A Well Worn Path

by

Jay H. Cravens

FIRST EDITION

Copyright 1994, by Jay H. Cravens
Library of Congress Catalog Card No: 93-94302
ISBN: 1-56002-418-6

UNIVERSITY EDITIONS, Inc.
59 Oak Lane, Spring Valley
Huntington, West Virginia 25704

Cover by Jacob Katari

Dedication

For **Gwen, Melissa & Cindy**
who made this book a reality

&

To **Joe Cravens** who asked for my story

* * * * *

I would also like to thank
Myrl & Gene Jenson, who read,
criticized, and re-read this book;
Rudy Gehn, who guided me through the
possibilities of publishing;
Cindy Johnson, who taught me to use a
word processor; and the many friends
and associates who gave me encouragement and advice.

CONTENTS

FOREWORD

A "Foreword" is usually written by a distinguished person whose name will add prestige to the undertaking. Why then, would the author ask an "unknown", to prepare an introduction? Perhaps the answer lies in our close family association for over fifty years, my own interest in natural resource management, and our parallel careers—his in forestry and mine in environment.

Forest management in the United States has a long and distinguished record, and it has produced both a profession and an organization. The Forest Service, unlike so many agencies of government, has enjoyed dedicated professional leadership, has adapted to changing demands on forests (and the lands upon which forest grow), has worked constructively with competitive user groups; and forged lasting working relationships with the other elements of our complex governmental spectrum.

The rate of change in forestry objectives, and therefore management practices, during the period covered by this book has probably exceeded that of any other historical period. In one short generation the public demands have shifted from an almost exclusive emphasis on "consumptive use" to a much broader set of demands including the "non-consumptive uses" of recreation in its many forms. Many new Federal statutes have been enacted in this period, and their objectives and constraints have been assimilated into forestry practices and administration.

The process of assimilation is neither painless or automatic, and it has placed unprecedented demands upon the many professionals who have guided the Forest Service through these changing times.

The author is one of the professional trained foresters who faced the task of carrying on under the old rules, while trying to understand, plan and adapt to the new demands placed upon his organization. His success in understanding and coping with these demands is reflected in his steady climb up the career ladder from a "beginning forester" to Acting Chief of the US Forest Service.

The author sensed, correctly I believe, that the key to the past successes of the Forest Service rested in formal training of foresters, and in a strong professional organization with worldwide connections.

His convictions lead to early retirement from the US Forest Service to accept, first, a professorship at a prestigious university, and second, to serve as President of the Society of American Foresters. At the time of writing the author was employed by a forestry consulting firm—the fourth leg in a long and distinguished career.

This book is the life story of a young man, interested in the

1

natural environment, who received degrees in forestry, and of his steady climb to the top of a career ladder.

This book is also the parallel story of the US Forest Service, and its response to the changing public and statutory demands which it encountered in this same period.

Talented, dedicated and well-educated people have given the US Forest Service its international reputation for excellence. But the organization has also had a profound impact on these same people, their wives and their children. The book illustrates, in very human terms, the interplay of these independent forces on people, organizations, and one of the great natural resources of the nation.

Gene Jensen, P. E.
Former Assistant Administrator/Water
US Environmental Protection Agency

PREFACE

I was first attracted to writing this story by my wife who conceived the idea of writing her family history. As she suggested I tried to write this story of my experiences over four decades as a forester as if I was telling a long lost friend or relative about my life and careers. She believed our children, grandchildren, people in the Forest Service and anyone seeking a natural resources career would be interested.

I went into the US Forest Service at a time when it was one of the most dynamic and highly respected Federal agencies. The Forest Service had evolved from a custodial, protective and land acquisition organization to one practicing environmentally sensitive management even before environment and ecology were common dinner table topics. Forest management practices utilized by the Forest Service were biologically sound, but some practices were generally unacceptable to the preservation groups. Our omnipotent attitude (we know what's best in the long run and we know what will result in a decade, or a score of years or a century or two) did not sell. We showed people our practices, then we listened to what they were saying, but we persisted in what we believed was sound management that cared for basic soil and water resources and provided a mix of uses and products for the public. We became an agency under fire.

My early love of the outdoors was fostered by my Father's teachings and fueled my ambition to become a forest ranger. I saw and grasped the brass ring and realized my goals. When I started with the Forest Service I had no idea where I would go, what I would do or how far I would advance. I was well trained academically, but woefully inexperienced in work practices. I was determined to learn and to perform whatever tasks were assigned to me, to the best of my ability. I learned as I went and I never had a job I did not enjoy. The struggle of newcomers to adapt and influence an agency's policy and direction was a challenge, as were the separate staff and line career ladders.

My travel down *A Well Worn Path* was exciting and the story goes through a war, or two, or three, falling in love with a Red Cross girl, being entranced by a Forest Ranger and obtaining an education in forestry. Then I learned the most important skill of a forester is that of working with people. Most young people enter forestry and related natural resources careers to work in the woods and the outdoors they love; working with people and getting things done through people is the last thing they expect to do or want to do. My progress from the bottom of the organization to near the top may provide a historic perspective to some of the inner workings of that proud organization. I had experience at every level of the Forest Service organization.

There are clues in my story on how to survive in the Federal bureaucracy, which are applicable to any organization. I devoted most of my attention and efforts to the job at hand and had some time left over for a sometimes tolerant wife and daughters.

My story is essentially about my life as a forester and deviates from an already exciting career when I was sent to South Vietnam during the peak of the war years to administer a forestry program in a war zone. Letters to my wife and daughters picture me living in South Vietnam and display my impatience, my exposure to constant danger and my insecurity and doubt about our country's motives. I tried to be graphic and convey the exciting aspects of my personal life and professional involvement in South Vietnam. This is what kept the adrenalin flowing and rendered the boredom of being far from home less debilitating. The book illustrates the excitement of a foreign assignment. At the beginning I was certain I could be a part of a huge civilian and military force that would make life better for all those that lived there. As the months passed my involvement in this intense action changed my outlook on what our country could and could not do. As the pages unfold the reader will note a reflective change in my attitude from "gung ho, damn the torpedoes, full speed ahead" to more mature thoughtful insight of the terrible abuse our government imposed on the people and land in this tiny corner of Southeast Asia. As I reflect on our country's involvement in Vietnam, I found we traveled the same well worn path as did the Chinese ruler Genghis Khan, the Japanese invaders and the French colonists. We became ensnared in the same diplomatic intrigue and fought over the same fetid rice paddies, river crossings, hills and forests. What a shame we were doomed to repeat those failures because we did not understand the Vietnamese or profit by the study of the thousands of years of history of that tiny country. In the end all of us, the Vietnamese and American people, military and civilians and the countries of South Vietnam, North Vietnam and the United States were changed. No serious student of history will deny that our involvement was a mistake.

My story in Vietnam does not pretend to cover everything that occurred during my tour of duty but there was enough excitement for me and those who shared being in that part of the world in 1967 and 1968 to fill a lifetime of memories. The reverse culture shock of returning to the mainstream of the Forest Service was mitigated by the wide circulation within the organization of my "Dear Gwen & Girls" letters. Contrasted to the experience of some of the military I was welcomed back to the United States and quickly resumed my career in the "real world".

On my return the scene was set for the "save a tree cult". Major battles erupted between the preservationists and "the tree

4

killers" as we were called. The dramatic pathway through the Federal courts took its toll on our people. As managers our job was to get work done through people. We had the task of seeing that our people were doing quality work and then back them up in the public arena. We lost cases in Federal District courts before environmentally inclined "hanging judges", but we won on appeal and the Supreme Court declined to hear any of our major environmental battles. Yes, we won the battles, only to lose the war. Laws were changed and only slowly did we learn the real value of alternatives. Working with people and helping sharpen the interpersonal skills of our people became paramount and began to pay dividends.

As my journey progresses it will become apparent how forestry evolved from the beginning and how forestry and I developed more of a global perspective. The history of forestry and the Forest Service is copiously chronicled in dozens of volumes to which little can be added. But there are few, if any, books of this nature. I provide a different historic perspective and a sense of identity of how we came together and traveled *A Well Worn Path*.

During the 1960s and 1970s, before I left the Forest Service, I saw many bright well educated young professionals and technicians enter the Forest Service. But they had two problems:

(1) they were less well trained in interpersonal and communication skills.

(2) many had difficulty surviving and advancing in the Federal bureaucracy.

It was to help prepare young people for natural resources careers that I took an early departure from the Forest Service and became a professor at a prestigious university which was to become during my tenure the largest undergraduate forestry and natural resources program in the country. This unique program taught integrated resource management and placed emphasis on forests as a renewable natural resource.

The Forest Service is adaptable and has survived over a century of change. It has become a global forestry organization. The strength of the Forest Service or any organization is in its people . . . people make the US Forest Service. The Forest Service develops talent. If the Forest Service people are properly trained, motivated and given the proper leadership both the people and the organization will prosper. In retrospect as I look at the span of my career and the succeeding processions of public administrations from President Harry S. Truman, to Eisenhower, to Kennedy, to Johnson, to Nixon, to Ford, to Carter, to Reagan, to Bush and to Clinton, I would like to think that one of these political cycles we will have an administration which will pay some genuine attention to the nation's natural resources which are capable of providing the quality resources, amenities, goods

5

and services our people and the world deserves. As Voltaire once said "in a democracy the people get the kind of leadership they deserve". Surely we deserve better.

Jay H. Cravens
Milwaukee, Wisconsin

BOOK ONE

Jay's Story

When the ancestors of my wife, Gwen Sanders Cravens, landed on the shore of Massachusetts in 1630 the forests appeared to be unlimited, full of Indians, wild animals and a resource to be pushed back for settlement and agriculture and to be cut for products essential to the building of a new nation. The Indians sold or gave some settlers vast tracts of forest land with no real sense of how much land or what consequences were involved (Pomfret, 1970).

Exploitation of the forest resources continued for the next three hundred years. It was during this period and long before I entered the world, Federal forestry work began. Books have been written to trace the early beginning of forestry in the United States and the National Forest System (USDA/USGPO, 1976). This book will trace it from my perspective.

Early in 1891 an act of Congress created the first forest reserves. Gifford Pinchot, one of America's most renowned conservation leaders became head of the Forestry Division in 1898 and Chief of the Forest Service when it was created in 1905.

The Society of American Foresters, a professional organization of technically trained foresters was founded in 1900. Gifford Pinchot was its first president.

During the tenure of Chief Henry S. Graves (1910-1920) and Chief William B. Greeley (1920-1928) Federal laws were enacted to strengthen National Forest administration, intensify protection of forest resources, provide for acquisition of forest lands to protect the flow of navigable streams, initiate management of forest resources and authorize cooperation with various states to help protect state and private forests from wildfire and improve forest land.

During my early childhood years the Forest Service visualized the wilderness concept and the first wilderness area was set aside administratively in the Gila National Forest in New Mexico. Later more millions of acres were dedicated to remain in their natural wild state to be visited by man only on horseback, foot or by canoe.

When I was a young jayhawker growing up in Kansas dust storms during the "dirty thirties", that greatest of all conservation programs the Civilian Conservation Corps was created. As a young man enjoying the northern Minnesota outdoors of the Superior National Forest I saw CCC corpsmen going to fight forest fires, plant trees, construct roads, trails and campgrounds and other improvements. Many of my early Forest Service supervisors and forest rangers obtained their start in the CCC program.

World wars, especially WW II made a heavy impact on the nation's forests. My early Forest Service mentors were called on to perform wide ranging special tasks such as timber harvest in Tennessee and producing rubber bearing plants in California.

While I was serving as a medic in Colorado's huge Fitsimmons Army Hospital the first black bear symbol appeared on a fire prevention poster. The Smokey Bear, who I later met on a forest fire in New Mexico, ranked with Coca-Cola, Ford and Chevrolet as one of the best known, worldwide advertising symbols. It was during this period that I had my first contact with a Forest Service district ranger and soon embarked on my life long career as a forester.

There were other Chiefs of the Forest Service who I did not meet. Robert Y. Stuart was Chief 1928-1933; Ferdinand A. Silcox was Chief 1933-1939; Earle H. Clapp was Acting Chief 1939-1943; Lyle Watts became Chief in 1943 and was in office until 1952. Chief Watts was only a signature as far as I was concerned. I never met the man, but through his leadership we improved our resource management activities.

Chapter I

In the Beginning

I was born a Hoosier in Bloomfield, Indiana on Easter Sunday, March 27, 1921. Helen Hampton Cravens was a domineering Mother and assumed from the beginning I would follow in my Grandfather's and my Father's footsteps, take over a real estate and insurance business and be around to take care of her for the rest of her life. My father, J. (James) Frank taught me to fish and love the woods.

At a very early age we visited Uncle "Bun" Cravens farm northeast of Bloomfield in the unglaciated hills of southern Indiana. We crossed a neat covered bridge to reach the farm. I remember the wood planks rattled out a tune as we drove across in our 1926 black Dodge sedan. This was a nice woods to explore and in which to hunt squirrels. My Father taught me the names of all the native hardwood trees. I learned which woods were good for lumber, furniture and other uses. He told me that one of his uncles once owned a sawmill in Tennessee. Father showed me which hickory trees produced sweet nuts and which trees were good for furniture. His brother Carl ran the Old Hickory Furniture factory in Martinsville, Indiana. I liked the chestnut tree best and especially enjoyed collecting and extracting those sweet nutmeats from those spiny burs.

I had a favorite Airedale dog named Buster. He enjoyed running with me through the woods. In the fall, if we could not get out to the farm, Buster would follow me as I gathered sweet chestnuts and beechnuts from any number of places around our small town. I did not know that chestnut blight would hit and wipe out those magnificent chestnut trees in our part of Indiana in the 1930s after we moved away. I learned how to tap the maple trees located in our yard during the late winter and then boiled the thin, sweet sap into delicious maple syrup.

Father made home-brew. This was during the prohibition years in the late 1920s and against the law. I thought it tasted terrible and could not understand what all the fuss was about. I remember the yeast and hops bubbling away in large earthenware crocks and then drawing off and bottling the beer. I could operate the bottler. One time two events took place at the same time. One batch of beer blew up in the cellar. It made an awful noise and a real mess of the fruit room where it was stored. My Mother frantically attempted to call my Father. But he was already on his way home because word was out, the "revenooers" were in town. Father destroyed the evidence as he sadly dumped his entire supply of beer down the drain. Of course no "revenooers" ever showed up and we were back in production in a few days.

We attended the Christian Church. I attended Sunday School and an occasional church service. My Mother pinched me when I nodded or was not paying attention to a sermon I did not understand or care about. I was scared when I was baptized, actually dunked, in a large tank of smelly water behind the pulpit. Up the street from Father's office was the Church of the Holy Rollers. We kids used to sneak up and watch their ranting, raving and rolling in the aisles. We stood there looking through the window with our mouths and eyes wide open until we were yelled at and chased away.

Late in 1931 we left Bloomfield. Father accepted a job with the Union Central Life Insurance Company in Topeka as a mortgage loan officer and I became a jayhawker. Kansas was not a happy place in the "dirty thirties". There was dust, unemployment and depression. People begged for dimes and sold apples on street corners. Many days street lights were turned on in the middle of the day due to the massive dust storms that blotted out the sun. Father traveled throughout the State and had a difficult job making decisions regarding numerous farm foreclosures.

In Kansas, at age 10, Father taught me to use a shotgun and hunt prairie chickens and ducks, when and where there was water. He stressed safety in the use of a gun of any kind and as soon as I had his approval of my performance he gave me his 16 gauge Belgium Browning automatic shotgun and purchased a 12 gauge of the same make for himself. After a bout with measles I became nearsighted and had to have my first pair of glasses. I vividly recall, when I first put on the glasses, I could see that those wonderful trees which I loved so much had individual leaves! It was a wonderful experience.

Father satisfied the Company, was promoted and transferred to Minneapolis, Minnesota. He became manager of the Company's North Central District. We lived well in Minneapolis. Father had a salary of $500 per month, plus a very liberal expense account, a real fortune during those years. The Company provided Father with a new Oldsmobile or Buick every two years. I was shielded from much of the poverty that existed around us in those years.

Father and I spent time together in the outdoors. We hunted deer in the north woods and pheasants in southwestern Minnesota. We had excellent fishing in Lake Minnetonka and Lake Mille Lacs. Each August my Mother and I went to Ely, Minnesota to find relief from my hay fever and asthma. There was good walleyed pike fishing with minnows and a June bug spinner below a nearby power dam. I learned to swim and canoe in what was to later become one of my major environmental challenges, the Superior National Forest's Boundary Waters Canoe Area. I had many canoe trips into the canoe country, where fishing and water quality were good. In those 1930s Ely's nearby Lake Shagawa was terribly polluted. It was not fit for fish or swimmers. Some of my trips to northern Minnesota were made as part of the Minneapolis YMCA program. Several trips were made in the rain. It could rain continuously day after day. We bailed out the water that collected on the floor of our tent, but it did not bother 12 year old kids. Our camp was at Half Moon Lake near Eveleth where I learned to canoe, shoot a .22 rifle, sail and ride horseback.

We had many friends in Ely and stayed at the Ely Hotel, owned and managed by Frances Moravitz. We became part of her family. Her kids, a boy, two girls, and I were good friends. We canoed, hunted and fished together. I went on a number of early deer hunts (before the season opened). It was said that the kids around Ely ate nothing but fish and venison until they were at least 21 years of age. I really enjoyed their Finnish fish stew, mojaka. I learned to make it out of a large northern pike, onions, milk, potatoes and pickling spices. The stew was started by boiling a large fish head, fins, tail and backbone with a gauze sack containing 2 or 3 ounces of pickling spices for an hour or so. This produced the broth. The bag of spices, fish head and other parts were discarded. The other ingredients plus chunks of fish were added. Really, the whole thing should have been thrown out, because at the early stage it looked like nasty, greasy, dirty dishwater! But the stew was delicious if you overlooked the broth stage.

In the Fall we returned to Ely for deer hunting. We were always successful on those hunts. While in the area we collected wild rice and cranberries. I loved the north woods.

At times during the summer I was a bellboy around the Ely Hotel and made a few nickels and dimes carrying baggage for the visitors. One of those visitors was a forester. This was my first real contact with forestry. I spent a day in the woods with this forester who worked for US Steel. He showed me some large magnificent white pines near Ely and demonstrated how to measure their diameters and heights and calculate board foot volume in the standing trees. Around the Ely area I saw Civilian Conservation Corps young men in their blue denim work clothes and floppy hats riding in the back of stake trucks on their way to plant trees, fight forest fires and build roads and trails. At the time I did not appreciate the significance of this greatest of all conservation programs, the CCC.

Many of the visitors to Ely in the mid-1930s were from Japan. The Japanese were interested in visiting mines in

Minnesota's Iron Range where they made purchases of iron ore for shipment back to Japan. Very few of them spoke any English. Of course they did not tell us they were stocking up for what was soon to become World War II.

I am uncertain as to the details, but the Union Central Insurance Company bosses in Cincinnati wanted Father to do something that he did not believe was proper. They asked him to do it or resign, but being very stubborn about his principles he said "fire me". And they did! That was a sad day. My Mother cried and carried on something awful. I had to leave my school, friends (no girl friends however . . . my Mother and Father completely neglected that taboo subject of sex education), a comfortable home, the land of lakes, fishing, hunting, swimming, ice skating, hockey, skiing, and boating. But we packed up and moved to Iowa. The family had a good salary and the good life one day and then nothing ahead but tough times and starting a new life in Iowa.

We moved to Cedar Rapids, Iowa in 1937. Father strongly believed there was a good future in Iowa farm real estate. Breaking into the real estate and insurance business on his own during the Great Depression was bleak. There was nothing "great" about it. Sales of insurance were slow and farm sales were few and far between. On many occasions Father returned home late at night after being out all day with a prospect looking at farms and telling us that he didn't sell one today, but things were improving. He never lost faith and had confidence in himself. He was always optimistic. My Mother cried a lot and was not very happy. The best thing in my life at that time was a black, curly haired Cocker Spaniel that I named "Smokey". I really loved that dog.

We did not have much to live on in those early days of the new and struggling business. We had enough plain food to eat. Some months it was difficult to come up with the rent money for both the house and office. There was no new car every two years. Both my Mother and I worked. She hated it, but she worked in grocery stores demonstrating food products and handing out samples. I worked at odd jobs . . . lawns, raking leaves, cleaning and hanging storm windows and screens, snow shoveling, and then "jerking sodas" at $0.35/hour. Later I really made progress when I worked weekends and after school in a dairy for $0.50 per hour! I operated and cleaned dairy equipment, made and packaged cottage cheese and drank gallons of chocolate milk that I mixed with whipping cream. It's a wonder I didn't explode from fat and cholesterol. But I worked long, hard hours and kept slim at 6 feet 2 inches and a trim 190 pounds. It was difficult performing the heavy work and cleaning equipment in the hot steamy dairy during the hay fever and asthma season. But although I wheezed and sneezed I was considered a good dependable worker. That job provided the beginning of my contributions to a new program called Social Security, and helped pay my tuition at Coe College.

My Father was a "died in the wool" Democrat. He was a great admirer of cowboy philosopher, Will Rogers, and used to describe how he and Rogers were dedicated Democrats! In Father's mind Democrats could do no wrong. He strongly believed even dead Republicans were no good! My Father experienced genuine grief in 1935 when Will Rogers was killed in an Alaska airplane crash. My Father's other heroes in those days were President Franklin D. Roosevelt and his Postmaster General James Farley.

Father was a great supporter of the Democratic Party. He did not donate much money to the party because he did not have it to give. Before every election he would go around town putting up posters for candidates and distributing pins and literature which supported Democratic candidates of his choice. On election day he worked from dawn to dark hauling, literally dragging people out of their sick beds and out of old people's homes, to polling places to vote. He was not very broad minded because he would ask people who they were going to vote for and if that turned out to be a Republican candidate he refused to give them a ride. Oh how he raged at the no good, crooked Republicans. He actually hated ex-President Herbert Hoover. Republican Presidents, and all Republicans as far as he was concerned, were dishonest, no good crooks and had come close to ruining the country. During his raving about Republicans his blood pressure would shoot up. This became a serious health problem. He solved it by losing weight, came down to an eighth of a ton and joined the Christian Science Church. He became a supporter of Mary Baker Eddy's church and claimed he cured his high blood pressure through Christian Science readings. When he was really ill his practical side won out and he would go to a medical doctor.

Chapter II

College & the War Years

I was not an exceptional student in high school and did not apply myself as well I could have, but did well enough to maintain a "B" average and was accepted at home town Coe College. There was a saying at that time "if you can't go to college, go to Coe". Coe was really a good liberal arts school and prepared me for my later forestry career. Courses in sociology and psychology helped me understand and work with people. I had a part-time job working in the Zoology and Chemistry departments cleaning up after lab experiments. That work along with my job at the dairy helped cover my expenses for books and tuition.

I had been on the track team in junior high and high school and continued at Coe. I high jumped over six feet, pole vaulted up to 12 feet and ran the high hurdles. I even had black cinders ground into my knees from falls on the track. I won a few medals in various track meets, including the Drake Relays where I met and was interviewed by a sports announcer for radio station WHO in Des Moines. His name was Ronald "Dutch" Reagan (later movie actor, Republican Governor of California and 40th President of the United States).

I took Reserve Officer Training Corps classes at Coe and marched in the 11 November 1939 Armistice Day parade. During the parade the temperatures plummeted and we just about froze before we finished. That day many duck hunters on the Mississippi River died as a result of that terrible Armistice Day storm.

Where was I on 7 December 1941? President Franklin D. Roosevelt called it "a date that would forever live in infamy"! That Sunday afternoon I was at home in Cedar Rapids, doing my chemistry homework and listening to my radio when the bombing raid on Pearl Harbor was announced. My Mother started crying when she heard the news and moaning that we would go to war and now she would lose me. I left the house and went for a walk. I could not visualize what the future held for me. My Mother had always wanted me to be a doctor. She pushed me in that direction and hoped I could get a draft exemption. She hoped my bad eyesight, 20/200 without glasses, and flat feet would get me classified "4F" and therefore unacceptable for military service. I had mixed emotions as many of my friends went off to war. Should I go? Or not go?

Nevertheless I put the war out of my mind and decided on medical school. I took chemistry, physics and zoology and qualified for medical school.

During this period Father and I enjoyed the outdoors together. We hunted pheasants and ducks and fished. Occasionally in August we returned to Ely for the hay fever season and good fishing. But most of the time I worked at the dairy to earn money to pay for college tuition and expenses and save up for medical school at the University of Iowa.

Business improved! Father became one of the leading realtors in the State. He was correct, the move was worthwhile and he enjoyed being his own boss. He was greatly disappointed when I showed no interest in joining him in the business. All along he assumed I would join him. My Mother continued to pressure me to become a doctor. I guess I wanted it and when accepted I went off to medical school in 1942 at the University of Iowa in Iowa City for a year.

I lived in a fraternity house with my good friend from high school days and Coe, Bill Franey. Bill was short and I was tall. People referred to us as "Mutt" and "Jeff", after the comic strip characters. At Medical School mainly it was all study and then work a few hours each day in the University Hospital business office for my meals.

I did well in biochemistry, human anatomy and histology. But I flunked neuroanatomy! That was the well-known "washout" course used to reduce enrollment. It took me out at the end of my freshman year! During that period I had a reserve officer commission as a 2nd Lt. in the Medical Administrative Corps (Serial No. 0-176559). When I flunked out it was a greater disappointment to my Mother than for me. I really did not mind, after all there was a war on and I was ready to go! My Mother did not agree with that either, how she cried and moaned. She said I would be exposed to those "bad women". Father said she acted the same way when he went off to World War I. Nevertheless the draft board quickly drafted me. I left family and friends and went off to the Army, passed my physical examination and was ready for World War II. My friend Bill stayed on in medical school, graduated, became a prominent Cedar Rapids physician and was my Mother's doctor for many years.

"What did you do in the war, Daddy?" My war experience was interesting. I was inducted into the US Army at Camp Dodge, Des Moines, Iowa on 6 April 1943 (Serial No. 37667965). I had two days of KP and attened illustrated lectures about sex and those "bad women" while the Army figured out what to do with a reserve officer draftee. That situation was quickly resolved and I was given an honorable discharge from my

15

commission on 9 April 1943. Fortunately my medical school experience surfaced on the day they were looking for a medical technician to work at the Air Corps Recruiting Station in downtown Des Moines. I was told to pack up, go look it over and if I liked the assignment and they were willing to take me, I could just stay. They told me to have the Recruiting Office inform them. I asked the Army people at Camp Dodge about the alternatives. They said it was likely to be Camp Polk, Louisiana for basic training in the infantry or whatever else came along. Needless to say the Air Corps assignment looked good and I was pleased to be accepted. There was no KP, latrine cleaning, no barracks life or basic training . . . rather an apartment with fellow recruiter Sgt. Joe Harbridge, per diem when traveling and flight pay added to my salary every month, provided I had eight hours of flight time.

I was promoted from Private, to Private First Class, to Corporal and to Sergeant, all within three months. As added attractions there were lots of WACs (Womens Army Corps) in Des Moines and lots of girls around Iowa. I worked in the Des Moines Recruiting Office for five days each month and traveled the remainder of the time. I could plan my schedule to be in Cedar Rapids on a recruiting visit to local high schools, see my folks and get away again. All in all it was a good life.

The recruiting assignment involved extensive travel with a combat officer by car and airplane, throughout the state. We covered all 99 Iowa counties and were very successful in meeting our quota of recruits. We talked about the aviation cadet program and combat experiences in every high school in the State. For about two years I gave written screening and basic physical examinations.

Army duty in Iowa was really hazardous! We should have received combat pay in addition to the other allowances. We had several near misses involving cars and airplanes. One day near Cedar Rapids a pilot and I walked away from a plane crash that completely demolished the aircraft. A frozen carburetor was identified as the cause of engine failure and the crash. I was a bit stiff for a day or two but recovered rapidly. Another time we were on a practice bombing run over South Dakota in a B-17 and the fire warning light came on. The pilot ordered us to put on our chest chutes and be prepared to jump. We lined up at the door but the problem was solved and we did not bail out. But the wheels could not be lowered, due to a failure in the hydraulic system and we were low on fuel. The mechanical method for lowering the wheels did not work either. So we had an exciting wheels-up landing. Sparks, dust and everything else flew through the air. The cabin was pierced with bits of flying metal, probably from the bent and broken props. The noise was deafening. But we made it, got in our flying time and went on the next challenge.

Those crazy combat pilots, navigators and bombardiers I worked with would drive our 1943 Plymouth staff cars at top speed, up to 85 miles per hour. Once we approached a busy 4-way stop crossroad. Cars were waiting in each direction for their turn to proceed. Captain Peterson roared through the junction at 80 miles per hour, honking, hand throttle pulled out all the way, waving both arms, feet up on the dash board and yelling "bombs away"! I about died then and there. I shut my eyes and we made it, leaving many angry motorists honking their horns and waving their fists at our rapidly departing car. There were many wild times for a single guy with no obligations and a questionable future.

After work we had a few drinks. We tried as often as possible to meet up with the other recruiting teams. Frequently I was the only one of the team sober enough and able to give all the talks the next day. I could talk to a bunch of starry-eyed high school students about the aviation cadet training program, physical and mental examination requirements and combat experiences. I had heard enough war stories from the European and Pacific war zones that I believed I had been on some of those flights and bomb runs. On one such occasion, after our day's schedule was completed, Captain Peterson and I started drinking to his buddies who had been shot down during bomb runs to Germany's ball bearing works at Regensburg near the Czechoslovakian border. Many of his friends were lost on those hazardous missions. After a big steak dinner and many toasts, in straight Canadian Club whiskey, we gave up and started back to the hotel. I was driving. Pete got sick, opened the car door when I stopped at a signal light and heaved away. The odor of his vomiting made me sick, I opened my door and heaved. My next cloudy, vague memories were of flashing red lights, going up in the air, hearing a door clang shut and the lights go out.

Several hours later I awoke to see vertical steel bars before my eyes and a moaning Air Corps Captain on a bare iron cot next to me. Hell's fire, we were locked up in jail! Then I really was sick. Looking out through the barred windows all I could see were the tops of trees below. My next recollection, after a few dry heaves, was a guard bringing food . . . more dry heaves. The guard said our Commanding Officer and the District Attorney were down below having a conference . . . more dry heaves. I have never been so sick in all my life. We were asked to get ourselves straightened up and follow the guard to the DA's office. We both were getting sicker by the minute. We sheepishly met the DA and our Major Miller and were given a stern lecture on the dangers of drinking and driving and were dismissed. We were never booked. We were placed in overnight protective custody. We were ordered to go to our car and stand by. The

Major arrived shortly, looking stern, and very angry. I could see the infantry, Camp Polk and Louisiana swamps looming over the gold oak leaves on his shoulders. The Major said it was a good thing the DA and he were brother Elks. They had a good visit and the Major got away from the office for a few hours to come to our rescue. The Major said the next time to invite him along when we were going to celebrate and now take him to the airport and get back to work. Fortunately it was a free day and we were able to clean out our stinking car, improve our appearances and go on to the next series of high school appointments. I gained great respect for the Benevolent Paternal Order of Elks that day.

From that time on I have never had a drop of Canadian Club whiskey. We did not go on the wagon completely. I stuck to "boilermakers", a drink of cheap whiskey with beer for a chaser. One night I had 15 of those things and was the only one able to drive the crew to the hotel and meet the next day's recruiting appointments. As I reflect on it now, our behavior was just horrible!

I continued to enjoy the outdoor life and being in the woods. Hunting was particularly good in Iowa in 1943 and 1944. There was good feed, cover and water. Pheasants and ducks were particularly plentiful and most of the hunters were off to the war. Restaurant operators gladly cooked game for the hungry "fly boys". We were treated well, sometimes too well. Everyone wanted to buy us drinks or meals. At that time war was raging on both sides of the world and we were providing it with raw material, aviation cadets. I remember D-Day and the invasion of Europe. Two recruiting teams had come together at Spirit Lake, Iowa. We were playing poker on the night of 6 June 1944 when the radio reported that the invasion was underway. More drinks and from then on recruit numbers greatly exceeded our quotas.

My life was directly influenced by a military decision made after the May 1944 battle of Cassino in Italy. After days of aerial bombing, a monastery full of entrenched German paratroops would not surrender or be driven out. The Air Force, as it was then called, received some bad press. It became obvious that airplanes alone would not win the war. The Air Force had grown too large. Within a few months most of us, including the combat officers, were transferred out. Only the Major and the two regular Army Master and Staff Sergeants remained at the Recruiting Office. The remainder of us were shipped out to various places. My friend and fellow recruiter Sgt. Virgil F. W. "Dick" Dickman was sent to Camp Carson, Colorado. I was transferred to Fitzsimmons Hospital in Denver as an Army medic. It was a trip to a new beginning, a new life.

I loved the West and the forests and mountains of Colorado. I left my hay fever and asthma behind forever. My first Army hospital assignment was in the psychiatric, locked ward. I cared for some very tragic cases. I learned how to help administer electric and ice water shock treatments and how to control and restrain violent patients in straight jackets. Some of those suffering from hysterical paralysis and the violent ones in padded cells were the sad results of war. One young man vividly remains in my memory. His name was Dwight Yost, an 18 year old tail gunner on a B-17 bomber. His plane had crashed after a combat mission and he heard the screams of his crew members as they died in the ensuing fire. He was rescued and suffered from hysterical paralysis. He could not communicate, feed or care for himself. I was assigned to care for him along with a number of other similar tragic patients.

My recreation consisted mainly of seeing many movies, visiting the USO, bowling and having a beer or two. I had tempered my drinking considerable from recruiting days. During the war Denver was a good town for GIs and I enjoyed it. My health was good and on my free days I went on hikes in the nearby mountains through the spruce and pine forests.

Fortunately the Army recognized the need for us to occasionally get away from duty in the locked wards. This provided me with a monthly opportunity to escort officer patients to other hospitals. My first trip was to Walter Reed Army Hospital in Washington, DC. I traveled by Pullman in a room next to my patient. That was really first class railway travel for an Army Sergeant. After my patient was signed into the hospital I had two days to take in the sights of the nation's capital. I ranged far and wide, from the Capitol Building, to the museums and art galleries, climbed the "umpteen" steps to the top of the Washington Monument, looked on in silent reverence at the Lincoln Memorial and at the Tomb of the Unknown in Arlington Cemetery.

My second escort trip was a flight to El Paso, Texas where I signed in my patient to the Veterans' Administration Hospital. I briefly visited El Paso before an Air Force DC-3 returned me to Denver. I enjoyed seeing and inhaling some of that arid southwest air. The next trip was by Pullman to Seattle where I escorted my patient by Army ambulance to Ft. Lewis Army Hospital. Back in Seattle I decided to spend my two days and take a bus ride to Vancouver, British Columbia. This was my first visit to the Pacific Northwest and Canada. I enjoyed the trip and will always remember the beauty of the huge Douglas-fir trees in Washington and Vancouver's Stanley Park. The return trips by train always switched escorts to coach. But I really enjoyed seeing more of the USA, especially the forests and mountains of the West.

I was a skier when I lived in Minnesota, even made a few jumps off of low structures. When I heard the ski troopers were looking for medics I spent almost every weekend in January and February of 1945 at Winter Park with Army buddies practicing and improving my downhill skiing skills. The 10th Mountain Division sounded like an exciting part of the Army to join. Army skis had "bear paw" binders which secured cables and strong springs rigidly to the ski boots. There was no quick release safety system and the boots did not fully protect the ankles.

All went well with my skiing until I fell on a steep slope. I made a high speed turn on a flat rock that had more ice than snow covering it. Army skis were designed to stay on if you fell in enemy territory. The theory was you could ski away and live to see another day. When I made the turn, I side-slipped on ice and went over a low rock ledge. When I fell, the crack I heard, that I thought was my ski, wasn't. It was my right leg just above the ankle. With considerable pain and some 12 hours later the ski patrol had me back at Fitzsimmons Hospital in the orthopedic ward, in traction. In the bed next to me was an officer who had been shot in the back of his neck on Attu in the Aleutians by a Japanese soldier firing a small caliber rifle. He was paralyzed from the neck down. I considered myself fortunate. While confined to bed I learned we really were not being considered as medics for the ski troopers but were prime candidates for assignments as medics in the South Pacific. Lucky break!?

As soon as I could get up and around on crutches I was placed in charge of a ward to care for the walking wounded. We treated and dressed some nasty, badly infected wounds. Most of the injuries were caused by bullets and shrapnel, the latter being by far the most damaging to muscles, nerves, bones and morale. These young soldiers came from both European and Pacific theaters of war. Victory in Europe (VE) Day was 8 May 1945 and the wounded arrived in increasing numbers and filled the Hospital to capacity.

I spent a wild VE Day in Denver with one of my patients, Leonard Stelmacowitz of St. Louis. He had received a severe arm wound that had to be dressed each day. We made quite a pair, me on crutches, leg in a cast and he with his shoulder cast. Everybody brought us drinks that day as we celebrated. We even kissed a few girls!

When I could use a cane and a walking cast, I requested 3-day passes and began hitchhiking around Colorado to see the beautiful forests, streams and mountains. During the War anyone in uniform, with a cast and a cane, had no problem securing rides. On one of those trips in June I was offered a ride by a Forest Service district ranger in the Gunnison, Colorado area. He drove me into the Gunnison National Forest and showed me some of his work. On the next day he had to hike into the backcountry. Since I could not accompany him he gave me his fly fishing outfit and plunked me down by his favorite fishing hole (full of large rainbow and German brown trout). From that point on I had conflicting dreams of going back to a medical or chemistry career or into the real estate and insurance business or becoming a forest ranger. I was away from home and really did not want to return and live in Cedar Rapids. As my Father used to say "Iowa is a good place to raise corn and hogs, but a hell of a place to live." I agreed! My recurring dream had me becoming a forest ranger and hunting, fishing, trapping and living in a cabin in the woods.

One of my trips, the summer of 1945, took me to Colorado Springs and the most significant event in my life. It was the beginning of an exciting new life. I was invited to visit my old recruiting buddy, Sgt. Dickman who was stationed at Camp Carson. Dick and I went to a Colorado Springs nightclub where he recognized and introduced me to a Gwen Sanders who he had met at Camp Carson. I asked Gwen for a dance and Dick danced with her sister Leela. These two young women came from a farm family near Clinton in western Minnesota. Gwen had graduated with a degree in English from Hamline University in St. Paul. Then she had become a social worker for Ramsey County in St. Paul. During the War she signed up as a hospital social worker with the American Red Cross and after training in Washington, DC was assigned to the Camp Carson Army Hospital.

Meeting Gwen was the beginning of a long and beautiful relationship. This gave me an opportunity to make several trips to Colorado Springs. Sister Leela worked in an office in Colorado Springs and was fun to be around. However, she had some weird spooky friends. I recall one and made it a point never to be cornered by her. Gwen worked at the hospital with Army Nurse, Lt. Myrl Halstead. Myrl and Gwen had gone to school in Minnesota and had lived together in a co-op house in St. Paul before the War. One of Myrl's patients was Lt. Eugene T. Jensen, a B-17 pilot with 35 missions out of England over Germany. Both of us arrived on the scene in pretty bad shape but between Myrl and Gwen, Gene and I both made rapid recovery. You might say neither of us could run fast enough to get away! Lucky for us! Myrl and Gwen ultimately caught us! Actually we really made no effort to get away. We were all very happy together.

I recall the first atomic bomb being dropped on Hiroshima on 5 August and the second on Nagasaki four days later. The Japanese acknowledged defeat and surrendered on board the USS Missouri on 2 September 1945. Victory in Japan (VJ) Day was a happy one in Denver. Years later I was asked by a young grade school student about my thoughts when I heard the bomb had been dropped and killed and injured many thousands. As I reflected, I recall thinking at the time that was a cruel, inhumane act for our country to commit, but it did bring a long drawn out, devastating war to an end. I had been hardened by seeing and treating so many young men whose minds and bodies were impacted and destroyed for the rest of their lives, that I did not have much sympathy for the Japanese. After all, this event meant that some day we could return to civilian life.

Chapter III

Falling in Love

I made spaghetti dinners at Gwen's and Leela's apartment in Colorado Springs. On one occasion I broke the glass top off of a bottle of Worchestershire sauce and it fell into a pan of excellent meat sauce. I was not about to throw it out. I was stirring and listening very carefully for the sound of glass scraping the bottom of the pan when Gene Jensen came in. He has been accusing me of feeding him ground glass ever since then . . . he still really loves my spaghetti sauce. My secret is to put a little bit of everything on hand into the pot, along with meat, tomatoes, onions and ground glass, which cuts through the grease! Gene always liked to sample what I was cooking. Once I offered him a chunk of cabbage from a salad I was preparing. I quickly substituted a big strong garlic clove. When he bit into to it he made a lot of commotion. I didn't think he would ever forgive me, or stop chasing me around the kitchen table.

Those were happy days for me. I had no Mother to boss me! I was in love with a beautiful, wonderful girl. I spent most of my free time away from Fitzsimmons Hospital in Colorado Springs with Gwen. I grew to love her dearly. She was bright, charming and fun to be with. We went on many hikes and picnics in the Pikes Peak and Garden of the Gods area. It required a lot of nerve for me to get up my courage to ask her to marry me. But I did and floated on clouds when she said "yes". I gave her an engagement ring with a small diamond and we made plans to be married at her home in Minnesota after she was released by the Red Cross and I was discharged from the Army.

Late in 1945 the flow of wounded into Fitzsimmons Hospital eased. After weeks of scuttlebutt, rumors and speculation as to who had enough points to get out of military service, I was given an honorable discharge as Sergeant from the US Army at Ft. Logan, Colorado on 13 January 1946. I enlisted for four years on the same date as a Corporal in the Army Medical Corps Reserve but was never called up for the next war, in Korea. I returned to Cedar Rapids for a short time and finished requirements for a Bachelor of Arts Degree in Chemistry and Zoology at Coe College. I was credited for the medical school classes I had passed.

Myrl and Gene Jensen had married at the end of the War and made a decision to go to school at Colorado A&M University in Fort Collins and they were the lucky ones to remain there. We all had fallen in love with Colorado. Of course, as to be expected, my Mother expected me to come home and work with my Father. The war years had stiffened my backbone, but I was not ready to go out on my own. I just ignored my nagging Mother and thought pleasant thoughts about my life ahead with Gwen, in Iowa or somewhere. During the letdown period of those post-war weeks the only bright spot in my life was to meet Gwen for a wonderful weekend in Des Moines. She was on her way home from Colorado.

Gwen and I were married on 30 March 1946 at the family (William Otis Sanders and Alice Hickson Sanders) farm near Clinton, (Big Stone County) Minnesota. Myrl and Gene Jensen were there, en route to Fort Collins. Gene was my best man. Many of Gwen's friends and relatives arrived at the farm for the ceremony. My Mother, Father, Aunt Madeline and my Army buddy Dickman were there. My Mother still did not like Gwen, was cool to her parents and hated being on the farm.

We were dressed for the wedding. The minister and guests were waiting. My Mother and Father called me into their room, where they shut the door and we had a big argument. My Mother tried to talk me out of marrying Gwen and insisted I return to medical school. My Father wanted an agreement from me to return to Iowa and go into business with him. My mind was in a turmoil. I reaffirmed my plans to marry Gwen and leave home. I realized we could never be happy in Cedar Rapids with my Mother feeling the way she did towards Gwen. At that moment of decision I remembered the Colorado forest ranger and his work. I made up my mind to become a forest ranger. Gwen and I would go to Fort Collins and I would enter the Colorado A&M forestry program, if they would accept me.

Those were four of my better decisions . . . (1) to marry Gwen, (2) give up the idea of medical school, (3) avoid the real estate business in Iowa, and (4) to leave home. Never once have I had any regrets in making those decisions. As I reflect now I had little to offer Gwen . . . no job, no money and a doubtful future. With that the wedding proceeded.

After the wedding Gwen and I went to nearby Ortonville for an overnight stay. Then we returned to the farm for a few days. During that period brother-in-law Billy and some of his friends put on a charivari ("a noisy mock serenade to newly-weds". Morris, 1975) one night at the expense of Gwen and me. It was embarrassing to be rolled out of bed in the middle of the night and be harassed in such a noisy manner.

It was cold and muddy that Spring but I was happy and warm and I did not feel the cold. I enjoyed tromping around in the mud helping Gwen's father and brother Billy with the chores. I was always comfortable and happy around the farm and being with Gwen's family. Gwen had a fine family, Mother, Father,

24

Billy and sisters Leela and Elizabeth. I have always been very fond and close to Gwen's entire family. They were good people and very kind to me.

Before leaving the farm I assured Gwen's parents that I would take good care of her. Our honeymoon was taken in company of friend Dickman, en route through the Black Hills and a heavy spring snow storm to Fort Collins. This was a little unusual, not every couple takes a wedding guest on the honeymoon, but Dick had the transportation and we needed a ride. Dick went on to Denver and stayed to begin a new life in Colorado. He was a natural born salesman and became one of Denver's leading realtors. He had wisely declined an offer made by my Father to work with him in Cedar Rapids. My Mother would have probably come to dominate him since I was gone. Gwen and I were optimistic . . . we had few clothes, little money and not much else. Gwen had cashed in an insurance policy and I had a little savings. While my Mother was not looking my Father slipped me a $100 bill after the wedding. Since we had so little, every little bit helped. We had no fear of the future, just the optimism of youth.

Chapter IV

On Becoming a Forester

Myrl and Gene Jensen were in Fort Collins when we arrived. She was a Larimer County Public Health nurse and Gene an electrical engineering student. We all loved Colorado and had a good life and many happy times together. I was accepted as a graduate student by Colorado A&M where I studied forestry under the GI Bill. I was paid $75 per month plus payment for all tuition, fees, books and supplies. Frequently I "damned" the Veterans Administration when the vitally needed check was delayed. For the final year my payment was increased to $100 per month. Gwen and I lived in the Walnut Apartments, our first home (one of 20+). It was across from the fire station and a second hand furniture store, our first source of furniture that later became valuable antiques.

Earlier Gwen had tried unsuccessfully to find work at the University. Then one day we heard about a notice posted in the Forestry and English buildings pleading "Would English graduate married to forestry student come to English Department as soon as possible". They had kept no record of Gwen's name during that previous interview, but remembered she was married to a forestry student. That contact resulted in Gwen being given a second interview and an appointment as an instructor teaching English to large groups of veterans. She loved it! "Fum" McGraw, a forestry student, was one of her typical students. His Colorado High School education had not prepared him to use English correctly. "Fum" was an excellent football player at Colorado A&M and later played for the Detroit Lions professional team.

The University gave me credit for many of my Coe College courses. After taking some of the forestry prerequisite courses and attending summer camp at Pingree Park I was granted a Bachelor of Science Degree in Forestry in 1947. Gwen, Myrl and Gene visited me at Pingree Park and we had some fine weekends hiking in the high mountains and eating fresh trout I caught in those cold, clear cirque lakes. I was a graduate student teaching assistant to Dr. Richard Preston. I corrected examinations and reports and taught dendrology and wood technology labs while doing my Master's thesis on wood preservation treatments of native Colorado woods.

During the summer Gwen accompanied me on trips into the nearby mountains to collect wood samples. We were very happy, we did well at the University, had a good family life, joined the Unitarian Church and saved a little money. We bought a used Cushman motor scooter and with Myrl and Gene rented a cabin up the beautiful Cache la Poudre River beyond the Water Works Hill and the settlement of Columbine. We went there on our motor scooters most every weekend; Jensens also had a Cushman. We four had a good life together. We hiked, hunted deer and spruce grouse, fished for trout, cut and burned many cords of fireplace wood. The cabin was a good place for Gene and I to study. We were good students and earned mostly "A" grades. Gwen found the cabin a good place to prepare lectures and correct examinations. There was no trapping but finally I had that opportunity to hunt, fish and live in a cabin in the woods! And there certainly was excellent pheasant and duck hunting along the irrigation ditches east of Fort Collins.

We moved out of Walnut Apartments into the Veterans' Village on Campus. We lived in a quarter-round Quonset hut with our second hand furniture and wooden orange crates for bookcases. We acquired our "first child", Huckleberry, a Dalmatian pup, with feet that were four sizes too large for his body. He was a real stumble bum. Huck was soon joined by Sammy, a stray golden Cocker Spaniel, who adopted us.

The motor scooter continued to provide reliable (although we had to walk it up steep hills) and inexpensive transportation. Occasionally Gwen ended up with bruises, skinned knees damaged stockings and clothes when we slipped on loose gravel or ice. Most of the time she could stick on to our hazardous vehicle with her long hair flying in the wind. She was a great wife and companion! Though she hated the hike Gwen accompanied Myrl, Gene and me almost to the top of 12,324 foot Flat Top Mountain. Gwen ran out of steam and huddled and shivered under a rock outcrop near the top while we three went on. She ate candy bars we had left for her. I loved Gwen dearly. And she got me away from a domineering Mother!

During the last semester students were interviewed by a number of prospective employers from industry, state and federal agencies. Classmates Steve Yurich went to the Forest Service's Region 2 Rocky Mountain Region and Alan Lamb and Chet Shields came to Region 3. (Our careers and retirement paths crossed many times.) I was interested only in the US Forest Service. I wanted to be a forest ranger. I researched and identified the needs of the Southwestern Region 3 (Arizona and New Mexico). When Personnel Officer Ralph Alskog, came to Fort Collins I convinced him that my studies and research in wood technology and preservation were just what they needed to get a wood preservation industry started. After all the Southwestern Region was a well known range management region. It was dominated by range conservationists and cowboys.

Subject to an acceptable score on the Federal Civil Service Examination, I was given an offer as a P-I (GS-5) Forester in Region 3 at a salary of $2,985.00 per annum.

My last semester at Colorado A&M was a good one academically. We had a seminar to help us study for the Civil Service examination. Studies covered every phase of forestry, wildlife, range, recreation, botany, entomology, pathology, soils, surveying, geology and related subjects. The fateful day arrived and we were given a four hour exam. To our horror it contained not one natural resource-related question. The difficult examination consisted of reading, reasoning, math and logic problems, spatial and visualization, history, English and subjects not even remotely related to forestry. I made a 72 + 5 points for being a World War II Veteran, for a total score of 77. Without my knowledge or permission, my Father wrote a letter to Democratic President Harry S. Truman urging him to give me an appointment. I was fit to be tied when I learned about that letter. I had earned the job through my own efforts and merit. The Federal Civil Service examination for foresters and other disciplines was discontinued a few years later because so few applicants were able to make a passing grade.

A week before graduation, forestry students were dispatched to a forest fire in the foothills above Fort Collins. In felling a dead snag with a cross cut saw, the tree kicked back and hit me on the knee and knocked me down the hill. I was laid up for several days with a badly bruised knee. I hobbled along in the graduation procession in my Masters of Forestry cap and gown. My folks came to the graduation ceremony. Gwen always tried to be nice to my Mother but the favor was not returned until almost 40 years later. For a few days, due to my injured knee, I was unable to drive our new (used) 2-wheel drive, 3-speed Willys Jeep metal station wagon. We had parted with the motor scooter for $75 and bought the Jeep for $700.

The day finally arrived when we loaded a small 2-wheeled trailer with all our earthly possessions, $1,000 in savings, Huckleberry and Sammy. We said goodbye to friends and optimistically headed for the Kaibab National Forest at Williams, Arizona for my first permanent job. Myrl and Gene Jensen stayed on in Colorado for a few more years, she as a public health nurse and he as a public health sanitary engineer. Our paths and lives were to cross many times.

BOOK TWO

About Forest Rangers

From 1948 to 1966 two dominant chiefs guided the Forest Service into the multiple-use era and influenced my early career. Both were outstanding leaders; they were people-oriented and they made prompt and well thought out decisions.

The eighth Chief, Richard E. McArdle, came up through the ranks from Forest Service Research and Cooperative Forestry. He projected the image of multiple-use forestry and helped the world community understand and appreciate the concept. Prior to passage of the Multiple Use-Sustained Yield Act in 1960, many groups wanted the same acres and pressure was on national forest administrators to yield in one direction. McArdle's job as Chief was to see that none of these pressures got out of line, that all were integrated and in his words "that all the user groups got a fair shake". Nevertheless, timber and range were dominant programs before 1960. Receipts from these activities returned huge dollar amounts to the US Treasury and to local governments, and in turn brought in the appropriations from Congress that enabled us to hire more foresters, range conservationists and engineers to construct a needed transportation system to protect and manage the timber and range resources. Other resource activities were managed, but these were not as well funded for the needed management. The MU-SY Act spelled out a needed authority and declared that the national forests are to be administered for the five major uses—wood production, use as watersheds, grazing by domestic livestock, habitat for wild game and fish, and outdoor recreation. While the Act recognized the established wildernesses, emphasis was on utilization, not preservation. Concerning wise use and balanced use, Dr. McArdle clearly stated the simple facts of life: "The question . . . is not preservation *or* use of resources but how best to achieve both preservation *and* use. If this be a sermon, make the most of it!" (McArdle, 1975)

The MU-SY concept endorsed by Chief McArdle resulted in improved management of the renewable resources in a combination that would best meet the needs of the American people. When Chief McArdle visited our ranger district on the Coconino National Forest he looked me in the eye and directed me to manage the Long Valley District and maintain a high level of output of forest resources without impairing the productivity of the land. That personal wise use and land stewardship message to me from the Chief was my paramount driving force for the balance of my career.

Edward P. Cliff was Chief from 1962 to 1972. He came up through the ranks. He knew the job of the district ranger, range staffman, forest supervisor, regional office specialist, researcher and state and private forester. He was a strong supporter of the total Forest Service mission . . . the National Forest System, Research and State and Private Forestry. He referred to this arrangement as the 3-legged stool. Chief Cliff worked smoothly with members of Congress, the White House and the public. He was well respected by everyone. We all shared in policy making and when he made the decisions we all supported him and went in the same direction. This resulted in a smooth working organization.

During these years the Forest Service helped focus attention on the global importance of forestry through symposia, the Fifth and Sixth World Forestry Congresses, supporting the United Nations Food and Agriculture Organization's Forestry Program and fielding Forest Service expertise to help other countries.

The environmental movement began in earnest. The Wilderness Act was passed in 1964 and 14 million acres of national forest land were classified as wilderness and primitive areas. The Forest Service was the first government agency to designate large areas of wilderness with specific restrictions aimed at preserving wilderness conditions. The effort to set aside such areas began on the White River National Forest in Colorado in 1919, Superior National Forest in Minnesota in 1924 and on the Gila National Forest in New Mexico in 1924. From the beginning wilderness preservation was tied to national forests set up administratively under Secretary of Agriculture regulations. The Wilderness Act legislatively established wilderness and Congress was to play a key role in the selection, designation and management.

Up to this time the Forest Service was dominated by foresters, range conservationists and engineers. We began to add more landscape architects, fish and wildlife biologists, hydrologists and other disciplines. While the Forest Service was quick to respond to various thrusts to improve the quality of the environment and natural beauty, we insisted that we knew what was biologically and silviculturally sound when it came to management of the timber resources. That inflexible attitude resulted in the public leading us to the "woodshed" and later to Federal court. The public looked at what we were doing and did not like it. But the "omnipotent foresters" of the Forest Service were not particularly good listeners and we missed the message. (Behan, 1966)

Chapter V

Learning the Tricks of the Trade

We arrived in Williams, Arizona on 1 July 1948 and reported to the Kaibab National Forest Supervisor's Office. Edward C. Groesbeck, Timber Staff Officer and Assistant Forest Supervisor, made us feel welcome and introduced us to the staff and my boss, Chalender District Ranger Vince Schroeder. Ed Groesbeck and his wife Blossom were to have a strong influence in the development of my career and family over the years. We were directed to Camp Clover, two miles west of Williams, and the house we were to live in temporarily. At that location was another ranger, Jewell Wyche and his wife Effie. They helped Gwen make the transition to this northern Arizona drought stricken world of red and black rocks and bushy juniper, pinyon and ponderosa pine . . . a world away from the green farmlands of the Midwest. Effie had been Gwen's grade school teacher in a one room country school near the family farm in Minnesota. Another old friend of Gwen's from back home, Kenneth Williams, was the lookout-fireman on nearby Bill Williams Mountain. Those friends helped Gwen a great deal since she was a long way from her family and friends.

The three bedroom house at Camp Clover was a disaster. The previous occupant moved out the previous fall and neglected to drain the water system. Pipes, water tank and the water heating chamber in the wood cook stove were split from end to end. That was where I received my first training in plumbing. It was a real chore and while I became a reasonably good, experienced plumber, I hated plumbing.

Working for Ranger Vince Schroeder, as his Assistant District Ranger, was a special challenge. His two previous new foresters had resigned or been discharged. He was a hard worker, a perfectionist, very demanding, but an excellent teacher for a young lad from Iowa with little or no experience, other than "book learning". He taught me timber cruising, log scaling, timber stand improvement, brush disposal, handling and caring for horses, range management, forest fire fighting, general maintenance and safety in all work practices. I had some difficult days trying to please Vince. He was also tough on the temporary fire guards and timber crews. I learned some of my best supervisory skills from Vince. I observed how others and I reacted to his direction and style and then acted in a "more positive" manner.

After we repaired the plumbing and made the house at Camp Clover livable, Vince moved us to a 3-room guard station at Spring Valley, about 25 miles northeast of Williams. Gwen will always believe he moved us just to be mean. That was the first time she said "damn the Forest Service", but it was not the last. I enjoyed my work and was gone for days at a time in the field or off on some other National Forest on forest fires or at a training session. Vince and I would work in the office on Monday. On Tuesday at daylight we loaded our food and gear and took off for the field and worked hard all day. At dark we made camp wherever we happened to be, ate and crawled exhausted into a bedroll. Sometimes I had trouble keeping up with Vince due to my war injury, but I did it. The ankle caused me trouble for about five years and then I got used to it. I wore out my right boot from dragging my foot over the rocks. We were up at dawn and worked all day. We accomplished much and I learned a great deal about national forest activities, including working with livestock men and loggers. I thought I did a good job but I never received one word of praise from Vince. I taught him how to read aerial photographs and how to pin point precise locations on the photos used to revise our district maps. He really took to that activity and when we were near a section corner we would stop, find the corner, pin point the location and go on to the next chore. Consequently we had the most accurate, up-to-date maps of any district in the Southwestern Region. Vince believed in and practiced progressive travel. Seldom was a mile wasted on the Chalender Ranger District. It was usually late on Friday when I returned home, completely exhausted.

Gwen did not drive and was lonesome at times for family, friends and the green fields of Minnesota. But she enjoyed Huckleberry, Sammy and the remote area at the foot of Sitgreaves Mountain. She cared for the horses, split wood, carried water, read, refinished more second hand furniture and took long walks in the woods. She could find her way around very well. Our toilet was a little building at the end of a path in the woods. She enjoyed her life at Spring Valley. She had her own personal Stevens Off-hand single shot .22 pistol and could use it and my revolver very effectively. I purchased a Smith and Wesson Model Military & Police .38 caliber revolver that I carried in the truck. Fortunately we never had to use them for anything but target practice.

Gwen went home on the train in October for her sister Beth's wedding. Spring Valley was a bleak deserted place without her. She enjoyed seeing family, friends and the corn fields of Minnesota. Gwen came back looking tired and worn out after sitting up for three days on an exhausting train trip. She looked good to me, and even more so, when she announced she was pregnant.

Night in Spring Valley could be blacker than the inside of a cow. We used a gasoline lantern and had a propane gas refrigerator, neither of which worked too well. It was a dark and lonely place with no nearby neighbors. I recall we were both spooked the first time we heard the bugle of a bull elk in the middle of the night just outside our bedroom window. We did not know what that high pitched sound was until we identified the source as a huge bull elk calling to his cows and warning other bulls in the vicinity to keep away.

Later that Fall we received word Gwen's Mother was in the hospital for a gall bladder operation and was critically ill. We packed up and drove non-stop to Minnesota. Sad to say, she died before we arrived. She had been given a blood transfusion and this resulted in severe hepatitis which caused her death. This was a great loss for Gwen, her family and me. Alice Sanders was a great lady and I have very fond memories of her. With heavy hearts we drove back to Arizona.

In November 1948 we moved out of Spring Valley in a fierce snow storm and into a rented one room and bath apartment in Williams. This rather bleak, cold home was a converted "tin" garage. Access was through an alley near the Forest Supervisor's Office. Gwen could get out to Williams' small library and visit with some of the people we had met. Local folks had been telling us how lucky we were to be out of that cold Colorado country, since Arizona winters were relatively mild. In the month of January 1949 alone, we received some 13 feet of snow. We spent much of that winter directing aerial hay drops and guiding Army, game biologists and sheriff's people to snowbound ranchers, livestock and starving deer, antelope and elk herds.

That winter was the first time I got back onto skis since I broke my leg during the War. I used the skis for the rescue work and once a month to take snow depth measurements and weigh snow cores for moisture content on a winter snow survey route. The snow settled to a depth of over six feet that winter. We heard no more talk about mild winters after that very severe one.

In the Spring of 1949 we moved first into a two room cabin, with an outside toilet and then into the modern two bedroom ranger's house at the Chalender Ranger Station, east of Williams in Pittman Valley. Melissa was born at the Flagstaff Hospital on 21 June 1949. We were both happy to have her arrive safely. Sister Leela came to Arizona to help Gwen for a short time. The Sanders' family was very supportive of one another. It was pleasant, though noisy at times, to have a baby in the house.

During this period our district office was in Williams and at least once a week, I worked there. On one occasion I was questioned by some Washington Office inspectors concerning my views of the Forest Service work planning system. They carefully reviewed my official daily diary and monthly work plans. I told them annual plans were a waste of time and any good manager could plan effectively on a monthly basis. That was the wrong thing to say! These inspectors were the originators of the annual work planning system. I received a lengthy lecture and probably a black mark on my record for expressing my views. (Some 25 years later I was in charge of policy planning for the Forest Service to the year 2020!) Much to my delight, most of my time that year was spent in the field.

My work became even more diversified. I was responsible for the total care of the horses, earned a blaster's certificate and successfully used many cases of dynamite on various projects, cleaned campground toilets and garbage pits, trained and supervised fire fighters and lookout firemen, maintained many miles of telephone line and back country trails, surveyed land and located section corners and property lines, worked with aerial photos, conducted wildlife surveys and handled the normal timber sale administration and range activities. I enjoyed the variety and could arrange my schedule to be home with my family evenings and most weekends.

I made a point of staying near the telephone or radio during the fire season. While this was extremely confining, and Gwen did not like it at times when she wanted to go somewhere shopping or visiting, I was always prepared to take prompt fire suppression action. Consequently we had a good fire control and fire prevention record. Unfortunately one of my contemporaries in the Region eased up and was away from his station at a critical time. A fire started and built up to a huge acreage with corresponding devastation of critical watershed and natural resources values and tremendous costs. His career never prospered from that point on. The Forest Service could not forget and was not very forgiving under such circumstances.

Vince became more satisfied with my work and let me carry on in his absence. During this period Ed Groesbeck devoted time to teach me more about cruising timber, scaling logs, laying out logging spur roads and working with loggers. Some of the loggers used new, heavy two-man chain saws, but others stayed with crosscut saws. Some operators used crawler tractors, but others used a team of horses to skid logs. Some loggers used mechanical, power driven log loaders, but many used poles and a cross-haul arrangement which were rigged in a tree. Work in the woods was very dangerous so I followed our rigid safety program very closely and insisted that others do likewise. This safety attitude paid off, for over the years I stressed safety and I never had a worker seriously injured or killed on the job. I found all this new work interesting and exciting. I recognized and respected the loggers and their job needs and I taught them our requirements. We got along very well. I quickly learned that without the loggers we could not carry out our management program.

Ed Groesbeck was an outstanding forester, a good teacher and very practical. Working with him was a special privilege and a challenge. Ed could not see or hear well in spite of thick glasses and a hearing aid. Our conversations were yelling matches. Ed smoked 15 to 20 Roi Tan cigars each day and those very strong cigars filled the air of the pickup truck with choking smoke. Ed was a fast driver and many times on loose gravel we skidded around curves and had a few near misses involving logging trucks. I was always grateful that Ed took the time to teach me the "ins" and "outs" of timber management and getting along with loggers and others.

Telephone line maintenance was interesting work. I learned how to service both single line grounded telephone system which was hung on trees and a double line metallic telephone system which was suspended by poles and cross arms. I used climbing spurs for climbing both poles and trees. The gaffs for climbing trees were much longer since they had to dig deeply into the bark. I slipped, fell and skinned a few trees, poles and myself before I learned the proper technique. But after awhile I could do it quite well. It was very difficult to climb trees four feet or greater in diameter to a height of 50 or 60 feet to repair or reinstall a telephone line that had broken or pulled out from a tree staple and split tree insulator. Some of those trees stood on the side of a steep mountain and there was nothing below except a rock slide. Needless to say I did not fall out of any of those trees. I did have poles break with me hanging on. The trick was to loosen the safety strap, kick the spurs free and get out of the way of the pole. I did not land on the bottom side of any of those broken poles. Fortunately within a few years all of the telephone lines came down and were replaced with 2-way radios. I truly had an interesting job.

Assistant Supervisor Sam Sowell was a grand old man. He taught me much about the practical aspects of range management and how to get along with livestock people. Among other things range management involved checking the range for trespass use by unauthorized horses and cattle. While I was not an experienced cowboy I rounded up several horses after we had given a rancher repeated warnings that he was in trespass. I notified him he could redeem the horses at the ranger station by paying the trespass fees. He came with a .30-30 rifle instead of the fee. I cited the rules for redemption. He said he would redeem them with his rifle. With that he levered a shell into the chamber and took his horses home without paying the fee. I used discretion and turned the law enforcement over to the local sheriff who disposed of the range trespass case by nailing the rancher's hide to the wall. In later years it became a federal crime to shoot a federal forester. Now that was a comforting thought!

That fall we moved back to Camp Clover for the winter. The Forest Service moved in a three room cabin for our use. Gwen and I spent many hours painting, repairing and making the place livable. On occasions we converted the bathroom into a darkroom and we began processing our own black and white film and photographs. We spent much of our own money in fixing up many Forest Service houses. I should note at this point that the annual salary of $2,985 did not go far. The $1,000 savings we brought with us from Fort Collins had long since disappeared, as we added more household furniture. We literally lived from pay check to pay check, even after I received a promotion to GS-7 and made over $3,000 per year.

We put in long work days. Vince and I continued to camp out, go to work at sunup and return to camp dead tired by dark. I put in 40 to 80 hours of non-compensable overtime work per 2-week pay period, that was over and above our 40 hours of paid work. Forest Supervisor Leonard Lessel told me about the 40-hour law and our entitlement to overtime pay, but warned me what happened to anyone that claimed overtime pay. And that would have not led to a long and successful career. He was a very conservative man and typical of many of those old time supervisors. When I asked for funds to buy paint and a few supplies to fix up one of the cabins on our own time he told me he started out in a one room log cabin with a dirt floor and what we had should be good enough for anybody. He also reminded me that I did not need to use all my annual leave and I could lose it rather than use it. I did not go for that suggestion and I either used all my annual leave or accumulated the maximum allowable carry over for later use or payment. I was fortunate in having no illness or injuries and was able to save all of my sick leave time. The lack of funds for paint and supplies and the talk of giving up annual leave did not go well with Gwen. I found myself between a rock and a hard place many times, as she said "damn the Forest Service!"

Supervisor Lessel also oriented me on the importance of doing whatever my bosses asked me to do. He said it really did not matter how well I did the field work if I did not get my reports in on time. In fact he said my career would be "dead" if I did not meet due dates. His supply clerk, Charlie Cartmill was even less flexible. Vince would send me in for two dozen pencils. Charlie would give me a lecture and only two pencils. He said two should be enough for any one. Bring the stubs in and he would provide replacements. Lose anything and you were in deep trouble with Charlie! I was really in trouble anyway when I went back to Vince with only two pencils!

I learned a great deal and had many interesting experiences during those early years on the Kaibab, like the time I put my hand behind me to lower myself to sit on the ground and there oiled a rattlesnake. Climbing tall trees to pin point forest fire locations was especially challenging. Then there was the time Huckleberry, Sammy and I were out on a timber sale area marking timber on Kendrick Mountain. The dogs starting chasing a new elk calf and its mother did not go for that foolishness. She started lashing out with her hoofs and chasing the dogs. They came running to me and I had to react quickly and climb a tree to get out of the action. Picture this, if you will, there was the elk chasing the barking dogs around the tree I was in and me yelling at all of them. It was exciting! But what worried me the most was the awful thought of having to take a handful of spotted dog home to Gwen and Melissa. Fortunately we all made it through the episode, wiser and without a scratch.

Let it be recorded here that one of my successors on the Chalender District as Ranger Vince Schroeder's assistant was as well prepared for the job as I was in the beginning, which was not saying much. But he did not last very long. He had a reputation for getting lost in the woods. Anyone who says they have never been lost in the woods is generally lying or has forgotten. It happens to every one, but most of us finally are able to become oriented and find our way out. But he could not keep from getting lost until he found a unique way to solve the problem. On one occasion someone went to his work area to check on his progress. They found his pickup truck along the side of a woods road. Attached to a door handle was a white string that disappeared into the forest. Following the meandering path of the string, which was laid out while he wandered through the woods marking and measuring individual trees for harvest, they found him with string feeding out of his backpack. He apparently retrieved the string and eventually wound his way back to his starting point.

His termination came before his probation year was completed and shortly after the deer season. Seems like he came through the Arizona Game Commission's deer checking station with the head of a nice buck and the body of a fat doe. There was no legal season for female deer. The buck's head was perceptively older than the carcass and putting the two necks together would have resulted in an abnormally long-necked deer. He was fined heavily for his violation of State game laws and left the Service shortly afterwards.

Another of Ranger Vince's assistants was actually too lazy to cut wood for the cook stove. One cold winter morning Vince found this forester waving a lighted newspaper through the frigid air in an effort to warm his one room cabin. I understand Vince blew his top! His face and bald head could take on the most vivid scarlet color. Vince apparently appreciated my work, since he told others I did a good job, but he never once told me. I was always grateful for what he taught me. Other valuable lessons on fire suppression, fire safety, advice on range management and how to get along with people were provided by Williams District Ranger Jewell Wyche.

In May I was offered a possible promotion to GS-9 and a transfer to the Coconino National Forest at Flagstaff, Arizona. We had lived in six different houses while on the Kaibab. It was good that we did not know at that time there would be 17 more houses to move into, out of and fix up, before retirement from the Forest Service.

First Turkey Happy Jack, Arizona 1952

Chapter VI

The "Mighty Coc"

My first job on the Coconino National Forest, the "Mighty Coc" as it was called, was a timber management planner. I was assigned to the Forest Supervisor's office and responsible for inventory and preparing a timber management plan for the 2 million acre forest. It was interesting work and began in the Flagstaff office. First I had to delineate timber types on aerial photographs and then statistically determine how many permanent 1/4 acre plots were required to sample the area to the desired accuracy.

During these first few preparatory weeks I was able to be at home every night except when I was called out on forest fires. We were assigned government quarters at the Elden Guard Station which was four miles northeast of Flagstaff. Gwen liked it there even though she was grounded and led what most people thought was a lonely life. Efforts to teach her to drive ended in near disaster. She ran the station wagon into the ditch on a practice drive and never had the inclination to learn after that. I guess she decided there were enough poor drivers on the road without adding another. That was the year we traded the Jeep station wagon for a new 1950 2-door Chevrolet sedan, which carried with it 36 months of payments. Gwen had Melissa, Huckleberry, chickens and two government horses to keep her company. Sammy, the Cocker Spaniel, became a drifter. He wandered off and must have adopted another family. We were happy there and had good times and vacations together. Cindy was born in the Flagstaff Hospital on 17 March 1951 and was brought home to Elden.

Elden was located in the fringe of the ponderosa pine timber type. In a short distance to the east the ponderosa pine was replaced by juniper and pinyon. The area generally had mild weather. There was some light accumulation of snow in the winter. From April to mid-July there was seldom any rain. June was particularly dry and the fire hazard was extreme. We had to be especially careful about fire when the temperature approached 90+ degrees, with winds of 20 to 30 MPH and relative humidity ranging as low as zero. It was so dry that you could smell the turpentine in the pine trees and the inside of my nose dried out and bled. Dry lightning and light scattered rain could be expected any time just before or after Flagstaff's 4th of July all Indian Pow Wow. Heavy rain and much lightning could be experienced in late July and August. Gwen enjoyed this variety. Occasionally she received rides from other Forest Service wives into Flagstaff and to social affairs. Mainly she enjoyed Elden, the children, Huckleberry and me.

Huckleberry did not like the milk man's truck and one day he ran too close and received a broken left hind leg when he was struck by the truck. Fortunately our nearby friendly veterinarian was able to splint his leg and he recovered with barely a perceptible limp. Huckleberry caused some excitement one day with his frantic barking near the front porch. He treed a small black bear. Gwen grabbed the children who were playing nearby and Huck and took them into the house. She was afraid mother bear would show up and not approve of the foolishness. After a while the cub came down and scrambled off into the woods.

We became acquainted with the southwestern Indians during this period. They worked for us on forest fires and we always went to the 4th of July Pow Wow parade and rodeo. It was a fascinating experience to go into one of the nearby Indian trading posts and look at the pawn jewelry. When the Indians needed money they pawned some of their silver and turquoise jewelry. Traders seldom sold any pawn to the public. They held it for years and eventually the Indian family would redeem it. The pawn was kept in a large walk-in safe and could be observed from a distance. It was a veritable treasure house full of bracelets, rings, buckles, squash blossom necklaces, and other fine pieces. We bought a few modest rings, bolo ties and bracelets that were made for sale. The jewelry as well as the rugs and woven baskets were very inexpensive, but we did not have much money to spare in those days. I purchased small Navajo rugs for saddle blankets, dog rugs and throw rugs for wiping our muddy feet. These sold for $6 to $8 each. Unfortunately we did not recognize the value of such common items until they were worn out or prices escalated beyond our reach.

While on a trip to New Mexico we purchased two small black pottery bowls made by an Indian woman by the name of Maria. We gave these two items to my parents for a gift. My Mother soon grew tired of having that "cheap Indian junk" around and sold it at a rummage sale for a couple of bucks. Years later a huge demand developed for Maria's pottery and prices sky rocketed. Small objects similar to what we had purchased were valued at hundreds of dollars.

I took the family with me on some of my field trips. The girls enjoyed the woods and splashing in the few shallow running streams. They did not enjoy Mack's Crossing. This access road had been started and half finished by a CCC crew in the 1930s. A carefully engineered and well constructed segment reached from the north rim to a crossing of East Clear Creek. The south side access was a pioneer road at best. It was an eight foot wide shelf that had been blasted into the side of a sheer cliff. At a number of points small rock slides removed parts of the road which were bridged with 4 by 12-inch timber planks. The road had an outslope and when wet passage over it could be a harrowing experience. Gwen always pulled the children and Huckleberry out of the pickup truck and they walked up the south side. I must admit it could be a bit scary even to walk it, let alone drive it, as I did many times to reach my work areas or forest fires.

Most of the time I was in the field working by myself and away from home for the entire week. As my field work progressed south I ended up over 100 miles from Flagstaff. I camped out with timber stand improvement (TSI) crews who were working in the area. I had purchased a Smith and Wesson K-.22 pistol. It was a honey and very accurate. Few people hunted the grey Abert squirrel in Arizona. There was the local popular belief that it lived in the pines and tasted like turpentine. In fact they were fat and delicious. During the squirrel season I helped keep the TSI crew at Buck Springs supplied with game. They supplied the beans, biscuits and bacon. We ate well and I became well acquainted with these skilled technicians. I knew, on a first name basis, most of the permanent and temporary forest workers. In time I became very familiar with the Forest from one end to the other.

1950 was a bad fire year. The dry season was prolonged and we had many man-caused fires before the lightning season started. I had become an experienced sector boss by this time. The first large fire that season started shortly after I arrived on the Forest. Fire danger was rated extreme. A Santa Fe Railroad welding crew was replacing standard rails with welded continuous rails three miles west of Flagstaff. The right-of-way was filled with highly inflammable vegetation and partially decayed ties that had been thrown aside with other debris the railroad crews had failed to remove or dispose of properly. This problem existed all across the Forest, from the eastern to the western boundary, some 35 miles. Repeated reminders and pleas from the Forest Service to the Santa Fe to remove the debris in accordance with their 19th Century right-of-way easement agreement went unanswered. Sometimes they would make a token start at clearing and then a train wreck somewhere along the line would occur and all the men would be pulled off the clean up job. They always conveniently forgot to return and finish the job. The welding crew started a fire on the north side of the double tracks and burned over 23,000 acres before we stopped it. I was a sector boss and my crews were on the head of this dangerous fire. The Santa Fe acknowledged blame, however months later the Department of Justice neglected to meet a deadline for filing a judgment and the Santa Fe had a free ride.

The fire did not cost them a penny, while it had cost the government thousands of dollars in fire suppression costs, fence replacement, reseeding and reforestation. After the fire Santa Fe cleanup efforts were still periodic or non-existent. All of this was recorded in the back of my mind for future reference.

That same Spring I was dispatched to a large forest fire on the Lincoln National Forest in New Mexico along with Harlow Yaeger and four other Coconino crew bosses. Harlow supervised project crews on the Elden District and had visited us at the Elden Guard Station. While he never talked about it, Harlow had been a Prisoner of War when he was captured by the Japanese in the Philippines during World War II. He was an excellent person to work with and get to know. He taught me a great deal. En route to New Mexico we saw a series of large fires burning on the Sitgreaves and Apache National Forests. Our fire on the Lincoln National Forest was known as the Capitan Gap fire. A very experienced forest officer by the name of Dean Earl was Fire Boss. Harlow and I were assigned separate sectors that burned hot all night and crowned and ran during the day. The temperature was very high, it was extremely dry and windy. The fire camp and our food was constantly buffeted by gale force winds. Wind blown debris and sand were in everything we ate. It was a very dangerous fire.

Harlow was in charge of a group of soldiers from Ft. Bliss and was assigned to the southwest sector running up on to Capitan Mountain. His company of soldiers were endangered by a crown fire and almost trapped before Harlow led them to safety on a rock slide. When the fire had passed around them one of the soldiers heard mournful crying and whimpering sounds. He found a tiny black bear cub clinging to the trunk of a burned tree. The cub's paws were badly burned and one of the soldiers carried him back to the fire camp. I was there when they arrived and the bear was handed to me. It was crying, whimpering and licking its burned paws. I recall it crying as it chewed my leather gloves. I did not know at the time this 10-inch ball of black fur I held in my hands was to become the original Smokey Bear. For years afterwards Melissa and Cindy asked me to tell them the true story of Smokey the Bear. Many people think that I found him. I did not find him, but I held the little creature for a short time before he was taken away to become that famous symbol for fire prevention.

My work on the timber management plan, including field work and office work, continued into 1951. I was assigned Rusty Richardson and some other valuable helpers and we finished the job in spite of the weather, getting stuck and being dispatched to numerous forest fires. Rusty was a graduate of Louisiana State University. He was a pleasure be around. He soon became a district ranger.

Rusty and I learned a valuable lesson on that job. Early in the Spring we drove a 4-wheel drive Dodge Power Wagon into a remote area near Barney Pasture. We had to straddle two foot deep ruts which were separated by a high center ridge. Rusty stopped and I got out of the vehicle to scout the route. I turned back and shouted "Rusty, are you stuck?" He replied "no" but as we watched, to our horror, the large power wagon truck slowly slipped into the ruts and was high-centered, but good! We built a lot of road that day and learned our lesson . . . "get stuck on government time, never on your own time!"

In May 1951 I was offered a promotion to GS-11 District Ranger and moved to the 425,000 acre Long Valley Ranger District. I knew the area well from my timber management field work. I knew most of the people and was aware of some of their morale and organization problems. One problem involved a conflict between the forestry technicians and the professional foresters. The previous district ranger had been unable to pull them together as a team. He worried a great deal and tried to do too much himself, rather than delegate work. Our family moved to district headquarters at Happy Jack, Arizona which was some 40 miles south of Flagstaff. We enjoyed our life there. I was able to be home almost every night. With the strong support of Forest Technicians Norman Johnson, Ken Cook, Homer German, Don Bassler, Henry Summerfield, Ciefredo Gutierrez, Alvin Teague, Bill Phillips, Henry Bolin, John Sanders and others, we developed a smooth working team. I gained great respect for these forest technicians. They were the real experts and I had learned a great deal from them about practical forestry and how to work with people. We had fine professional foresters at Happy Jack, such as Frank Carroll, Bob Bates, Ted Ingersoll, Frank Bell and many others. Marge Bates was my secretary in the Happy Jack office. Ken Cook's son, Jim, went on to become a highly respected reporter for Phoenix's *Arizona Republic*. Norman Johnson's son, Norman, Jr. went on to earn a Ph.D. and became Vice-President for the Weyerheuser Company's forestry research program. Dave German became one of the Region's top timber staff men.

The Coconino had more fires than any other national forest in the United States. And the Long Valley District had the highest number on the Forest. Most of these fires came before and during the rainy season in the form of lightning fires. Those that came in May and June were the dry variety of lightning and occurred without a drop of rain. These were the most troublesome, particularly following a prolonged dry Spring. The day I arrived on the District a "lightning bust", resulting in multiple fires, was predicted. Experienced district personnel ordered 100 Navajo Indians to be transported to Long Valley and were available on standby for dispatch as needed to fires. The Southwest Forest Industries' loggers and support personnel along with large trucks and bulldozers were available for fire suppression action. Sure enough the dry lightning storm rolled up over the Mogollon Rim from the hot, dry Tonto National Forest country below and lightning literally stood on end. Before it stopped we had 125 fire starts in less than three hours and not a drop of rain. The fire lookout stations were smoked in and blinded. But under the direction of Ken Cook, the experienced District people transported and dropped off 5-man Indian crews whenever and wherever they found a fire. Through such organization and prompt action, not one of those fires spread to over 10 acres in size. I did not interfere or give any orders. I let the organization function and it did and did well. I believe my behavior on that day endeared me to the men and helped strengthen the organization. From then on I saw to it we were well prepared, trained and equipped and then let the organization work as it was designed to function. This gave us all confidence and a high espirit de corps. Before I left the district we had one "fire bust" that resulted in over 300 fires from one dry lightning storm. Results were the same.

Long Valley was a heavy work load forest. In addition to large timber and fire programs we had a large road design and construction program, and major range, wildlife, recreation, soil and watershed programs. The work load and pressures were intense at times. As the line officer I was responsible, but we had good crews and I learned to delegate authority to others to accomplish quality work done and on time. We had a good team and a good production record. Field work continued well into winter. But finally we were forced to move out and Happy Jack closed for the winter. We rented a small house in Flagstaff on Elm Street which was within walking distance to the office, shopping and the library. Gwen became an assistant librarian and enjoyed the work for the remainder of our stay in Flagstaff. Our winter office was in the basement of the Post Office, where we practically sat on one another's laps. We all developed "cabin fever" and were glad to see winter pass and move back to Happy Jack in the Spring. We rented the Elm Street house year long so we would have a place to stay any time we came to town.

Work pressures finally got to me and I developed a pain in the center of my chest. Our physician was Dr. Dan Bright in Cottonwood. He diagnosed my problem as "Coconinoitis" and advised me to ease up, relax or I would develop ulcers or cardiac problems. I followed his advice and more fully used our fine district organization to get the work done. The crew responded well to their added authority and my health improved. Dr. Bright was also the girls' doctor. We worried about Melissa who was very slender and had arms and legs like a little sparrow. I called her "Bitsy". Dr. Bright told us to relax and assured us that she would grow up to be a fine young lady, which she did.

Gwen, the girls and I enjoyed life at Happy Jack. By 1953 the Chevy sedan was paid for and we traded for a new 1953, all metal, brown and tan Chevrolet, 4-door Chevrolet Station Wagon. This one cost $2,950 and carried another 36 months of payments. (It was part of the family and with us for 30 years. With mixed emotions I sold it in 1983 for over $3,000.) Gwen became a good friend of Betty Carroll. This beautiful friendship ended tragically in 1978 when Betty died of breast cancer. Gwen and Betty seemed to think alike and enjoyed many of the same activities, talking, reading and travel. Against my better judgment Betty gave Melissa, on her third birthday, Shadrack, a long haired gray cat. Betty and Gwen liked to tell the Forest Service officials what was wrong with the outfit and what they should do about it. Huckleberry also helped in that regard. He bit Forest Supervisor Ken Kenney every time he came near. It is a wonder my career ever survived Mr. Kenney!

Generally Supervisor Kenney was supportive. He was stiff, unyielding and a former merchant marine officer during the War. He could never relax and forget he had been an officer. It showed in his stiff military posture and in his behavior towards all of his employees and the public. I recall being under his gun during one fire episode. We had a very long, hot and dry fire season. Our performance and results could not be questioned. We were good! We kept the burned areas small. At that time the policy was to control a fire by 10AM the next day and keep the fire manned until it was declared out, with no exceptions! By late August we had experienced well over 500 fires, the rains had come, vegetation was greened up, it rained hard part of every day and the fire danger was very low. Ten or so lightning caused fires cropped up each day and our crews were under great pressure. Some had worked for weeks without a decent break from continuous fire control work.

Late one afternoon the Minter brothers, Roy and Charlie, two of the most experienced fire fighters on the entire Forest, called in on the radio to Fire Dispatcher Bill Phillips to report the status of their fire. It was in a single dead snag they had felled. They had a line around it and water filled the fire line. It was raining and expected to continue all night. There was still some fire deep inside this extremely large dead ponderosa pine, but in their judgment it was safe and would not go anywhere. They asked if they could be relieved to return to their station for dry clothes, a hot meal and some much needed rest before returning to the fire the next day at daylight. They were exhausted having been on forest fires continuously for over two days.

I was in another part of the District when Phillips called me. I had heard the request over my radio and since we had no relief crews I directed him to release them with instructions to return at daylight. Radios have "big ears" and Supervisor Kenney heard the same dialogue. Following our discussion he radioed Dispatcher Phillips and asked for Ranger Cravens to call him over the telephone when I returned to Happy Jack. I called as directed and expected to be commended for the good work being done on the fires. He asked me how I liked being a district ranger. I said I liked it fine and had excellent crews that I was very proud of. His crisp concluding military-like remarks were to the effect that if I enjoyed being a district ranger and if I expected to continue as a district ranger I would see that the fires were manned until they were dead out. End of message! Bang went the receiver in my ear! Even before I could say "yes, sir!" So much for exercising good management and cost effective judgment. At that moment I had a wicked thought. I wished Huckleberry could have taken him by the neck and shaken some sense into him. I do not recall how we did it but that fire and all others were manned until they were declared dead out and followed by a check 24 hours later. It cost, but it was effective. We never had a fire escape after it was controlled. That policy continued in effect for almost two decades and then was eased. Action and control was then based on cost effectiveness and values protected. That may explain why some large fires in the West have run out of control and cost millions of dollars. I contend that experience and judgment make the difference. Unfortunately in recent years computers and simulations have taken the place of experience and judgment. The results are a design for disaster and losses have been tremendous.

In dealing with loggers and grazing permittees I did not try to show off my "vast experience", education or authority. I treated them with respect and consequently learned a great deal about the practical applications of timber and range management. I knew that if we did not have the cooperation of these groups we could not function effectively. The logging camp was across the road from our Happy Jack Forest Service site. Logging crews went by picturesque names such Dink, Houndjaw, Buttercup, Jellybean and a Henry. Dink was the very cooperative woods boss. Houndjaw was a truck driver who rolled in with a D-8 bulldozer on a trailer to save us on number of threatening forest fires. Buttercup and Jellybean were operators of huge D-8 Caterpillar tractors and were highly effective on forest fires. The skidding operators with their huge "Cats" and arches could make or break a good forestry program. They could obtain good utilization by picking up every merchantable log or carelessly leave logs behind for us to find in a utilization check of the area. They could help us thin the thickets by going through the most appropriate places, knocking down trees we could spare and leave the best crop trees. They loved to irritate Supervisor Kenney. They did not like his superiority attitude especially when he ordered them to do something for the benefit of Regional or Washington Office inspectors. Kenney about had apoplexy when they would show off in front of him and the inspectors and knock down good young trees.

When I came to the district I found the loggers were not fully cooperating when it came to forest fire suppression. The previous ranger had shut them down when the fire danger reached "extreme". Of course with no work they would all take off and go fishing. Then we lost a reliable source of manpower and operators of dozers and tractor-trailers to transport them. On the advice of the forest technicians we soon changed that practice. It was better to have a fire conscious source of manpower on the district than rely on outsiders. They never started a fire as the result of their operations on one of those extreme fire danger days. In dealing with loggers there was a need to release them for return to their regular work as soon as a fire was controlled. Loggers, like every one else, hated the dirty mop-up work required to put a fire dead out.

And who was Henry? Henry Ryberg was the head cook and along with his brother Emil, ran the logging camp. Henry's word was law. So keeping on his good side provided us with complete cooperation and the most delicious full course meals, both in the mess hall and on the fire lines. His peach pie was food for the gods! Gwen and the girls enjoyed eating an occasional meal at the mess hall. Henry even waived his rigid, almost unbendable rule of *"No talking during meals"*. He liked to come out of the kitchen himself and visit with Melissa, Cindy and Gwen. We had many of Henry's delicious buffets for the square dances we held in the loggers' meeting room. We enjoyed a special relationship with the loggers and local ranchers.

Livestock men were a special breed. Ernest Chilson had the largest grazing permit in the entire National Forest System. Irvin Walker started roping and breaking wild horses in the 1880s before there was a Forest Reserve or a Coconino National Forest. Since we had no government horses at the time I made my first horseback range inspection of his grazing allotment on one of Irvin's horses. It was the slowest, oneriest and roughest gaited horse that ever existed. Irvin was on a fleet, white stocking-footed mule. That horse just about beat me to death and spurs did little to change it out of low gear. Irvin told me years later that my action that day gave him a good opinion of me. He knew I was suffering. I never tried to be a cowboy and show off like some of the young foresters did. I learned much from that old man. Another rancher, Bruce Brocket, had been in the area almost as long as his neighbor, Irvin Walker. Bruce and his son-in-law Bill Sullivan frequently told me that it would not be the Forest Service rules and regulations that put them out of business; in their opinion it would be the recreationists. And over the years I have seen selfish, misinformed, special interest groups use heavy handed methods to deprive land owners and the public users of needed resources. Indeed a wide variety of surrogate measures were used to block timber harvesting and range operations over the years.

Long Valley was the place I was able to fulfill my wartime dreams of real hunting. I harvested elk, deer and antelope each year. I acquired two .30-06 rifles through the National Rifle Association and the Director of Civilian Marksmanship. Both were bolt action models. One was a Springfield and the other an Enfield. I traded the Springfield for a K-4 scope, had the "ears" ground down and barrel shortened on the Enfield so I could carry it in a scabbard when I was on horseback. Later I had it bedded in glass on a sporter stock to ensure more accuracy.

I enjoyed tracking and hunting elk late in the season, preferably during a snow storm. One season I was given the assignment of escorting one of the *Outdoor Life* editors, Charles Elliott, on an elk hunt in the Mogollon Rim area. Unfortunately my elk hunting permit was in a different part of the district. I acted as a guide for almost a week while he observed our management programs. He visited with hunters and took many photographs. The elk he finally shot was a huge bull, but it had antlers on only one side. Nevertheless he was proud of it. After we delivered it to a processing plant for cutting, wrapping, freezing and transport to Atlanta Charlie went on to the North Kaibab for a deer hunt. I was then free to go hunting. My crew was sympathetic and took over while I went hunting on the last day of the season.

I asked Alvin Teague and Ken Cook where to hunt. They both suggested I go down and have a cup of coffee at daylight with Irvin Walker on the "poor farm". That's what Irvin called his log cabin headquarters. It was snowing and had accumulated about a foot of fresh snow. I visited with Irvin about his cattle and plans for the winter, had coffee and then asked him where I could find a bull elk. He suggested I cross the fence out his back door and walk along the fence until I found fresh tracks. It was just breaking daylight and off I went. The snow was in my face as I hunted up wind. I had gone about 300 yards when I came across the tracks of a large bull elk. The tracks were so fresh they were literally "smoking", and heading into the wind in the direction I was going. I went another 500 yards and through the snow I saw a man and woman coming towards me. They were back tracking the elk. They asked if I had seen it, I crossed my fingers and said "no" and wished them luck in tracking it down. I was not about to give them a lesson in tracking on that last day of the season. Within another two hundred yards I saw where these two people had crawled through the fence and started following the tracks. Snow fall was increasing and daylight had arrived. In a short distance the elk circled to the west and climbed a low hill. The tracks disappeared into a young pine thicket. I hunkered down and scoped the area and finally located the elk and dropped it with one shot at less than 100 yards away. I never shot a deer or elk that was more than 100 yards away. The two hunters I had met earlier came running up to me and asked "how did that elk get over here"? Then I gave them a tracking lesson. They went away bewildered and scratching their heads.

One deer hunt was made with Roy Barnes, our timber stand improvement foreman. We hunted on the adjacent Mormon Lake District where the ponderosa pine merged with the pinyon-juniper type. We attempted to keep one another in sight in dense juniper. I jumped two large mule deer with racks like an elk and took a shot. The brush was dense but I had the cross hairs behind one deer's left front shoulder and fired. The other one ran towards Roy. Mine never appeared to slow down and I found no sign of blood. About then another large buck came busting out of the junipers. I sighted, fired and the deer tumbled head over heels. I put my deer tag on an antler and then field dressed it. Still no sign of Roy. I got to thinking how could I have missed that first deer and I started circling around for sign. I found it folded up and partially obscured under a juniper not too far from where I had first sighted it. I field dressed it but had no tag, in the land of many game rangers! Still no Roy! I spent a couple of frantic hours locating him and led him back to tag the deer. He wanted the meat and was glad to have that chore done. That relieved my "guilty conscious". All this brought to mind the transgressions of Ranger Vince's assistant on the Kaibab.

During dove, duck and goose seasons I had good jump shooting on livestock ponds. Once in two shots, I dropped more than my legal limit of ducks and had to call on Bill Phillips to come to my rescue. I still remember his "yuk, yuk, yuk". Duck hunting moved to the lower country in the Verde Valley when the ponds above 7,000 feet froze. In that lower area there was good Gambel quail hunting. Earlier I commented on my success in squirrel hunting. In the ponderosa pine forest the turkey season was also good. We used a light center fire turkey load. Shotguns were not permitted. I brought down my first turkey with a shot that connected at the hip joint. None of that 25 pound gobbler was damaged. The shock of the bullet dropped him on the spot. My next turkey was a young six pound hen that I spotted at a livestock tank. I jumped out of the pickup truck, crossed into the woods, loading a shell as I went. Unfortunately I had a 220 grain "Silver Tip" elk load in my jacket pocket and loaded it instead of a 110 grain turkey load. I hit a small hen dead center and feathers, blood and guts flew in every direction. I took home a small wing and drumstick from that mistake.

Elk hunts took place in late November, usually after we had burned the accumulated logging slash piles. We finished burning one Friday in a light rain. Elk hunters were coming into the area while we were burning in advance of the opening of the season the next day. Many came from Phoenix and were not well dressed for the high Mogollon Rim Country. From the Rim the ridges extend north to lower elevations. The only way out was back to the Rim road. On Saturday a light rain turned to heavy snow on the Rim. By Sunday night there was a good 30 inches of snow in the area. The weather cleared and the temperature plunged to below zero. We had moved to Flagstaff for the Winter. Sunday night I received a call from the Sheriff's Office . . . we had over 1,000 hunters stranded in the Rim Country and down under the Rim on the Tonto. Monday we started a full scale rescue operation, involving bulldozers, snow plows, 4-wheeled equipment, airplanes and rescuers on skis and snowshoes. The Sheriff was in charge and his efforts were coordinated with the Forest Service, State Game Commission and Army. Our headquarters was on the Rim at General Springs Cabin. A *Life* magazine photographer and writer recorded our rescue efforts. Within a week we had most everyone located and out. Hunters tried to return as fast we could extract the stranded ones. It was a real mess! Five hunters perished, four in the Rim country and one on the Tonto. All who stayed with their camps were safe. Those five had tried to walk out or were unable to find their camps and never made it.

1953 was my last year on the Coconino as District Ranger. The year ended while I was on a large fire on the Angeles National Forest in California. I was dispatched on Christmas Day along with a large group of Forest Service and Indian fire fighters. We spent New Year's Day fighting fire on Monrovia Peak and watching the Rose Bowl football game, from some 20 miles away. That fire threatened TV and observatory facilities on nearby Mt. Wilson, which we saved. Grateful TV stations provided fire camps with TV sets. It was the first time the Indians and many of us had seen TV. During the day the fire crowned and spread all over the mountainside. At night it slowed down when the temperatures dropped to freezing. The fire camp kitchen was manned by a religious cult of robed, barefoot believers. They were good cooks but those serving us had frozen, red and dripping noses which did little to enhance the quality of the fire camp food or our frozen bodies.

The fire area contained vast stands of poison oak. Indian fire fighters fell victim to the poison which was spread by the very dense smoke. I transported many to a nearby hospital. One of the foresters, Dennis Grassi, was terribly allergic to poison oak. Later I removed him from the hospital and transported him home by air. He was a festered sore from one end to the other. He was not a pretty sight to turn over to his wife. Getting the Indians and Dennis home was an assignment that finally returned me to Flagstaff at the end of the first week of January where I found I had no job! My position had been abolished while I was gone! The Long Valley District and four other Coconino districts were abolished effective 1 January 1954.

It took a while to recover from the fact that my job had been abolished. A delightful two week family camping vacation in Guaymas, Mexico made us all feel better. On return I still felt like the leader of a country who took a trip abroad and a coup occurred while he was gone and he was replaced. The Coconino National Forest reorganization created the Flagstaff Administrative Unit. Basically this eliminated five ranger districts from north of Flagstaff to the Mogollon Rim. Only the Sedona and Beaver Creek Districts remained essentially intact to carry on business as usual. The Flagstaff Unit was essentially one large ranger district managed by the Forest Supervisor and staffed by a number of resource specialists. This meant that the range staff person and his crew would operate out of Flagstaff and handle range management on the five former districts. Timber, soil, water, wildlife, recreation, lands and fire activities were handled in the same fashion. Engineering and construction and maintenance services were provided as needed, or essentially with no change. I worked on the transition and briefed the specialists about our former district programs. I could see this was an obvious design for disaster. We would have a group of specialists with blinders on heading out from Flagstaff to do, for example, a range job. They would fail to see the sign that needed to be replaced, the timber or occupancy trespass, a campground that needed attention, a timber salvage need, etc. This range specialist would not bother to talk to the special use permittee or the logger, his job was strictly range. It was clear to me that communications with the respective forest user groups would be muddled or lost. Fortunately an opening developed for me and I escaped the short life of that disastrous organizational experiment. (I returned later as Assistant Forest Supervisor to put the Forest back together again.) I was offered and I accepted the position of Fredonia District Ranger on the north unit of the Kaibab National Forest. Before leaving Flagstaff we had to break trail through deep snow drifts to extract the furniture we had stored at Happy Jack for the winter.

Chapter VII

The Arizona Strip

Fredonia, Arizona was 197 miles north of Flagstaff, 75 miles from the North Rim of the Grand Canyon and seven miles south of Kanab, Utah, in the so-called Arizona Strip. Fredonia was an isolated Mormon community. Previous district rangers had their office in Kanab. The Forest Service had recently decided to move the district headquarters to Fredonia. They purchased a 3-bedroom house in town and constructed a 2-car garage for the ranger. A 3-bedroom house, garages and warehouse were constructed in a flood plain at the south edge of town at a place affectionately called "Tanglefoot". Office space was rented in a newly constructed county building on US Highway 89 in Fredonia.

Some of my predecessors and their families had not been happy in this isolated corner of Arizona. Fredonia had less than 300 people, of which our family and one other were non-Mormon. The town supported a garage, two filling stations, cafe, small general store, post office and a new Mormon Church. I can say, without any reservations whatsoever, we were happier in this little town than any place we lived in before or since. We became part of the community. Gwen worked with the Women's Relief Society. Melissa and Cindy were active in Sunday School. We attribute the girls' self-confidence and ability to speak well as adults to their start in the Mormon Sunday School. At the respective ages of 3 and 5 they learned to stand up before a class and give talks to other students and adults.

Each month during our three years in Fredonia Mormon missionaries would call on us for one hour evening sessions. All these young missionaries come from the same mold. They come in pairs, two clean-cut, fresh-faced young men wearing conservative clothing and carrying small blue books in black brief cases. They knew the *Bible* better than any other religious group. They were dedicated to their founder, Joseph Smith and his beliefs which were recorded in the *Pearl of Great Price* and the *Book of Mormon*. They devoted much time educating us about their Church. We agreed with most of their beliefs except those concerning Joseph Smith. They told us that without Joseph Smith they would have no religion. They never pressured us to join the Church but they kept educating us all during our time at Fredonia and for many years afterwards. Gwen can still spot Mormon missionaries from a mile away . . . all project the same image. Our only real concern with the Church was the wonderful suppers that were held once each week. The food was absolutely the best ever. A problem occurred when it came to eating the most delectable deserts. Mormons do not drink coffee or tea, we were served cold punch. I did not mind but Gwen hated cold punch since she was a tea and coffee person. The same thing occurred following our regular square dances . . . delicious deserts, followed by ice cold punch.

Gwen enjoyed her life at Fredonia. I was gone much of the time since the district was 35 to over 100 miles away. During the fire season I was dispatched to forest fires on other national forests south of the Grand Canyon or in New Mexico and California. Gwen was a good wife and mother. She had the girls for companions and spent much time with them. She read stories to them, sewed and helped them fix up the playhouse I built for them in the yard. Looking out of our kitchen window we could see them and their many friends sitting on the low roof like so many little birds. They were very fond of Huckleberry and he guarded them well, sometimes too well. We had to warn people about approaching the girls when he was on guard. He continued to bite forest supervisors at every opportunity.

Gwen was our gardener and horse tender. The property in town had water rights. Our share of irrigation water was usually available for one hour between 0200 and 0300 hours. The local watermaster turned water into our ditch and Gwen would divert it to her garden and the horse pasture. She raised a wonderful garden on the site of an old corral. The soil was rich and very productive. We kept a government horse and mule on the property part of the year. These two were inseparable. In fact if I took the horse and left Jakie the mule behind he could jump the fence and come trotting and braying, as he tried to catch up. I either had to tie him up or take him along. When he was packing gear or supplies he would follow at my heels like a faithful dog. He never required a bridle or lead rope. Both were good to enter a horse trailer. However the mule was mean to certain people and could not be fully trusted. But he permitted the girls to crawl all over and under him. He liked Huckleberry too. Both animals were kept in good condition since they had good pasture and grain rations. An old Mormon cowboy told me one time when we were out riding, that in his reincarnation he wanted to be the ranger's horse. I am not sure that was meant to be a compliment.

Fredonia was the place we adopted, or vice versa, "Blackbird", a stray young black cat that hung around the nearby filling station until the girls found him. The garage man gave him to the girls and they tied a piece of rope around his neck dragged him home. Gwen and the girls loved that cat and nursed him for 15 years through many critical diseases, a kidney stone operation and tended his wounds after fights. I had been brought up to hate cats and told Gwen and the girls that "either the cat goes or I do!" Well we have had cats ever since then, and that shows who was boss in our family. Confidentially, I admit I was fond of all of the succeeding generations of cats we possessed. Among others we had Shadrack a longhaired gray, Calico and Patsy her kitten, black and white Pinky, saved on the day she was born by Melissa before people she was baby sitting for could flush her down the toilet, Laura a longhaired stray, pumpkin-colored Christopher a stray and the best of the lot in my judgment, tiger-striped Emily, Rose who was rescued after being caught in a fence, Gus the friendly gray and white biter and One-Ear a black and white stray.

Gwen did her shopping at the tiny local general store and in Kanab when I could drive her across the line into Utah. Frequently she could get a ride with her good neighbor Shirley Swapp or one of her many other Fredonia friends. Usually when I was called to the forest supervisor's office in Williams she and the girls would go with me and shop and visit friends in Williams and Flagstaff. It was during this time of our lives that Gwen lost her longing for the green farm fields of the Midwest. She loved the vast expanses of House Rock Valley and the Navajo Reservation which we passed through en route to Flagstaff. Then there were our own towering red cliffs filling the skyline north of Fredonia. To the south of Fredonia were the magnificent red canyons of Kanab Creek and the vast Grand Canyon. Further north we had ready access to scenic Bryce Canyon and Zion National Parks. It was beautiful, empty country and we loved it.

After normal working hours Gwen was my principal contact and relay point for radio messages from the Forest. She was very effective and efficient in using the 2-way radio in our living room. She was a real professional in using "10-4" language. She loved keeping in contact and relaying messages to and from the supervisor's office, local secretary Delores Riggs, assistant Bill Anderson, timber officers John Churches and Paul McCormick and cowboys Billy Swapp, Veldon Judd and Jense McCormick. In fact she enjoyed all of the Forest Service people. She and the girls enjoyed trips to the adjacent North Rim of the Grand Canyon National Park, the Kaibab Forest at Jacob Lake, Big Springs and the Dry Park Guard Station. She will always have a warm spot in her heart for Dry Park. It was in that little cabin we visited a few years earlier on a deer hunt that it got so cold that Cindy's diaper froze on her in bed. I am not so sure that is a heart warming thought for Cindy.

Gwen became an assistant to Dr. Phillip Fulstow in Kanab. For a period of time she worked several days each week in his office as a receptionist and laboratory assistant. While she still did not drive she could always find someone going to Kanab. She enjoyed the work and it was a very sad day when Dr. Fulstow was killed in an automobile accident.

As for my work on the Kaibab, the place was a paradise for a District Ranger. The Forest Supervisor was 230 miles away in Williams and he and his staff could not afford to take the time for the round trip to visit us very often. We were much on our own. We were responsible for 730,000 acres of federal land, with only 600 acres of private in holdings. I had outstanding people and a tremendous timber, wildlife and range and recreation program. A modest special use permit program involved two State and federal highway systems, resorts at Jacob Lake and VT Park and a telephone line. Fire did not amount to much in comparison with the Long Valley District. One year on the Fredonia District we had 75 fires. We had to hurry to reach some of them before the rain put them out. We averaged 20 to 30 fires per year. The few fires that occurred during the very dry months of May and June were pounced on by our experienced fire crews and held to less than an acre or very few acres at the most. The only real concern we had about forest fires were the National Park Service's fires. They were not too well trained and not very prompt in hitting fires early with the necessary resources. On many of the Park fires the Park Service personnel frequently found them manned by our crews when they arrived.

Fire season was generally over by the late deer season and our three fire lookouts closed. However one year dry conditions persisted and we hired an old gentleman from Panguitch, Utah to provide coverage for the balance of the fire season. He said he knew the forest like the back of his hand. Assistant Bill Anderson drove him to the 110-foot Dry Park Lookout Tower and gave him training on the use of the fire finder and radio. He demonstrated he did know the forest. All went well until a small fire was reported to be just south of Dry Park. The smoke was detected by the Jacob Lake Lookout and one on Bill Williams Mountain south of the Grand Canyon and near Williams. The fire was quickly controlled. I was near Dry Park and drove to the tower. Our faithful lookout man was on the job. Indeed he had seen the fire, but froze when he picked up the radio microphone. I asked the nice old man why he had not reported the fire. He said he just could not remember how to use the radio. I told him it was just like a telephone, press the button, talk, release the button and listen. The response of that 65 year old man was unforgettable. He said "Mr. Cravens, I have never, ever, used a telephone". So much for modern technology!

Frequently we were able to take law enforcement action on man-caused fires, littering, timber trespass, game law violations and vandalism. Most of the cases involved responsibility for starting a fire that spread out of control. Individuals were taken before the Justice of Peace in Fredonia. The routine before the

JP never varied. Both sides of a case were presented and the JP would beckon me to come with him into the next room. With the door closed he would ask "is this defendant really guilty?" If I said "yes, sir" we returned and the defendant was judged "guilty" and fined. There was no appeal. We had a record of 100% convictions. Often I thought about the merits of the legal system and the blindfolded statue balancing the scales of justice.

We had an annual timber harvest of 44 million board feet of timber. All of it going to the Kaibab Lumber Company mill in Fredonia. The loggers were excellent, careful workers. Some of them were from the polygamy settlement of Short Creek which was a few miles west of Fredonia. Having four or five wives did not interfere with their work. Contrasted to many of the national forests south of the Grand Canyon the north Kaibab had only a few light timber harvests, principally for the control of the Black Hills bark beetle back in the 1930s. Our light selection cuts left the ponderosa pine forest aesthetically pleasing and very productive. Contrasted to the vast forest areas south of the Grand Canyon we had only very small areas of dwarf mistletoe or diseases affecting the pine. Following the timber harvests we had the funds to take care of the logging slash and timber stand improvement, including any mistletoe or disease problems that could be detected. Erosion was no problem on top of the relatively level Kaibab Plateau.

We used a number of locals and forestry students to do the TSI which consisted of thinning and pruning. Forester Ned Jackson supervised the crews and their camp at Big Springs. One Fall the regular cook had to leave early and the mess hall was turned over to Ned's wife, Joanne. She was a good cook and kept the crews well fed and happy. I took the family out to Big Springs, some 35 miles from Fredonia, for Thanksgiving dinner. The turkey and all the trimmings could not have been better, everything was delicious. Then came beautiful looking pumpkin pie covered with whipped cream. We all took a bite of pie at about the same time and you never saw so many puzzled expressions and mouths full of something we did not know what to do with it. Seems like Joanne had never made pumpkin pie before so she just emptied the contents of several cans of pumpkin into pie shells, with no flavoring or sugar, just plain pumpkin and baked it. You never tasted anything so awful! On the side Gwen gave her advice for the next time. Joanne was so embarrassed.

Assistant Bill Anderson was from the old school with 30+ years of service. He handled special use permits, fire prevention and suppression, maintenance and improvement crews and supervised the deer checking stations. He was good and knew how to work with people. He lived alone and paid high rent for the 3-bedroom house at "Tanglefoot". His many stories about the

Prescott National Forest and the Groom Creek area were legends in their own right. The one I liked best was about the old tobacco chewing cowboy and the practical jokers. All cowboys liked to pull tricks on one another. Seems like the old-timer chewed from the time he got up in the morning until he went to bed in the bunkhouse. Each night he would take out his chaw of tobacco and place it on a shelf by his bunk. Then in the morning he would pick it up and continue to extract the remaining juices. One night the jokers substituted a large cat dropping for the tobacco cud. Next morning at daylight the jokers peeked out of their bed rolls with one eye and expected to see the old timer have a fit and explode when he tasted the cat's dropping. But he didn't. He put it in his mouth, chewed and walked away as normal as could be and left the disappointed practical jokers shaking their heads with disbelief.

The wildlife program on the North Kaibab was a tremendous opportunity for me to follow in the footsteps of President Theodore Roosevelt, "Doc" Rassmusen and many other famous wildlife biologists. The area was also known as the Grand Canyon National Game Preserve and had a long history that had been studied by many generations of students and frustrated Forest Service and Arizona State game managers and wildlife biologists. Deer populations had exploded in the 1920s and devastated available food supplies. Attempts to drive surplus deer across the Colorado River, by way of House Rock Valley and Little Saddle, to the south rim area of the Grand Canyon had failed when many cowboys and Forest Service riders found the deer were not about to go into that big, deep canyon and ford the Colorado River. One old timer told me when the deer approached the inner gorge of the Grand Canyon they exploded all over, around and under the horsemen and headed back to the Kaibab and starvation that awaited them.

By the mid-1940s the habitat had improved somewhat with the increase in logging by the Whiting Brothers' Kaibab Lumber Company and the condition of the deer improved. In fact on my first hunt on the Kaibab in 1949, with friends from Williams, we killed four mule deer that weighed from 185 to 215 pounds field dressed. They were fat and in good condition. By 1954 when I was transferred to the North Kaibab deer numbers had sky rocketed and the habitat was again being eaten into the ground. Deer ate all available vegetation. Spruce trees were hedged up as far as deer could reach . . . an indicator of severe overuse and starvation. In fact the deer stood around waiting for the loggers to fell an aspen and then consumed all the twigs and leaves in a short time. Seeding for site restoration on skid trails and closed logging roads was unsuccessful since the deer slicked it off as soon as the grass and clover shoots appeared. We could count several hundred undernourished deer at one time in some of the overgrazed open parklike areas on the upper summer range. In these areas the grasses and other vegetation were fully consumed. The critical winter range at lower elevations was hit heavily when the deer arrived in late fall. Only the intermediate range between the top of the Kaibab Plateau and the lower winter range was in fair shape. Deer being creatures of habit did not pause in their migration from the summer range on top, including those coming from the Grand Canyon National Park, until they reached the lower elevations above Kanab Creek, House Rock Valley and the Grand Canyon. Adding to the problem was over half a century of overgrazing by cattle, horses and sheep. The average field dressed weight of bucks harvested the fall of the year I arrived averaged 129 pounds and all were in poor condition. Does, which normally produced twins or triplets, produced none or only one which usually could not survive through the winter. These were some of the challenges we faced in the mid-1950s.

I worked with our Forest Service and Arizona State wildlife biologists on deer management. We made deer counts in the summer and winter. As on the Coconino, big game counts involved being on a survey route before dawn and coming in late at night. We also measured the amount of vegetation utilized on summer, intermediate and winter ranges. The Forest Service was responsible for operating 24-hour deer checking stations during an early and a late hunt. We weighed, aged, recorded sex and location of where each deer was harvested. We designed and constructed roads to improve hunter access, installed pipelines, developed springs, constructed earthen stock tanks, deer guzzlers (covered rain water catchments), cleared and burned low value vegetation, planted deer browse and reseeded thousands of acres with crested wheat grass and other suitable species. This work was accomplished through the use of cooperative funds coming from a deposit paid by each deer hunter. As with all other resource programs, and before funds were allocated, we prepared detailed work plans, budgets and made presentations to the Forest Supervisor and staff.

We conducted many show me trips for hunters and State people to demonstrate our management program and the seriousness of the over population of deer. I made many trips to Phoenix with the Forest Supervisor Flick Hodgin in unsuccessful efforts to try and convince the State Game Commissioners to issue more permits and increase hunting pressure. We told them the consequences would be further destruction of the range resources and a winter dieoff. The State feared the political consequences and refused to increase the number of permits. They wanted improved and reduced livestock use, more water improvements, reseeding and the Fish and Wildlife Service to set up a predator control program. By working with the grazing permittees we obtained reduction in livestock numbers and improved management. We continued a program and installation of water developments and reseeding. But we refused to permit a predator control program, including the establishment of poison (Compound 1080) bait stations to kill coyotes and mountain lions.

The deer were in poor condition when they migrated from the summer range. Our predictions came true that winter. We lost thousands of deer on the winter range. Deer carcasses could be found in every canyon and deer browse was fully consumed. With the large dieoff and a significant reduction of deer numbers, acceleration of reseeding and water developments, improved range management and continued timber harvest, the range and deer responded. Calves were fatter, does produced healthy twins and the average weight of male deer increased. The Kaibab was famous for its trophy deer. Antlers were huge, up to a 40-inch spread and almost the size of elk. When I saw coyotes I was pleased to think we had not been a party to poisoning them. After all they helped eliminate the weaker deer and keep the numbers under control. I felt particularly good early one morning when I saw a mother mountain lion and two spotted cubs cross a woods road in front of me.

Deer hunts also brought the VIPs. We had outdoor writers and photographers to show around. Brass from the Washington and Regional Office usually showed up and had to be put up at the historic Jacob Lake Forest Service Cabin. Most of our limited building improvement funds went to keep that place in good shape for our guests. We had a laugh when we were invariably asked to point out Jacob Lake. It was just a low spot containing an earthen livestock tank on private land. The Assistant Regional Forester in charge of Wildlife Management, always came to one of the hunts. He did not hunt, but he loved to cook and drink Jim Beam whiskey. And he did exceptionally well at both tasks. I returned one evening with a group of other VIPs and there he was, very intoxicated and sweeping the floor. He had just spilled a large skillet of gravy on the floor and was using a broom to clean it up. What a mess! Cleaning up after some of those people was not one of our fun jobs!

We had another interesting feature on the Kaibab . . . a herd of a few hundred buffalo. This use was permitted to the State of Arizona on one of our grazing allotments. Managing buffalo is a chore. If an animal decided that it wanted to go somewhere there was no fence, or any man-made structure, or any amount of effort that could stop it. There were times that these animals drifted miles out of their allotted range and had to be moved back. The State conducted "hunts" by once-in-a-lifetime permits issued only to Arizona residents. Hunters assembled and were led to a small enclosed pasture and told which animal to shoot. Some were dropped in their tracks but a small wounded cow could take off and go for miles. Again nothing could stop it, except death. Earlier I had shot a 1,000 pound bull on the Coconino where the State had another buffalo herd. I traded the magnificent head and three quarters for the tanning of the hide. We lived at Elden at that time and Myrl and Gene Jensen were visiting us. Gene and I butchered the front quarter on the kitchen table. The bones were about eight inches in diameter. What a task! We had some whiskey to aid us, however a chainsaw would have been more useful. But the meat was delicious. The girls were still loving that hide to death in their playhouse in Fredonia.

I learned much on the Kaibab. We had a heavy public relations program. There were hunters, recreationists, conservation and hunting groups, local people and visitors from as far away as India. I enjoyed this, as I did working with our various forest workers. I learned a lot from cowboys such as Billy Swapp. He taught me how to construct something very useful out of a little of nothing and how to make baking powder biscuits without a bowl. He mixed them in the top of a sack of flour and baked them in a Dutch oven covered with just the proper amount of red hot coals. Billy always had starter for delicious sourdough biscuits. He taught me how to pack, tie a diamond hitch and care for the horses and Jakie the mule. He taught me how to rig and use a "lizard". What's a lizard? What is it used for? At times we had to transport by trail an 18-inch by 10 foot water trough to a spring development. We had no helicopters and the trails were too narrow or steep for our 4-wheel drive equipment. We made a "lizard" by cutting a small forked tree. The tank was lashed on, the front end hitched up to the mule Jakie and with the fork in the rear it was dragged to the work site. It worked well and the "lizard" could be discarded when the job was done. We also packed in pipe, dynamite, caps and tools on pack boards or panniers (large saddle baskets).

I hunted deer every year I was on the Kaibab. Most of my hunting was done near the end of the second, late season. By that time hunters had thinned out and the snow had normally driven the deer to the lower elevations. I enjoyed hunting open sagebrush ridges and canyons. I learned from the local people that you could find the big ones in those areas. And I always did! As on the Coconino, I kept our freezer full of venison, mutton and beef. I could purchase a dressed, whole sheep for $10 and a half of beef for $50. The girls would eat beef but they refused to eat "Bambi" or "Bo Peep's" sheep.

We had turkeys on the Forest. Some 40 wild turkeys were flown in by District Ranger Pat Murray from the Apache National Forest in 1948. They did well and by 1955 they had multiplied and spread all over the Kaibab Plateau, including the Grand Canyon National Park. The Park Superintendent was heard to complain about the presence of exotic birds that had never historically been known to inhabit the area. In confidence he admitted enjoying seeing these majestic birds.

One of our early deer hunts coincided with the turkey hunt. I had a turkey permit, but it was a busy fire season and I did not have time to hunt. On the last day of the season we were putting out a small forest fire and one of the fire fighters called to me and pointed to a flock of turkeys passing through the burned area. It was just about daylight and the light played tricks. The turkeys looked huge as I picked out and shot what appeared to be one of the largest. It turned out to be a young three pound hen! It was not much to be proud about.

After the fire season I did well fishing for trout. John Churches, Pat Murray and I usually went to Idaho and fished the Salmon River for steelhead trout and the Big Lost and Little Big Lost Rivers for brook and rainbow trout. We had some great, exciting times and made many new friends in Idaho. John was an excellent German brown trout fisherman. He knew their habits and taught me the secret of catching these wily trout. The trick consisted of casting a fly or spinning lure into a likely spot, usually in a pool under an over hanging bank on a small stream, as many as 20 to 30 times . . . always in the exact same place! Most trout fishermen will tire of that practice, decide after a few casts there are no fish and move on. But repeated casts in the same location finally make the elusive German brown trout mad and it will finally lunge at your lure. I fished Asay Creek which is 42 miles north of Kanab, Utah. This small stream heads in the Dixie National Forest and leaves the forest boundary just before it crosses US 89 Highway. The area was heavily grazed by sheep and had the living daylights beat out of it all summer by fly fisherman. I could go there after work, about dark in late September and come home with a limit of German brown trout as long as my arm. Of course I told everybody I caught them out of a lake up on the Dixie. That also was part of John Church's training . . . be elusive like the trout and don't give away all your secrets! Never give away the location of your favorite fishing holes, except to friends, or like that friendly ranger on the Gunnison National Forest in Colorado did for a crippled soldier back during the war. Hunt, trap and fish, live in a cabin in the woods . . . I had it made, except I never went for the trapping part.

Much of the Kaibab was inaccessible to vehicles so we necessarily did a great deal of horseback work in connection with range and wildlife activities. Many months I spent 20 days on horseback. Some deer surveys were made on horseback to check winter use and deer survival. It required much riding with grazing permittees to learn about their needs, problems and range conditions. This knowledge helped us when it came to requiring improved management and reduction in livestock numbers. Most of the grazing permittees had been in the area for most of their lifetime. Some were in their 70's and 80's and quick to criticize the Forest Service for its management of the deer herd over the years. Most of them viewed the increased public interest in the Kaibab as support for the deer program and opposed to livestock use. It was our job to secure balanced use. We recognized on winter ranges in particular that cattle grazing kept the grass density and competition down and prepared a suitable seed bed for reproduction of vital winter deer browse species, such as cliff rose, bitterbrush and mountain mahogany.

One grazing permittee was an unforgettable character. Eddie Hatch was a World War I veteran. His lungs had been damaged by German poison gas and he wheezed with each breath. His summer place was located on one of the limited tracts of private land within the Forest and adjacent to Jacob Lake. He had a grazing permit and used two allotments on the Kaibab, one near Jacob Lake and one in Snake Gulch and Kanab Creek. His winter place was at White Sage Flat which was located on the main access road between Fredonia and our Big Springs Station. Most of his winter grazing took place on the Bureau of Land Management's Public Domain land adjacent to the Forest. Eddie practiced more intensive range management than any other grazing permittee I had met up to that time or since. Each of his cows had a pet name and they knew him well. He either drove them by horseback or led them on foot to choice grass. Some of these grazing areas were very remote and located on isolated ledges beside a steep canyon.

I was just another "gov'mint" bureaucrat when I first came to the Kaibab, but after several visits Eddie Hatch trusted me and taught me a great deal about handling livestock. He invited me to ride down Snake Gulch and into Kanab Creek to see his "pets". Some of this area was inaccessible, even to horses, but it was extremely scenic. He pointed out a cow here and there as we dropped into the canyon. I would swear a bird would have trouble landing on some of those narrow ledges, let alone finding it. But he led his cows into these almost inaccessible areas to graze and then helped them turn around and get out, they trusted him. Normally much of his winter range on the Forest in Kanab Creek would have been designated as "unsuitable range", but he used it and used it properly. Water holes for the cattle, and for us, in that area consisted of stagnant pools on the red rocks and water that collected in cattle tracks. When you get so dry your mouth has absolutely no moisture, anything would do. Even the trick of sucking on small pebbles will not do the trick. So you grit your teeth and drink whatever is available.

Staying overnight with Eddie Hatch at his shack at the lower end of Snake Gulch was above and beyond the call of duty. The place was overrun with pack rats and their droppings were everywhere. We arrived after dark, late one night. We were tired and dry. A single, nearly used up candle was our only source of light. I looked and felt around for water. Eddie directed me to the rain barrel around back of the shack. The water in the 50 gallon oil drum was hot and tasted strange, but it was wet and satisfied some of my thirst. Eddie cooked slab bacon and made biscuits and gravy. That was all he ever ate and it did taste good, particularly the biscuits which he mixed right in the flour bin. Sleeping with Eddie wheezing and pack rats running across me and screeching would keep most people awake all night. But I was dead tired after a long hot, dry ride and slept soundly.

Next morning it was the same routine, same food. The only problem was I could see what I was eating and drinking. The rain barrel was full of hot, rusty and stagnant water. I did not mind the mosquito larvae wiggling in the water but the floating dead rat that had lost most of its fur and pulsated with a body full of maggots was not inspiring. Those delicious biscuits contained some of the same water and the flour bin seemed to be alive with crawling weevils and ants. The bacon had a sickly green color. Other than that it was a great trip and I survived. Our local cowboys were quick to ask about my trip and overnight stay with Eddie. I stole their thunder when I told them everything, including the food, was great. I guess I came up a notch in everybody's book after surviving that test. Eddie even respected my judgment when we later asked him to reduce his permitted numbers of livestock.

Another horseback trip took me into the Red Rocks area with the Forest Supervisor and the Forest's Grazing Advisory Committee. Most were from the South Kaibab, with three from our area. They wanted to see the Red Rocks and Thunder River. The very scenic Red Rocks are reached by dropping several thousand feet from Big Saddle off the rim of the Kaibab Plateau. The Red Rocks are on a bench of several thousand acres between the Kaibab rim and the rim of the inner gorge of Colorado River in the Grand Canyon. Spectacular isolated buttes and mesas rise above the area. The famous western writer Zane Grey described the area in one of his books where the Mormon Bishop blocked the trail and hid out from a posse. That would have been very easy to do in any number of locations. Each winter a number of local cowboys would go into this remote area to rope and lead out some very mean and very wild 20+ year old cows and steers. They were veritable longhorns. Usually on one of those trips someone would get a broken arm or leg or they would have to shoot one of their horses which suffered a broken leg. It was rough country!

There were a dozen of us on this particular trip. While Billy Swapp packed in the gear and set up camp on the Red Rocks, I led the group on a side trip into Thunder River. The Kaibab

Plateau consists of a very porous "karst" limestone formation. Most of the moisture falling on the area sinks into the ground. There is seldom if ever any surface runoff. The resulting river is actually a huge spring gushing out of a cavern in the side of a sheer cliff. It then drops rapidly to the Colorado River. The access trail to Thunder River was very steep and I asked an experienced local rancher to go first. I brought up the rear. Before starting down the very steep trail, with a sheer drop off of several hundred feet on one side and sheer cliffs along the uphill side, I tightened the cinch and adjusted my saddle. The March weather in the area grew hotter as we rapidly dropped in elevation. The temperature must have been nearly a 100 degrees and all of our horses were sweating profusely, even going slow, slipping and sliding down hill. As each horse came to Thunder River it stepped into the water for a drink. By the time I reached the River the more accessible drinking areas were occupied by the proceeding 11 horses. My horse had to have a drink so he set all four feet and slid about six feet down to the River. That was fine for him, but he had sweated and lost considerable weight on the last bit of trail. I knew the cinch was loose but on that steep, narrow trail I could not get off and tighten it again. The saddle slipped and over I went, right between the horse's ears. I landed right side up, didn't even loose my cowboy hat, but the River was so swift that I was carried over rocks and rapids to where the first horseman could grab me by the collar as I went by. I took a real beating but my heavy chaps kept me from injuring my knees. But the big shock was going from 100+ degree air into the 40 degree or colder river water. It was a while before I could breathe normally. Naturally this provided good entertainment for the others and I had the distinction of getting wet first. My little run in the River was the subject of many conversations by the local cowboys at the Buckskin Tavern. News traveled fast on the cowboys' communication system.

On another trip two of us walked in to the Red Rocks area and climbed several hundred feet to the top of the very isolated Fishtail Mesa. This reconnaissance trip was made to compare the use of an area that historically had been used only by deer, with areas had been used for many years by sheep, cattle, horses and deer. The climb was almost vertical and only one place contained a gap in the sheer cliff that could have possibly been used as a route for the deer and us to reach the top. It was absolutely impossible for any livestock to have reached the top area. On top the views up and down the Grand Canyon and Kanab Creek were absolutely spectacular. Vegetation consisted of juniper, sparse old sagebrush and decadent cliffrose plants and a very heavy stand of perennial bunch grasses. It was obvious the dense grass cover limited reproduction of the brushy species that were important deer browse. There were very large deer tracks, some huge antlers shed over the years but absolutely no sign of livestock use . . . no tracks, droppings or bones. Elsewhere on the Forest livestock use limited grass cover and the bare soil between grass plants made a good seedbed for the seed of brushy species to germinate and become established. This information served us well and helped hold on to livestock use on the Kaibab when there were pressures to eliminate all livestock use.

We had good working relations with the National Park Service people at Grand Canyon. They worked with us on deer census and attended meetings with the State and sportsmen's groups. They supported our recommendations for larger hunts, only on the Forest of course, since the deer were overusing their portion of the summer range. They also urged us to continue opposition to a predator control program. As noted before the Park Service cooperated with us on fire prevention and suppression. They relied on our judgment since we were more experienced. Many times some of Billy Swapp's fire control crews made the first attack on Park fires. Some of Billy's colorful language must have polluted the pristine atmosphere of the National Park and the virgin ears of park rangers.

Park rangers were nice people. On the North Rim they had a tough life. They spent six months on the North Rim and then moved out to the South Rim, Florida or another more moderate site for winter. They had to move back early in the spring to prepare for summer use and then move out before the roads were closed in by an early snowfall. Sometimes the top of the Plateau received over 15 feet of snow. One fall they did not get out soon enough and we had to go in by 4-wheel drive and a State snow plow to rescue Ranger Clyde Maxey, the same one I sold my motor scooter to back in Fort Collins. He was embarrassed about being rescued by the Forest Service, but grateful.

The National Park Service considered the National Forest to be a good buffer zone along the Park boundary. They were critical and looked down their noses at our timber harvesting and wildlife improvement projects. But each year they would come with pointed hat in hand, to request a wood permit. They cut their firewood from dead trees on the Kaibab. We refused to give them permits for garbage dumps on the Forest. They did not like it when I told them they had a big canyon in which to dump their trash. On one occasion we talked to the Superintendent of the Grand Canyon National Park about exchanging land. We offered them some of the National Forest in the inner gorge of the Grand Canyon on the House Rock Valley side and in the Red Rocks, including part of Kanab Creek in exchange for a tier of timbered sections in the Park which were on top of the Plateau and adjacent to the National Forest boundary. The Superintendent wanted the National Forest land so bad he could taste it but refused to even consider giving up an acre of the

Park. Later the Grand Canyon National Park was expanded by law to include these same National Forest lands. The Forest Service acquired nothing in exchange.

One day I received a call from the federal aviation people. Had we seen or heard any four engine airplanes over the south end of the Forest near the Grand Canyon? We contacted our lookouts and work crews. Someone had seen a 4-engine Constellation over Jacob Lake earlier that day. That was not the one they were seeking. Later they discovered the crash site of two 4-engine planes, an American Airlines DC6 and a TWA Constellation which had collided over the Grand Canyon. Everyone in both planes was killed. We were relieved when we were not called in to help the Park Service pack out over 100 bodies from the south side of the Grand Canyon, near the confluence of the Colorado River and the Little Colorado River.

We enjoyed good working relations with livestock permittees and cooperators at the Mt. Trumbull Unit of the Kaibab. This relatively small 17,500 acre area was about 50 miles west of Fredonia, over roads which were frequently impassable, and in an extremely remote area. This isolated peak on the north side of the Colorado River was adjacent to the Grand Canyon National Monument and Public Domain land administered by the Bureau of Land Management. The area supported an almost insect and disease free stand of ponderosa pine. It contained one area of two or three acres of dwarf mistletoe, probably from seed brought in by a bird. One of our TSI crews cleaned that up in a short time. Other than that the only major cutting had been done in the late 1800s when large timbers were cut and transported by teams of oxen to St. George, Utah for use in building the huge Mormon Temple. Grazing on that portion of the National Forest was administered by the Bureau of Land Management and the National Park Service handled our forest fires. We had established excellent relations with both agencies and they reported back to us on the respective activities.

One summer National Park Service Ranger John Riffey sent us a report on a Class E fire (over 100 acres) that he had handled. We had not even detected smoke in the area and knew nothing about the fire, until we received his report. His fire suppression costs were less than $100. When the fire report and the bill hit our supervisor's office in Williams I received a reprimand because I had not reported the fire or had even been involved in the action. The pressure eased when we reported no resources were damaged and that if we had sent in fire crews from Fredonia that the cost could have amounted to thousands of dollars. The delay in moving a crew in would have resulted in the burning of many more acres.

I visited the Mt. Trumbull area once or twice a year. It was completely removed from the real world and even further away from the forest supervisor's office. From the top of Mt. Trumbull one could see Lake Mead to the west, St. George to the north, the Kaibab Plateau to the East and the Grand Canyon and mountain peaks near Williams and Flagstaff to the south. From the rim near Riffey's Ranger Station you could look up and down the magnificent Colorado River for miles and look straight down several thousand feet to the Lava Falls. These Falls were one of the big drops and a major obstacle for rafters on the River. In the rim area we frequently observed desert big horn sheep, wild burros and bandtail pigeons.

I took the Forest Supervisor, Regional Forester Otto Lindh and Personnel Chief Allen McCutchen to Mt. Trumbull. I remember the trip well since they cornered me beside a campfire and tried unsuccessfully to talk me into accepting a transfer and promotion to an Assistant Forest Supervisor timber position on the Apache National Forest. I turned it down after Gwen and I visited Springerville and could find only a former dentist's derelict office to rent for a home. My colleagues said I was dead for turning down the Regional Forester. That occasion was one of the few times during my Forest Service career that the Forest Service came second to the family's needs. But I survived rejection of the offer and spent another delightful and productive year in Fredonia.

On another occasion, while I was stationed at Fredonia, I attended a regional training session for rangers at Ft. Valley and the Regional Forester emphatically told us to vote for a specific national candidate or the Forest Service would go down the drain. He committed the cardinal sin of publicly endorsing a political candidate and trying to influence our vote. He did not last long after that and retired shortly after being forced to transfer. From that time on I cautioned our people about endorsing specific candidates for political office and reminded them about the restrictive provisions of the Hatch Act which limited political activities of federal employees.

Billy Swapp and I organized a great watershed management trip for Forest Service people from Albuquerque and the Rocky Mountain Region. We spent several days with them on the Forest and in Monument Valley. We traveled in the Navajo Indian Reservation by using 4-wheel drive equipment. We reached the spectacular country back and beyond the mesas, mittens and buttes that are so well photographed. Swapp was a good camp cook and story teller. Everyone enjoyed the trip into that colorful, isolated, windswept, semi-desert country. We urged people to sleep out under the magnificent stars and warned them about sleeping in bug infested, abandoned Navajo Indian hogans. But two of our party thought it would be great to sleep in a hogan. They picked up bed bugs as a result. No one would go near that pair and we made them sit in the bed of the power wagon until the trip was over.

I made many trips to our Flagstaff sub-regional equipment shop, fire dispatcher's office and to the forest supervisor's office in Williams for various meetings, to turn in equipment and to pick up supplies. It was a long trip across the Navajo Reservation. I made trips in pickup trucks, 4-wheel drive Dodge power wagons, Jeeps and five ton tilt-bed fire trucks. When Gwen and the girls were not with me I usually traveled at night in order to be in Williams by 0800. On a return trip from Williams one dry day in June, I stopped by the fire dispatcher's office in Flagstaff. I found they were looking for me and were ready to dispatch me and four others to a large project forest fire on the Tonto National Forest's Pinal Peak. My assignment as communications officer had already been cleared with the Williams office which had in turn notified Fredonia I would be gone for a while. I had a problem. I had clothes for only two or three days and I wore only cowboy boots. The latter worried me most when I thought about my feet and the fact I did not have enough money to buy new boots. I did not think about accessing our checking account which had been in the First National Bank of Arizona, Flagstaff Branch since my pay checks had been going there for direct deposit in 1949. (Over 40 years later my retirement checks are going there!) I called my friend Frank Carroll in Flagstaff and asked if he could loan me a pair of boots. He had a pair of Sears logger boots that he graciously let me use. That was very kind of him at the time. After I had worn them for two weeks they about "killed" my feet. I was ready to wring Frank's neck when I later returned the boots. He thought it was a great joke since they had always hurt his feet too.

The Tonto fire was large, but was under control in three days and we were released. It was good experience, but I was ready to go home. On the return trip we spotted smoke boiling up on the Chevelon District of the Sitgreaves National Forest. It could be seen from a hundred miles away. We debated. Should we or should we not? In the end dedication won out and we checked in to the forest supervisor's office at Holbrook and were given fire assignments that they had ready for us. If we had gone on to Flagstaff we would have been turned around and sent to the fire. After five days on the fire line we were released. Oh, how my feet ached! Fortunately Frank Carroll was not home when I went through Flagstaff and returned his boots. I promised Betty Carroll I would return and break his neck. She thought it was funny too! Frank still thinks it was poetic justice. I returned to Fredonia two weeks after I had departed. I was tired but wiser. As I recall Gwen made me undress and take a sponge bath in the garage before she would let such a smelly creature into her home.

There were two sad events to report concerning our life in

Fredonia. First, while we lived there the US Government conducted surface firings in Nevada to test atomic bombs. The scheduled times of the frequent test firings were announced over the radio (we had no TV). Each time, the government spokesman announced the tests were carefully controlled and monitored and there was absolutely no danger to people in surrounding areas. The explosions took place exactly on time. Looking out of our kitchen window, several hundred miles away to the west, we could see the brilliant flash, which was brighter than the sun, in the pre-dawn sky. Several minutes later the sound of the explosion reached us. Windows rattled, the house shook and these events were followed later by a perceptible rush of wind. Most of our prevailing winds came from the direction of Nevada. On several occasions and hours after the tests I recall seeing particles of white material drifting down like very fine snow. The wind whipped up frequent dust storms that were probably highly radioactive. As a result several residents of Fredonia developed leukemia and died. One of Melissa's and Cindy's school mates sickened and died. One of our grazing permittees developed burned spots from the fallout on his face and died. More from the area have developed radiation sickness and many have died since we moved away. So much for trusting what our government says! So far we "downwinders" have been lucky and survived that awful experience.

The second sad event involved ordering and installing a bronze plaque on a boulder designating this as the site of President Theodore Roosevelt's hunting cabin back near the turn of the century. Teddy loved the area for its famous deer and mountain lion hunting. Instead of restoring a deteriorated log cabin, near the old Ryan Ranger Station, one of my predecessors had burned it down and cleaned up the site. Unfortunately in those days, throughout the Region we were a burnup, cleanup and get out bunch. Many old historic structures were eliminated in this way.

During our stay in Fredonia we made several trips into scenic Utah. We grew very fond of that state and its people. We made a shopping trip to Salt Lake City and were awed by the wide streets, Mormon Temple and the surrounding mountains. On that trip we added a beautiful, 350 pound Franklin stove to our growing collection of antique furniture.

While at Fredonia I made several trips to Lake Mead with John Churches and other friends from Kanab. In early March the fishing for large mouth bass and crappies was fabulous. Again this was a very scenic, remote place. Access was over non-existent roads so we generally had our camp site by the lake to ourselves. We usually fished near camp on a Fork of the Virgin River. If we had been so inclined we could have traveled over a hundred miles on the surface of this huge lake to Boulder

(Hoover) Dam.

What was my most embarrassing moment? Most of our family trips were made outside the field and fire season. One year we drove to Las Vegas which was a short day's drive from Fredonia. We stayed in one of the inexpensive downtown hotels. We had breakfast in a restaurant just off the casino. Melissa and Cindy had each been given a roll of nickels by a friend and they took money out of their piggy banks to give to me to play in the slot machines. They had heard this was the place to make money grow. Since the girls were underage and could not enter the casino area they stood in the entry way and expectantly watched me. When I won and coins dropped and rattled in the tray, the girls would clap their hands and squeal with joy. However after awhile the machines gobbled up the coins and failed to payout. Cindy detected this, tears came to her eyes and in her shrill little voice yelled, "Daddy, daddy, don't lose all our nickels!" Usually a casino is a noisy place, but at that moment it was very quiet. People looked over at me from their slot machines, blackjack, roulette and crap tables with disapproving stares. I could have crawled under the carpet. Gwen about exploded with laughter when she saw my predicament. At least the girls learned that Las Vegas was not the place to make money grow.

Years later we visited Las Vegas as a family. The girls were beautiful and in their 20s. Cindy was a good crap shooter and during one session she threw seven straight sevens. There really was excitement at our crap table that night. People were yelling, "Come on, Cindy! Do it again!" I made money that night off of Cindy's efforts. Some people made thousand's. I could have broken the bank if I had faith in my daughter and guts enough to stay and double up each time. When we left the table I was aware of all the whispers, grins and sneers directed our way when I walked away with two tall, good looking 20-year young women holding on to my arms. The best part of Las Vegas for Gwen has always been sitting beside the pool or in a big comfortable bed reading a good English detective story.

Chapter VIII

San Francisco Peaks

The time finally arrived in 1957 when we were faced with critical family and career decisions. Either we remained in Fredonia for an unpredictable length of time or we accepted an offer of a possible promotion and transfer back to Flagstaff as Assistant Forest Supervisor. We were very happy in Fredonia. We had not transferred for three years. Being a District Ranger was the best job in the Forest Service. On the other hand a move, if like all the others, would cost us a $1,000 or more and eat up our modest savings account. Remaining in Fredonia would mean that the girls would finish grade school in a tiny town, which now having reached 500 was incorporated, and finishing high school in Kanab. So thinking about the better and larger schools, more outlets for Gwen and the girls and a broader career opportunity for me, we accepted the offer and with mixed emotions packed our belongings, Huckleberry and Blackbird and returned to Flagstaff. Ray Housley succeeded me as District Ranger in Fredonia. He followed my career route, became an Assistant Forest Supervisor on the Coconino, then to the Regional Office, return to Flagstaff as Forest Supervisor and later to the Washington Office where he became a Deputy Chief.

At the base of the San Francisco Peaks, Flagstaff had not changed much in the three years we had been away. Of course we had kept in touch and returned frequently for shopping and visiting en route from Fredonia to and from Williams. We moved into a large red stone house on Beaver Street. Living costs were higher and we paid $135 per month for rent. We obtained some relief from this high rent by renting out two rooms to Forest Engineer John Wakenigg and one of the office people. We were even able to have a modest savings account and purchase an occasional small savings bond. Federal salaries were low and we did not have many luxuries. Our home was convenient to schools, shopping and was half way between my office in the downtown Post Office and our Forest Service facilities at Knob Hill. We were able to walk most everywhere we needed to go.

The girls were enrolled in Northern Arizona University's grade school where they were happy and did well. Gwen had read to them since they were babies. We had no TV and they became good readers. They made new friends to replace those they left behind in Fredonia. Transfers became a greater problem as the girls grew older. They hated to leave friends and move. On weekends I took Gwen and the girls with me on trips into the Forest. They became very familiar with the Coconino National Forest. They particularly enjoyed trips to the Snow Bowl on the San Francisco Peaks, Cinder Hills, Happy Jack, the Mogollon Rim, General Springs, Battleground Ridge, Dirty Neck Canyon, Pivot Rock, Oak Creek, Sedona and Red Rock Crossing. We made numerous trips to Montezuma's Castle and Well. At that time the Park Service permitted visitors to climb ladders and explore the cliff dwellings. We also enjoyed trips to the old mining town of Jerome and over the Mingus Mountains to Prescott. We traveled together as a family when I went to meetings in Phoenix and Albuquerque. We made many long, hot trips down US 66, past the trading posts and snake pits (tourist traps) to Albuquerque. Gwen and the girls did well on their own during my frequent absences on field trips, training sessions and forest fires.

Gwen renewed acquaintances with old friends and made new ones. Most of these were not from Forest Service families. She enjoyed University affairs. She became Assistant Librarian at the Flagstaff Public Library. We participated in Forest Service activities and attended picnics when people like Chief Richard McArdle came to town. He was a great Chief and a very fine person. He had an effective system for remembering people's names. At one Forest Service family picnic he called me by name although he had not seen me since I was Ranger on the Long Valley District. From that point on I worked out my own system for remembering names. I became quite good at it too.

We enjoyed the Museum of Northern Arizona and the Lowell Observatory and the special programs they held from time to time. Then of course there was the All Indian Pow Wow and rodeo on the 4th of July. We became quite active in the community.

Gwen's father, William Otis Sanders, passed away in 1957. That was a great loss to all of us. He was a great man. Gwen and the girls went back for the funeral. By this time both of Gwen's parents were gone. We saw my parents occasionally. They came through northern Arizona on one occasion to visit friends in Phoenix. They did not even stay overnight. My Father enjoyed the girls but my Mother never showed any interest in them or Gwen at that time.

My work on the Coconino was interesting. I had moved without a promotion, as I did for every one of the jobs I held in the Forest Service. Promotions always seemed to come after I had been on the job for awhile. I was one of three Assistant Supervisors and was responsible for timber, recreation and lands. The others were responsible for soil, water, range, wildlife and fire programs. The Forest Engineer was responsible for the transportation system and all construction and maintenance projects. All were large programs on that heavily used national forest and our group operated well as a team. The new Forest Supervisor was Ralph Crawford. He had transferred to Flagstaff from Bend, Oregon where he had been Supervisor of the Descutes National Forest for 17 years. He had many stories about that forest and apparently it had many similarities with the Coconino. When we had a problem facing us he dredged a crystal ball, covered with a velvet cloth, out of his desk drawer and attempted to predict the future.

The greatest task we faced was putting the Forest back together. The Flagstaff Administrative Unit had been a total disaster and when failure was evident, former Supervisor Ken Kenney gave up and was transferred to Missoula. Not even the good Lord could have made that thing work! We selected and oriented five new rangers and moved some people around to accommodate their personal and organizational needs. We held many meetings with timber, wildlife, recreation, special use and range permittees and water users groups. We made much effort to meet with the press, city, county and state officials and representatives of other federal agencies. The Supervisor delegated a great deal of authority to me and I enjoyed it. I had the advantage of knowing most of the people we contacted including those outside the service. We helped new rangers establish contacts with their forest users.

I had an excellent crew of timber, lands and recreation specialists and enjoyed my work as a staffman. That was a big change since I had been a line officer, District Ranger, for almost six years. We conducted a large timber program and had one large mill and three smaller ones to supply. The Coconino was designated as the Flagstaff Federal Sustained Yield Unit, we dealt with the same mills all the time. We had different rules and regulations and the timber had to be carefully appraised and with the exception of set asides for the small mills, all timber sold and harvested went to the large Southwest Forest Industries Mill in Flagstaff. I was responsible for reviewing and inspecting all timber activities. We had some of the same great forestry technicians I had worked with before. They conducted most of the timber sales and trained the new foresters as they were transferred in. All in all it was a smooth transition.

Water shortages have existed in Arizona since pre-historic times. It was during my stint as Assistant Forest Supervisor that emphasis was placed on the importance of Arizona's forests in respect to the yield of water from the State's high altitude forest watersheds. Proposals by water-hungry Phoenix and Tucson water-user groups and academics in the southern part of the State advocated project proposals that ranged from the very sound and rational to the absurd. One group advocated combinations of clearcut, burn, pave and paint it green. Our participation in state-wide planning involved many meetings in Phoenix, Tucson and Albuquerque as well as soliciting ideas from local interests. Our public relations efforts were intensive.

There was much politics involved in the water program but we approached it in a rational, inter-disciplinary professional manner. We worked closely with counterparts in Forest Service Research. Our proposals for calibrating water yield on treated and untreated watersheds and vegetative manipulation ranging from clearcuts to various degrees of thinning attracted professional and political support. This was the beginning of the Arizona Watershed Program. Our portion was the Beaver Creek Watershed which covered over 250,000 acres. As soon as our detailed plans were approved the dollars rolled in and continued. Employment reached high levels, particularly among highly skilled Hopi forest workers; 100 worked out of Happy Jack. We had construction crews installing water measuring stations, bulldozers clearing pinyon-juniper and crews reseeding and applying many varieties of herbicides and conducting thinning of ponderosa pine stands. We became experts in applying 2-4-D, 2-4-5-T, and other plant killers. One of my tasks was to demonstrate to our crews how to use these toxic chemicals safely and according to label instructions. While we were extremely careful this was the beginning of my experience with toxic chemicals.

We made many timber sales ranging from selective cutting to clearcuts. I continued to raise a "critical question" . . . since the original law, the 1897 Organic Act stated that we could cut only large, old growth, dead and dying trees, how could we legally cut and thin immature healthy trees. I was always told this was an old-fashioned concept and we had many newer laws on the books and much scientific research to back up our current activities. I had raised this issue when I was first introduced to the Forest Service Manual in 1948. The legal reference clearly restricted our cutting practices. (More on this a decade later!) Anyway the Forest was a beehive of activity and we kept busy.

We had foreign visitors from Germany and Britain. The Germans were not too impressed by our pine stands particularly those in fringe areas. They suggested we should clearcut all limby, short pines and start over again. We even had royalty visit us. Lord Dalkeith from Scotland spent several days with us. He was a real gentleman and gave us some good suggestions for improving parts of the forest. Years later I had an opportunity to see his forest land and he certainly practiced what he preached. We usually invited these foreign guests to our home for dinner. Gwen was a gracious hostess and was most interested in anyone from the United Kingdom.

We had over two million recreation visitor-days of use on the Forest. I devoted a great deal of time to this activity. I reviewed development plans, public safety programs and visited with the public to learn what they liked and what they wanted. As I reflect on this concept, I spent almost 30 years providing the public with what I thought they wanted. But the biggest problem I had was determining just what they wanted. Ideas, concepts and various suggestions were sometimes 180 degrees apart.

It was during this period that we recognized the need to upgrade ski area facilities on the slopes of the San Francisco Peaks. A small area started by Ed Groesbeck, when he occupied my position on the Coconino in the 1930s, had slowly grown like "Topsy" over the years, but it was still inadequate in meeting critical public needs. We brought in Forest Service ski experts from Colorado to help with our planning. I necessarily, but cautiously, found myself back on skis. But I had no mishaps this time. I spent many weekends on the ski slopes at our Arizona Snow Bowl observing use and talking to users. This was the beginning of much needed expansion and improving our winter sports resources.

The lands program consisted of several hundred special use permits and land adjustment. Special use permits covered such uses as bee hives, transmission lines, roads, gravel pits, mines, fences, radio transmission sites, a ski area, garbage dumps, railroads, buildings and summer homes, to name a few. Fees were paid for most of these uses. Once early in my career I had asked why do we permit all those summer homes on public land. They occupied and cluttered choice areas needed for public use and were often a headache for the district rangers when they tried to obtain compliance with permit conditions. The answer I received to my question was that it was this permitted use that would discourage the National Park Service from ever wanting to take over that part of the National Forest. There was good cooperation between the National Park Service and the Forest Service at the field level, but there was much rivalry and competition for recognition and land at the national level.

The land adjustment program was large. It involved classifying which public lands should be retained in national forest ownership, which should be exchanged and which private or other public land should be acquired by the Forest Service. We made many land exchanges and traded off public land in exchange for private land. In the area within the city limits of Flagstaff we had a very favorable exchange ratio. In some cases, for each acre we traded off, we received as many as 50 acres in exchange. The lands we traded were such places as the city dump and isolated tracts that were impossible to manage. Some of those 40 acre tracts we exchanged had as many as 25 special use permits or rights-of way encumbering the land. Most of these were in the recently expanded city limits. Logging, grazing and escaped fires from garbage dumps just did not mix with gardens and petunias, so we phased out of that urban area. All appraisals were subject to very rigid standards and critical review by several levels in the Forest Service. We had to verify that the exchange was to the advantage of the government. I became very familiar with land values and which choice private lands were available for sale or exchange. Some private land was available for very low prices compared to what they ultimately were sold for years later. This was particularly true in the Sedona, Red Rocks Crossing and Verde River areas. I refrained from any personal purchase of anything within or adjacent to the Forest. I was very sensitive about creating any perception of a conflict of interest. That really was not too difficult since we had very little money at that time.

We had many forest fires as usual but I recall no particular problems. We had well-trained people and we operated as a team. The Santa Fe Railway continued business as usual and ignored repeated warnings to clean up the right-of-way, but there were no major railroad fires.

My involvement with the Society of American Foresters continued to grow. I had joined our professional forestry organization when I was a new Forest Service employee. I vividly recall the first meeting of our Arizona-New Mexico Section at Grand Canyon, Arizona in the Spring of 1949. Executive Vice President Henry Clepper was there from Washington, DC. Henry was a real gentleman, an ardent fisherman and excellent at motivating people. He collared me and said . . . "young man I want you not only to be a member but be involved. It will help you grow professionally. You can help your profession grow stronger". By the late 1950s I had chaired some committees and been elected secretary-treasurer, vice-chair and chair of our two-state section. We rotated meetings between Arizona and New Mexico. Gwen and the girls usually went to these meetings with me to see old friends. We attended an exciting national meeting in Atlanta the year I was section chairman. I recall Atlanta being a place where we had TV in the room. The girls, Huckleberry and the cats sat glued to TV and ate French fries and juicy roast beef sandwiches while Gwen and I participated in Society affairs.

Lake Mead was a good place to visit in March or early April; fishing for large mouth bass and crappies was fabulous. I went there many times with some of the local people, Equipment Shop Foreman Ken Butler, Timber Staffman Gordon Bade from the Kaibab and Ed Groesbeck from Albuquerque. We had great times and caught many fish. I have many rich memories of the Lake, like the time a herd of burros and a wind storm hit our lakeside camp at midnight. The burros did not make a sound until they were passing through our camp. Their braying and knocking over everything in their path created havoc. It scared the daylights out of all of us. Even without his hearing aid Ed Groesbeck heard all the commotion. His classic comment was "what the hell goes on?" While we were putting things back in order the wind hit and a strong gust blew Gordon Bade's hat and cot into the lake. We couldn't help but laugh when Gordon was running around in his skivvies and jumping into one of our boats to retrieve his floating hat and cot. Fortunately Ed did not lose or step on his hearing aid or thick glasses.

On another occasion the same group of fishing buddies, plus Lee Wang from the Regional Office Division of Operations, had tremendous fishing success. As all of us, with the exception of Lee were sweating and cleaning fish just before we departed Lake Mead for the long drive home, a friendly black man walked by. He looked at us working away on the fish and then he looked at Lee who was sitting on one of the boats and smoking his pipe and said, "You the supervisor or something?" Ed and Gordon about busted a gusset with laughter. Lee did have a reputation for being a little lazy when it came to doing chores around camp.

The highlight of every fishing trip to Lake Mead with Ed Groesbeck was to sit around the campfire at night and hear him recite a limitless amount of delightful poetry. His mind was a precious reservoir of poetry. Paul Bunyan's "Round River Drive" in the Lake States, "The Cremation of Sam McGee", "The Shooting of Dan McGrew" and other tales of the Yukon by Robert Service were my favorites. (Service, 1940)

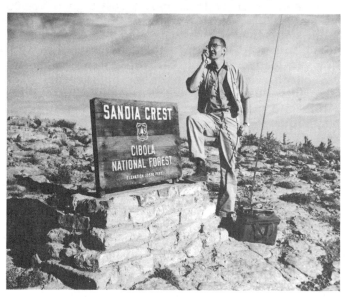

Directing Insect Control Project 1959 (U.S. Forest Service Photo)

Chapter IX

The Land of Enchantment

Again, all good things came to an end, or at least changed in 1958. I was transferred to the Regional Office in Albuquerque, New Mexico, the so-called Land of Enchantment. Again the move was difficult for Gwen and the girls. The tears flowed freely on this move. We had to pack furniture and pets, leave a good school, friends and the library and the excellent Mexican food at the Spanish Inn south of the tracks. As with all the past and future moves our allowance for moving was grossly inadequate to cover expenses. Packing meant haunting grocery, liquor and department stores for cartons and saving newspapers to have enough packing materials. While we never liked it, we became very good at packing. Nothing was ever broken. The one good feature about frequent moves was we tended to rid ourselves of some of our surplus junk. However we kept accumulating, buying, and finding second hand furniture, like the black walnut chest we saved from the bulldozer and fire at the Flagstaff city dump, and the girls began to accumulate their own treasures. We were always over our weight allowance. As previously noted each of these moves cost a good $1,000 out of our own pockets. This move was to cost even more!

In the summer of 1958 we headed for Albuquerque . . . a new home and a new job. We took off with the 1953 Chevy Station over-loaded from a car top carrier to bottom with precious cargo. We carried phonograph records, photograph albums, plants, guns and eats. The remainder of our household goods were in a moving van. Our objective was to beat the van to Albuquerque and the house we had rented on Girard Avenue Northeast. Driving to Albuquerque we had a good tail wind, even if it was in a terrible (normal) dust storm. It was the same old US Route 66 made famous in song and TV series. Scenic vista points were frequently screened by an almost continuous array of bill boards, tourist traps, wild animal displays, snake pits, junked vehicles of one sort or another, filling stations, run-down motels, Indian trading posts and roadside stands. It was never an impressive route and my passengers were not in the frame of mind to be impressed, motivated or even humored. Seemed like the girls fought and bickered more than usual in the back seat. And the pets were not too happy in the terrible heat and dust. At a time like that I kept my silence and drove on for hours.

We beat the moving van to Albuquerque, but we weren't

even fully unpacked and settled in the new home when the owner served notice that he wanted it back. We had to move again in 30 days! That went over real great with the family. We decided to look for a house to buy. We had a few thousand in Government Savings Bonds, a small savings account, started with Gwen's family inheritance, and an equally small checking account. We thought we would not be required to transfer for another three or four years. I contacted my folks to see if they could help us with the down payment which we would pay back, with interest. At that time the down payment and the 5% interest and principal payments on an $18,000 house were modest. True to form, my Mother forbid my Father from giving us a dime. So we plunged in and along with one of the local banks we purchased a 3-bedroom Spanish-style home on Madison Northeast, across the street from the Ed Groesbeck's. It was a lovely home compared to what we had lived in since we had been married. It had a large family room with fireplace, enclosed backyard and attached garage. We enjoyed the house, especially the backyard living and the family room. The only problem was the flat roof, it required much maintenance to keep it from leaking.

The girls enjoyed their school, made new friends and made a start in learning Spanish. We still did not have TV so much reading took place in our house. The girls were well ahead of their Albuquerque classmates since they attended such a good school in Flagstaff. Gwen enjoyed the house and visiting across the street with Blossom Groesbeck . . . a great lady. Gwen was a good seamstress and provided most of the clothes for the girls and herself. She planted flowers and had a small garden. In Albuquerque plants thrived as long as they were watered. The pets enjoyed the shady backyard and we added female Calico who quickly attracted boy friends and provided us with even more cats. Albuquerque provided unlimited access to Mexican food and by that time the girls had switched from hamburgers and French fries to the good hot stuff. The library, shopping and bus service were convenient. We all enjoyed Albuquerque.

On a number of occasions I was able to take the family along on field trips. They particularly enjoyed the Upper Pecos Ranger District where we had the use of a large Forest Service log cabin, horses and stream adjacent to the Pecos Wilderness. We all were very fond of Taos and the Philmont Scout Ranch, since we had spent some time there on a training detail when the girls were just babies.

The Regional Office was in downtown Albuquerque. I was in a car pool with Regional Forester Fred Kennedy, Ed Groesbeck and my boss Dahl Kirkpatrick. They were a congenial group. Frequently our departure for home was delayed by the Regional Forester's involvement in some crisis situation. I was well informed on what was going on in the Forest Service. This was good training for me.

I was a branch chief in Timber Management, a counterpart to Ed Groesbeck who was in charge of timber sale appraisals and administration. I was responsible for silviculture, timber management plans, timber stand improvement, slash disposal, along with insect and disease control. My job as a staffman was to provide services to the national forests, train, inspect, ensure that work was done to standards and coordinated with other resource programs. I worked very closely with Groesbeck and he continued to provide me sound guidance and training. He even stopped me a few times before I went off the deep end. This job required much travel to all of the 13 national forests in Arizona and New Mexico. And as in the past I continued to seek out training opportunities to prepare me for the future. I enjoyed attending training courses in supervision, personnel and management.

I conducted many timber stand improvement inspections, especially in New Mexico and northern Arizona. The crews were usually composed of Spanish-Americans or Indians from one of the many tribes, along with a few college students. On one occasion I inspected a large TSI operation being conducted by Spanish-Americans on the Cuba District of the Santa Fe National Forest. The crew boss invited me to eat with them and spend the night. I heard him tell the cook in Spanish this "Gringo likes hot food so fix him up!" They did not know I spoke and understood a little Spanish, but I comprehended what they were talking about. I had learned Spanish from Basque sheep herders in Arizona. And "fix me up" they did! But I enjoyed hot Mexican food and they beamed when I asked for more hot chile, salsa picante, tortillas and frejoles, for both supper and breakfast. The Spanish-Americans and Native Americans were great people, good workers and excellent fire fighters. I enjoyed my close association with them.

My involvement in forest fire suppression activities continued. On one occasion I was sent to Winslow, Arizona with orders to recruit all the Navajo and Hopi fire fighters that I could sign up for a series of very large fires in California. Working through the State employment agencies of Arizona and New Mexico, we brought scores of Indians into Winslow and bused them to Southern California. After five long days and nights I called the Regional Fire Dispatcher and told him, with a sigh of relief, we had signed up 1,000 Indians and the last bus load was ready to leave. I asked him "what do I do now?" He said "it has been cleared with your boss and the Regional Forester and you should get on the bus and go with them." Three weeks later I returned home. Fortunately by this time of our lives Gwen and the girls were getting along fine without me. I had learned my lesson and during the fire season I carried along a fire pack containing fire clothes and good, comfortable boots.

Forest insect and disease control was interesting work. We had a good detection and reporting system and between Forest Service Research and my office we were able to take prompt control action following detection of an outbreak and identification of the cause. Disease consisted mainly of dwarf mistletoe. The prescription for control was to prune out sources of infection or fell and remove the tree if it was merchantable. We had good technical descriptions on how to do the work. Our job was to follow up to see that the work was done properly.

In the case of insect outbreaks it was vital to identify the insect causing the outbreak. We had bark beetles that caused minor damage, but others, such as Black Hills or spruce bark beetles, were deadly and required prompt removal or disposal by fire and chemical treatments such as EDB (a very toxic poison) that was hand sprayed on the trunks of the infested trees. These were the days before we had heard of Rachel Carson or her *Silent Spring* message. (Carson, 1962) But safety was paramount; we used extreme care in administering chemical treatments and projects were carefully monitored to be certain none of the chemicals affected trout streams, orchards or humans. We successfully treated one outbreak of Engelmann spruce bark beetles on a large tract of private land in northern New Mexico. That particular tract had a meandering trout stream running through a mountain meadow. The water was cold, deep and not over a foot wide. The magnificent trout were so large they had difficulty in turning around in that small stream. The grateful landowner gave me an invitation to return and fish whenever I liked, but unfortunately I never returned to that paradise.

Our largest insect control project involved isolated areas totaling about 1,000 acres in New Mexico and Arizona. The insect in this case was the Douglas-fir tussock moth, a voracious defoliator. One area was on the Sandia Mountains in the Cibola National Forest just outside Albuquerque and the other two were in Arizona on the Sierra Ancha Experimental Forest and near Globe in the Tonto National Forest. This was a major project and timing to control the insect was critical. We had to prepare contracts, advertise and select a qualified, experienced aerial applicator, award the contract, secure DDT that met our specifications, notify the public, states and adjacent landowners and put out monitoring stations. We had sensitive areas, such as a fish hatchery, trout streams, orchards and farm lands near these projects.

Any flying we did on insect control projects was in small observation planes since we were not permitted to fly in the aerial tankers. My job was to coordinate the flights of the tanker crew, observation plane, crews at the air base, ground crews and observers on the treated areas. We had a good team and our contractor, Abe Sellers, flew a converted World War II B-25, twin engine bomber and delivered the chemical right on target and in the proper amount. This work required proper weather conditions, not too windy and low temperatures and the insect at the proper stage of development. As soon as the temperature or the wind increased we shut down for the day. Flights began in the early morning hours, as soon as it was light enough to fly safely in the mountains and hit target areas. As soon as the plane landed we loaded the tanks rapidly and had it on its way again. Time on the ground was as brief as possible. We selected airfields near the areas to be treated. One was just west of the Rio Grande River at the edge of Albuquerque. The other was a crop duster field near Coolidge, Arizona. The latter was by far the best location since the land owner adjacent to the airfield let us help ourselves to the sweet, juicy watermelons as long as we did not drip any DDT on his melons. While we were soaked with DDT a few times during the loading operations all went well and we had no mishaps.

In 1948 my pre-employment research indicated the Southwestern Region was range management dominated. Over the years I had constant reminders of that fact. The Region had millions of acres of pinyon-juniper, from the North Kaibab to the Lincoln and the Carson to the Coronado. This was a significant resource and provided wood, cover and food for many species of wildlife and benefits for livestock. During the 1920-1930s, forests such as the Prescott and Kaibab had intensive timber management plans for the pinyon-juniper timber type. Plans covered regeneration, harvesting levels and management techniques. But in the late 1940s and 1950s decisions were made to eradicate it, since the trees out-competed grass and browse species and used precious water resources. From a public relations standpoint "eradication" did not play well and the practice was renamed "juniper control". However removal practices did not change and if anything they were intensified. Control consisted of using two D-8 Caterpillar (or equivalent) tractors to drag massive battleship anchor chains or cables through stands and uprooting the trees. The hula dozer, a hydraulically controlled blade, invented by our Forest Service Equipment Engineer Hank Mullins was used effectively to uproot individual trees. Fire and herbicides were widely used to kill the trees. But the juniper was a tough old tree and the seeds regenerated 10 trees where only one had existed before. Our recommendations in Timber Management to go slow, retain, protect and manage the pinyon-juniper resource fell on deaf ears and we lost out to the range and water interests. Years later, I am happy to say, the value of the pinyon-juniper resources are recognized and being managed.

Here I must record a near mishap. Just before the holidays I was selected to run the regional office turkey raffle. We put on a big push to sell tickets to benefit our Forest Service Welfare Fund. The entire regional office staff cooperated and purchased tickets. We found a cooperative supplier and obtained a good price for the birds. Now comes the near mishap . . . Gwen and I were invited to a cocktail party at Engineer Max Ilfeld's home before the banquet. He served only martinis. Gwen was smart enough to refrain. But I did not know much about this gin, vermouth and olive concoction and in a short time I consumed eight of the "derned" things. I thought these were good, mild lady-like drinks and I liked the olives. But was I wrong! Wham, it hit me! I could not recall the ride to the banquet place, the drawing for the turkeys or the dinner. And I had to stand up at the head table and be the master of ceremonies and conduct the drawing. What a headache! Fortunately we rode with someone else and the Regional Forester won a turkey. Gwen was terribly embarrassed.

We had a visitor from the Chief's Office in Washington, DC come to the regional office one day. I was told to go meet with a Mr. Lyman and answer his questions. So I did. I did not know who he was nor did I bother to ask. Most of his questions pertained to personnel management. By that time in my career I thought I knew absolutely everything wrong with personnel management, particularly the arbitrary decisions made by classification officers. I unloaded on Mr. Lyman and told him everything negative I could think of concerning the Forest Service's personnel management program. I had nothing positive to offer. He thanked me and I returned to my office. I asked my boss who that guy was. I could have died on the spot when I learned Mr. Hy Lyman was the Chief of the Forest Service's Personnel Program in Washington, DC.

Chapter X

The Railroad

Travel eased over the holidays and after the 1960 New Year's I reviewed timber management plans and worked on project and travel plans for the coming season. Again the fateful day arrived. I was called into Dahl Kirkpatrick's office. The Regional Forester and his Chief of Personnel Management Allen McCutchen were there. Good old "Mac", he seemed to be around everytime they decided to move me. I thought, now what have I done? I recalled my conversation with Mr. Lyman and thought, here it comes . . . maybe I will be sent to that timber stand improvement crew on the Santa Fe for the rest of my career. So it did come as a shock when I was given not one, but two job offers. One offer involved moving to Washington, DC and becoming the Assistant Director of Personnel and the other was to become Supervisor of the Coconino National Forest. I was overwhelmed since I had only been in the regional office for a little over a year and I thought my career was doomed after that conversation with Hy Lyman. But he apparently liked what I had said and wanted me. After discussing the offer and opportunities with Gwen and the girls (they did not want to move anywhere) we decided on Flagstaff. I called and thanked Mr. Lyman for the offer and declined it. I told him I believed I could do more for Personnel Management in the Forest Service as a line officer than as a personnel staffman. He accepted my explanation, but later invited me to a number of national meetings to further express my strong views on personnel management.

Now we had a real family crisis! We had a house to sell. Normally that would have been no great problem. Albuquerque was expanding and new people were moving into New Mexico. However the timing of our move was atrocious since Sandia Air Force Base had just completed hundreds of new housing units on the Base and the Commanding Officer had given orders to their military personnel to move onto the Base. That flooded the market with over 500 homes for sale. We contracted with the same real estate firm that had sold us the house because we liked the agent. He had a few suggestions for improving the resale value. So we patched, painted, fixed this and that and prepared for the beginning of many open house visitations. My transfer to Flagstaff was effective in March 1960. Between open houses we all drove to Flagstaff and found a place to rent on Leroux. Gwen and the girls stayed behind in Albuquerque to sell the house and complete the spring term of the girls' school.

In March I took off alone for Flagstaff to begin my fourth job on the "Mighty Coc". I lived alone in the house on Leroux and dined on Kraft dinners and kipper snacks. In my spare time, mostly evenings, I worked on the job list Gwen had prepared. As usual the house needed many repairs and much painting. I slept on a folding cot and did not have much in the way of any furniture. Other than returning to Albuquerque once a month to conduct business in the Regional Office and see the family, I was able to devote most of my time including weekends to working in the office or being out in the Forest. I was expected to become a member of Rotary International since every forest supervisor since 1930 had been a member. I developed many friends and key community and business contacts through these weekly Rotary meetings. During one of my earlier assignments in Flagstaff I had been a member of Toastmasters. Both of these affiliations helped improve my public speaking ability and self confidence.

I was within easy walking distance to my office in the downtown post office and our Knob Hill warehouse, equipment shop and fire dispatcher's office. I had a real advantage in returning to this Forest. I knew the area and most of the people since I had worked in and around it for a decade. My successor, Ralph Crawford, had successfully completed the transition from the Flagstaff Administrative Unit back to ranger districts and transferred to Prescott where he wished to retire within a few years. Ralph had it all figured out. He had seen too many people work all their careers in an area and then pick up stakes and move to a strange new area. He said most of them were lost souls. He wanted to become established before he retired. And this he did. He deserved it! Unfortunately he did not leave me his crystal ball. I could have used it on many occasions.

We had a top quality staff in the Supervisor's office and the seven ranger districts were working well, that is with the exception of one. The Beaver Creek District was a heavy range management unit. The Ranger was an expert timber person but he did not know much about livestock and had problems in dealing with some of the old hard core grazing permittees. For the good of the district and the Ranger we decided we needed an experienced range type. We put in a requisition to the Regional Office and within a short time we had a top quality rangeman transferred in from the Intermountain Region. He was a few days late in arriving but we were delighted to have him. The former Ranger was transferred to a heavy timber district and did well in that position.

We breathed a little easier until someone notified the Regional Office that our new Ranger had moved south with his secretary and had left his wife behind. That raised a few eyebrows and the Regional Forester asked me to investigate. Indeed he had separated from his wife and brought his secretary along with him. But that was not the real problem. Soon I began receiving reports from some of the livestock people that the new ranger was great one minute and then he would act like a complete greenhorn the next. Some of the grazing staff spent time with him and found that he could be the most brilliant rangeman one moment and then a complete novice the next. It took us a while to identify the problem, which we suspected was alcohol.

I verified the cause when I left home early one morning and walked in on the Ranger in his barn one morning before 0500 hours. He was very surprised to see me that early for the range inspection we were scheduled to make. From that moment on I stuck to his back like a wet shirt. Sure enough while we were out on horseback, at about 1500 hours, he had an attack of DTs (Delirium tremens). He had been unable to take a desperately needed drink as long as I stuck so close and he suffered the consequences, including the shakes. After gaining his confidence we found he had been drinking a full quart of whiskey every day for many years. He even carried some in his pack when he went off to large forest fires. He kept convenient caches of liquor in the refrigerator, the barn, his pickup, briefcase and on his person. Reluctantly he acknowledged to us that he had a problem. The problem had existed before he was transferred and en route to the Coconino he had a relapse that resulted in a delay of several days in his arrival. That cleared up an earlier question on why he was late in arriving. We put him on sick leave, provided him medical attention, rehabilitation and a transfer to a district in New Mexico. He did well for a number of years and then began drinking and died as a result. That was my first experience with an alcoholic.

I enjoyed my work and we had excellent people on the staff and on the districts. We all took much pride in our work. One inspection of our public relations (Information & Education) activities brought out the fact that the inspectors had never seen a unit that was so close to the forest users and the public. They reported that we had the local and state press eating out of our hand. Another thing they said was the "morale on the forest was, if anything, too high". That made us all feel good because we were a "can do" organization that could get the job done and done well.

In August, some six months after I had transferred, I brought Gwen, the girls and the pets to Flagstaff. Our house was sold to a neighbor who wanted it for an investment. We assumed a second mortgage and lost money on the deal. Holding open houses and having people march through the house was not a pleasant experience for Gwen. It tainted us all when it came to buying houses. We liked the Albuquerque house but then we hated it when it drained away most of our meager savings. At that time most everyone else purchased a house, kept it for a few years, sold it for a profit, moved and bought a bigger, better house. We had little to show for our experience. So much for home ownership. Between the transfer and the problem we had in disposing of the house I was not very popular with Gwen, the girls, Huckleberry or the cats for some time.

The transition from New Mexico to Arizona schools was not easy for the girls. They had to work hard to catch up, since the Albuquerque school was far behind Flagstaff's. Again I was gone most of the time on field trips but Gwen was good to help them with their home work and they were soon up to speed. Having no TV to distract from studies was helpful.

I was selected as one of the Region's representatives to attend the 5th World Forestry Congress which was held in Seattle in October 1960. We had a good trip across the country, visiting San Francisco en route for the first time. The girls enjoyed the trip, cable cars and motel swimming pools.

I attended many of the impressive Congress sessions from the opening ceremony to the closing. The theme was "Multiple-Use of the World's Forests". Chief McArdle was Chairman and again he remembered my name. When Gwen attended social affairs the girls stayed in their room with Huckleberry, Blackbird and Calico and ate French fries and roast beef sandwiches and watched TV. We all enjoyed a boat ride on Puget Sound one evening with most of the delegation. We had a pleasant visit with the interpreter for the Russian delegation. She gave the girls a set of nested wooden dolls. While on the boat trip the Russians were quick to detect and point out with pride the reflection of the first satellite in space, their "Sputnik".

Life returned to normal when we returned to Flagstaff. We were all busy with one thing or another. I spent considerable time that first year in providing key individuals with information on some pending legislation, known as the "Multiple-Use Sustained Yield Act". We had to be extremely careful and gave out only information. Really, it was lobbying, pure and simple. We Forest Supervisors were carefully briefed by the Regional Forester on how to go about selling this piece of pending legislation and soliciting support. Then we were warned that if we got caught at it we were completely on our own. It was not a comfortable feeling, but I did what was expected and did not get caught. The bill passed and our region quickly developed multiple-use plans and improved integration of all of our resource management programs.

Information obtained in making multiple-use surveys of proposed projects enabled us to do a better job of timber harvesting, road design and construction and locating major transmission facility sites. We thereby avoided impacting critical resources and scenic areas. In one case we disapproved part of a Bureau of Public Roads Federal Highway project and saved critical habitat of the tiny spiny dace (minnow). This was years before the Endangered Species Act was passed by the Congress. The decisions we made on Bureau of Public Roads highways and other major projects, including the relocation of right-of-ways for major power lines to be built by the Bureau of Reclamation from Glen Canyon Dam on the Colorado River to Phoenix, were frequently appealed to the Chief or to the Agriculture Secretary's office in Washington. However our decisions were sustained because they were based on the results of good analytical work done by skilled staff people using the multiple-use impact survey techniques.

Fire seasons continued to be critical, especially in May and June. One potentially devastating fire started on the north edge of East Flagstaff. Fortunately we were well organized and had the manpower and aerial tankers to stop it before it climbed more than a few hundred feet up the south slope of Mt. Elden.

In spite of numerous meetings the rangers, fire staffman Dean Earl and I had with the Santa Fe Railroad, they were generally uncooperative. They made only token efforts to clean up the right-of-way, until one day in June, when we got their attention! The temperature was in the 90's, humidity at mid-day was near zero and a dry southwest wind was gusting 20 to 30 MPH. Under those extreme fire danger conditions we had ground and air patrols out, each of 20 or so work crews had a person or radio standby, heavy equipment, fire plows and pumpers were loaded and operators were on standby, 50 Indian fire fighters were on standby in Flagstaff, aerial tanker bases were on alert and planes were loaded and ready to roll. Local city fire department units and US Army Navajo Depot fire crews were alert and ready for action as needed. The Supervisor's office and district staffs were ready to implement well prepared fire prevention and suppression plans.

Under these extremely high fire danger conditions the Santa Fe Railway hooked up a switch engine as a helper to assist a long, heavy, west bound freight train up the steep grade west of Flagstaff. When the throttle was opened on the helper engine it started blowing sparks onto tinder dry inflammable material on the north side of the tracks. Several things happened at once. The Mt. Elden fire lookout and the patrol plane detected smoke immediately and reported it to the dispatcher. Patrolmen on Highway 66 also spotted it and drove to where they could take initial action. Aerial tankers en route to another small fire on the south end of the Forest were diverted to the railroad fire. Heavy equipment, work crews and fire departments were dispatched. The fire was whipped up by high winds and headed northeast. In its path were the wood structures of the famous Lowell Observatory and the City of Flagstaff. Everything fell into place and we were feeling confident until the patrol plane and lookouts reported new fires being started by the train as it slowly worked its way up the grade towards the divide near the Coconino-Kaibab boundary. Our fire dispatcher called the railroad dispatcher and asked him to stop the train, since it was starting more fires. The reply was that only the General Manager in Los Angeles could stop the train and he did not want to be bothered. We continued to pour in manpower and equipment and kept the fires from spreading, but with 500 men on the fires or on their way with additional equipment, we were rapidly over extending our resources. The Mayor considered ordering an evacuation of Flagstaff. It was a critical situation!

Where was I while all this was going on? I was on a district south of town and was pleased to listen over my car radio to the prompt action being taken. Everything was falling into place just like we planned. But I headed for Flagstaff and ordered our fire dispatcher over the radio "to get that train stopped". Keith Hunter kept trying but was unsuccessful. I drove to the Flagstaff railway station and parked my nice, brand new, green Chevrolet sedan across the double tracks and locked it. I went into the train dispatcher's office and ordered him to stop the train. He said he couldn't do it since another freight due in 10 minutes. I pointed out the window to my car sitting straddle the tracks and said it would not look good to have a car and train wreck in the middle of Flagstaff and they would be responsible for not only destroying government property but they were also racking up fire suppression costs at the rate of well over $10,000 per hour. Then, in no uncertain terms, I ordered him to call the General Manager and get the train stopped! The train dispatcher just about had a heart attack on the spot.

Well that did it! The power of suggestion is a wonderful thing. The train was stopped within a couple of minutes. That resulted in trains being stopped and backed up on that busy main line from Kansas City to California. We had a 10 mile long trafficjam on Highway 66. Most of the tourists, highway patrol and sheriff's deputies and city police had never seen aerial tankers dropping fire retardant so nearby and they enjoyed the air show. Our great fire dispatcher, Keith Hunter, received one complaint from a home owner in Flagstaff who did not like to have pink colored fire retardant drift onto his wooden shake shingles. The owner was referred to the Mayor who refused to give him the "time of day". The train started 10 fires before it was stopped and all were controlled at less than 10 acres in size. The press and other media had a field day. Our people were the heroes of the day. It was a pleasure to work with such an outstanding group of people. Bottom line for the railroad was a citation and a fire suppression bill of well over $100,000, plus damages. And we collected this time! The cause of the fire was a build-up of carbon in the switch engine and when it was used as a helper it blew red hot carbon particles out the stack. After that railway crews were assigned to dispose of inflammable material in the right-of-way.

My work routine changed radically when Gwen went into the Flagstaff Hospital for a hysterectomy. She came through the operation in fine shape. But waiting for the results of the biopsy was stressful for the entire family. The results were somewhat inconclusive and contained some "maybes" and some "perhaps". But the end result was she recovered rapidly and kept busy at the library and with the girls. Gwen had several good friends, the closest being Francis Babbitt across the street from us and Mary Sweitzer behind us. The girls were good friends of Christine Babbitt and they became acquainted with all the brothers except Bruce, who was away at Oxford.

Each Thanksgiving week we usually traveled to the Frank Carroll's in Luna, New Mexico, or wherever they happened to be stationed at the time, for a reunion. Also with us on those occasions were Natalie and Pat Pattison and Shirley and Wally Gallaher. The eight of us had worked together at one time or another. On one of these trips we visited an old abandoned mining site near Mogollon, New Mexico. The operation had been closed abruptly and junk furniture and artifacts of the mining days had been abandoned. Naturally we returned to Flagstaff with more "precious junk". Gwen and the girls were real "pack rats".

Work kept me busy and the pace was lively. There were always local meetings with County Board of Supervisors, and Forest Supervisors' meetings with the Regional Forester and his staff in Albuquerque or elsewhere in the Region. I was frequently invited to speak to National Park Service trainees at the Grand Canyon Albright Training Center. I described the role of the Forest Service in administering the National Forest System, State and Private Forestry and Research programs. These efforts helped facilitate our coordination and cooperation with the Park Service. The Beaver Creek Watershed program was active and continued to grow.

We were provided funds and given frequent assignments to try something different, to innovate. One such effort was to practice unit area control in managing ponderosa pine. The method had worked well in California so we were directed to try it in Arizona. This involved group selection, which was really a series of small clearcuts followed by commercial thinning in the residual stands. In theory it was good. But it failed and the technique was abandoned after only one timber sale. Many of the group areas cut contained a dense cover of a bunch grass called

Arizona fescue. Cutting released the grass and density increased to the level that pine seed falling on the area failed to germinate or the seedlings perished due to lack of moisture or growth inhibitors produced by the grass roots. The logging company had a surplus of pulpwood for their Snowflake Mill and failed to make the necessary thinnings. While it was not a good example of the best timber management, livestock permittee Hy Kennedy, at Lost Eden, was delighted with the increase in forage production. Through his improved management and the increase in forage he was able to obtain approval for an increase in permitted numbers of yearlings.

Fortunately most of the timber harvests made on the Coconino up to that time were light selection cuts which removed 35 to 50% of the merchantable timber. The forest looked good and was very productive, for timber and other resources. In some areas we averaged five dead trees per acre. Most of these trees had been damaged and weakened by lightning and then killed by bark beetles. Rather than leave the dead trees for wildlife habitat we felled the snags on each timber sale area to reduce the safety and fire hazards.

The Forest had an active slash disposal and prescribed burning program. Slash disposal was achieved by requiring the logging operators to do the work or make cooperative payments into a slash disposal fund for us to perform the work for them according to timber sale contract provisions. I had helped pioneer one aspect of this activity when I was on the timber staff. I was sent to the Southern Region and near a place called Monk's Corner on the Francis Marion National Forest. I was trained in how to conduct broadcast prescribed burning in pine stands. There was a slight difference between northern Arizona and South Carolina in that burning was done in loblolly pine stands which stood in a foot or so of water. Burning of longleaf pine was done on so called ridges where the water was only ankle deep. Bringing those techniques back to dry Arizona we quickly demonstrated we could kill ponderosa pine either during the rainy season or late in the fall. That practice of trying to burn large blocks of logging slash was quickly abandoned. We did not have research recommendations or the experience to pursue prescribed burning. Our most successful prescribed burning for fire hazard reduction was done with machine bunched logging slash piles. We successfully burned many miles of slash along logging roads each fall. Unfortunately we had limited funds to experiment and learn how to do prescribed burning safely. Funds for fighting forest fires were unlimited. I believed we were wrong in putting out all wild fires regardless of the values involved. Under favorable conditions fire could be used to reduce accumulated fuels and make fire control easier. But the policy was to put them all out as quickly as possible and our leaders held us to rigid requirements.

Our next major effort at prescribed burning was for range and wildlife habitat improvement. Projects such as this required intensive surveys and preparation of detailed plans which were approved only at the Regional Forester's level. The idea was to remove competing vegetation and encourage the growth of desirable range and wildlife plants. Such a plan was ready for implementation east of Flagstaff at a place called Cosnino. Funds were provided for the project with the understanding if the fire spread out of control we could use fire fighting funds for control. The area contained a heavy stand of pinyon and juniper. The day prescribed was rapidly approaching and Ranger Milo Jene Hassell alerted us. We invited Regional Forester Fred Kennedy and his staff to come and observe our team at work. The day arrived and the manpower, equipment, Regional Office brass and the press were on hand. There was only one "fly in the ointment". That involved an incomplete weather forecast. We had every weather factor that was predictable at that time except information on the jet stream. We had pleaded with the Weather Bureau for years to provide us good fire weather forecasting, including jet stream information. That was still missing, but everything else looked good so we torched it off according to the approved plan. For about an hour the fire behaved beautifully and then it happened! The "jet stream" influence dipped close enough to the surface to produce intense winds and caused the fire to crown and start running in every direction. It was terrible! Crown fires in pinyon pine resulted in extremely dense black smoke and severe damage. Fortunately we were able to stop the fire within a few hours. When the smoke cleared and we could assess the damage we were appalled as were others. The bottom line looked like this:

1. We burned over one-half of the land area (several hundred acres) of Walnut Canyon National Monument. To say the least, the National Park Service was not happy!

2. We burned and seriously damaged a transcontinental AT&T telephone, telegraph and television line. The fire was so intense that sections, several feet in length, were burned out of the middle of individual poles. Those sagging wires on damaged cross arms and the burned poles did not make a pretty picture.

3. The 169-KV power line that served most of northern Arizona was put out of commission for almost 24 hours. Parts of Flagstaff, Winslow and Holbrook were out of power.

4. US Highway 66 and the Santa Fe Railroad traversed the fire area. The fire and dense smoke created a monumental traffic jam on US Highway 66 and a failure in the signal system resulted in trains being backed up from Kansas City to California. We could not blame this fire on the Santa Fe.

5. Local and state press had a field day and came out with

spectacular photographs of the fire along with red headlines.

The situation was indeed bleak. Hassell could see himself being sent back to the North Kaibab, where as a student he had pruned trees and rolled rocks behind a road grader. I did not know what would happen to me. But again it was fortunate we had invited the Regional Forester, who along with his staff, had no quarrel with our action. We followed his approved plan to the letter. All our prescribed burning plans from that time on included a reference to the jet stream. Hassell went on to become Regional Forester in Albuquerque and later State Land Commissioner of Arizona.

Fortunately, there was one bright spot. Fire Staff Officer Dean Earl had briefed the press, highway patrol and railroad ahead of time about our plans, including our concern about the lack of information on the jet stream. We had followed Chief Gifford Pinchot's earlier advice to "use the press first, last and always if you want to reach the public." In the minds of the press we were the good guys and the failure was blamed on the Weather Bureau. The Weather Bureau promptly obtained the funds and started providing the Forest Service with more complete fire weather information.

That was not to be the end of the Cosnino Burn. At our next Rotary meeting the manager of Arizona Public Service told me his people were calculating the costs resulting from 24 hours of interrupted power and compiling claims being submitted by merchants for the loss of ice cream and other perishable products. I swallowed, breathed deeply and took a stupid calculated risk. I told him that our people were devoting much time, effort and attention to Arizona Public Service applications for many miles of new powerline rights-of-way across the Forest. But if we had to stop and evaluate all these claims he was planning to make that it could possibly be months before we could return to processing his special use applications. He looked me in the eye and with a smile said "why you so-and-so!" That was the last we heard of any claims. The applications and approval of his permits were expedited. If I had not been on good terms with that manager through the Rotary International I could have been reported to the Regional Forester. Then I would have been the one pruning trees and rolling rocks behind a road grader.

The next event related to the Cosnino Burn came two years later. The grazing permittee had fully cooperated with the plan for improved management, called deferred-rest-rotation grazing. The burn and reseeding were so successful in securing increased forage production that we recommended an increase in permitted numbers of livestock. This had to be approved by the Regional Forester. When our recommendation reached the Regional Office a team was sent out to investigate what we were up to. Such a recommendation was perceived to be unwarranted since most grazing permits at that time were being drastically reduced in numbers. The investigators found we were conservative in our recommendations and approved the request.

Another interesting feature of our job was administering sheep grazing. These operations were owned by fine people who were very cooperative and maintained the range in good condition. Sheep herders were imported from other countries. They were given a work permit to enter the United States for two or three years and then returned to their homes. These hard working people lived out of sheep wagons and traveled with the herd. They spent the winter months in the desert country of southern Arizona. In the late spring the herders slowly drove the sheep over designated sheep driveways to the high country on the Coconino Plateau. Our district people met them at the National Forest Boundary and counted the permitted number of sheep on to the summer range.

On an extremely high fire danger day in June a fire popped up on the east slope of the San Francisco Peaks in a sheep grazing allotment. Fire control action was prompt. Patrolmen, work crews, aerial tankers and heavy equipment were promptly dispatched. In such situations a fire investigator and a Spanish-American interpreter were part of the initial attack team. It was determined the fire had started at a stock tank which had been used that day by a large band of sheep. They found the remains of a burned wooden match and a fresh set of horse tracks at the source of the fire. The investigators tracked down the herder, interrogated him and told him they would return and obtain his signed, witnessed statement as soon as the fire was under control. The herder acknowledged he had lit a cigarette while waiting for his sheep to water and had thrown the match down. I was in another part of the Forest and heard the investigator report his findings to the fire dispatcher. I got on the radio and ordered the investigator and interpreter to return immediately and obtain a statement from the herder. This they did. The fire burned and damaged several hundred acres before it was controlled. Our investigation clearly identified the responsible party and the grazing permittee was billed for costly suppression and site restoration expenses. Shortly after receiving the bill the permittee came to my office and said "You so-and-so, if you had not got in the act I would have had that sheep herder on his way home before the fire was controlled!"

I had a perennial visitor call on me at my office. Hazel Carter had a special use permit to operate a lodge at Mormon Lake. Her family had been in the area for many years. She believed all conditions of her Forest Service permit were foolish and unnecessary. And she was quick to give the Ranger and the Forest Supervisor her unflattering views of the government bureaucracy. My secretary, Mrs. Black, would politely usher Mrs. Carter into my office, close the door and leave the building. For the next 20 minutes I could expect to hear everything that was wrong with the Forest Service and the federal government. Her voice carried far beyond the closed door and throughout the office. Her language would have made the roughest loggers turn pale and cover their ears. After she said her piece she abruptly departed, slamming the door behind her. My secretary would meekly return and ask me why I took all that abuse. I told her Mrs. Carter always felt better after telling me and the entire government off. And following her session with me she always returned to her resort and fully complied with the terms of her permit. I added that Mrs. Carter was a lifelong friend of Senator Carl Hayden, the powerful Chairman of the US Senate's Appropriations Committee and a strong supporter of the Forest Service. If anything, after a session with Mrs. Carter, our appropriations for the Beaver Creek Watershed Project and the entire Forest budget increased.

Another event involved the Mormon Lake and Oak Creek Canyon areas. We were ordered to revise summer home fees based on the value of the national forest land they occupied under special use permits. At that time fees ranged from $25 to $30 per year. Land values had skyrocketed and reappraisals based on fair market rental values resulted in permit increases of well over $1,000 per year. We obtained approval of the Regional Office to make adjustments over a five year period. Then we had a real problem. How do we notify the permit holders? We decided to do it through their group associations and invite members to meet with us. This we did. The rangers, my staff and I met with as many as could attend the meetings. They were appalled at the amount of the increase. But they listened and gave us their views and recommendations. They did not like it, but our explanation was well thought out and was on a sound basis. Following the meetings we sent a personal letter, which I signed, to each of the several hundred people. We did not receive one appeal. A neighboring forest had the same situation, only they held no meetings and notified their summer home permittees by form letter. That resulted in several hundred appeals which required years to settle.

We had many visitors to the Forest, especially to see the very active, innovative water yield programs on the Beaver Creek Watershed Project. Between the Forest visitors and foreign students who came to Rotary we entertained and made many friends. We still have close relations with Sue Passmore and her family in Wales. On 22 November 1963 Richard Costly, Assistant to the Chief of the Forest Service and I were inspecting Beaver Creek Watershed Projects. On that day it was announced to the world, including a report over our Forest radio network, that

President John F. Kennedy had been shot. That created an empty feeling in all of our hearts and minds. On the day of JFK's funeral we closed Forest Service offices. Our family went next door and watched the solemn ceremony on a neighbor's TV.

I remained active in the Society of American Foresters. We formed the San Francisco Peaks Chapter, which required members to live within sight of the Peaks. It has been said that on a clear day (or was it night?) that the Peaks could be seen from San Francisco. I checked out that theory one time when I was on the top of the Mark Hopkins Hotel in San Francisco and I could not see the Peaks. Nevertheless the requirement stuck. Our major accomplishment was to sponsor the establishment of a forestry program at Northern Arizona University. This came to pass and we became a strong supporter of the program at NAU and the first Dean, Chuck Minor.

I continued to successfully hunt deer, turkey and elk. Until lakes in the high country froze, duck hunting was particularly good. My favorite spot was Marshall Lake which was about 30 minutes out of Flagstaff. If the weather turned stormy we could leave the office and return with a limit of ducks in a couple of hours. Every time I hunted there I thought about the time Fire Dispatcher Bill Phillips and I took Forest Supervisor Ken Kenney and Administrative Officer Joe Lucero duck hunting. Bill's small duck boat was just right for Marshall Lake. Bill had on a body cast from a recent back operation. Bill, Joe and Ken were hunkered down in the boat in a dense stand of cattails. I was on opposite side of the small lake. A flock of ducks came over, much too high to hit, but Joe and Ken both stood up on the same side of the tipsy little boat and fired. I could hear Bill shouting "sit down, sit down". Of course they did not hit anything, but the recoil of both of them shooting at the same time and the unbalanced boat resulted in all of them being flipped over into a very frigid four feet of water. They came up snorting, covered with cat tails and muck from the bottom of the lake. I about exploded with laughter, but kept it inside and helped the freezing duck hunters home. I did not hear Bill's familiar "yuk, yuk, yuk" that day. You might say we harvested three mud hens that day.

One day we received a very appealing, heart rending letter from a 90 year old grandson of a US Army Indian War soldier. He described how his grandfather had died and was buried beside the old military road on the Mogollon Rim. The grandson wanted to erect an Army grave stone at the grave site, but was not able to travel and do it himself. The Army records gave a very precise location of the grave. Staffman Doug Morrison and I decided to help. We searched out the grave site on our own time and found it within sight of the road. As noted in the records it was marked and protected with a pile of rocks. It had remained undisturbed for about 100 years. Fortunately road reconstruction during the CCC program of the 1930s and subsequent improvement and maintenance of the road had not disturbed the grave. The grandson had the Army Graves Registration Office send us a headstone and we set it in concrete. I do not recall going through special use permit application procedures, but after all the old soldier had been there long before there was a Forest Reserve or a Coconino National Forest.

1964 was a sad year. That was the year we lost Huckleberry. He became partly paralyzed in his hind quarters. We all loved that dog, Gwen's dog. After all he had been with us for 16 years. Gwen cared for him tenderly when he had almost died of leptospirosis and suffered from a broken leg. He had guarded the girls all their lives. He had bitten supervisors, landlords and even me. During his final days we cared for him tenderly, fed him what little food and water he could take from our hands and carried him outside to do his business. Seeing the suffering in his eyes we made a family decision to have him put to sleep. I took him to the vet and before he received his final injection he licked my hand. I wrapped him in a Navajo saddle blanket and buried him along side the road to the Mt. Elden Lookout. I carved "HUCK 1948-1964" on a large boulder. Then I kneeled and wept over his grave. I loved that dog like one of the family, we all did. I must return one day to northern Arizona to visit the old soldier's grave and pay my respects to Huck. I know when I find Huck's grave that tears will come to my eyes, then as they are now, as I record these thoughts.

In 1964 I received my next transfer offer. I was to join the "Battle of the Potomac" . . . the title resulting from the interagency battles for power within the Federal bureaucracy, among agencies responsible for resource management activities. By this time in my career I realized these really were not offers of a transfer but orders to move! This next move was to be to the Washington Office of the Chief of the Forest Service. It was a lateral transfer from a GS-14 to a GS-14 position. I was doomed to again move without a promotion. Gwen and the girls wondered why then was it worth all the bother to move again? That was a good question, but it was all in line with serving the Forest Service and a career move to gain experience at different levels. It appeared my career ladder was pointed towards a Washington Office assignment and then, if I was lucky, I could escape back to a field assignment. As we collectively considered this latest offer some thoughts concerning people came to my mind.

Two of the Forest's outstanding GS-462 Grade 11 Forestry Technicians were Norman Johnson and Homer German. They had been my strong timber staff leaders when I was a District Ranger at Happy Jack. They continued in timber work and annually received outstanding performance ratings. I had gone to "bat" literally and figuratively with the Regional Office Personnel Officer and we put together a case for converting these fine individuals to professional Forester GS-460 Grade 12 and the actions were approved. Norman became Assistant Forest Supervisor in charge of timber and Homer became the Forest check scaler. Approval of these classification actions set the pattern for other similar conversions and promotions within the Region.

My other personnel efforts involved bloody, knock down, drag out battles with the Region's Personnel Classification Officer concerning the grade level for women. On two occasions the Regional Forester intervened to separate us. I was especially provoked when we recommended promotion of an outstanding woman in our business management section. I knew the classification rules and had learned some of the "tricks" of the trade. We recommended reclassification and promotion to Grade 9. The Classification Officer told us this could not be done for women because they did not warrant as high a grade as men. That touched a tender nerve and I took on the whole personnel system. And we won! That too set a precedent and helped other women advance.

I recall when one of the national federal employee unions came to Flagstaff. They wanted to unionize the Forest as an example for the entire region to follow. They directed us to set up a series of information meetings in Flagstaff and on the outlying districts. We made them feel welcome and agreed that employees could meet with them any day outside of official work hours. We reviewed the rules concerning relationships with unions. We sent out notices to each employee and scheduled the meetings. Flagstaff was the site of the first meeting. There was little interest in union membership and only three people from the Flagstaff area districts, my secretary and myself attended the meeting. Only the respective rangers attended the meetings on the districts. We listened to what unions could do and heard about the dues structure. When there were no questions from the participants the union organizer handed me an application and said, "sign here and become the first member". I had been instructed that as the principal line officer not to join a union, so I politely declined. This angered the union man and he promptly filed a complaint with the Secretary of Agriculture concerning my lack of cooperation. The appeal of course went nowhere. When we asked the employees why they were not interested in a union they said they were satisfied that management took care of their needs. This made us all feel good!

While I was on the Coconino we had a strong and effective safety program and a reasonably good safety record. There were the usual, frustrating, nicks and sprains, but no serious accidents.

I was always grateful I did not have to call on any children, wives or parents and inform them that a loved one had been killed by a falling snag or involved in a fatal car or airplane crash. However we were indirectly involved in other serious accidents.

Once we received a report about a Navy jet fighter plane crashing in the Mogollon Rim area. We located the crash site which was 10 miles southwest of Happy Jack and very near permittee Irvin Walker's barn. The pilot had bailed out at a high altitude. He was not located until a search plane saw the top of a small pine move. The pilot was about two miles from the crash site and had suffered a fractured spine but was able to pull on the shrouds of his parachute and cause the tree top to move. We helped pack the poor guy out. The Navy expressed their gratitude to the Forest personnel who were involved in the search and rescue efforts.

The next tragedy was below the Mogollon Rim on the nearby Tonto National Forest. During periods of extreme fire danger we were always a bit nervous when the Tonto had fires just below us. Old scars in the Rim area were evidence that fires from a thousand or so feet below had crowned and blown up over the top. Dean Earl, other fire people and I took turns on observation flights. On some of those hot June days the air over the Rim was so rough that it tossed the plane up, down and from side to side, and my guts along with it. Observe and heave, heave and observe, that was what I went through on a number of occasions. One day I was "riding shotgun" on one of these flights and we observed a nearby single engine World War II TBM tanker plane make a run to drop slurry (fire retardant) on a Tonto fire. To our horror we saw a puff of smoke come from the plane as it dove on the fire and in two or three seconds it crashed into the ground. We saw no survivor. I had to repeat that story several times to federal investigators who determined the tanker plane had blown a piston, lost power and crashed. On another occasion in the same general area two observation planes collided over a fire and the occupants were killed.

On one of my last official trips around the Forest I was returning on the newly opened Interstate Highway 17. It was a well designed road and reflected much team work and coordination. My staff and I had spent many days walking the proposed alignment from the Verde River to Flagstaff with State and Federal Highway design engineers. It was our task to specify and approve such items as clearing limits, timber salvage, debris burning, interchanges needed to serve future recreation and timber harvesting plans, forest fire control access, fence design to control livestock and facilitate wildlife migrations, underpass structures to serve as passageways for livestock and roadside rests. As I was driving along behind an Arkansas furniture manufacturer's semi-truck I saw a bright red Coke can go sailing out of the truck's right side window onto the clean, uncluttered roadside. In a couple of minutes another Coke can came sailing out the driver's side into the clean medium. That made me angry and I radioed the fire dispatcher to have a sheriff's deputy intercept the truck and I would pick up the evidence and file a complaint for littering. When all was said and done the truck driver and his assistant were fined $50, plus their lost time, each for littering. We notified the media and headlines in the papers announced that littering in the Coconino National Forest cost $50 per can!

We enjoyed our life in Flagstaff and the Southwest. This was a good period for the entire family. We all carried with us fond memories of the Southwest and the people we met along the way, after all it had been home for almost 15 years. I had served as a Forest Service line officer on two ranger districts and one forest supervisor office and gained experience as a staffman in national forest and regional offices. As I was about to leave the Southwestern Region I reminded myself that the knowledge I had about wood preservation got me there in the first place, but I never used it. Reluctantly we accepted the offer to move. Folks on the Forest presented us with a beautiful framed color photograph of the San Francisco Peaks. Many of the local citizens, city and county officials and the newspaper editor, Platt Cline, thanked us for our contributions to the community and wished us well.

BOOK THREE
The Foreign Fantasy

USDA and Forest Service support of world agriculture and forestry increased, especially in Southeast Asia. When the US Agency for International Development (USAID) could not obtain agriculture expertise needed for its expanded South Vietnam pacification Program the USDA entered into a Participating Agency Service Agreement (PASA) and provided crop specialists, irrigation engineers, agricultural cooperative experts, economists and foresters. Along with the military thrust into Southeast Asia USAID had the backing of President Lyndon B. Johnson to move ahead with the civilian programs. The forestry program was well supported by the Forest Service's International Program people and backstopped by the entire National Forest System, Research and State & Private Forestry organizations.

Meanwhile back in the real world recreation use of National Forestry System lands was skyrocketing. Other resource programs continued producing wood, water, improved wildlife and fish habitat and range resources. Programs to enhance air and water quality were given priority as the Forest Service became more involved in environmentally popular programs such as natural beauty, wilderness, a wild and scenic recreational river system and a national trail system.

Chapter XI

The Washington Scene

Our accumulation of household goods continued to grow. After we had packed all the dishes and fragile items we called United Van Lines to provide one of their largest vans to load and transport all our "treasures" to Bethesda, Maryland. On an earlier house hunting trip we had located and rented a split level, 3-bedroom house on Danbury Road. The thought of home ownership still left a bitter taste in our mouths and pocketbook. We loaded the Chevy station wagon with plants, records, camera gear, guns and cats. In addition to Blackbird and Calico, we now had Calico's daughter Patricia and Laura, a longhaired stray who had adopted us in Flagstaff. The long trip across country required frequent stops to water plants and cats and especially to air cats (phew!). The girls were their own "sweet selves" and continued to bicker and argue. Anyway it took their minds off the sadness of leaving good schools, friends, Mexican food and familiar places they had known all their lives. We all missed Huckleberry on that trip and he was in all of our minds.

Of all the good, less expensive areas to live in, why did we move to Bethesda? Why Myrl and Gene Jensen were there to offer us the valuable companionship of dear old friends. They had visited us at least once each year since we had left Colorado. Gene was an officer in the Public Health Service and in charge of the shell fish program. We lived about half a mile apart and spent many hours together. We enjoyed lobster boils and bushels of steamed clams at their nearby PHS Officer Club.

The girls found Bethesda a real change. They moved from a modest western community where everyone knew everyone else to the most affluent county in the United States. Everything from food to clothes was more expensive. Maryland taxes were high enough but Montgomery County imposed a surtax on top of that. The girls attended Walter Johnson High School, where students wore better clothes and drove more expensive cars than the faculty. Fortunately Gwen had taught them well and the girls were appreciative of her frugality, sewing and making their clothes. The progressive Flagstaff schools had them well prepared for eastern schools.

After we were settled we enjoyed many of the free amenities of Washington, DC. Most weekends found our family visiting museums, art galleries and places that were rich in history and free. We had frequent family trips and vacations. In the summer we made short trips to West Virginia's Seneca Rock in the Monongahela National Forest or Chesapeake Bay. When we had more time we camped out at our favorite Small Point Beach near Bath, Maine. I fished for delicious flounder. On one such trip to Maine, Gwen and the girls went off to an auction and Melissa made a successful bid of $2.50 for a tall Hoosier kitchen cabinet. I was fit to be tied and puzzled as to where we could load it, but we packed it in our rented popup trailer and hauled it home. It turned out to be a very valuable antique. We visited Canada where Quebec City became our favorite place. Gwen loved the old part of Quebec City and spent hours following and photographing beautiful French speaking children.

Spring vacations were spent in the Charlestown, South Carolina area. We enjoyed camping on the nearby Isle of Palms and visiting and photographing some of the beautiful, stately, antebellum homes and plantations. I enjoyed seeing Hampton Plantation. My Mother always told me that was where General Wade Hampton, her great-great grandfather lived. Whether or not that was true, it gave her a good status symbol and something to brag about.

The Danbury house was rather new and lacked the charm of the older kind of houses we preferred. It just did not fit our life style. We paid $325 a month rent which was more than twice what we had paid in Flagstaff. It did not take long before we realized that our cost of living had skyrocketed. We again had to watch our expenses to get by from pay check to pay check. Just before our year's lease on the house was due to expire the owner said she wanted it back. We could have purchased it for $38,500, but we were still gun shy about home ownership and short of money, so we declined to make an offer. (22 years later that same house was sold for $385,000 . . . so much for our ability to make good investments).

We found an older house more to our liking on Green Tree Road. The area had no sidewalks. We had a driveway but most of the neighbors parked on the street. It was a mess that winter when 30 inches of snow came in one January storm. All we could see of the cars on the street were the tops of their antennas. We were snowed in for several days because DC and the suburbs had no snow removal equipment. This house was within easy walking distance of Myrl and Gene's home and to high school. The girls liked the house and by then they had made friends with a number of girls and that is where boys came into the picture. We finally purchased our first TV and I could watch my favorite professional football team, the Washington Redskins.

My new job was in the Division of Watershed Management. I would usually leave home at 0700 hours, catch a bus and be at work in the South Agriculture Building just before 0900. I would leave at 1700 and be home by 1900 hours. It made no difference whether I traveled by bus or car pool, the commute was long and tiresome. I preferred the bus since I could read and relax and watch frustrated car drivers fight the traffic, to and from downtown Washington.

I was one of two Assistant Directors in the Division of Watershed Management. I was responsible for service-wide preparation of multiple-use impact surveys for major water resource projects. Vern Hamre, the other Assistant Director, was responsible for all other National Forest soil and watershed programs. Our boss was Byron B. Beattie, a very intelligent and demanding leader. He was a perfectionist and an outstanding writer. Our work had to be perfect or he returned it to us to do over. I was frustrated in that job. When I looked around for staff assistance to do a job, there was no one there, just me. In Flagstaff I had been one of the respected community leaders; in Washington I was a non-entity, a small fish in a great big ocean, as far as the community was concerned. These two factors were a rude awakening for an ex-forest supervisor. Vern had just come in after being Supervisor of the Helena National Forest in Montana and we shared the same feelings. I had always required good, complete staff work and now I had to do it myself; there was just me and no one to call on for assistance.

I learned I had been selected for the job as the result of the excellent multiple-use plans and impact surveys that we had prepared and implemented on the Coconino. Now I had the over sight task to perform concerning every large water resource development project impacting or occupying National Forest System land. I reviewed multiple-use impact surveys prepared by all of the Forest Service regions and prepared the Chief's response. These ranged from projects in the Yukon, Pacific Northwest, Northern Rocky Mountains, California, Southwest, Midwest, East, New England and the South. Frequently I traveled to these areas to provide training for the people involved in project evaluation and to gain firsthand knowledge for myself and the Chief's Office. As soon as we sized up a proposal we made presentations to the Chief and staff and obtained their views on the region's and our recommendations. Most of my meetings with Chief Ed Cliff were at these staff meetings or when we talked in the men's room. Ed Cliff was a tremendous leader and we all worked our hearts out for him. He provided us clear policy and direction. At that time the entire Forest Service worked well together as a team. After obtaining clearance from the Chief we worked with the construction agencies to see that they recognized our interests and National Forest resources. Among the many federal agencies we had to negotiate with were the Soil Conservation Service, Corps of Engineers, Federal Power Commission, Bureau of Reclamation, Bureau of Outdoor Recreation, National Park Service and the Water Resource

Council. Our objective was to ensure that the design, construction, operation and maintenance of a major dam or other facility provided proper conservation water levels and down stream water releases to provide suitable habitat for fish and wildlife, recreation use and aesthetics, and to relocate or provide needed transportation facilities to provide for public use and management of national forest resources, etc. It was complicated and demanding work. As I gained experience and the confidence of Director Beattie I was given more responsibility to carry out the program.

The Forest Service was a member of the interagency Wild Rivers Study Team. The role of this group was to identify, study and recommend selected rivers and streams for retention in a free flowing condition and participate in the preparation of proposed legislation. Many waterways had good potential for recreation and other uses. This effort developed an effective legislative device for stopping the dam builders, who wanted to eventually control every free flowing stream in the country. This project required multiple-use surveys and we had to have detailed facts about each river. We made numerous presentations to the Bureau of Outdoor Recreation and its very critical Director Ed Crafts. I knew I had passed the extremely rigid demands of this ex-Forest Service Deputy Chief when he would praise my presentations and criticize his own staff because they had not done as well as the Forest Service.

I enjoyed the Wild Rivers work, especially on the occasion when I was called into the Chief's office one Friday morning and told that he and the Secretary of Agriculture needed detailed information on the nature of the improvements on private lands which were intermingled with the Federal land adjacent to six rivers which were candidates for inclusion in the initial Wild Rivers System. They needed this information to testify before Congress. I had ten calendar days from that time to personally visit the rivers, map, photograph and identify significant features on these rivers and return to brief them on my findings. I called the regions and arranged for maps, aerial photographs, guides and transportation. The next week I traveled by commercial airlines, small fixed wing planes, helicopters, jet boat and car. I evaluated the Eleven Point River in Missouri, Salmon, Middle Fork of the Salmon, Middle Fork of Clearwater, Lochsa and Selway in Idaho and Rogue River in Oregon. It was an exciting assignment. The days were long and it was an extremely demanding week. I had excellent cooperation everywhere I traveled and by the end of the week I obtained the desired information. My only regret was that I could not stop for a while in each of those lovely places just to watch the water flow, fish, or observe the abundant bald eagles, elk, deer or mountain sheep.

I returned with maps and photographs of each of the

proposed wild rivers. My briefing for Chief Cliff and Secretary of Agriculture Orville Freeman took place on the floor of the Secretary's office. We crawled around and looked at maps and matched photographs with significant improvements on private lands. In summary I pointed out locations and described houses, corrals, shacks, resorts, mines, orchards, crop land, saw mills, plywood mill and one place I referred to as a multiple-use establishment. There was a place on the Middle Fork of the Clearwater River locally known as Maggie's Bend. This was a motel type of building that projected out over the River. Secretary Freeman asked me to describe exactly what I meant by a "multiple-use establishment". I told him that as an ex-Marine he would have identified it as a "house of ill repute". The Secretary roared with laughter and Chief Cliff almost bit off the stem of his long curved pipe. With that conclusion I figured it was time to roll up my maps and photographs and get out. The Chief was still puffing, snorting and shaking his head as I departed. The information was useful to those two as they testified the next day. I received an expression of appreciation from the Secretary and thanks from the Chief when all of the candidate rivers became "instant wild rivers" in the Wild and Scenic Rivers Act.

In 1966 I moved across the hall to the Forest Service's State & Private Forestry program. This was a promotion to Director of Flood Prevention and River Basin Programs at Grade GS-15. This program had been administered for 25+ years by the same Director. He had been in the Forest Service for over 40 years. Unfortunately he had fallen into a pattern of checking with the Chief on what he should do about this or that activity and then passed the directions on to his staff. Naturally that management style did not go over well with the Chief or the program staff. The staff essentially worked on day-to-day work plans and under these circumstances they were not highly motivated or productive. My greatest challenge was my lack of knowledge of the Forest Service's State & Private Forestry Program. I had no experience with State and Private Forestry, Public Law 566, the Small Watershed Program or River Basin Planning. I had to study the respective laws, Forest Service Manual and Handbooks and consult with the Soil Conservation Service and other more experienced State & Private Forestry people. My staff and everyone were very helpful in getting me up to speed.

The Forest Service had staff people in each regional office who handled our program activities. I met with the program people in all regions and visited their major and minor projects. This involved traveling from New England to California and throughout the South. We were involved in watershed planning for protection of major cities such as Los Angeles and St. Louis. I found the work interesting and my staff very experienced and supportive. Instead of asking the Chief what to do, we told him what to do and our program took on new life and visibility.

As principal USDA liaison with the multi-agency Water Resources Council and the Soil Conservation Service I attended many meetings in Washington and throughout the country. I became an active member of the Soil Conservation Society of America. I was privileged to receive Departmental approval to accompany SCS, officers of SCSA and Mrs. Hugh Bennett, the wife of the founder of the SCS, to the First Pan American Soil Conservation Congress in Sao Paulo, Brazil. The long Pan American Airline flight was broken up with a welcome stop at the new modernistic Capital of Brasilia. Except for the shacks of the poor people that surrounded this federal enclave, the awesome, artistic (?) place seemed to be deserted. I later learned the Brazilian politicians and foreign embassy representatives did not want to leave the amenities and beauty of Rio de Janerio for the "bush", hot climate and red soil of the interior in Brasilia.

I had the honor to speak to the Congress on "Multiple-Use Management of the National Forest System". This was my first real exposure to foreign travel. I loved it and met people from all over the World. One of our field trips carried us into the western part of Sao Paulo State where we visited farms, coffee plantations, eucalyptus groves (planted in 1860's) and ranches. We were treated to barbecues, cafe zinia (tiny cups of very sweet, concentrated coffee) and powerful pinque which was distilled from sugar cane. In one research laboratory we were each given a large beaker of a product they had manufactured out of corn. It had an alcohol proof equal to or greater (150%+) than the "white lightning" I had been introduced to a few years earlier at Monk's Corner, South Carolina. This diversion added interest to research tours. My later suggestion to Chief Cliff that this innovation be added to Forest Service Research was rejected flat out.

At the conclusion of the Congress in Sao Paulo I traveled to Rio de Janeiro. Avoiding the dog droppings, I walked the interesting serpentine, black and white sidewalks near my hotel along the beautiful, terribly polluted, Rio beaches. I took the tram to the top of Sugar Loaf Mountain and viewed the statue Christ of the Andes. From there I could see the beauty of the city, the Atlantic Ocean and the terrible favelas (slums) surrounding the city. One afternoon I took bus No. 4 to the ferry station and visited a nearby island to see the sights. I had a delightful time. By then I could speak enough Portuguese to find my way around. On return to the mainland it was nearly dark and I boarded the No. 4 bus which the driver said returned to the beach area. The further we went, the higher we climbed and the darker it became. Abruptly in the middle of the roughest of the favelas on the steep mountainside above Rio the bus halted, the driver said "end of the line" and turned out the lights. Out of the bus window I could see some of the toughest looking characters I had ever seen. Frankly I was scared! The driver beckoned me out and slammed the bus door shut behind me. There I stood in the midst of what looked like a bunch of cut-throats longing for new blood, my blood. After what seemed like an hour, but which was probably closer to five minutes, a very large tough looking black man walked up and held out his hand and introduced himself. He was a 7th Day Adventist preacher from New York City and was in the Rio area as a missionary. He had noted my apprehension and assured me a new bus driver would be along in an hour or so and that I would be safe as long as I remained with him. I trailed him through back alleys and trails for an hour while he preached the gospel. The bus was about to leave when we returned. Sure enough No. 4 bus delivered me to the front of my hotel about midnight. That was an exciting evening but it was a big relief to be back to the relative security of the beach area. Since then I have always had a kind word for the 7th Day Adventist Church.

From Rio I flew north to Sao Salvador. This capital of the state of Bahia was an Atlantic sea port, fishing town and market center. The African influence was very strong in both the people and the hot, spicy food. I enjoyed visiting the smelly fish markets, colorful fruit and vegetable markets. A funicular carried people from the lower level of the town to the upper level. The simultaneously ascending and descending cars counterbalanced one another. I rode up and down the steep incline of this cable railway a number of times and took some interesting photos.

Next I flew on a Brazilian airline, Cruzerio de Sol, to northern Brazil and Belem, capital of Para State on the Rio Para. The plane stopped several times en route and in the dark I could not see much or identify where I was, but the attendants knew where I was going and told me when we finally landed at midnight in hot and muggy Belem on a tributary of the Amazon River. I went by taxi to the Grand Hotel. My reservation had been made by one of the Congress foresters. The Grand Hotel was the sort of a place where you would expect to see Humphrey Bogart. It had been grand in its day but now it was just a sleepy, run down, charming place. The high ceiling fans barely turned and did little to stir the hot, humid tropical air. I had a beer in the Hotel bar and listened to the piano player play a tune about a city on a bay, far away on another planet . . . in beautiful, cool and clean San Francisco.

The next day was a full one. I had breakfast of lush tropical fruits and cafe zinia. By then I was addicted to that sweet, strong coffee. I walked to the fascinating water front and snapped many pictures of boats and the produce and people they hauled. Some of those boats looked like Egyptian sail boats and others were very similar to the "African Queen". I took short rides up and down what looked like a great river. It was very wide and mingled with the water of the majestic Amazon.

A walk around Belem was equally fascinating. The sights, sounds and smells of the tropics were all there to challenge my senses. People were friendly, especially the merchants and children who wanted to sell me something. Then there were the young women who leaned out of the half doors of the cribs and tried unsuccessfully to entice me to enter their places of business. The markets contained colorful fruits, vegetables and flowers. I visited a curio store and purchased small rubber figurines of people. I was intrigued by the sights and smells to be found in a witch doctor's shop. He sold bat wings, bark, leaves and bits and pieces of everything imaginable. He had a cure for everything that affected man and beast.

I would have enjoyed staying longer in the Amazon region but I had an appointment with a US Forest Service watershed planner who was on river basin planning study in Venezuela. I had a good look at Caracas, the north central and the oil producing part of the country. There I saw the same conditions that I had found elsewhere in Mexico and Latin America. There were the very rich and the very, very poor. After five weeks I had seen enough of South America to make me want to return and help these people.

On my return to Washington I gave a series of slide talks to the Chief and staff and to the National Capitol Section of the Society of American Foresters. As a reminder of my trip I continued to successfully take anti-malaria drugs for three weeks after my return. It was good to return to Gwen and the girls but I had been badly bitten by the foreign travel bug. It really infected me.

1966 was the year Melissa started driving. She had taken lessons in High School and we purchased a new 1966 Volkswagon Beetle. That spared the 1953 Chevrolet Station wagon which continued to serve us well. The girls loved that little VW car and now Gwen had a chauffeur. Melissa and Cindy both became excellent drivers, probably because they learned from an expert driver, rather than me. Gwen went on a number of trips with the girls and their friend Sue Maynard.

Once Gwen, Mrs. Maynard and I had to rescue Melissa and Sue from the Bethesda Police Headquarters. The two girls had the not-too-bright idea of selling tickets to their own turkey raffle. The problem was they had no turkeys, no intention to get any and no charitable cause except their own. One of their customers had called the cops who hauled the girls into the Police Station. The girls were very meek and much wiser when they were turned over to their parents. Mrs. Maynard believed her daughter got into trouble because of her association with bad company. Gwen and the girls, along with friend Sue, enjoyed the longhaired hippies and their protest songs and music at Du Pont Circle. At that time the Vietnam War was being opposed by the hippies and other more rational people.

My work continued as before and the entire Forest Service worked well together as a team. It was a privilege to be a contributing member of that team and be involved in the revitalized, respected State and Forestry Program.

In the spring of 1966, the Dwyer Mission was appointed by President Lyndon Baines Johnson and sent to the Republic of Vietnam to investigate opportunities for developing short and long term programs for the country's forests and forest industries. Robert F. Dwyer, a lumberman from Portland, Oregon, was mission chief. Team members consisted of Forest Service representatives Chief Edward P. Cliff and Rufus Page, Branch Chief of Utilization and Marketing, Thomas Mainwaring, Senior Economist of the Stanford Research Institute and Mackay Bryan, US Agency for International Development (USAID) Forestry Advisor to Nepal. As a result of their efforts and recommendations, a forestry technical assistance program was approved as part of USAID's Pacification Program in South Vietnam.

We Forest Service Division Directors were asked to recommend candidates for the Vietnam forestry team leader and other positions. I named our Forest Service watershed planner in Venezuela and Mark Johanneson, an outstanding young man in the Division of Timber Management. Mark was selected as the team leader and went off for physicals, inoculations, security clearance and five weeks of training provided by the State Department, USAID and USDA's International Agriculture people. Mark was a GS-13 and was promised a promotion to a GS-15. Again my work continued as before.

Shortly after the New Year holiday break in 1967 we were called to the Chief's office for an emergency staff meeting. Mark had finished his training, received a Top Secret security clearance, met all the other requirements and was ticketed to fly to Saigon on Friday of that week for a two year assignment. At that late date someone discovered Mark could not be promoted two grades to a GS-15 and he declined the appointment and transfer. This resulted from his learning that status was all important and powerful in State Department overseas operations and a GS-14 and the equivalent Foreign Service grade did not warrant any status or the privileges that went with rank. Mark sincerely believed that he could not effectively carry out the assigned mission. That decision caused all kinds of consternation since the Chief was under pressure from the Secretary's Office, which was under pressure from the White House and President Johnson personally, to move and move fast, to support the USAID South Vietnam Pacification Program. We were again asked to recommend candidates. I told the Chief we had selected one of my candidates in the first place and our forester in Venezuela had a change in family status and was unavailable. The other Directors, Deputy Chiefs and Associates had only a few recommendations. No one was really interested in going to South Vietnam at that time. News reports of terrorist attacks and a widespread war made the place sound rather unfriendly for foresters or anyone else. The thought was also expressed that anyone going overseas for that long would be out of the main stream of career promotions and forgotten. As we were being dismissed from the staff meeting Chief Cliff asked me to remain. I did and he asked if I was interested in going to South Vietnam. That caught me up short and I told him I would have to think about it until Monday and discuss it with Gwen and the girls.

We had a family discussion about my leaving for one to two years. Gwen and the girls were strongly against the Vietnam war and shared my interest in wanting to help that little country of South Vietnam in some peaceful way. Gwen said she could get along and she would look after the girls and I was free to accept the assignment. She was taking on quite a challenge to watch over two beautiful teenage girls in those days of wide spread drug use, boys and teen age pregnancies. Soon both girls would be licensed and learn to drive a car. This was the hippie period when both boys and girls had long hair and a strong spirit of freedom of expression. Then what if the girls decided to have another turkey raffle? I had mixed emotions about leaving, but I wanted to go. All these thoughts raced through my mind, but after all Gwen had brought up the girls successfully with minimal input from me. On the following Monday I told the Chief I accepted the offer to go to Southeast Asia. He then worried for an entire week about sending a family man into that dangerous war-ravaged country. He almost changed his mind about sending me but in the end he had so many pressures from the Secretary and the White House that he reluctantly agreed to approve my promotion to Grade 16 and loan me to USAID for up to two years.

Chapter XII

H-E-L-L-O, V-I-E-T-N-A-M!

The next few weeks were busy ones as I prepared for a prolonged overseas assignment. It dawned on me I would be in Vietnam for 22 months with three one month home leaves. Gwen would have her hands full trying to keep track of the girls without help from me, but she had done well for 18 years during my frequent absences. I had my State & Private Forestry Program to administer, plus a search for replacement candidates for my position. The next few weeks were a blur of feverish, sometimes hectic activities.

Among the normal Vietnam forestry activities my work would involve sensitive Top Secret matters. To obtain this level of security clearance I brought my Standard Form 171, "Federal Employment Application" up-to-date. This had not been done since I had applied for federal employment in 1948. I also had the difficult task of answering questions on a Security Investigation for Sensitive Position form and listing the address of every school I attended and places I had lived and worked since birth. I would not have been surprised to be asked the name of who is buried in the Tomb of the Unknown Soldier at Arlington. Since Vietnam was high in President LBJ's list of priorities the Federal Bureau of Investigation and Civil Service Commission investigators went to work immediately checking with former employers, neighbors, universities, college and current and former associates. In a short time I received calls and letters from some of these sources. They wanted to know why the FBI was after me! Within a month I was cleared and ready for classified briefing from the State Department, USAID and Office of International Agriculture. Generally I rated most of the security measures as paranoid, however some information was sensitive, and all was fully respected.

The State Department and USAID briefing by personnel with experience in Vietnam were the most valuable. These helped prepare us for the Vietnam experience. I can truthfully say when I arrived in South Vietnam I did not see, hear, feel or smell anything I had not been fully prepared for, culture shock was no problem. One series of lectures concerned communism. I was startled when we were told about the merits and glories of communism. At first I thought I was in the wrong place, but as the briefing unfolded we began to see communism in the worst light and how in time the human spirit was completely subordinated to the state. I suppose this was done to be sure we were not attracted to or taken in by communist propaganda. We were told to inform the Embassy Security Officer immediately if we ever came in contact with a representative of the International Control Commission, Russia or one of the Eastern European countries. That included riding in a hotel elevator with a Polish member of the ICC. The ICC was set up in 1954 to enforce the Demilitarized Zone (DMZ) between North and South Vietnam. As I later observed the ICC representatives played a lot of tennis and the DMZ was as militarized as any strip of land on earth.

Preparation for working and living in Vietnam involved exhaustive physical examinations, health lectures and many inoculations. It was scary to hear about the health risks. It sounded like I was going back in time to the Middle Ages. We had smallpox vaccinations and shots for yellow fever, plague, cholera, typhus, typhoid, tetanus, diphtheria, influenza and polio. Three weeks before departure we started taking once-a-week malaria prophylactic. The last week in the US we had massive, painful 5cc shots of gamma globulin in the hip for prevention of hepatitis, to be repeated every five to six months.

Lecturers made brief indirect references to the dangers we were about to face. These were the only safety warnings we received from USAID during my entire Vietnam tour of duty. But this impressed us with the reality we were going into Saigon, a city full of terrorists and a country which was a war zone from one end to the other. It was desirable to take out more life insurance and prepare a will and power of attorney. I stored these documents in our safe deposit box, told Gwen where they were located, but did not bother her with all of the details.

During my orientation and training period I scheduled an official trip and flew to Florida. While there I was able to spend a few days with my Mother and Father who were wintering with relatives. As usual my Mother moaned about me going to war. And what would she do if anything happened to my Father or to me? My Father thought it was great I was going over to help that little country. I was convinced he favored the effort only because it was supported by a Democratic President! I spent several hours with my Father, including taking him on a motor boat trip from Silver Springs down the beautiful Oklawaha River, which was the focus of my official business in Florida. I gathered information on the resources of the River and the nature of the impact of the proposed Cross-Florida Barge Canal. (That first hand knowledge served us well a few years later when we convinced President Richard Nixon to stop the Canal project.) During that field trip my Father and I had one of the best talks I could ever remember. He suggested I had inherited my love for the timber business from his grandfather Thomas R. Williams and his four sons. They operated a large sawmill adjacent to the old Monon Railway Depot near Bloomfield, Indiana and moved to Jackson, Tennessee to cut virgin timber. They thought Green County, Indiana was all done. He told me Thomas could not get help to shingle his roof in 1920 and when he did the work himself he fell off his roof at age 92 and broke his hip. He spent the remaining two years of his life on crutches. Regretfully that bit of family history came too late for me to explore further with my Father. I am thankful I made the effort to see him for the last time, before I departed for Vietnam.

During the last week of February I completed all my arrangements. I was promoted from a GS-15 to an FC-2 (Foreign Service classification equivalent to a GS-16). This meant I was classified at the Senior level and could participate in high level Embassy and military briefing. This was a major concern of the first proposed team leader. In addition I was given a 25% differential for hazardous duty pay, separate maintenance allowance (SMA) for me and also one for my family. Gwen and the girls would not have been permitted to accompany me and live in Saigon even if they had wanted to go with me, due to the security situation. They could have gone to Hong Kong, Bangkok, Thailand or Manila if they had wished to and I could have visited them once each month. But for obvious reasons and using their good judgment they chose to remain in the States where I could visit them every six months.

With mixed emotions I said good-bye to Gwen, the girls and Myrl and Gene Jensen at Washington's Dulles Airport. As I departed on Pan Am Flight 2, I realized for the first time I would soon be 12,000 miles away and on the other side of the world from my family and friends. Still I was very excited to think about what I would see and do. I had a one day rest stop scheduled for Hawaii. After a night's rest in Honolulu I walked the beaches and rented a car for a drive around the island. As I wrote on my postcard to Gwen and the girls, "it was beautiful, wish you were here". At midnight I reboarded Pan Am #2 and headed west. A few hours later we stopped at Guam for refueling and I breathed in and enjoyed the hot moist tropical air. Our next stop was Manila. There the hot tropical breeze carried some of the odors we had been told about. Then we flew across the South China Sea to Vietnam. At the first, faint sighting of the mountains, I thought to myself, "Land ho!" From 35,000 feet the country looked stunningly beautiful. There were white sandy beaches, ranges of green hills and mountains, rice paddies, rubber plantations and other crop lands. With my nose pressed against the Boeing 707's small window and looking closer, I could see holes in the forest and craters in the crop lands. Only the rubber plantations appeared to be undisturbed by the war.

Still at a very high altitude we approached Saigon and the huge Tan Son Nhut Airport. Instead of the normal let down, line of approach, we came in high and spiraled down over the so-called "safe area" and the beginning of my "Asian Adventure". A delta-winged fighter aborted a landing and was waved off to permit our heavily laden airplane to come in for a landing. In every direction I could see more jet fighters and propeller driven bombers circling around in the sky, each waiting for their turn to land, one after the other. A scary thought came to me as we touched down . . . *What am I doing here?* But it was too late. *I was here!*

As I walked down the ramp of my Pan Am Clipper the stifling, breathless, suffocating heat and odors of Saigon told me I was in the Orient. I was met by Rufus Page of the Forest Service and USAID officials. They helped me pass through customs and immigration in a short time. Whisked into the center of Saigon I began to see and smell some of the things I had been prepared for. As we neared the downtown area and my hotel we passed the huge Saigon Cathedral. Between the spires and a nearby lofty radio tower I could see a US Army L-19 "bird dog" airplane drift slowly through the skies on a routine patrol. I could not yet comprehend what was in store for me. After all my mind was dulled from 25 hours of jet travel to the other side of the world.

A shadow passed over me and I looked up to see a Vietnam flag flutter in the hot tropical breeze. I was proud to be here and living a part of the history being made. The flag was a bright yellow banner crossed by three red bands representing the hope for a united South, Central and North . . . the flag of the Republic of Vietnam. On that day, 4 March 1967, representatives in the National Assembly set their signatures to the Constitution of the RVN.

At this point in my story I bring in a series of letters. I call these my *Dear Gwen & Girls* letters. These letters were prepared for two purposes: 1) to keep my family informed, daily and weekly, about my activities and 2) to keep the Forest Service people informed and stimulate their interest in my work. The reporting aspects of my job were demanding. USAID required periodic and special activity reports. USDA and the Forest Service were provided monthly reports. During my preparation for the Vietnam assignment I was warned by friends in the Chief's office that I was taking a great risk by leaving the mainstream of the Forest Service and I would be forgotten and my career sidetracked. The (dated) *Dear Gwen & Girls* accomplished the dual purpose of keeping the family informed and creating an interest on the part of the Chief and staff in the everyday, exciting, rewarding, frustrating and dangerous aspects of my life and work in Vietnam. In rereading those letters, which Gwen saved, I found they contained a wealth of material and observations about my work, the Vietnamese people, their customs and the perils of conducting a forestry program in a foreign country in the middle of a war zone and in the midst of terrorist activities. Some of my work was classified as Top Secret and I could not write about what I was doing at the time. But with the passage of time I am able to comment on those sensitive activities. In those letters I note a change in my attitude and views toward Vietnam and US involvement in Vietnam. I went from a "gung ho, John Wayne mentality, give 'em hell, we can whup em" philosophy, to one of more mature thought and questioning our motives and reasons for being in Southeast Asia. As Regional Forester Bill Hurst wrote me from Albuquerque during my first few months in Vietnam, "the letters are fascinating but the details must cause your wife great concern for your safety".

* * *

USAID/AGRICULTURE
C/O American Embassy
APO 96243 San Francisco
5 March 1967 1700 hours
(Saigon Time)

Dear Gwen & Girls:
I was deeply shocked today when I received the following telegram from the Secretary of State through the US Ambassador's office:
FM SECSTATE WASHDCTO
AMEMBASSY SAIGON PRIORITY
STATE GR NC BT
UNCLAS STATE 149094
AIDAC AIDTO 85 19
SUBJECT: J.H. CRAVENS
FOR VAN HAEFTIN PASS TO CRAVENS USDA PASA TEAM:
FATHER OF J.H. CRAVENS PASSED AWAY THIS MORNING OF HEART ATTACK. FUNERAL ARRANGEMENTS UNDETERMINED. MOTHER AND AUNT BEARING UP WELL. JERRY DRIVING FAMILY CAR TO IOWA WHERE IT WILL BE HELD. WIFE AND DAUGHTERS WILL FLY THERE. FAMILY SUGGESTS HE NOT PLAN TO RETURN UNLESS OTHERWISE ADVISED. HE MAY BE MORE HELPFUL AT LATER DATE. PLEASE ADVISE IF ANY PLANS ARE MADE FOR JAY TO RETURN HOME. WILL PASS INFORMATION TO FAMILY.
RUSK

116

The USAID people had a difficult time finding me since it was Sunday and they did not expect to find me working at the office. I appreciate receiving word that you and the girls will be flying to Cedar Rapids for the funeral. I wrote Mother and told her I will plan to come home about the first of July. Please do what you can to help her in the meantime. I know Dad would have wanted her to exercise care in selling the house and ranch property and not give it away. These are valuable assets and he wanted her to have the full benefit of the values. He was seriously considering selling and moving her to Florida. I know this will be difficult for you, but offer her what advice you can.

I am exhausted from the long flight—15 hours to travel from Hawaii to Saigon. The city of Saigon and the country of South Vietnam are everything I expected. I will write more details when I catch my breath and get over Dad's death; the shock and realization he is gone were a great impact. That is a helpless feeling here on the other side of the world.

I will not need an alarm clock because this morning at 0500 hours the nearby artillery opened up and they have been pounding away all day. At one time during the day B-52 bombers unloaded hundreds of bombs ("arc-light strike") over towards the Cambodian border. The noise and concussion were great even though the activity appears to be over 50 miles or so away.

I shall take care of myself and keep you posted. Write as you find time. I am sending a copy of this letter to Bethesda and Cedar Rapids to be sure you receive it.

Love, Jay

* * *

As the leader and first member of the forestry program to be recruited, I was part of an 80-man Department of Agriculture team fielded and lent to USAID under a Participating Agency Service Agreement (PASA) to support the USAID agriculture program. In a short time we recruited Barry Flamm a management forester, Martin Syverson a utilization and marketing specialist and Lewis Metcalf and Walter Pierce two logging and sawmill experts. While we were selecting our team members, USAID was recruiting 600 additional employees for its Vietnam mission. Most needed for the pacification projects were doctors, nurses, agricultural specialists, secretaries, administrative personnel and the like. Under proposed plans the USAID mission, the largest of all missions in Vietnam, was to be built up from 1,400 to 2,000. Due to dangerous conditions USAID had difficulty in recruiting specialists and convincing their regular employees to volunteer for second tours of duty. That was one of the reasons USAID and USDA entered into an agreement whereby USDA provided the agricultural specialists. Tours ranged from 18 months for those with families in the States to 24 months for those who took their families to Hong Kong, Manila or Bangkok.

Working with USAID was valuable experience. USAID had many top quality people, however the tremendous overnight buildup of USAID/Vietnam programs brought in some people who were not well trained. Many of these were journeyman level people who had acquired their experience and training elsewhere. On arrival in Vietnam there was no time for training and no trainers. And the personnel system existing in USAID at that time was absolutely mediocre. We really had deep sympathy for the permanent USAID employee. This produced a mixture of people who were not well organized, incapable of producing good staff work and lacked cohesiveness. Fortunately we had been trained by the highly efficient Forest Service and we PASA people received our support and payments from USDA.

Our efforts were directed toward improving logging practices, increasing lumber output, lowering production costs and prices and manufacturing better lumber. We were also to assist primary and secondary wood industries with other technical problems, particularly in the field of marketing, wood seasoning, lumber storage, wood preservation, charcoal production and furniture manufacture. We were encouraged to establish a sawmill demonstration project where we could illustrate the importance of proper mill layout, alignment, log handling, lumber seasoning and use of metal detectors to identify bullets and shrapnel before the saw hit them. (Page, 1967)

The forestry program included the important task of training Vietnamese foresters and technicians to conduct a comprehensive survey of the country's almost 14 million acres of upland pine, hardwoods in the dry tropical forest and the wet tropical (rain forest), mangrove and flooded forests (over 1,500 species of which only 20 were being utilized) and develop management plans for the national forest areas. The Directorate also had responsibility for wildlife which included deer, elephants, tigers, monkeys, birds and other creatures. The forestry team worked very closely with the Directorate of Forest Affairs people in putting this program into operation, since the objective of the program was to assist the Directorate rather than supplement it. Our approach was to become familiar with their methods and then help them improve what they had, rather than providing them up with expensive, sophisticated US instruments and machinery they would have trouble operating and more trouble maintaining. At a very early stage we found it necessary to provide technical assistance to the US military forces since they provided security and significant transportation for us and our Vietnamese counterparts. Commanding General William

118

Westmoreland supported our efforts and gave priority in providing security for logging operations and lumber transport. (Westmoreland, 1976)

USAID pressured us to plan a forestry program and propose a budget of millions of dollars to import US sawmills and industrial equipment. We resisted and helped the Vietnamese improve the installation and operation of their own locally manufactured machinery. Certainly it was less efficient than sawmills manufactured in USA, but the Vietnamese could build their own and manufacture replacement parts. The French designed, Model CD-4, horizontal bandsaw was labor intensive, but millions of people desperately needed work. It did not take us long to come to this conclusion because we saw vast quantities of rusted, neglected, idle and broken down made-in-the-USA equipment that USAID and its predecessor agency had imported into the country from the US prior to our arrival. Replacement parts, knowledge and technical assistance simply had not been available to the Vietnamese for utilization of this sophisticated equipment.

USAID further encouraged us to select promising Vietnamese for training in the US. We resisted this also. The Directorate of Forest Affairs had many well trained foresters with degrees from Yale, Michigan and other US colleges and universities. But on their return to Vietnam these highly motivated foresters were given meaningless jobs. They had no authority or responsibility and little future in a bureaucracy based on seniority and power. Power meaning family status or money to buy key positions where through various means they could recoup their investment in a short time. Furthermore most of the potential candidates for training abroad did not speak English well enough to study in the US. So we trained them and the technicians and transported them out of Saigon and into the forest, and to many places they did not particularly like . . . they were uncomfortable in the field and it was dangerous.

Through my studies and experience I learned Vietnam was a cultural melting pot. The Vietnamese originate from about one hundred or so interrelated families. Then there were the ethnic Chinese, the darker skinned Cambodians (traditional enemies of the Vietnamese) and the Montagnard (a French term for a race that migrated into Indochina long before the Vietnamese) highland people consisting of Muongs, Nungs and Tays. It took a while to even begin learning about the pragmatic Asian mind. I am not certain any Americans fully understood the Vietnamese. These people had been exposed to imperialism, colonialism, communism, Marxism, Leninism, socialism, democracy and the French and Vietnamese bureaucracies. But the power of Saigon and the authority of all other sources stopped at the village gate. The real historical power of the country was vested in the village hierarchy. Therefore the war, the creation of millions of refugees and our presence seriously disrupted village life and created long lasting problems of major significance. (Sheehan, 1988)

While I was far from being an expert, I learned to speak and understand the Vietnamese language. Few Americans made the effort. A handful had taken a few lessons and could ask and give directions. Most transactions, civilian and military, took place through interpreters. And most interpreters were incompetent. This further compounded the problem of trying to get through to and understand the Vietnamese mind.

Vietnam looked like a land of opportunity. It had strong people and a wealth of soil, water and forest resources, and all appeared to be crying for help. Vietnam was also a land of death, disease and sadness. My letters did not linger on these later aspects nor will this story. But I vividly recall one of my early letters from Gwen. Aside from providing welcome news from home and about our teen-age daughters and their friends' activities, she told me two little five year old neighbor boys came running over to her one day. They were crying and sadly asked, "Did poor Mr. Cravens die in Vietnam too?" She answered them, "Of course not! Why do you ask?" One replied through his tears, "But my daddy said you went to his funeral". She told them it was my father who died. They threw up their little hands, yelled "Yah!" and ran off apparently satisfied the report of my death had been greatly exaggerated. As I reflected on these childish utterances, it is sad, but at that time even 5-year olds associated Vietnam with dying.

Within a few days of my arrival we suffered along with the Vietnamese, one of the bloodiest weeks of the war. Some 272 Americans were killed and over one thousand wounded in military actions from the DMZ to the Delta. Seven members of a University of Wisconsin education task force perished with their USAID education advisor and Air America pilot as their plane slammed into the highlands near Danang. Hundreds of Vietnamese, both northerners and southerners, were killed and wounded. These tragedies continued throughout the 18 months I was in Vietnam and for seven years following my departure.

Saigon, the Pearl of the Orient, as the French called it, was a city of contrasts. There were the lovely French-colonial mansions and public building set back from the tree-lined streets. At the time of my arrival there were over one million refugees crowded into a city designed for 200,000. This number doubled before I departed. I lived in hotels for several months in the center of Saigon with its infamous Tu Do, a street lined with bars nightclubs, massage parlors and brothels which ran till dawn.

It was sad to see hundreds of young teen-age GIs, in Saigon Bien Hoa, Danang and everywhere I traveled in the 44 provinces of Vietnam from the DMZ to the Gulf of Siam, searching for something to relieve both the continuous boredom and the threat of imminent death in the jungles and swamps of Vietnam or on the streets of Saigon. They found release in drugs, alcohol and the prostitute resources of places like Saigon's Tu Do Street. Take a young American teen-ager, far from home for the first time, give him a powerful automatic weapon, the assets of the Tu Do Street, merchants in the city markets providing cheap heroin, marijuana and cocaine, low cost PX whiskey at $2 per liter and 20 cent beer and you have the perfect formula for producing scrambled brains. The culture shock and post traumatic stress syndrome these former GIs have experienced over the years comes from that experience and the fact they had no preparation for being dumped, overnight, back into America, or as they called it then, the land of the big PX . . . a land which did not welcome them back or consider them to be war heroes. No one who has not gone through a tour of duty in Vietnam can fully appreciate and recognize what those young American soldiers felt.

We civilians experienced the stress of Vietnam. We saw people killed and wounded on both sides, but I believe our maturity and previous Forest Service training helped us to absorb the shock and shielded us from lasting impacts. We learned about escape routes in our Forest Service fire fighting experience. We knew how to recognize hazards, traps and patterns that should be avoided in our work, travel and daily habits. We met an American who visited a Buddhist orphanage in Saigon every Wednesday night at 2000 hours to teach the children English. Terrorists detected his travel pattern within a few weeks and shot him dead one dark night as he headed back to his apartment. Forester Barry Flamm taught at one of the orphanages for two nights each week until we heard of the other American's death and then he changed his pattern.

* * *

8 March 1967

Again I want to thank all of you for taking care of my family obligations. This must have been a most difficult week for all of you. Since Dad had to go I am thankful that it was not a lingering illness. I understand that he had a myocardial infarction, and everything was over in the blink of an eye. I have notified USAID I want to return to the States at the first opportunity which will be about 1 July. Please advise me what needs to be done and what you plan to do during the time I am on leave.

With this letter I will begin a series of letters on my observations concerning the Republic of Vietnam (South Vietnam, SVN, RVN or VN). With each one will be a personal family letter and a general one that can be circulated, if you wish, to family and friends. Please retain the letters and at some time in the future I may write about my experience, if you judge it to be worthwhile. In addition to my official reports to the Forest Service I will send the Chief of the Forest Service a copy of the general letter. One of my peers warned me about going overseas for an extended period. He said you will be out of the mainstream and they will forget about you and your career. I do not intend for that to happen. Perhaps my general letters will stimulate their interest in what I am doing and help keep me in mind.

After work today I accompanied my boss, Assistant Director Jerry Overby and his driver to his apartment. As you noted the Americans are not exactly living under primitive conditions. He has a three room apartment, completely furnished. Furniture is locally manufactured from beautiful, heavy, dark grained wood. An overhead fan circulates air in the living room. An air conditioner in the bedroom does a good job of cooling. The kitchen is modern. A gas stove, which has a gas supply tank sitting on the floor of the kitchen, would not meet Forest Service safety requirements. The electric refrigerator has a large frozen food compartment and is well stocked with frozen steaks, beer and all the goodies you said the Americans would not be without. I will tell you about the Post Exchange (PX) and Commissary later. The bathroom is modern but the fixtures are uncommon as are all the electrical fixtures. These are all of French design. The French must be very backward. The shower has a flexible, metal hose that fastens to the wall or can be taken down to wash hair and private parts. Everything seems to work well except the electric power. It fluctuates or goes off frequently as the load becomes too great or the Viet Cong (VC) cuts the line or blows up generators or transformers.

The driver waited while Overby changed clothes. I shake my head and wonder when I see all the chauffeurs and personal use of government cars. We were so strict in the Forest Service when it came to using vehicles for personal use. Here it is done for security purposes, but I do fine walking to and from work. I change my routes frequently to avoid getting a grenade tossed at me. I will tell you about my walks in future letters. From Overby's we went to the Australian Embassy to attend a reception, the purpose being to welcome members of the Australian forestry mission who arrive tomorrow. Their flight schedules were changed, but the Ambassador decided to hold the reception regardless. Real nice man, the Ambassador, he was very friendly. He talked to me about forestry in Vietnam, the US and Australia. He knew some of their foresters I had met in the States. Trays of hors d'oeuvers were passed around continuously as were drinks. I stayed with beer, since it is safer in many respects and satisfies my thirst. About 25 people attended the reception, including four from the SVN Forest Service. I did well mixing with the group and talking. This job should be good training for me and will provide me with much to talk about for many years.

The Ambassador's residence was a lovely old French-style home, as are most of them here. The French lived very well in Saigon which they called Pearl of the Orient. This mansion was well maintained, had high ceilings, white walls, beautiful dark wood floors, large heavy wood furniture and attractive cut glass light fixtures. The place was well protected by two armed guards who stood on either side of a large steel gate. The grounds were surrounded by high walls containing glass fragments embedded in concrete on the top.

From the reception I was driven by the Vietnamese Director of Forest Affairs, three of his staff members and chauffeur to my hotel. The car was the Director's pride and joy. It was a bright, black, four-door 1937 Ford sedan. You could actually see your reflection in the paint. The streets contained an endless stream of bicycles, Honda motor bikes, motorized bicycles, scooters, boxy leg-powered pedicabs, cyclos (which are like a motorized wheelbarrow), occasional Harley-Davidson motorcycles, small blue and tan Renault taxicabs, Army Jeeps, large military trucks and an assortment of other vehicles such as vintage Buick, Chevrolet, Citroen and Ford. These flowed along in both directions and resulted in a continuous stream of traffic. At times the activity was obscured by dense exhaust smoke. I am amazed by the skill, or luck, that prevents more mass slaughter on the streets and highways; the only salvation seems to be speed . . . the streets are so crowded that high speed is impossible.

After being let out at my hotel I walked to the nearby Rex BOQ (Bachelor Officers' Quarters) for a beer on the rooftop patio, where I ducked bats which were feeding on insects attracted to the lights. Civilians working for USAD have use of BOQ's, PX's and Commissary. To enter the BOQ, or any military establishment, we pass through a barricade of oil drums filled with concrete. These are placed to keep unauthorized cars out, particularly those loaded with explosives, and serve as shields for the guards in case of trouble. Then we show our ID card to Military Police (MPs) who stand behind a concrete, sand bagged revetment. They are vigilant in their flak jackets and steel helmets and with their automatic rifles held at ready. Inside the building a line forms for elevators. The French also fouled up the world with the design and function of their elevators. All elevators in Saigon hold no more than four people and the doors will get you sooner or later. The doors open and unless someone jumps in and pushes the hold button the door slams shut. The doors do not have the benefit of a safety button. You start at "RC", which is ground level, the next floor is 1, then 2, etc. "Arret" means stop and "Marche" means go. I do not savvy the RC bit, but it indicates the ground floor and carries you there. At the top, after a very slow ride, you jump out, before the door gets you, take a ticket, display your BOQ card and proceed to the dining room. Tonight I paid 10 cents for a large tasty dish of vanilla ice cream, made from reconstituted milk. I visited with a Colonel who works for Commanding General Westmoreland and a Captain in the Navy. The Colonel has been here for nine months and longs for his home and family in Annandale, VA and the Captain was from Georgetown. In addition to talking about the hazards of commuting to work in the DC area we talked about our respective jobs. This was my introduction to the military role in SVN.

The Colonel works 14 to 16 hours a day, goes at a dead run and keeps this up for seven days a week, week after week. He has strong feelings about what has been done to this lovely land, both by the VC and our military operations. He says some areas of the forest look like the craters on the moon. He told me the General is very interested in forestry and wants to do something to salvage the down timber and put people to work. That sounds promising for my mission! He favors restoring the deforested sites when peace comes. The Colonel sees no end to the war but believes we are making progress.

The Navy Captain has been working on plans to salvage barges and ships that have sunk in the Saigon and Mekong rivers. His major project consists of refloating and repairing a dredge that was recently damaged and sunk by mines in the Delta area.

So it goes—the room is full of officers and civilians, like me, who are enjoying a few moments of peace, quiet and good food. The food is good, inexpensive and includes very American items such as fried chicken, roast beef, steaks, ham and eggs, milk, wonderful uncontaminated ice, coffee, hot or ice tea and delicious salad vegetables. The vegetables are safe here since they are disinfected and washed very carefully under supervision. So far I have had no stomach problems.

In US facilities we pay in Military Payment Certificates (MPC) which are in dollars and cents and look like "monopoly" money. The local currency, dong or piasters (US$1=VN$118), is used for purchases of items from the Vietnamese. It is illegal to possess or use green dollars in Vietnam. On entry at the airport we were required to exchange our green for these two kinds of money.

On the walk to my hotel I worked my way through crowds of walkers, motor scooters, etc., past stands on the street which display everything imaginable, including items which are unobtainable in the PX. This is part of the very active, illegal black market. The time was 2100 hours and still three hours to curfew and the city was very much alive. Except for an occasional jet streaking for the other side of the Saigon River or a helicopter patrolling over the city, the sounds of war are shut out for the moment.

Back in my room I turned on two slowly moving overhead fans and I was quite comfortable. Periodically the sky lights up from bombing which was 30 to 40 miles away or by "spookey" a C-130 Hercules flare ship that turns night into day as the flares drift ever so slowly on parachutes to the ground. C-130s serve as transports, flareships and gunships. When equipped with an arsenal of miniguns the firepower unleashed evokes a sense of awe. While writing this letter the big guns have been booming off towards the Cambodian border. Their dull "whumps" are reminders of the nearby war.

That's all for now and soon to bed. Sleeping is very comfortable. My bed has a solid base, Hollywood-type, with a firm rubber foam mattress. Windows have no screens and by using repellent I am only bitten once or twice a night by mosquitoes. We have constant reminders, in the form of posters, advising us to "take your pill". I am sure they mean malaria suppressants because there seems to be little concern on the part of the GI's or many of the civilians to prevent anything else. Well so to bed. The next sound I will hear may be one of several things—a rooster crowing, since he never fails to sound off just before daylight, the big howitzers "whumping" in the distance, nearby mortar explosions, isolated rifle shots, B-52 bombers unloading a string of bombs which sound like a series of rapid whumps or a glow from a fire in the city. We seldom hear what causes these activities, however I did read in the paper this morning that 10,000 people, mostly laborers, were left homeless as the result of a fire near the docks last night.

I will write another letter tomorrow night. Rufus Page from the Chief's office departed for Washington today, so at the moment I am the total forestry mission. I am sleepy now and will end this letter and say good night.

10 March 1967

My letter tonight will describe a day in Saigon. I am also sending one to our friends in Flagstaff, Arizona on the Coconino National Forest who were so kind as to send me my box of forestry reference books. They arrived yesterday in good condition.

As I write this letter there are a few interruptions. The guns are booming about 10 miles out across the Saigon River. A flare ship dropped a light just a few minutes ago. The sound of the guns is closer tonight and they seem to be spaced about every 15 seconds. This shakes the building windows and I can see where they are striking.

This morning my rooster neighbor and I were awakened just before 0400 hours by a series of rifle shots. It sounded like opening day of deer season on Jumpup Point on the Kaibab National Forest. At 0400 hours the curfew goes off and the city quickly comes to life. I am on the 7th floor of the Oscar Hotel. That's actually the 8th because the French choose to call the ground floor "RC". Speaking of the French it is a wonder they lasted for a 100 years in this lovely country, they botched up the electrical, sewer and water systems and the elevators must have caught more of them than the Viet Minh. As I noted before, the elevators are tricky and you have to move fast. To ride in one is taking a calculated risk in view of the power outages and mechanical failures. Under most circumstances, it would be most lonesome to sit out the war in a stalled French elevator. You can be certain there are no valid safety inspection certificates.

On the ground floor, each morning, I am met by a news boy who sells me the *Saigon Post*. This is a very good English language paper which brings us up-to-date on the war and Congressman Adam Clayton Powell's carrying-on's with girl friends in Bimini. As I pass the hotel desk I read the instructions, quote:

Notice—Prostitutes (Taxi Girls) are strictly forbidden in the hotel. Customers are requested to sign the police form for their girl friends if they bring in a girl.

The Management

The Oscar is a respectable hotel, very clean and has pleasant workers who leave your gear strictly alone. USAID leases and pays for our living quarters and is selective in their choices. I suppose that is why the management puts up the sign. I do not see similar instructions in other non-USAID hotels.

I have breakfast at the Brinks BOQ which is just across the way. Access is through staggered concrete filled barrels and past the ever alert guards. It is nice to have them on your side. I hear many US civilians bitching because SVN is under strict military law as far as civilians are concerned. Breakfast at Brinks is like breakfast in any first class hotel back home, except for the presence of the military. Today I sat looking down the barrel of a carbine looped over the back of a Colonel's chair. That was a surprise since signs warn us to "check your guns" at the door. Anyway the ammunition clip was removed and I trust the chamber was empty. Breakfast costs 30 cents (MPC) for eggs, toast, fruit and coffee. From the roof deck after breakfast I watched jets streak and scream off toward the Cambodian border. Bomb shock waves could be felt and heard a few minutes later. Looking down from the roof top of Brinks I could see the city streets were swarming with people.

Saigon heats up early in the day. Heavy humidity hangs in

the air. Your hands, face and body feel wet at all times. You said I had a crazy thermostat! However I enjoy the weather which is about like Phoenix in July or August, only the humidity is much higher. People going to and from work mostly ride. They ride bicycles, pedicabs and cyclos. The latter must be run with very little gasoline and much diesel or oil because they smoke something awful and go pretty fast. In fact it appears to me it would be like riding on the front seat of a roller coaster. Mixed in with all these vehicles are motor scooters and motor driven bicycles, these number in the thousands and carry from one to four people. There are Lambretta carriers which have a small covered box on the rear of a scooter and carry up to 12 people or four or five people and a load of produce. The small Renault taxis have very loud, excellent horns and are good for a drag race any time they can find space to get up speed. Throw in a few hundred Jeeps driven by GIs and Vietnamese soldiers, plus some very large 6x6 trucks and you have a design for disaster.

From Brinks I walk about 3/4 mile to work. I find a smile directed at the Vietnamese still does a lot, but in some cases that does not work. These people have had 3,967 years of hardship, starting before Genghis Khan invaded the country in the 12th century and was repelled. Then there were 1,000 years of Chinese domination and 100 years of French "high grading" all they could pack off to France. So the local people are suspicious of the "round eyes" or "long noses"; I stand out on both of these counts, plus I am more than a foot taller than most of the Vietnamese. Most of the Vietnamese are very thin, emaciated and diminutive. We still have to prove ourselves and our motives to these people. And that will be extremely difficult since the average American has little or no rapport with the Vietnamese. The Colonel whom I ate with the other night summed it up in a few words . . . "it's a shame what's being done to this lovely country".

My walk to work is through a veritable obstacle course. I have to be careful about traffic, holes in the street and sidewalk, concrete bunkers, sand bags, piles of garbage and numerous other obstacles. In most of the sidewalks the French planted trees and utility poles which must be avoided. I admire these tree-lined streets. I have been advised never to kick a can or a box lying on the sidewalk or in the street since it may be a bobby trap and explode in my face. I have a walk lined out that permits daily changes in my route. Following a definite route at the same time each day can lead to trouble. Sooner or later you could be ambushed by a terrorist. My routes follow our Forest Service fire fighting instructions . . . "always have an escape route". I could use some of the above items to duck behind in case of trouble. This walk and alternates take me past several BOQ's (I still have not figured out where the enlisted men live and eat), RVN government offices, a bowling alley and the Korean Embassy. That Embassy is an interesting place. The four guards out front are young and very tough looking. If a car stops they blow a whistle and their automatic rifles are leveled to cover driver and occupants. They are all business and I am glad they are on our side. I understand the VC fear Koreans more than any other group. The Filipinos are also all business and look as hard as nails. By the way my smile works on the Korean guards and now they smile and salute me each morning as I pass. I feel secure in this area.

Crossing Le Van Duyet (La Van You-Wet) Street is a good trick! The traffic is solid in both directions. You step out into the brief intervals when there is an opening and work your way across. Melissa, remember I told you in your driving never to do anything unexpected? If I walked too fast or jumped back I would end up on somebody's radiator ornament or lap. And speaking of laps, these young Vietnamese girls sitting on a motor scooter with their long black hair flying and in their flowing ao dais (ow zi) are a picture of grace. Around the office these Vietnamese women are good workers and very helpful and considerate. I limit my contacts just to looking, both in the office and elsewhere.

Our office, referred to as USAID No. 1, is in a series of tall buildings surrounded by coils of 4-barbed wire, a high fence and a number of guards. Each car passing through the gate receives a thorough inspection. Guards move around the car looking into a mirror mounted on a long handle and inspect the underside. They tell me fastening explosives on the bottom of a vehicle is one of the ways for getting explosives into the compound. Speaking of explosives the building next door has no windows, a large hole in the roof and no second floor. The Army used it as a store house until last December when the VC tossed in a satchel charge (of dynamite) and blew it up. This blasted all the windows out of our building.

To enter our building I show my pass to a Marine guard who has a .45 pistol on his hip and a grease gun (.45 Cal. automatic weapon) on his desk. Then by taking two steps at a time, I can reach the 6th floor (remember +1). That is the only way to get there because there is always a long line of people waiting to take the elevator, at the rate of four passengers at a load. This is a so-called modern French office building. My office reminds me of the type of office Henry Summerfield might have had at the old Bly Ranger Station or the General Springs Guard Cabin back in the late '20s or early '30s or even old Ranger Dan Judd's cabin on Jumpup point on the North Kaibab. Remember? He was the one who started a fire in the cook stove and had a black cat explode out of the oven when it got too hot. My office furniture consists either of heavy, rusty metal desks or is made out of wood. The wood is extremely heavy and was put together when it was green. The wood is split and the drawers stick. That is one thing we hope to get across to the sawmill operators—the importance of proper seasoning and storage of wood to avoid twisting and splitting. Today, to add to the confusion, we moved up one floor. Nine Vietnamese workers arrived and started planning how to move the furniture. I am only on lesson number five in my language training but I detected they could not agree on how, what or where.

About this time I was handed a Top Secret file by a courier. I was asked to read the file (about 1/2-inch thick) and brief my supervisor on the contents. Now that was an interesting assignment. In my security training I was told never to lay a classified document on a desk if anyone was around and not to permit anyone to see the contents. Get the picture? With this crew of movers chattering and having several arguments, plus several loud crashes and the lights going out, followed by nearby explosions (we never hear what happens and where) and all this accompanied by the continuous whump, whump, whump of B-52 bombers unloading their bombs, it makes for an interesting, unrelaxed atmosphere in which to review documents, or do anything else.

My new office occupied what formerly was a library. There was a slight problem when they moved in five desks and files for my crew and did not take out the tables, or chairs, or book cases or the accumulation of at least 50 years of French technical manuals and literature. Finally I took over, got the crew of Vietnamese movers together and got everything pretty well cleared out and organized. Behind the books and bookcases was about a bushel of broken glass that resulted from windows being blasted out last December. They said USAID was a real mess at that time. I am not sure it has improved a great deal since then!

People in USAID and the USDA people like me, are dedicated to their work and go about it, as I have described, under rather challenging circumstances. People stationed in Saigon feel safe here and are nervous when they go out into the provinces where the war is going on. People in the provinces are said to be spooked about coming to Saigon with all the terrorist activity. It all depends on your frame of reference. Our crew will work in both areas.

Next I was invited to sit in on a briefing session for Australian foresters and construction people who arrived at Premier Nguyen Cao Ky's invitation. Near the conclusion of the meeting one of the participants reminded Agriculture Deputy Director Van Haeftin not to forget the USAID forestry program. I smiled at the guy for his comment. Later I learned he is the Director of USAID and well aware of our program. I was told later that I would get to meet him one of these days.

Work hours and all other hours are recorded in military time. 0800 hours is 8:00 AM, 1200 hours is noon, 1330 hours is 1:30 PM 2300 hours is 11:00 PM and 2400 hours midnight. Our hours are from 0830 to 1230, 1430 to 1800 hours, Monday-Friday and 0830 to 1230 hours on Saturday. When it is time to leave work a very loud bell rings one short ring, followed by another short ring. With that everybody heads out into the traffic. My routine will be somewhat different since I have signed up for a Foreign Service Institute language course which will run from 0730 to 0830 hours and from 1330 to 1430 hours Monday through Friday. That will keep me busy and off the streets. The 50 or so words of Vietnamese I do know are very helpful and seem to be appreciated by the Vietnamese. I have been told by the older Vietnamese employees of USAID that the French were here for 100 years and they never attempted to learn the Vietnamese language. Vietnamese had to learn French if they wanted to succeed. It should be worthwhile to try and learn Vietnamese. French would be much more useful for me in the future, but since I am in Vietnam to do a job, Vietnamese should be helpful.

On the way to my USAID hotel through heavy traffic I observed the US and RVN MP's from MACV (Military Assistance Command Vietnam) and ARVN (Army Republic of Vietnam) pick up two Vietnamese who apparently were impersonating officers. The MPs were not very gentle. This place has a language of its own and all agencies, civilian, military and foreign use acronyms to abbreviate their respective agency's name. It certainly makes for a very strange language.

The next event involved a very large GI gasoline semi-tanker truck speeding down the street with its air horn blasting out a path. People scattered in all directions. It was scary. I doubt there was an emergency, only another example of the complete lack of consideration the US military gives to others. That is not what wins the hearts and minds of the local people, rather it is the image of the ugly American. However in general most of the Americans seem to be considerate and aware that we all have to work together to get the job done.

Before I sign off for the evening I want to record a few words about the local forestry people. They are very fine dedicated people and as Director Nguyen Van Tan (When Van Tun) says he has been waiting for his advisor for two years and he is happy now that I am here. During our first meeting in his beautiful, wood paneled office I told him I once was supervisor of the Coconino National Forest. He jumped up like he had been shot and said "See that photograph? Chief Cliff gave me that in Madrid at the World Forestry Congress." It was a framed color photo of a scene in the White Mountain National Forest. "That was my favorite until now!" With that he snatched it down and went off into a side alcove and took down a beautiful color photo of the Coconino's Oak Creek Canyon and proudly hung it in a place of honor. He said, "this is in honor of my advisor who I welcome to the Republic of Vietnam". Makes you feel pretty good and choke up a little. It made me want to work a little harder to help these troubled people. Good night to you all.

12 March 1967

It is Sunday afternoon. Everyone has been sympathetic about the loss of my Father. News travels fast. This morning I tried to go to church but all I could find were very crowded Catholic services. I will start going to the Episcopal Church next Sunday. I listened to a Mormon service and the Tabernacle choir over the American Forces Radio. That radio station operates around the clock and keeps us up-to-date on happenings back home, plus rock & roll, classical and folk music. The announcer just dedicated one with words that went something like this: "Big hearted Charlie, Goodtime Charlie. He's such a wonderful guy". Since "Charlie" is what we call the Viet Cong or VC, the enemy, that probably raised some eyebrows and hackles. But that seems to be the objective of this radio station. The announcer cries out in the morning with a very loud, distinctive, strung out greeting of "G-o-o-d m-o-r-n-i-n-g, V-i-e-t-n-a-m!"

I had breakfast on the roof top patio of the nearby Astor Hotel. It consisted of papaya, a piece of dried toast and black coffee. The cost was 66 piasters, about 55 cents. After breakfast I looked out over the city. Red tile roofs, with no chimneys, stretch as far as the eye can see, in three directions. In the other direction you can look across the Saigon River. The Port of Saigon is very busy at all hours, day and night, unloading supplies for peaceful purposes and the war. There are all sizes and shapes of ships from large freighters to small sampans. Between the wharves are the shacks of the very poor homeless people and the black markets. Beyond the River I can see some of the lovely green fields, rice paddies and a few scattered patches of forest. There is some mangrove forest along the River. Yesterday a battle raged from 0700 hours to well past dark just 14 miles out across the river. Several Americans were killed by "friendly fire" when one of our shells fell short. The VC also caused a few casualties. Today it looked like the peaceful rice fields in the delta country of Louisiana and Mississippi.

After breakfast I walked to the office. Traffic was lighter today as very few people were working, that is office and construction workers. Others, like the small peddlers with their tiny sidewalk stands, are there every day, in the same place. The women set up and operate most of the sidewalk eating stands. They carry supplies in two baskets suspended from a pole carried on their shoulder. This piece of wood is not much over 3/8-inch thick by two inches wide and is about three feet long. This supports loads that must weigh up to 50 pounds. These women prepare quite a variety of food and have a steady supply of customers. One was cooking rice over a charcoal brazier, along with meat and something green looking, perhaps it was spinach or something that had turned green. She chopped green onions into the mixture, poured on some smelly (stinking) fish sauce and the customers were shoveling it in with chop sticks. While I do not eat at these places I am now glad I learned to use chopsticks before I left home, since everyone else, outside of the BOQs uses them. That fish sauce is called nuoc mam and is eaten by all the Vietnamese. You can smell it on their breath when they are 10 feet away.

One of the other sidewalk peddlers was serving various kinds of meat. Some of it looked quite tasty, but I would avoid the plate of pig tails, another of pigs' ears, a plate of feet and a plate that looked like chicken guts which attracted several customers. do not believe these items were cooked very well, or not at all The customers either hunker down on their heels, like all the Vietnamese do when they are resting, or sit on little two or three legged stools which are about five inches high.

Other peddlers sell tea, iced drinks of various colors. I watched one lady prepare a drink for small children. She took parts of several different concoctions and put them in a glass with a few bits of ice. The kids loved it. When I grinned and licked my lips, the children laughed and offered me a glass Naturally I politely declined, in Vietnamese. That stuff would have turned me inside out before sundown.

You may wonder about the dish washing. Gwen, you will remember my description of the peddlers washing their dishes in the Amazon River. In Saigon it is done differently. The city has a water distribution system and outlets at each street corner Outlets are two-inch open pipes set flush with the sidewalk. The water runs continuously to flush the sidewalk and gutter. To fill a bucket or pan you hold your hand over the opening and the pressure forces a stream up about 12 inches and it arches into the container. The street peddlers splash water from the flowing pipe or a pan onto the dirty dishes. Of course they use no soap!

Now about some of the children on the streets of downtown Saigon . . . some are clean, dirty, healthy, sickly or just about any combination. The very young ones play, lay and crawl about on the dirty sidewalks as naked as "jay-birds". They love to play in those water outlets. I have seen some of them squat over the flowing water and do their business. That is not the worst thing that happens to the water distribution system. The water comes into the system from deep, clean wells and is chlorinated, but once into the antiquated French distribution system it becomes contaminated from seepage and cross connections. It seems the French were not successful in teaching the workmen the

fference between water and sewer systems. USAID is working
this problem and has provided funds and materials for some
miles of completed system and for training people on how to
e and maintain the water system and how to keep it separate
om the sewage system. They certainly have a long way to go.

The children are like little kids everywhere. They play
mes. One game looks like "tit-tat-toe", they pitch one piaster
ins to see who can come the closest to a line. They play soldier
ith small wooden swords and guns and they play dead. Saigon
s a few teeter-totters, but I have seen no swings. I have only
en one child begging and he was pleading with GIs. I find the
ietnamese are very strict regarding the behavior of their
ildren and do not permit them, knowingly, to beg. They say
lo not make beggars out of our children". These are strong
illed people and proud of their heritage. With an average life
an of 37 years these people lead a precarious existence.

After working at the office for about five hours I made
ogress in getting things organized. On the way back to my
tel I passed one of the shops where operators bring their
oken down scooters. Workmen repair engines, transmissions,
les, brakes, tires and just about any part. The forge is an open
re and they straighten or bend parts by pounding them out on a
ece of iron or by inserting one end into a crack in the sidewalk
in the fork of a tree. I see many such shops around the city.

The next shop repairs bicycles. The shop consists of a piece
canvas supported overhead to keep out the blistering sun. The
w tools and tire repair items are laid out on canvas. The two
en operating this shop smile and wave at me every time I go
y. At times they seem to do a thriving business. I have seen
me bicycle tires that have been patched many times, even the
tches have patches. Our throw-away society could profit from
e lesson to be learned here. Everything is used, reused and
used again until it disintegrates. I have seen no junk yards.
ven the refuse and garbage piles are worked over and then
worked for any useable material. Last night I saw a dirty old
oman, in very ragged clothes, sorting through a trash can and a
le of construction rubble in front of the Hotel. She appeared to
looking for pieces of wood which she placed in a tattered
sket slung from her shoulder. She apparently needed wood to
ok a meal, makes you wonder what the meal would consist of,
esn't it? She went about her scavenging apparently oblivious of
e relative affluent society around her. She was not begging,
st searching, as so many people are in this troubled country.

As I continued my walk from the office I looked down
wards one of the busy intersections where 4-way traffic
tempted to occupy the same place at the same time, in complete
sregard of one the basic laws of physics. I witnessed my first
sualties and blood in Vietnam. A Volkswagon bus and a

Lambretta carrier collided. Vehicles and passengers flew off in opposite directions. The Lambretta had six passengers and a large load of sugar cane stalks tied on top. The street was littered with sugar cane, parts of the Lambretta and bloody bodies. I pulled my camera out of my pocket and took a photo of the disaster scene. I stood there momentarily as a crowd quickly gathered. recalled we had been warned to avoid crowds and disturbances. Cars, scooters, Army Jeeps and people on bicycles kept streaming by. Nobody was doing anything for the injured people lying in the busy street, so I dove in and helped them. I would not have felt right to ignore their injuries and just walk away. These people needed attention. US Army soldiers with their first aid training ignored the people or honked their horns so they could get by on their way to do whatever they had to do in town. stopped the bleeding on one man's throat. In my limited Vietnamese I instructed the other people who came to their aid to let the woman who was stretched out to remain where she was lying. She did not seem to have anything broken unless it was her ribs or other internal injuries. She could move both of her legs. These people thanked me through their eyes and expressions. It made my efforts worthwhile. Finally one of the Vietnamese men thanked me and said "okay now". I left wondering what would happen next to these battered people. heard the sound of an approaching ambulance as I headed down the street to my hotel.

Along with Elwood Harder, a Soil Conservation Service man who works for USAID, I went to the Rex BOQ for the cookout. This is a regular Sunday evening affair. For $2.50 (MPC) you select a steak which can be a one pound New York cut, or a one and a 1/2 pound T-bone, or a one pound filet mignon or a two pound sirloin and throw it on a large charcoal grill. There were four of these grills, made out of half an oil drum with heavy wire grating on top. I selected the T-bone. Makes you wonder to see all of us with those huge, thick steaks hanging over the edge of eight-inch plates, when you stop and think of all the under nourished, hungry people who try to survive in the slums of Saigon. As I ate the steak I recalled reading in the newspaper this morning that 100,000 tons of rice has arrived from the US for distribution in Saigon, starting tomorrow. In addition to steak we each had a huge baked potato, green beans, sliced tomatoes, potato salad, lettuce, pickled beets, cucumbers, carrots, salad dressing, ice tea and coffee. The food was well prepared, even the steak that I cooked. To finish it off I had a huge bowl of fresh pineapple and papaya. This made up for the meager breakfast I had several hours earlier. The rooftop of the Rex is famous for the bats that dip and dive overhead as they seek insects which are attracted to the lights.

After the big meal and a short walk, I am back in my hotel

room typing this letter. The only distractions are the sounds of an overhead fan which is slowly stirring the hot tropical air, classical music from American Forces Radio and the pounding of artillery in the distance. I am quite comfortable sitting here in my shorts.

13 March 1967

Gwen, thanks to you and your sister Leela for going to Cedar Rapids and Dad's funeral. Sorry to hear you had so much trouble with travel. I am glad you were able to say something about how much Dad loved his "43 Ranch". I regret hearing Mother turned in Dad's large collection of old silver dollars to the bank for green at 1:1. What a windfall that was for the teller who got her hands on that bag full of coins! I am glad to hear my Mother was calm. I did not figure she would be.

I received my last shot today from the 17th Field Hospital. It was for the last of the plague series. There was a long line of soldiers and civilians waiting at the hospital sick call. Some looked very ill. I hope I can keep out of that waiting room. I was in line behind a Negro Colonel and we were given our shots and were on our way in a very short time. There is a large number of Negro officers and non-commissioned officers in SVN. They are very good according to what I hear and read in the *Stars & Stripes* newspaper. Their rate of reenlistment is very high.

I have several other newsy topics for you but will work on them for tomorrow night's letter. After dinner tonight I went to the top of our Hotel Oscar and watched the sun set. It was very beautiful and the war seemed far away. As it grew dark I could see flares floating down and across the Saigon River. One area was illuminated continuously. It appeared to be adjacent to the River. I suppose our troops were in a fight with VC that have been harassing shipping that passes through the dangerous 50 miles between the South China Sea and Saigon. Looking down the River I could see a large fire. I believe it was in one of the areas of the shacks I wrote about earlier. Off towards the huge Tan Son Nhut Airport I could see flares being dropped continuously so I suppose there was trouble there also. Today we heard several large explosions in the vicinity of the airport. We seldom hear what all the activity is about. I guess that comes under the subject of withholding information that would give comfort to the enemy.

Today during our staff meeting there was a large explosion that shook the building and the lights went out. The large generator supplying power to our building exploded. The engine may have thrown a rod and blown up. Anyway that sound and the "whoosh" of CO-2 fire extinguishers and all the related shouts made quite a commotion. It was difficult to hear about all the progress being made in corn, rice and hog programs. What

happened to the generator? We never heard. It probably just gave
up due to poor maintenance or lack of lubrication or a pump
failure and over heating. Again we were without lights, air
conditioners, electric typewriters and elevators. The Forestry
Branch has all manual typewriters, four of them, and with only
me to operate them, we made out okay. We means "me" because
that's all there is of the Forestry Branch of USAID-AGR at the
time.

Tomorrow I meet with General Westmoreland's science
advisor, who is also the General's roommate. That should help me
initiate some new projects. If they are classified Top Secret, I
will not be able to write to you about them.

14 March 196

Our staff meetings are really something out of "Alice in
Wonderland". You recall I wrote you last night about yesterday's
staff meeting and the generator blowing up in the middle of the
meeting. Today two spare generators were moved in and were
operating during our staff meeting. However the generators did
not produce enough power to run elevators, lights or electric
typewriters. It reminded me of our Forest Service set up at
Happy Jack, Arizona. In fact the staff meeting was about like
holding it in the Happy Jack generator building, with the
generator running. But there all similarity ceased. I mean the
peace and serenity of the "lip reading" sessions (I could not hear
a word) were continually interrupted by screaming jets, sonic
booms, howitzers operating just beyond the River and the roar of
the usual choppers (helicopters) which ride shotgun over the city
at all hours. Looking out through a tear in the plastic window
covering I could see a machine gunner peering down on us. And
as if that wasn't enough noise, a vehicle with a conventional
siren went screaming by every few minutes and that was
followed by one with European type sirens, the ones with two
tones, high and low blasts. What was the staff meeting about? It
is difficult to say but I believe it concerned the need for a
system to record policy, standards and procedures. Apparently
USAID has been operating by the "seat of their pants". While
most of the regular USAID people are dedicated, they are not as
well organized as we were in the Forest Service. After the
meeting I outlined what I could recall of the Forest Service
directives system, the Manual and Handbook and forwarded it to
the brass. It will be interesting to receive their response. Either
they will say it won't work or, horrors, they might put me in
charge of the directives system! Now that is an awful thought! I
should have remembered my Army training which amounted to
"never volunteer for anything".

My next job for the day was to document the status, plans,
prospects and future of the forestry program for a scheduled

136

conference. The Ambassador, USAID Director and General Westmoreland will meet with President Lyndon B. Johnson and staff shortly on Guam. I summarized the forestry program and reported that our principal problems involved VC harassment of logging areas and transportation of forest products to the markets. On the bright side the Army is working on the VC interdiction of logging and lumber trucks and appears to be making progress. I had a very productive session with General Westmoreland's science advisor today. I plan to support military operations as requested because we need the military's cooperation to carry out the forestry program. We also need them to provide security, support, transportation and occasional supplies.

Security at our office has tightened. Guards are giving the cars a thorough inspection as they pass through the gate to our USAID compound. Marine guards check all of us and our briefcases, while Vietnamese guards check and search the Vietnamese. No high level SVN officials will come to the USAID offices because of the searches. In all my contacts I go to the Directorate of Forests and Waters (Nha Lam Vu) to see Director Tan.

Tonight I worked at the office until 1900 hours. When I walked down to the ground floor I could not get out. I was locked in! Since the elevators have been out of order for two days I could see myself walking up seven stories to my floor and then across and down through the adjoining USAID building. But I searched my way through a maze, wandered through a series of offices and finally found my way to another entrance. There I waited along with the Director of USAID until the Marine guard returned and unlocked the outer door. The Director and I talked about the forestry program. The two Marines had been off on a routine search of the building. This includes looking for any classified secret documents that someone may have left on top of their desk or stuck inside to get it out of sight. Such violations are serious but infrequent.

Tonight at Brinks BOQ I had dinner with a Forward Air Control (FAC) officer. He was in Saigon for rest. He flies out of Pleiku, with the 1st Cavalry in the Central Highlands. He flies three missions each day for a total of about eight or nine hours. His job is to locate suspicious targets or draw fire, then he calls in jets "to tear up some of the timber". It sounded like flying for forest fires on the Coconino National Forest, except that frequently the "unfriendlies" shoot back. He said if he ever has to bail out he would try and land in the top of the tall trees and hope a rescue helicopter could snatch him up with its hoist. He was glad to be away from the action for awhile and rest up in Saigon.

Speaking of rest in Saigon. Those poor guys at Tan Son Nhut

Airport, at the edge of the city, never have any rest. They drop flares from a flare ship and patrol the area at all hours. Large lights from a light ship turn night at the air base into day. I can see it tonight as I type this letter. I can hear mortar shells exploding but we are too far away to hear any of the other shooting. Down on the River a huge search light comes on at frequent, irregular intervals and searches both ways, up and down the River.

Tomorrow I will make an effort to locate some of the supplies we have on order for the forestry program. With 80 some ships in the Saigon Port either unloading or waiting to unload I imagine our small order will be difficult to locate. My footlocker arrived yesterday. It was 12 days en route and the quickest anyone here has ever seen an air shipment come through. It came TWA to San Francisco and Pan American to Saigon. It received a few bruises and it appears the bottom landed on a sharp edge because it almost broke through in one area; it was nothing serious. The padlock must have been removed and then locked up a different position, but nothing was disturbed. I suppose Customs had to take a peak. I was pleased to receive it and especially my "survival items", i.e. spare tooth brushes, Halazone, etc. I always remember to use treated water to brush my teeth since I do not trust the water distribution system.

Since a copy of this letter will go to Ray Housley, I have a request. Ray, please send me some literature and specifications on the herbicide Fenuron. I have designed a project for the Vietnamese Forest Service that calls for elimination of woody vegetation. I also need design plans of the heavy equipment we use to remove pinyon-juniper. Today I visited the University of Saigon, College of Agriculture, Department of Forestry Library. It was absolutely appalling! The only forestry related book they had was one *1937 USDA Agriculture Yearbook*. There was nothing else! If you, or any of the SAF members or other foresters have any forestry books not being used, please consider donating them to this poor library facility. Please ship them to me in small boxes, of four or five pounds each, with the Official Postage Paid. Have them well taped. One box of books I received survived only because it was well taped. That is all the news for tonight.

15 March 1967

Today it is 94 degrees with 95%+ humidity. My Perma-Prest, 3-M Scotch-guard treated, (Sears) pants are right for this climate. The crease is perfect and no wrinkles are created by sitting. When I walk to work in the morning it is very hot. I could commute on an air conditioned USAID bus which comes to our hotel three times each day. The bus is enclosed in a heavy wire mesh which is supposed to discourage the tossing of hand grenades or bombs. I think it is really designed to keep the USAID inmates in their place!

The regulars are complaining about the heat, no operating air conditioners or elevators. These things do not bother me and I usually stay in my office during the lunch break. Today our office became the language classroom. I listened in on the lesson a Vietnamese lady teacher was giving to two students. I have not yet been approved for taking the formal lessons, so I just sat in and listened.

The regular old time USAID employees are an interesting group. I use the term regular since they are the full time, appointed USAID employees. The 60 of us from USDA Forest Service, Farmers Home Administration and Soil Conservation Service are PASA people and are loaned to USAID for a limited period. Some USAID regulars got out of the Army, here or in Europe, received an appointment and came to work. However most of them have been in USAID or its predecessor agencies for many years. Some have a strange outlook, which may explain USAID's successes and failures. For example, I was reviewing the forestry files we inherited to pick up some of the background and history of forestry in Vietnam and to try and offer some continuity and improve on past performance where possible. One of the regular USAID specialists asked me what I was doing. When I told him he said "throw it all out, and start over from scratch. What you are doing is a waste of time!" I suspect much time, effort and millions of dollars have been spent as the result of such attitudes. They are willing to repeat mistakes and start over from scratch, but the problem is they do not appear to know where to start or where to end. There are no well defined goals and objectives.

Recently I prepared a good, brief staff paper for the Guam Conference. It had an introduction, situation, conclusion and recommendations. With an end to the hostilities I visualize an opportunity to turn our material over to professionals in my field. We have very specific goals and objectives and I am working with my counterparts in the Vietnamese Forest Service with that understanding and agreement.

On my way to work today I discovered a lost Soil Conservation Service engineer who had just arrived in country and could not remember the location of the USAID office. I took him in tow and showed him around. I recall meeting him in Washington, DC where he was also lost. Coming from Lincoln, Nebraska he has difficulty in coping with the large cities. I predict he will have continuous problems in this country. He is like many of the USAID and PASA people, they come and go at all times and tell no one where they are going or when they will return. I asked them if they have a system for signing out and keeping track of one another. They showed me their sign out book, which no one had used for months. All day we answer one another's phones "where is so and so? When will he return?" I usually do not have the foggiest idea where they are or when they will return. After a number of such calls I stopped answering their phone. I have set up a sign out sheet and I tell people where I am going and when I expect to return and to alert my boss if I do not return. I believe that is the only safe and businesslike way to operate.

This afternoon I mounted a good map of South Vietnam on my office wall. Out of the hundreds of offices in our building I have not seen one decent map of the country. In fact there are only one or two of the offices that have any kind of map at all. I can not operate without a map and learning the geography of my work areas. Most people have little knowledge of place names and locations. I have already had some of the old timers come to my map to find places. They say what a good map that is. Where did I get it and could they get one? I tell them I obtained it from the USAID supply room and they just mutter and drift off.

The suggestion I made yesterday concerning a system for recording policy, procedures and guidelines was accepted by USAID. In fact the Assistant Director of USAID came by to thank me and tell me what a good suggestion it was. That's how you get the attention of the front office, establish rapport with the hierarchy and obtain support for your program. Fortunately I was not asked to refine and manage the system.

Today at the office one of the Americans was feeding two of the Vietnamese administrative assistants, Mr. Long and Mr. Quann, criticism about Saigon's polluted air and dirty streets. I could not resist entering the conversation and telling them about the many sections of Washington, DC and our other cities that have much dirtier streets and more polluted air and water. In our beautiful country we have created a threat to humans and other creatures. These two men respond to a little kindness and consideration. They seem to like me because I treat them with respect and try out my Vietnamese language on them. They in turn are helpful in correcting my accent, providing helpful advice concerning Vietnamese customs and advising me how to avoid trouble.

Tonight on my return to the hotel I waded through dense scooter, bicycle, pedicab, cyclo and automobile traffic. It was terrific. The air was blue from exhaust smoke. I saw accidents where bicycles and scooters had locked horns. There were a few ambulances and fire trucks racing around in circles. The radio announced the VC blew up a Korean home about the time I was returning to the hotel, but I did not see or hear any explosion. This is a good reminder for me to keep alert as I walk around Saigon.

An SCS engineer friend arrived today with over US$300 (green dollars) in cash in his pocket. I suggested he had better get rid of it or find himself in trouble. The only place he could pass it would be on the black market and that is very risky. The person who met him at the airport today neglected to tell him to cash it at the airport for piasters or MPCs. I took him to a nearby BOQ and they agreed to exchange only US$100 in each transaction. He had difficulty in understanding the three kinds of money and where it could be used legally and where it could not be. Then I took him to the Air Force Post Office (APO 96243) to obtain a US$200 money order to send home. But the PO would not accept green dollars. Twice he returned to a long line at the BOQ to cash dollars for MPCs. Then he returned to the PO for the money order to be paid for in MPCs. As if that was not enough of a hassle, he had promised his wife in Phoenix, who works for the Tonto National Forest, that he would wire her when he arrived. That meant a trip to the PTT (Vietnamese symbol for Postal, Telephone & Telegraph service). Our next problem was to locate the PTT. I did not know where it was and my friend had lost the map showing its location. So with my broken Vietnamese and blind reckoning we located the PTT. An eight word message cost him VN$895 (piasters) or about US$8.50. I then took him back to his hotel to clean up. Then we went to the Rex BOQ for two 20 cent cans of Schlitz beer and supper. My friend relaxed, probably for the first time since leaving home some 36 hours earlier. I felt sorry for him because for a day or two after traveling so many miles, and going through so many time changes, including gaining a day, and seeing so many strange sights it is real culture shock to arrive.

As you note in my letters, I keep soaking up the sights and sounds and I will try to describe them to you. You can see I am enthused about the prospects for this Nation. And since I came to help them, I'll do my best.

16 March 1967

Today was payday. Checks come to the USAID office from the Embassy, where they are prepared. I did not expect anything except my transfer allowance, which is a one time payment to cover expenses, such as tropical clothing and other special items. I was told I would receive a check today but mine did not arrive. Normal pay checks arrived for those whose regular salary is paid here. The local Vietnamese must go to the cashier's window to receive their checks. It was quite a sight to see all the men and women lined up for their pay. Most of the women wear the colorful silk ao dais. A few wear western style mini-skirts, tight dresses and 3-inch heels which makes them look cheap. They believe it is stylish to wear American-style clothes.

It is interesting to watch Vietnamese women handle the long

flowing tails of their ao dais on bicycles, motor scooters or just walking down the street. There are no fat Vietnamese women, they all appear to be very slim. Some must weigh about 85 pounds and the top weight would not be over 115. They walk erect. Some of them are as tall as you girls. I recall one I saw riding on a motor scooter today. She had on a beautiful brown and tan silk ao dais and rode on the jump seat, like you used to do, Gwen. She held a colorful purple and yellow silk parasol over the head of her driver.

Today I attended another meeting with forestry delegation from Australia. This is part of their AID team. I listened to a description of their forestry program and they listened to my presentation about the USAID Forestry Program. I offered to coordinate our efforts with theirs. They seem uncertain how to implement their program and this may be the last we see or hear from them. After the meeting I walked to one of the BOQs for lunch. I passed the Korean Embassy as usual where my saluting friends, the Korean guards, were on duty in the blazing sun. Temperature was reported to be 95 and the humidity 99%. I admit it was hot and sticky. The guards had their usual smile, bow and salute for me as I passed. I do not see them doing that for others, but then no one smiles to them and calls out their names, Kim Lee, Hy and Yuu, as I do. The guards appeared to be extremely alert today. As usual they held their carbines ready for action, but their eyes were constantly sweeping the street and sidewalks. The bombing of the Korean or Chinese (not sure which) residency yesterday was not far from this Embassy. One woman was killed in that attack. I do feel safer walking on this particular route because there are 10 armed guards along the way. On the other hand they could be considered prime targets. Who knows?

After a lunch of cold cuts, ice tea and vanilla ice cream I decided to walk back to the office by a different route. The streets were quiet because it was the time for "only mad dogs and Englishmen" to be out in the noonday sun. Two Vietnamese policemen were standing in the shade on a corner holding hands, which is a common custom in this country. We see many men and groups of young boys walking hand in hand or with their arms around one another. This practice is socially acceptable. But I never see a male with his arm around a woman, like in our country. I saw another policeman squatting and repairing his bicycle and another reading a newspaper. The police wear white caps and are called "white mice". They lead a rough existence according to one of the public safety men who I met in Washington. He is here now conducting training at the police academy. Most of the RVN police are young, since they have no opportunity to age in police work. They are paid the equivalent of $30 per month. Last week 28 were killed by the VC. They also respond to a smile and a friendly greeting. Most of the Vietnamese are surprised when one of us "round eyes" speaks to them in their language. As I was walking I almost blundered into a roll of uncoiled 4-barbed wire. Those barbs are fierce and are not intended for cattle control. My walk had brought me to the new palace which is surrounded by a tall iron grill fence and a large guard force of tough looking elite RVN Marines. No one occupies the palace at this time but it is a carefully guarded symbol.

There are many beautiful trees on the palace grounds and in the sidewalks of Saigon. Some day I hope to be able to identify these trees, after all there are only 1,500 species of trees! As an indication of the tremendous heat the leaves of the shrubs have rolled up edges and the grass appears to be dead. Some trees are losing leaves and have dried epiphyte growths on the trunks and branches. Some of these will become beautiful orchids during the rainy season.

On the far side of the 1/3 mile square palace grounds were small refreshment carts. I now know where the soft drink peddlers wash the glasses. They have a small half-round can fastened to the backside, which holds about half gallon of water in which they rinse the glasses—that's all they do, rinse them. They sell green, brown and water colored drinks, plus bottles of Coke, orange, lime and lemon.

The next item of interest was a truck loaded with chickens and ducks. There were hundreds of heads stuck out through the slats. As a policeman stopped the driver to examine his papers, the poor desperate birds had their heads drooping, beaks wide open and tongues sticking out in the boiling hot sun. Checking identification papers is a regular procedure. That is one way to control who comes and goes. The next vehicle I noticed was a motor scooter which appeared to be alive with ducks. Some 200 of the birds were neatly tied on and almost completely surrounded the driver as he went whizzing by. Watching means of transportation is good entertainment. Next was a huge load (of something?) pulled by very small horses. They trotted along, very smartly, with a bobbing feather arrangement on top of their collars.

Next I saw a motor scooter buzzing along with a large sack of rice tied to the rear and another sack on a bicycle. I have seen many of these loads the past few days since a large shipment of rice arrived on one of the Saigon River ships. Distribution started Monday. These are sold to the local people at controlled prices, which is about 17 cents a pound. Printing on the sacks shows they come from Houston, Texas. I understand the few provinces in the Mekong Delta can produce more rice than the entire United States but the VC seriously harass and curtail harvests, planting and distribution. They collect a large amount for "VC taxes". Some of the farmers hold their rice for an increase in prices. That too is a problem due to the lack of storage facilities, heat, moisture and rats. I have seen a few rats scampering around a construction project next door to my hotel. The rats are as large as our cat Pinky!

When I arrived at the office I looked in on the broken down generator. There sat the maintenance man, sound asleep, with his feet propped on it. It does not appear we are making much progress in restoring power. We continue using the steps, which is difficult for some of those who are overweight and older. "RC" plus 10 stories, in the dark stairwell, is a long climb. In our offices there are no lights to work by and no air conditioners. That's noontime in Vietnam!

I enclose my $75 transfer allowance check. Every two weeks I will be receiving $126 SMA (separate maintenance allowance). That should more than take care of my expenses.

17 March 1967

I thought I was not going to be able to write to you tonight, because just as I sat down to begin typing the lights went out. Trouble was brewing when I returned from dinner with my SCS friend, George. The elevator had difficulty coming down to pick me up and then it struggled back up to the 10th floor (remember "RC" + 10). From the 10th I walked up a flight to the roof for the spectacular view of the city.

On the way from my floor to the roof I pass a family's home. It is a small storage room, 3 feet by 5 feet and similar to the one found on each floor, but this particular one is occupied. The family consists of a young husband and wife. She is always in the vicinity day and night and I see him working around the hotel during the day. Be it ever so humble, this is their home. It is spotless and dry. Each of them carefully leave their shower-type slippers in a neat row just outside the door.

One evening she was cooking their meal over a small kerosene burner. The meat was in delicious looking brown gravy sauce and smelled very tasty and spicy. I always greet him first and then her, which is proper protocol. They are becoming friendly towards me. The other day when I was on the roof she was picking a chicken. Tonight after dark she was in one corner of the roof washing clothes. The door to their home was closed so I presume he was there. I see him on the roof weekends reading a Vietnamese translation of one of Huxley's books. He and I have spent some time together helping one another learn our respective languages.

There is a similar family living in a much smaller room in our office building. The lady is pregnant and sits there knitting all the time. I never see her husband, I presume he works in the building. I note some of his clothes hung out to dry.

Speaking of clothes hanging out to dry, I am doing quite well with mine. I do a shirt each night, plus socks and shorts and always have clean clothes. When the rainy season starts I may have trouble drying them. The short sleeve blue shirt is my favorite, it comes out looking good. The short sleeve white one comes out fairly well. Fortunately I do not have to iron anything. That would present a problem since I do not have an adapter to fit the crazy French electrical sockets, however I could purchase one on the black market. They have everything!

I want to tell you about Mr. Hac. He is a waiter at the Rex BOQ. He is a small thin, almond-eyed man, as they all are. He is the hardest working little fellow you ever saw. When he sees me or one of his regulars and he waves us over to his tables, one seats four and the other six. Before getting there I usually stop at the salad bar for a plate of cucumbers, carrots, green onions, olives, potato salad, cottage cheese, cole slaw and fruit which consists of bananas, fresh pineapple and papaya. If you wonder how I pile all that on a six-inch plate, it is the training I have had from watching Colonels and Majors pile it on. Around here those two military ranks are as thick as hair on a dog. There are very few Captains and 1st Lieutenants and they usually come in looking haggard, sweaty, dirty and most of them carry a weapon. Except for the patrolling MPs special permits are required to carry a weapon in Saigon. I have never seen a 2nd Lt. in Saigon. They must be out fighting the war.

Back to Mr. Hac . . . he smiles and says "tood ening" and pushes the chair under you so quick, it is quite a trick to keep from hitting your chin on the salad you have just collected. Then he jams the menu under your nose and says "ouu ike solethi nlic tonlight?" I fill out my order for the "Special—MPC$1". Then let the service begin! Ice tea is his favorite and you can not drink over two swallows before he snatches the glass away and fills it, again and again. As soon as the ice melts down he says "ouu ik mo lice". Off goes Mr. Hac with the glass for more ice. Before you can see the bottom of the ice tea glass you are ready to float away. You can retain the glass only if you have a better hold than he has.

Before the salad is finished he asks "ouu ike oup?" and before you can answer, off he goes and comes running back with a bowl of hot soup. You better have the deck cleared when he comes in for a landing with the soup because he never bothers to move any of the dishes aside. He is just programmed to deliver, not to remove. Then he brings a tray of fresh hot rolls and practically before you can eat one with your soup, he brings another tray of hot rolls. Tonight at the table for six we had seven trays of rolls. Next he brings the main dish. There is no room for it, but he pushes the soup bowl away just as you are spooning out the last bit and slams down the plate. At that point he asks "ouu ike mo oup?" I say "no thank you" and before I can grab the ice tea glass it disappears again for about 10 seconds and reappears with more ice tea. The sugar bowl is refilled while all this is taking place and you are offered more sugar whether you use it or not. As I finished the last bite of my turkey and sat back to relax, Mr. Hac appears and asks "ouu ike lice keam"? A "yes" gets you a heaping bowl of vanilla "lice keam" floating in chocolate sauce. Before that is gone he is at your side with a great big grin and asks "ouu ike mo lice keam"?" A "no" gets you a cup of coffee whether you want it or not. Needless to say if you do not call a halt, you will never see the bottom of the coffee cup again. By this time you are ready to pop, float away or go into hysterics. But the next step is very slow, that is getting your check. I do not understand the procedure but I believe his possession of your check proves you are his customer. Finally by holding up a 20 piaster note (17 cents) he will quickly surrender it to you. Then you enter the extra items he has brought, which adds up to a total of MPC$I.15. I do not always sit at one of his tables because the poor guy just over does the service bit and tries so hard to please that you tend to overeat.

The generator was working this morning when we arrived at work, but it shut down several times before the day was over. During the few days that I have been here I have made repeated efforts to determine the status of the sawmill support equipment Rufus Page ordered in January. Today I located the order on the desk of a man in the logistics section. It seems with all of the home leave, R&R (rest and recreation) trips that employees with no family get once each year and TDY (temporary duty) somewhere. Our order was set aside and buried at the bottom of a pile of a papers on the desk of a man who was on a family visitation trip to the Philippines. So I started the process all over again. I completed an eight page form in "octupulet". When I prepared the final forestry statement for briefing the President in Guam this weekend, I identified, among other problems, VC control of the forests and roads and US Government procurement practices. We may be able to defeat the VC, although their supply system far exceeds ours and is a secret weapon baffling the experts, but I do not know about our procurement procedures. I will try to beat the US procurement and logistic people at their own game. One of my accomplishments today was to defeat a damnable French-style stapler and I did! I opened a hole in the plastic window and threw it as far as I could into the depths of the bombed-out building next door. If it comes back I will follow it out the window myself!

Are your checks arriving regularly? You said they were becoming larger. The first increase should come to about $500 since I canceled the bond deduction. The next one should have a $611 increase and then there should be one that includes the differential (hazard pay) plus the retroactive portion. The amount should then remain constant. Remember the SMA (separate maintenance allowance) for you was to be $3,300? When I arrived here the payroll people said there was a provision that if the family expenses exceed your allowance you can receive an increase of 20% up to a maximum increase of $660. (When I included the items they base the calculations on (housing, utilities, food and car expense) it more than exceeded $3,300. Now your SMA will be $3,960 per year. Is it any wonder that foreign aid costs so much? I believe it is too much but the office here insisted I file for it. Now you have confirmation of your suspicions that I am paid more than I am worth! On this end I will be receiving a check for $151.90 every other Thursday. I will bank most of that and we will use it for something special. My living expenses cost between $4 and $5 per day, while the allowance is something like $10 per day.

I have recovered from a siege of the "Ho Chi Minh trots", which is much worse than Mexico's "Montezuma's revenge". It could have been caused by the malaria pills. The medics say that sometimes these can result in an upset stomach. Some people have to change medication. I plan to take the pills religiously because this strain of malaria is a bad one. Most of the cases result from those who fail to take their pills once each week. They have a special hospital in Vietnam for GI and civilian patients which provides 30 days of prescribed treatment (includes filling sand bags) and then they have the "privilege" of added time before completing their tours of duty. For obvious reasons no one is sent back to the states for treatment of malaria. Otherwise many would forego their weekly pill just to get out of here.

Good night for now.

18 March 1967

Today was a work day and I requisitioned the sawmill parts we thought were on order. *Biw tge kuggts gave gibe iyt agaub, si U'kk tr y fir a wguke ub tge darjk.* Whoops! The power went off and I kept typing . . . please note there were a few more typographical errors than usual. This happens too frequently. This time it was probably an overload on the hotel generator. Most of the major hotels and Army BOQs have large generators roaring at all hours. I am glad I do not have a room near one of those noisy monsters.

After work I went with my boss Jerry Overby to the Cholon (the City's Chinatown) Commissary and purchased kipper snacks, crackers, Tide and toilet paper. The TP provided here is one grade removed from corn cobs. I will put these items in my travel bag because I understand many of the places we will be staying in as we travel around the country do not even provide cobs. I never cease to be amazed by the amounts GIs pack out of the Commissary and PX. Many of the items they purchase, such as chile, enchiladas, cheese and ice cream, are not available in their camps. Most items are inexpensive, brands are familiar, but the stock varies according to the shipments received during the week. T-bones, rump roasts, pork chops and other frozen meats are less expensive than at home. The meats come from Australia and US.

When I move into an apartment I believe I can do fairly well. But I am advised it will be two or three months before an apartment is available. So far I enjoy the GI food at the BOQs. However we hear housing for officers will be available soon at Tan Son Nhut Airport and they will be required to move. This may result in closing some of the BOQ facilities and could limit the availability of eating places.

Speaking of construction, a new hotel is being constructed next door to mine. My outside room is relatively cool and has a good view of the activity. The construction project utilizes about 50 workers. Supervisors appear to be Chinese. The hard working laborers are women. They mix and carry concrete in two buckets balanced from the end of a pole placed over their shoulders. One group excavates muck which is piled on the edge of the hole and another group carries it away. They work like a colony of ants. They carry heavy tile and bricks up a flimsy ladder and bamboo scaffolding to the masons. Heavy structural steel is bent into proper shape and carried to where it is fastened in place. Some steel and concrete appear to have been designated for another project. Cement bags carry RMK-RBJ trade marks, which designates the largest consortium of contractors in the world. RMK-RBJ handles US Government contracts. Cement is made in Taiwan. I understand many supplies disappear before they reach one of Uncle Sam's projects. When noon arrives all workers fall down completely exhausted. I never see them move until a gong sounds. The gong beat starts out with a beat every second and then increases until it is constant. The workers jump up and go at it again. Laborers work 14 hours each day and make less than a dollar a day.

I had a can of Sego for lunch while sitting on the roof soaking up some rays. The sun is not quite overhead but it really bears down. I did not stay out over 1/2 hour. After some exercise I did my laundry, had a shower and took a short nap, before going for a walk. Waiting outside my room were the maid and two workmen who were there to finish the floor tile. I gave them permission to enter and the maid stood in the doorway watching my possessions while I departed.

I am mailing you two rolls of black and white film to develop. You will be able to see some of the sights I describe in my letters. Some will show the area between my hotel and the aigon River docks, some will be of women, children and workers along the way, a guard with a shotgun between his knees, bicycle and scooter repair shops and a series along the river. You will see a fisherman and heavy ship traffic. Ships are anchored wherever they can tie up. Many of the ships look and sound like the African Queen. One was a tugboat out of Los Angeles and flew SVN, US and Confederate flags. The 50 mile river passage from here to the South China Sea is very dangerous. At least two ships were attacked by the VC last week. Considerable damage resulted and there were casualties.

The fisherman in the photos had a small bamboo pole with monofilament line. He cleverly mounted a beer can on the pole and used it like a reel. He could wind his line around the can and then throw it out as far as I could have done with an expensive spinning outfit. I sat beside him for a while and talked to a small boy who wanted to practice English. He in turn helped me with my Vietnamese language. I noticed the fisherman was using a bent pin for a hook and was having trouble hooking fish. I walked to my room three blocks away and returned with six hooks. He was all smiles when I gave them to him.

While I was at the river front a tiny old woman came, set up a charcoal brazier and began preparing the evening meal. The woman obtained her supply of water for cooking and washing dishes by wading out into the river. She had to carefully swish the bucket around to avoid floating debris and crud. Next she washed the dishes in this muddy water. She had no soap but under the circumstances that would not have helped. A small girl split a small bundle of wood and built a fire under the charcoal. Dirty looking eggs were put in a pot to boil. When these were done her customers sat on their heels and cracked the top off the eggs to get at the goodies inside. Sure enough, as I had heard, they were eating either unhatched chickens or ducks that were curled up in the shells. The diners seemed to enjoy this food. The sight and thought of what they were eating delayed my supper for a while.

On my return I crossed the street in front of my hotel and worked my way through an interesting crowded mass of small sidewalk shops. These had a good supply of black market items, which were made in America and Japan. This is where stolen cameras and radios end up. While in this area I went into an electrical shop and purchased you a Vietnamese insulator. It is made of white porcelain and should make a good addition to our collection. It is too heavy to mail, so I will bring it when I come home. I did pick up a nice birthday present for Cindy and some small items for Melissa. Happy birthday (yesterday), Cindy! I will collect a few more items before I have a box worth mailing.

You can judge from the limited range of my walks that I

have only touched the surface of Saigon. I have only skirted the edge of the huge central market. It will take weeks for me to cover all the possibilities.

22 March 1967

Since you are on your spring vacation I will start this letter and just add to it.

I am glad you girls can not see me just now because I have a really short haircut. It is shorter than any I have ever had before. I went to the Brinks BOQ barbershop. This is a one chair operation, complete with Vietnamese barber and manicurist. After waiting for my turn and seeing him produce fairly good long and short hair cuts I sat in the chair. He asked, "How ou ike?" I suggested medium on the sides and flat on top. He looked at my head, gestured and made a few remarks which I apparently misunderstood. I noticed quite a lot of hair was falling. I did not get a manicure but I really got clipped otherwise. The top is less than 1/4-inch long and the sides are all skin. The cost was VN$35 or 29 cents. I gave him VN$50 (42 cents) for this bargain. We settled on the price before he started because some of my friends have been paying VN$200 to 300. I guess you get what you pay for! I can assure you I got what I paid for and this cut will last much longer than any previous one.

After lunch today I crossed the street to the US Army Library to use a tape recorder and a Vietnamese language tape. A young Vietnamese woman signed out tapes and earphones to users. When my turn came I told her what I wanted and she perked up and said "You study Vietnamese?" I said, yes, and asked her to show me how to use the machine. She did this while GIs and civilians impatiently waited in line. The machines are good and accommodate large reels at two speeds. Perhaps you could fill a tape and send it to me. One with your familiar voices would be most welcome. After using the tape for an hour I learned a few more useful phrases like "that is very fine work. Which way is it to the church, post office, bank, hospital, etc.?" As I was leaving the recording/listening booth, the same Vietnamese woman was walking by and I greeted her in Vietnamese. She smiled and said "you speak Vietnamese very well".

Tonight after work I accompanied SCS irrigation project leader, Arnold Snowden, to the Vietnamese-American Association. This is where he spends some of his spare time learning Vietnamese and attending some of their social affairs. Many classes are in session in this large building. I plan to return when their office is open and join. The objective of the Association is to promote better understanding between our two people. I recognized this part of town since the VAA is just across the street from the Directorate of Forests.

Today I went on my first field trip which was to a match factory and shipyard in a part of Saigon I had not been in before. I saw many interesting sights along the way as I traveled with three local forestry people.

We passed several massive street markets, not the large central market, but one with hundreds of stalls along the street. This was the largest concentration of black markets I have seen. They had just about every item found in the PXs and the military and commercial warehouses. Well dressed, local people were making many purchases. The Vietnamese tell me they can purchase imported products such as Cokes, cigarettes and liquor cheaper in these street markets than anywhere else. The black market is said to be the engine for Vietnam's economic growth.

We crossed a small side channel of the Saigon River that presented an interesting sight. The stream was jammed with sampans of every possible size and condition. I do not believe that some of them could float but since the tide was out these were resting on mud flats. I did not have an opportunity to see if they floated. People, entire families, were jammed on the boats not much larger than a canoe. Living conditions were indescribable. On the adjacent shore were shacks of those who did not have a boat. Shacks were made out of cardboard, crating, packing boxes, sheet iron, galvanized and corrugated metal, thatch, palm fronds and bark from trees. Those who did not have any of the above used a few rags to cover the openings of three foot diameter culverts which were stored alongside the road and lived in these.

I saw miles of shacks and people living as described above until we came to the more permanent homes made of wood, masonry and palm thatch set on poles. Many of these had green, brackish fish ponds out front. I could see ducks and a few fish breaking the surface of the water. As thick as the water was I do not see how fish survived. But the fish, like the people, must be a hardy lot and able to survive under harsh conditions. At the edge of each pond or perched out over it was a small structure that looked like a child's tree house. These are toilets. People have learned the value of fertilizing fish ponds to produce more fish to supplement their diet of rice and provide much needed protein. The sad thing to remember is before all these people became refugees they were once happy farmers and woods workers from all parts of Vietnam, including Hanoi in North Vietnam. Many of these refugees fled south in 1954 after the French were defeated and before the Communists took over and are still homeless.

The match factory was interesting and its problems were typical of what most wood using industries are experiencing. All were suffering from a serious shortage of raw material. Available wood is very expensive. In addition to the logger's normal costs he has to pay the VC a tax just to work in the woods. The log hauler has his normal expenses and before he can make a profit or be blasted into the next world by a land mine he has to pay a VC tax. Wood using industries must finance all these costs and still make a profit. The match factory was operating at about 50% capacity since choice softwood was about exhausted. As a substitute they were using logs from rubber trees. Defect and waste from the very cross grained rubber wood runs as high as 70%. Added to these costs was the damage to the peeler knives resulting from bullets, shrapnel, nails and other pieces of metal found in the logs. These industries need a dependable wood supply and metal detectors.

Our next stop was a shipyard. This was a fascinating place. I could have spent hours watching these craftsmen build junks, sampans, large ocean going fishing and cargo ships. A large sailing ship was being constructed for a rich American at a cost of VN$2,000,000. This one was a beauty displacing seven to ten metric tons and 14 meters in length. The wood workers were experts. These small men would move a huge support timber into place, another skilled boat maker would step back, squint, step up and move the timber a fraction of an inch and then drive a steel pin into place. Sawyers scribed a curved design on sections of logs. Then they cut the pattern out with weird, crude saws that cut a very narrow kerf (groove made by the saw). Two other sawyers cut a huge, two foot square timber which had been squared by hand. These men were using a larger saw but one of the same design. One stood on top of the huge timber and one below in a pit. Lumber has been produced by primitive peoples using this pit saw method for centuries. I admired these craftsmen for their precision work. I learned these two sawyers are the highest paid workers in the shipyard and receive VN$800 for cutting two 30 meter lines.

Workmen had rice for lunch. Their homes were shacks in the shipyard. In some I could see the floor covered with a woven mat. There was little else other than a few blackened and battered utensils. Their women and babies kept in the background. Men were dressed in black pajamas (uniforms of the peasants and VC) and they looked fierce. But I greeted them in Vietnamese and told them they were doing a good job. I asked to take their photographs and they broke into broad grins which showed stained, battered gold capped and missing teeth. As I said good-bye they all stood and bowed to me. I hope if I ever meet any of them out in the forest, when they are serving in an other role (VC?), they will remember me and be friendly.

When I returned to the office one of the Vietnamese administrative assistants greeted me and inquired about my trip. They were interested in my observations of their country. When they learned where I had been they said that was a very

152

dangerous place. But I knew the local forestry men had checked and found the "signs" were favorable. I wonder what takes place on those crowded, waterways in the dark of night? Verification of that situation is not in my work plans. No way!

(continued)   24 March 1967
Today I went to a furniture factory. The operator of this facility processes logs and produces lumber on a horizontal band sawmill. This product is French designed and known as a CD-4 sawmill. Pins fasten the log between tracks and a frame, supporting engine and saw, is pushed along the tracks to cut into the log. Resaw of huge boards is done in the same manner. It is slow, but it works. This factory produced beautiful furniture but the stock pile of lumber appeared to have random thickness, widths and lengths. They need help with log procurement which is delayed by VC tax collectors and other problems. The operator recently received an order for 7,000 crutches. The products looked suitable to me, but the operator said they were all rejected because they were a millimeter off in one dimension. He said he solved the problem by paying a bribe of VN$70,000 and the order was accepted. I understand bribery permeates every activity throughout the country.

I visited a pencil factory employing 50 people. The small factory was in the owner's back yard. The greatest problems facing this operation are an uncertain wood supply and competition from Cambodia which has no problem with wood procurement. This man's wood was imported from California at high cost. Quality control and the product looked good.

Meanwhile back at the office work goes on in spite of red tape. At the moment I am caught in a bind between US procurement procedures and Vietnamese political sensitivity. You recall I located the unprocessed supply order, but I could not get anyone to sign it. So I started the procedure again, signed it myself, then my boss and the program officer signed. Now I have to overcome the political fears of the Director of Forest Affairs and the Minister of Agriculture. I am surprised someone does not require the VC to sign orders. After all they have a great interest in the wood business. I have appointments with the SVN officials and hope to obtain final signatures tomorrow. What a way to run a railroad!

(continued)   25 March 1967
Back at work I drafted a letter to resolve my procurement problems in both USAID and Government of Vietnam (GVN) offices. But then more US-generated red tape slowed the process. By the time I obtained the necessary USAID signatures Mr. Tan was unavailable. So goes it in Vietnam. Looks like it will be a long war.

To finish out the morning I began training my new Administrative Assistant, Mr. Tran Cao Thuong ("Two-ong"). He put in a request to work for me. The other Vietnamese men working for USAID had recommended me as a "No. 1 boss". Mr. Thuong is an emaciated young man of 31. At this time he has a draft deferment. His history is typical of many people working for USAID. He fled south from Hanoi in 1954 to escape the Communists. He left his family behind while he searched for a free new life. For a while he taught Vietnamese literature in the old imperial capital of Hue (Whay) at the University of Hue. Speaking good English, he was hired by our Special Forces advisors and served them along the DMZ line, which is the dividing line on the 17th Parallel between North and South Vietnam. The DMZ is a joke . . . it is as militarized as it could possibly be. Thuong had a rough time and was involved in heavy fighting. This resulted in a nervous disorder which helped him secure a draft deferment. He went to work for USAID in 1964, where he has maintained a good employment record. He provides us with an experienced person. He remained after work today to help me with my Vietnamese language lesson.

After Thuong departed I finished a four page letter to my men. I provided advice on what they should bring and what to expect. The last crew member to be selected is Barry Flamm. He is an old friend from the Southwestern Region and currently Forest Supervisor of the Shoshone National Forest in Wyoming. He will report to Washington on 4 April to begin training. I would like for him to bring a few things. I will ask him to contact you before he leaves Washington. I want him to bring my field glasses. The war is just beyond the edge of the city and there is much to see of bombers and helicopters in action. Then there is the busy harbor to observe. Watching the stream of ships from all over the world is interesting.

We received sad news this week, perhaps you saw it in your newspaper. Nine persons including a seven man study team from Wisconsin State University and other colleges and universities were killed Thursday afternoon when their Air America airplane crashed into a mountain near Hue. The team was led by President James Albertson of Stevens Point, Wisconsin, wherever that is, and was under contract to USAID to study higher education in South Vietnam. I met with Dr. Albertson last Tuesday and discussed our forestry program and the needs of the Forestry Department at the University of Saigon, College of Agriculture. In addition to Dr. Albertson, others killed were the President of Bemidji State College, a Dean and a Science Department representative from Wisconsin University, a Dean from Gustavus Adolphus, two Ph.Ds. from Harvard and the University of Illinois, a USAID education advisor and the pilot. This was a real tragedy and deeply felt by those of us who had met with the team.

This afternoon I exercised on the hotel roof. The hot sun helped me work up a good sweat, but it made me feel good. I have trimmed down from the 210 pounds I weighed in early December to 190. The walking and exercise help keep my weight down. I eat no candy and drink only treated water. I am enjoying good health. I have no gastric disturbances but almost every one else around me has suffered from serious ailments such as amoebic dysentery and hepatitis. My SCS friend, George, has acquired amoebic. Two US secretaries, who arrived only last weekend and live in my hotel, are both ill from amoebic and hepatitis. These women hate bugs, dirt and poverty and the war frightens them. Vietnam is not what they expected. One is from St. Paul and the other is from Louisiana.

This afternoon I walked to the Vietnamese-American Association. But it was closed and I was unable to join. It looks like an interesting set up. They have concerts, lectures, classes in Vietnamese and English. I plan to return next week and sign up for a membership.

Next Wednesday during my lunch period I start my formal language training under a professional teacher, Mr. Phuong, who is hired by the Embassy. The training will consist of a minimum of 100 hours and I am looking forward to it.

(continued)   26 March 1967
                                              Easter Sunday

Last night some of the fellows in my hotel had a poker game and invited me. It was very enjoyable, especially since I won MPC$8. These MPCs look like play money but they are accepted by everyone. MPCs have been in use for over 20 years in foreign countries where the US military operates. Around here it is accepted by everyone.

I was faced with a decision as to where I would go to church today. I could have gone to the Rex BOQ and participated in a military Easter Service or I could have gone to one of the local Protestant churches where most of the Embassy and USAID people attend. Instead I went to the Saigon Catholic Church. It is the largest in Vietnam and a beauty. There are many Catholics in this area and they hate and fear the Communists. I find it strange that many Vietnamese are both Catholic and Buddhist. It appears religious beliefs are blended with the political system for personal convenience.

The Bishop of Saigon, looking splendid in his red outfit, conducted high mass in Vietnamese and Latin. I did not understand much, but the service was impressive. Attending were over 3,000 well-groomed people. I stood along one side of the church with hundreds of others. I was amazed that the numerous small children were very attentive, sat, bowed, jumped up, kneeled and prayed at the proper time. There was no fidgeting or whispering. The colorful assortment of silk ao dais made the affair look like a grand Easter parade. After the service I took a number of photos of people leaving church.

I am using High Speed Ektachrome film in the camera. At a PX cost of 80 cents for 35mm, 36 exposures film, including processing, I am able to shoot many photos and record scenes of people, the country and some paintings in an art show I visited. The Vietnamese have talented artists. I will purchase some paintings when I have a place to hang them and later ship them home. They are done in oil and water colors and include scenes of farms, rice fields, villages, sea, boats, water buffalo and Montagnards (highland people). I prefer the latter. I also took photos of singers and a band at the Rex BOQ.

The band at BOQ is there each evening. The volume of their music sounds about like the girls' boy friends' band, The Scoundrels. The group consists of alto sax, trumpet, lead electric guitar, piano, drums and three singers. I fear the volume will bring down the ceiling. They play for two hours without a break and after 15 minutes are back at it again. They can harmonize and are a big hit with all the military and civilians. Songs included "Green Berets, Let the World Go Away, Together Again, Poor Little Rich Girl, Come on Baby Be My Man, When the Saints Go Marching By" and a number of scoundrel-type songs. Some were so loud that I could not understand the words. Today there were a number of US civilians wives around. They were permitted to enter Vietnam for the Easter period. These dependents must return to their "safe haven" residences, presently provided in Hong Kong, Taipei, Bangkok, Manila, Baguio, Singapore, and Kuala Lumpur, on Tuesday.

I had my usual Sunday evening steak fry, which included a two pound T-bone and the works. I was hungry because I had only papaya, dry toast and black coffee for breakfast and a can of Sego for lunch.

This afternoon, after sitting and exercising in the sun on the roof, I went for a walk through the street markets, both legitimate and black markets. I noted huge supplies of PX and Army goods for sale. Prices are higher than the PX and the merchants will not bargain. At one stand I saw more new, unused Craftsman tools than our local Sears Store stocks. Other items included hair spray, Dial and Zest soap, cameras, radios, tape recorders, whiskey, Levis, Dickies, Munsingwear, etc. If the Army Quartermaster Corps runs short of new GI clothes, I can advise them where to find more. These items must be stolen. I presume bribes support the black market process.

Food, vegetables and meats in the markets were colorful and fascinating. Food vendors served rice, fish and other meat using their normal food handling practices. It makes me shudder when

I see people eating everything imaginable and drinking all varieties of liquids. Just one sip or taste of any of that would put us "My's" (Americans) in the hospital.

Take care of yourselves. I worry when I do not hear from you. Our newspapers report on the robberies and violence taking place in the DC area. I wish you could see this country. Parts of it, such as the coast near here are lovely. I have not been there for swimming because it is now off limits due to a black plague epidemic. Sounds like the Middle Ages, but this side of the world is very primitive in many respects. It is difficult to imagine how the Vietnamese have survived for 3,000 years as a separate and distinct ethnic group, but the answer must be found in the strength of these strong willed people. Perhaps I can help them find some of the answers that can help lead to progress and solutions to their problems. I strongly believe we can improve their ways and means of doing things, while not necessarily imposing the American-way on them.

Girls, take good care of yourselves and your mother. I love you all dearly and miss you.

27 March 1967

I received your letters (3/1, 8, 12 & 16) and my birthday card today, on my birthday! That made it a happy birthday. Finally the mail is coming through. Mail brings a touch of the real world and I welcome your letters. I save them and will read them again and again.

It was 90+ degrees again today, as it will be tomorrow, the next day and the next day. I am glad you are enjoying the snow, frankly I do not miss it. There has been no rain in this area for months. Cindy writes about Steve Fields' brother and the rain. He must be in the northern mountains along the border with Laos and North Vietnam. I suppose the reason they are required to pay for rain gear replacement is to remind the GIs to take care of their gear and keep it out of the black market. Yesterday I noticed the black market had a good supply of rain gear and shelter halves. These items were for sale and used as shelters for the sidewalk stands.

I am glad you received the money order. When I am paid on Thursday I will purchase a money order and mail it to USDA Credit Union for the final car payment. They will send you the title showing the car loan has been paid. Please place the title in the safe deposit box for safekeeping. With the check you received last week you should be solvent in a short time. Please remember the Montgomery County Income Tax is due by 15 April. Paying Federal, State and County Income taxes is a burden.

Cindy, it was good to receive your newsy letter. How is your sister Melissa? Has she lost her pen or broken her left arm? I regret missing your "sweet" 16 birthday. We will have a birthday party for all of you when I return this summer. No, I have not met General Westmoreland, just his room mate. But there are many one, two and three star generals around here. You can be assured they support this war because it justifies their rank and gives them something to do. Generals always prefer live wars rather than planning for future wars.

The Vietnamese women who work for USAID are in general good looking and neatly dressed. Most of them are just over five feet tall, thin as a board and weigh from 85 to 110 pounds. One of the secretaries named Miss Onah delivered your letters to me today and said "Mr. Cravens, I wish I could work for you. You are so nice to all of us." Even the three American secretaries in Agriculture appreciate someone being courteous and talking to them. You recall the Korean guards, well tonight I rode the USAID bus to the hotel since it was so hot and sticky. The guards spotted me in the bus and gave me a big salute. My friends on the bus noted I really had good rapport with those guards. That made me feel good.

One of the American secretaries accompanied me to the Directorate of Forest Affairs to do some work. At noon Mr. Tan announced he was taking his advisor and Miss Weixel to lunch in honor of my birthday. Mr. Tan asked my age and birthdate when I arrived and he had remembered. This is the third occasion he has taken me out for a meal. Lunch today was at L'A'miels, a small French restaurant near my hotel. I had not eaten there because I was shocked by the prices posted in the window. Mr. Tan placed the order in what sounded like perfect French. First we had small loaves of delicious French bread, then five spears of tender, pale asparagus. The main course was crab in a half shell mixed with mushrooms, small onions and luscious herbs. This was followed by five kinds of cheese, a large bottle of champagne, French pastry and strong black coffee. Needless to say it was a welcome change from my usual lunch of a can of Sego or a bowl of soup.

I did not write you about my first evening meal with Mr. Tan and Rufus Page. First we went to the luxurious Caravelle Hotel where we had a drink and hors d'oeuvres. A spectacular panorama is afforded by the ninth floor restaurant. While there we watched a street demonstration by Lambretta drivers who were chanting "Death to De Gaulle". At sundown we went to Maxims which is said to be the finest and most expensive restaurant in Saigon. We started with delicious onion soup which had floating stringy cheese. The main course was a tender, well prepared fillet mignon and small creamed onions. The Chinese fruit lichee was the best fruit I have ever tasted. These are green and about the size of Queen Anne cherries. For desert we had French pastry and coffee while we watched a very good floor show which featured French-Apache dancers, singers and a good orchestra. This was no "girly" show, just high class entertainment.

For the next meal, I accompanied Mr. Tan and the Australian AID delegation to "la Peprika", a combination Spanish-French restaurant. The restaurant was located on top of an apartment building in the northwest part of Saigon. The first course was served on the patio. It consisted of Scotch and soda and fat little sausages on a toothpick. The sausages were as hot as any Mexican food I have ever tasted. Next came a skewer of delicious meat . . . it may have been veal, mutton, snake or dog, but it was marinated in "ambrosia" and cooked to perfection. The technique for eating these bits of meat was to grab hold of a piece of meat with your teeth and pull out the skewer. Dinner was served at a beautifully set table for eight. The main dish was a delicious rice dish containing a chicken leg, a lobster and three other delicious meats which I did not attempt to identify. All of this was covered with excellent Spanish sauce. To wash this down we had two large bowls of iced red wine, containing slices of orange and other fruit. Desert was a delicious fruit sauce and good black coffee.

In spite of all this fine food I have lost eight pounds since I arrived. I feel very well and do not have a sign of a sniffle or sneeze. I have a tan you girls would envy. The sun is powerful at this tropical latitude. I do not stay out in it very long but just walking along the streets to and from work provides enough sun to brown your face and arms.

The doctor that talked to you girls at school reported Vietnam health conditions correctly and compares with what I have seen and heard. Two sections of the country are presently closed to entry due to plague. I am grateful I had those many inoculations. Concerning malaria, I use mosquito repellent every night, but still get one or two bites. There are bottles of malaria pills at all of the BOQs and placed at other strategic locations around American installations. With all the warning signs and hourly announcements over the American Forces radio there is little excuse for coming down with malaria. I take two pills each Sunday morning when I get up. We are advised to wear long sleeves, button the collar, use repellent and bug bombs and sleep under mosquito netting. Most of the malaria is found in the central highlands and some coastal areas. Pills must be taken for eight weeks after leaving a malaria infested area. If malaria parasites are in your blood the malaria pills merely suppress them and you have no symptoms. To knock it out completely requires massive doses of medication which is administered only after leaving Vietnam. Treatment is not given routinely in Vietnam because the malaria returns. I hope to be one of the lucky ones and escape this ailment.

Tomorrow I go to a memorial service for the Stevens Point College President, his team of college and university officials, the USAID education advisor and the pilot. Artillery guns are booming steadily tonight to remind us there is a war going on.

I will be mailing you some items of interest!?

28 March 1967

Your Easter and birthday cards are posted on my wall. I appreciate the thoughts they express. My cleaning man and woman have also admired the cards.

This afternoon I went to the US Embassy to meet with Ambassador Lodge's deputy. The meeting concerned certain phases of the forestry program which interest the Ambassador. The Embassy building is downtown near the river and on the edge of the sidewalk. There is no wall or setback zone. The place is surrounded by armed guards, concrete barricades and rolls and rolls of coiled barbed wire. You get the impression that certain groups are not welcome. The building shows the scars of a previous bombing attack. A new US Embassy building is under construction in an area with more space and security.

I was driven to the Embassy by one of our car pool drivers. We waited for another passenger who had signed for the car originally. This other guy was late for a meeting and was very rude to the driver, Mr. De. I had ridden with Mr. De before and we had talked in Vietnamese. He enjoyed that conversation. Although the US Embassy was in the opposite direction of where the other passenger wanted to go, I was delivered first. The other guy was fuming and fit to be tied. I told the driver not to return for me since I would walk back to USAID.

On my walk back to the office I saw a very old man selling something to children. They crowded around and reached through a trap door into a small box. Guess what was in it? There were giant roaches, about three inches long! These cost 20 piasters (17 cents). Giant insects are used in races and spectators bet money on which one will win.

Next I saw what looked like an old Buddhist lady with a saffron colored robe. She had a shaved head, was stripped to the waist and was squatting down and washing herself with water running in the gutter. I see many people washing in gutter water. They wash themselves, their dishes, bicycles, scooters, pedicabs and dogs. There are many kinds of dogs, some are fat and sleek and others are skin and bones and missing most of their hair. Many people walk their dogs on a leash and make no attempt to clean up after them. The markets I passed were selling puppies, monkeys, birds of many varieties, fish, ducks, chickens, gold fish, guppies and a fish that looks like an eel with external gills and is said to thrive in any kind of water and can stay alive under wet leaves.

Four of us from the office attended the memorial service for the nine men who were killed in the plane crash last weekend. It was held in the very modern International Protestant Church. Some 50% of the audience was Vietnamese and the remainder Americans, Australians, British and French. A Dr. Clevenger read from one of Whittier's poems. He had worked with the team of educators while they were here and had known many of them since their college days together.

After the service my boss, Overby, invited me to his apartment for dinner. His cook prepared Campbell's chicken and rice soup and salami sandwiches. When and if I get a cook, I will have rice, Vietnamese dishes and lots of fruit. The secret will be getting a cook who can prepare food properly.

While eating I told Overby I believed the proposed increase in the differential (hazard pay) for those of us working in the Saigon area did not seem reasonable. I said those working in the provinces were in more dangerous zones. He reminded me of the memorial service we had just attended and emphasized there are risks everywhere. If that passes as proposed in the President's bill and approved by Congress the differential will be increased from the present 25% to 50% of the total salary. That should help you get further out of the financial hole I left you in. Enjoy your vacation.

Chapter XIII

Around Saigon

After all my precautions I picked up a stomach bug. Normally on Sunday afternoon I would be out walking. But I am staying close to my bathroom today and curtailing my side trips and picture taking excursions. Now I do not know anyone who has not had stomach trouble. In spite of all my precautions the primitive conditions finally got me.

It may be the dirty rats. They are all over the place. I just looked out the window of my hotel and could see three big fat rats, in the construction area next door, scampering about. Yesterday as I started up the steps to one of the BOQs a large rat ran across in front of me and disappeared into the kitchen. It is no wonder that the city is so infested with rats. The few garbage trucks can not begin to clean up all the garbage that is thrown into open barrels, boxes, wicker baskets or on the sidewalk or street. The slums and refugee centers are indescribable in that respect.

The Vietnamese have many illnesses. The night man on my floor had severe stomach, incapacitating pains yesterday. For treatment he placed a piece of tape on his stomach. Yesterday morning on my way to work I saw a very old woman squatting down on the sidewalk beside a street vendor. Everyone assumes that position to rest; they are just like rabbits, when they stop running they squat, flat footed and rest. This old woman had the back of her blouse raised and the "healer", another old lady was applying suction cups to her back. She was being bled just like in medieval times. She had 12 cups, 1/2 inch in diameter and about two inches long fastened to her back. I could see them filling with blood and some seeping down her back. The poor thing did not look like she had any blood to spare. She looked very ill.

The next bit of personal hygiene I observed was a woman washing her baby's mouth and nose with water from the gutter. Babies have to be tough to survive. These people have a critical need for public health educators and medical people. At a recent briefing session given by a USAID physician we saw slides of some of the critical conditions, such as three people in a hospital bed, and children who had been severely burned when braziers of charcoal had overturned and set fire to shacks and houseboats. One showed a baby that was badly burned on its head and hands by a candle which had set fire to bedding. The child had been treated with Mercurochrome. Gangrene had set in and affected both hands and the child was to die soon. A nun was shown pleading with the mother to bring the child to the hospital. Many amputations result from the lack of training or concern for the patient. This sounds and looks like our Civil War times.

As I type this letter I now recall something is missing from the roofs, there are no chimneys. That could explain the many fires and use of so many dangerous kerosene stoves and charcoal braziers.

You will be pleased to learn that my penmanship has improved. It had to improve so the Vietnamese staff could read my drafts. It embarrasses me to think how my previous secretaries must have struggled to decipher my writing. I now do better and my Administrative Assistant Mr. Thuong can read my drafts.

Our boss, Mr. Overby has been charged with a security violation. He left a document marked Administratively Confidential on his desk where a Marine guard discovered it. It was not a leak of vital information or anything that would embarrass the United States, but regardless, this violation goes on his record. He had to go before the Embassy Security Officer for a stern lecture. We heard all about it and were advised not to let it happen to us.

The large contingent of Marines guarding our office and the Embassy are all specially trained for this duty. They are well trained physically and mentally to cope with anything. We hear many rumors, some may be true. Two weeks ago on a Saturday morning there was a report that one of the Marine guards shot a terrorist at USAID 2. Just today we heard that a terrorist had leaned a bicycle filled with plastic explosive up against a tennis court fence. When it exploded a number of people were injured. Another report concerns the VC using poison darts. There has been nothing about these events in the newspaper so we do not know if these reports are true. However, they do serve to keep us on our toes.

USAID has developed a system to beat the black market. Any items imported for resale carry the USAID symbol. Imports are certainly needed to provide people with the supplies they desperately need. But I do not see how this will stop the thefts and bribes which keep the black market very much alive. We shall see.

Our coffee set up at the office is interesting. I contributed MPC$1 to the fund, then I could not find a cup and I refused to use a community shared cup. I purchased a cup on the black market for VN$80. A sign on the coffee room door says "Phong Ve-Sinh Dan-Ba" which translates to "Ladies Room" which is on through in the adjacent room.

I find enjoyment in the people I see and greet on my walks

to and from work. This Saturday morning I walked by a school. On the way to school children stop at nearby sidewalk peddlers and purchase sticky rice wrapped in a banana leaf, or a piece of bread, or rice cakes, or a piece of fruit, or stuff on a stick that looks like a frozen popsickle. Small boys frequently hold hands or have their arms around one another's shoulders. But I never see little girls do this. I dropped into the school yard and looked in several of the school rooms. I counted along with the children as they sang out their numbers and the alphabet. Fortunately in the 16th Century French missionaries introduced the Vietnamese to Arabic numerals and the Roman alphabet, otherwise they would still be using Chinese characters.

There is a Japanese-American in my Vietnamese class by the name of Roy Yamamoto. He is a CPA and speaks English better that most of us. He finds this "sing-song" Vietnamese, as he refers to it, very difficult to master. He tells me Japanese is strictly a monotone language. Vietnamese, for example, has many tones such as level, rising, falling, rising-falling and falling-rising. A two letter word can have as many as five different, completely unrelated meanings, depending of how it is pronounced. It is an interesting language but very difficult to master. I spend a great deal of time studying.

While it harms the local economy I found a way to defeat the numerous shoe shine boys. I purchased a pair of rough skin Hush Puppies and the boys stopped bothering me. On my walks after dinner I have been approached by rough looking characters wanting to exchange piasters for green dollars, at the rate of 88 over the going rate. Besides being illegal I have none. The other offers are for a "nice girl". No sale on that either. Another thing I can do without is TV and that is not a difficult decision. The uncertain power and the abundance of TV programs about Batman and Vietnamese drama make that an easy decision.

Please do not worry about me. I am being very careful. I believe the work to be accomplished will make it all worthwhile.

4 April 1967

You could call today and tonight a textbook description of April in South Vietnam. The daytime temperature was well above 100 and reached 100% humidity. It is real cozy! This temperature reminds me of that day in August 1948 when we left Williams, Arizona dressed in our finest for Phoenix and driving a Forest Service 1940 International to be turned in for a new Ford pickup You will remember the old pickup boiled going down Yarnell Hill and blew the radiator cap over the edge of the cliff to join the wrecks scattered below. En route we stopped for water at every one of the few and far between filling stations and in between we filled the radiator from our desert water bags. We reached Phoenix completely wilted. You hated that weather and you would hate this too. I must be used to it because I found today quite comfortable. Everybody here agrees with you and believes I am either crazy, or have a faulty thermostat or both. I now climb steps two at a time and keep in good shape. Those individuals that are over weight are suffering in this heat and humidity. To continue with the weather report . . . it feels like rain but never quite makes it. Tonight after dinner I went up on the roof and watched the lightning flashing in the clouds over the South China Sea. It was a calm and peaceful scene, until broken by the flares being dropped around the City. I understand they are searching for infiltrators who are probing the perimeter defenses of Tan Son Nhut Airport and other strategic areas.

Today on the way to work I walked by the large central market. Parked there in the hot sun were two meat trucks. One was a 3/4 ton pickup and the other a ton and 1/2 open stakeside truck. Regardless of the temperature the vehicles are always piled high with freshly processed quarters of beef. A man with his pants rolled up was standing on the beef in each truck in dirty, blood stained shoes. They hook the quarters and pass them to skinny little men on the ground, who have a piece of dirty canvas draped over their heads and shoulders, and off they go at dead run into the market building. They make the round trip in about three minutes. The quarters of beef must weigh two or 1/2 times their weight. This takes place each morning and I see those trucks regardless of which route I take to work.

I walked by the lady "physician's" place of business. It is beside a Hindu Temple which I suppose imparts spiritual values to the treatments. Today a young woman sat in front of her and was receiving foot treatments. This consisted of making a small incision and attaching suction cups to each foot, just above the toes. One cup was full of blood and the contents were being emptied into a small can. The lady had her cups, dirty knives, robes and medications spread out on the dirty sidewalk. I did not go too close to examine them. The horrible thought occurred to me that if I suffered a fainting spell or was injured, I did not want to be in her vicinity and receive any of her TLC.

Speaking of injuries, Moui, the small girl who is the maid on my floor, usually sits on the floor just outside my room. I believe she sits there so she can listen to my radio when I have it turned to the American Forces Radio. Occasionally I tune in the local Vietnamese station to listen to the "sing-song" music. Recently Moui received a deep cut on her cheek which I treated with Merthiolate and covered it with a band aid. I gave her some of the black and white photos I took of her and her small girl helper, Chine. She was pleased to receive them. This Saturday afternoon Moui was not at her usual place. At 1900 hours I finished writing an article and was just leaving for dinner. Moui showed up with a bandage wrapped around her head and neck and under her chin. There was blood all over her blouse and she was about to cry. I could make out enough words to learn she had been helping clean on the next floor and had slipped and fallen on the slick waxed floor and struck her head on the sharp curb under the railing. Could it possibly have been child abuse? She had been taken to the nearby lady "physician" for treatment, which consisted of more bleeding. She told me she would not return to work for five days. Before she left I should have given her some aspirin for what must have been a major headache. The fact that I was able to learn what had happened demonstrates the level of my language proficiency. It is improving!

Speaking of language training, it goes well. The first series covers 100 hours and prepares me to ask and give directions, go on a trip and make the proper travel arrangements, order meals, find my way to hotels, ask prices and a few other similar items. If I can do that with a degree of proficiency I will be entitled to a step increase in my salary. However the greatest reward is to see the surprised expressions on people's faces when I talk to them. They are pleased and go all out to help me. For example when the USAID car dispatcher sees me coming he always provides me with a car, even without the required reservation or advance notice. Occasionally he will give me a car when others are waiting. He announces that I had one reserved. He tells me the drivers like to drive for me because I talk to them and treat them well. I avoid talking to drivers when they are in a tight spot in traffic and that is most of the time. To drive in Saigon requires much horn and a little brake. I call the game "chicken". All drivers seem to enjoy the game . . . usually at the last moment someone yields an inch and miraculously we miss one another.

A few days ago I had an appointment at the JUSPAO Office (Joint US Public Affairs Office . . . jargon for "propaganda office). One of the USAID staff members, Peter Newcomer, said he was going there and he would show me where to get my official black & white photographs processed. I agreed to ride with him on the back of a large Honda motorcycle. Gwen, I now have much more respect for you and those here who ride on the rear seat. He wove in and out of traffic and showed me how fast he could accelerate and bluff out the Vietnamese scooter operators. I suggested he was overdoing the offensive driving bit and deserted him when we arrived at our destination. Yesterday Peter proved his point and the "ugly American" in him came out. He tried to see how close he could come to a policeman directing traffic in the middle of a busy intersection. He hit the policeman and knocked him down and then took off to get away from the scene of the accident. Unfortunately for Peter, traffic stalled as the result of incapacitating the policeman and the officer jumped up and ran to Peter and arrested him. Peter wa taken to the police station where in addition to the charge of hit-and-run it was discovered that the Honda had been purchased from someone who did not own it. He could not produce the proper ownership papers and was required to cool his heels in a Vietnamese jail for a few hours until the Embassy obtained his release. In my judgment they should have left him there. I told him later, that as USAID's reports officer he should now be able to write a good, eyewitness report on what it is like to be held in a Vietnamese jail. He did not appreciate my suggestion, but I am certain the Vietnamese do not appreciate his kind. Please do not misunderstand me, the Vietnamese drivers are horrible or even worse, but I rationalize, after all, this is their country and we are their guests.

This afternoon I had an interesting trip into some of the back alleys and waterways of Cholon, Saigon's sister city and home of many Chinese. I would not care to be in this area after dark. I visited a sawmill with the local forestry people. The mill was in a small sheet iron building which was poorly lighted and ventilated. In this building were three horizontal band saw head rigs and three crews cutting huge, very heavy, oily Dau ("Zow" in North Vietnamese and "Yow" if from the South) logs. Four men operated each saw by pushing it on tracks as it cut through the log. Women carried the heavy green lumber out of the building and loaded it on trucks parked on the street. This is a highly inefficient way to operate and the boards have random thickness and widths. Women do all the heavy work of moving logs into place and turning them as needed. A steam winch is used to pull logs out of the adjacent stream to the saw. The women and children exposed themselves to danger as they crawled around the saw and salvaged sawdust and bits of wood. The wood and sawdust are used for fuel and joss sticks, which are used by Buddhists in their religious ceremonies.

From the mill the owner took us to the streamside. He wanted to show us logs on the other side of the wide stream. There was no boat or barge. The tide was coming in swiftly, but the men ahead of me took off and ran down the length of logs which were strung out across the stream. The logs were so heavy that they just barely floated. The workers were bare footed and sure footed. I had on crepe sole Hush Puppies and off I went. I ran down the length of a slowly turning log, stepped across to another, jumped on to a raft of logs tied to a sampan, up onto and across the sampan, then off onto another raft of logs fastened to the other side of the sampan. To the applause of the onlookers I finally and gratefully reached solid ground on the other side. It was an exciting trip! The streamside was lined with refugee shacks supported on poles in the water. The water was filthy and smelled badly. The tide refreshes the stream but it carries as much sewage in, as it carries out. But I made it to the

other side to examine huge logs being unloaded off trucks. Whil
we were there the operator dispatched one of his motorize
sampans, with logs lashed to each side, to the Delta, some 10
miles away. Escorting this raft of logs was an assortment o
derelicts, sampans, junks and two SVN Navy patrol boats. Th
patrol boats looked like porcupines with many .30 and .50 calibe
machine guns sticking out in all directions. All the sailors wer
well armed with submachine guns and automatic rifles. That i
the way convoys are formed to travel the waterways.

Trucks transporting logs to the Cholon/Saigon area trave
over 150km over roads controlled mostly by the VC, from wood
that are controlled by VC. The VC collect at least two sets o
taxes, one in the woods and at least one and sometimes mor
along the roads to Cholon. I have some feelers out to obtain a V(
tax receipt. I hope to get it second hand. The VC would probabl
overcharge me.

In view of the heavy water traffic the sawmill owne
instructed our driver to drive to a nearby bridge and pick us u
I was grateful I did not have to make a round trip across th
stream. While waiting for the car we watched workers on the lo
landing move huge logs to the water as the tide rose. Some of th
logs were at least 40 inches in diameter on the small end an
were at least 10 meters long. I calculated a load of four log
would scale better than 12,000 board feet and must have weighe
well over 20,000 pounds. As the logs hit the water about 15 men
women and children scrambled over them as they turned in th
water. Quickly they removed bark with short bark spuds. The
fought over possession of the bark which was used as fuel. Som
of them took off carrying huge loads of bark on their backs t
homes along the stream.

All of the hundreds of little children in this area wer
running around naked. Some as young as two or three years o
age were swimming in the dirty stream. It was frightening to se
them scramble around and under logging trucks being unloaded
While it gave me the creeps, I took color slides of the activity.

Here's a word about the delightful weather. In the tropics th
sun appears to rise and set about the same hour each day. As
type this letter, sitting in my shorts with the sweat pouring of
me, the overhead fan barely stirs the hot humid air . . . yo
might say it is quite cozy. Before going to bed I must study m
language lesson for tomorrow.

(continued)   5 April 196
It is now Wednesday evening, Tuesday morning where yo
are, and the girls must be back in school. Mail is now comin
through quickly. I am glad you dropped the American Embass
out of the address. USAID/ADGR, APO 96243, San Francisc
provides direct delivery service.

Around here we hear many tales about the lazy Vietnamese, how they overcharge us and how dirty their streets and country are, etc. This talk disturbs me and I remind people there is a war going on and millions of people have been driven from their homes and have no place to go. That is anything but normal. Take the forestry program for example. It is far from being normal. It is not running as efficiently as our Forest Service and I do not expect to see it ever operating at that level. They have real problems. They were trained and operated under an archaic French administrative system. I can not comprehend how the French held out here, or any place else, for almost a 100 years. Many key people in the Vietnamese Forest Service have been drafted, killed or crippled. There are many amputees because their medical people have not been trained to save limbs. From one end of the country to the other the forests are full of VC, mines and booby traps. Some forestry officials are threatened and forced to do certain things for fear of their lives and those of their families.

I hear and read in the newspapers the Air Force is defoliating hundreds of thousands of acres of timber. This will either kill or weaken the trees to the point where insects, diseases and decay may destroy the tree. The dead trees will then probably be replaced with worthless "jungle" sprouts. I am also told by people who have been out on field trips that there are forest fires raging from the 17th Parallel to the Delta. This will likely continue until the rains come. The forestry people have fire fighting forces but all they can do is suppress fires around cities, safe villages and hamlets. The VC will not let them enter the woods.

Added to our forestry program problems is the fact that the Vietnamese Security Agency recently learned that the owner who had been selected to set up and operate our demonstration sawmill had made a contribution to the VC and given them a radio in 1965. Now he is blacklisted. Added to the confusion is the fact the US Army just gave him a sawmill to cut lumber for their use. USAID just gave him a large contract to supply lumber for the pacification effort. These are the sort of events will keep this job from becoming dull.

I am pleased to hear our financial picture has improved. I will send more of my expense money to the bank to help build up the account. My expenses are low and my needs are few. Since Saturday I have not spent anything on food. My "innards" are upset. While I have no aches or pains, it is a great nuisance. I went to sick call today along with other broken down civilians and wounded soldiers. A checkup revealed no parasites or ova in my system. The doctor said it was just mild food poisoning. It really was not very mild! Now I am drinking much water and swallowing small white pills.

My language study goes well. I derive pleasure out of it. I am convinced the best place to learn a foreign language is in the country where it can be used every day. I now have Vietnamese employees of USAID greeting me when they see me out in the city. I enjoy working for their country and they know it.

In this wrap up paragraph I will cover a number of items. I regret missing your cousin Jo Ann's husband Don. He was only in town for an hour and came to my office while I was locked up in the maximum security briefing room. I will be sending you a box of gifts and some slides soon. This ailment, whatever it is, has weakened me to the point that I now take the bus to and from work. I have not been taking the stairs two at a time lately. I checked with the Embassy housing people today and it appears it will be months before I get out of this hotel room. The walls are beginning to close in on me. There are many new apartments under construction but many people who came here in November are still living in hotels. It is now 2130 hours and I will crawl into bed. I have been pretty tired since I caught whatever I have.

6 April 1967

Shortly after I finished my last letter I recovered from the "scourge of the Orient". I hope to enjoy good health, since it is really much more pleasant.

Today I had some interesting experiences with my language training and practice. The Vietnamese training is becoming more complicated each day. We are now putting more and more of the words and expressions learned in previous lessons together into more meaningful phrases and sentences. My Vietnamese friends say I am doing well and do not have much of an accent. I doubt that, since these people have either a northern, southern or middle (Hue accent). Since I am from the Midwest, naturally my accent more nearly approaches the middle. On my way to the hotel a US Army officer asked me the way to the Hotel Splendid. I answered him in Vietnamese and he went on down the road muttering to himself about "all the damned foreigners". The next contact was a sailor with Begay stenciled on his shirt. I greeted him in Navajo "Yat ta hay". He replied, "Boy, it is good to hear someone from home! Are you from Arizona or New Mexico?" I told him both and we had a good visit before we parted.

I had planned carefully for my next contact. Recently I had talked to the Korean Agriculture team and I learned how to greet the Embassy guards. One of the guards had been particularly friendly as I passed him each day. So tonight on the way to my hotel he was on duty and I stepped up to him and said "An yong, ha ship ni ka!" I hoped the other Koreans hadn't set me up since the guard had an AR-16 automatic rifle instead of a carbine. He gave me a big broad smile and said to me in perfect, unaccented English, "Good evening also to you, my friend. You speak Korean very well".

My next observation was a Vietnamese "lady" who I heard telling a construction worker in a hard hat, "that will be US$10 please", as they stepped into a taxi. Yes, it was a peculiar evening with the Navajo, the Korean and the Vietnamese "lady" all speaking in English. My other contacts were in Vietnamese with shoe shine boys, beggars, orphans of the street, sidewalk vendors and the people around the hotel who provided me with an opportunity to practice.

Today I wrote the manual for the forestry technical assistance program. It is much shorter than our US Forest Service Manual, which is about 14 feet of books. Mine for Vietnam is much clearer. I recall one of the old Forest Service manuals that had all the rules and regulations in a small book of about 1/2" thick . . . so much for the good old days. Our forestry manual for Vietnam covers eight pages and I will resist attempts to expand it.

Tonight I ate dinner with an interesting Negro Captain from Alabama. He was sitting alone at a table and I joined him. He does not like the Army too well, likes Mrs. George Wallace and her husband even less, and has decided to make a career of the Army. He is typical of many of the officers I have talked to in Saigon. They count the days remaining in their one year tour of duty in Vietnam. They shudder when I tell them the length of my assignment. It helps their morale to hear someone will be here longer. But they have many gripes . . . they are being reduced from a food-subsistence allowance of $67 back to $48; their quarters are crowded and dirty; even Lt. Colonels have to live with as many as five to a room and one bathroom; and most of them work 12 to 16 hours each day, day in and day out, with no break in the routine except when chasing or being chased by the VC.

The VC have been giving the Saigon area fits the past few days. There have been continuous nighttime probes of defenses and attacks on police check points. The sky has been illuminated continuously the past few nights by flares.

After arriving back at my hotel tonight I dropped by to visit one of my friends. He and three other men were watching "Batman" on a new TV set he had just purchased at the PX. The determined, intent look in their eyes as they concentrated on every move made by Robin and his faithful friend gave me an idea on how to beat the VC in one final battle. All we have to do is scatter battery operated TV sets out in the forest and let the VC watch American TV programs. They will soon decide that anyone with the stamina to stomach all that junk is much too formidable an enemy and they will throw down their arms and come in, leaving their TV sets behind.

Tomorrow I go on a dendrology trip to help Vietnamese

foresters identify some of their trees. Many of the professional foresters, like Mr. Le Viet Du, Mr. Mo and Mr. Hy, graduated from forestry schools in the States and they know all the US trees in the Lake States and the South, but they have had little recent experience with their own country's trees. I do not know them well either, but I have a French taxonomy book that keys out the chief characteristics of the trees. It is written in French with scientific names in Latin. A much better device will be to take along a Vietnamese botanist who is reported to be an expert. The forest we will be visiting is the Saigon Botanical Gardens and Zoo. While it is in a less secure area near the edge of the city it is reported to be surrounded by guards and miles of barbed wire. Saturday morning I will go to a place called Bien Hoa (Ben Whaa) to visit sawmills, a pulp and paper mill and a veneer plant. The road to that area is safe during the day but after dark it belongs to the VC. Police check points in that area were attacked this week and many police and VC were killed.

That is all the news for tonight. Since the mosquitoes are hungry, I will put on some "Off" and hit the sack. Good night.

9 April 1967

Good morning. It is Sunday and a cool 90 degrees. I will catch up on my writing and ID my slides. I will send you two boxes which will take you on a trip around Saigon. This will remove the slides from these hot, humid conditions.

My trip to the Zoo with the botanist and foresters Mr. Hy and Mr. Du (also pronounced "Zoo" in the northern accent or "You" in the southern) went well. The French established the well maintained Zoo and Botanical Gardens over 75 years ago. The area was carved out of tropical forest, with selected trees and vegetation being retained. Other native and exotic trees were planted. It is an interesting area for study. The trees are much different than ours. The botanist pointed some of the trees that had been killed or damaged by spray drifting in from nearby herbicide projects.

The Zoo adjoined one of the many waterways and thus had its share of armed guards. While there looking at the trees I had an opportunity to see some of Vietnam's magnificent tigers. There are many of them in the forest, and being man-eaters and scavengers, they do well with a war in progress. I understand they have learned to go to the sound of explosions where they may find a supply of food.

My field trips are increasing as I become more organized and make more contacts. These are essential in providing me with an overview of the situation. Saturday I went to Bien Hoa to visit a District Forest Office, a plywood mill and a pulpmill. The District Office looked a little worse for wear, about like the old Ryan Station on the North Kaibab during the dismantling process. The yard contained odds and ends of logs, pulpwood and chunks of wood. These had been confiscated as illegal products, i.e. harvested without a permit. These items did not represent much value but showed some effort being made to control theft and fraud. Most of the high value logs move at night under VC sanction and are never picked up. This is a serious problem even in secure areas like Bien Hoa.

The two forest industries had many things in common. Both were expanding. The plywood mill was being improved with the addition of new glue machines and dryers. The products look good and would make excellent wall paneling. Both native and imported woods from other SE Asian countries have spectacular color and grain. Prices are very reasonable with a sheet of 1.5 meters by 3.5 meters costing just over a dollar. But again the big problem is a dependable supply of wood. The Australian who operates this mill pays heavy taxes to both the VC and GVN for the privilege of operating. However he is optimistic about the future and is expanding.

The pulpmill is owned by the GVN and Whittmore-Parsons, a New York company. They too are expanding by adding another paper making machine. Their products look good. Quality control is achieved by continuous monitoring in a modern laboratory. The chief engineer, Duong Trung Hung, had his training in North Carolina. He was indoctrinated in Washington, DC by Mr. Hukenphaler of the US Forest Service. He recalled the kind and courteous treatment he received from "Huck". This mill also has raw material supply problems. Most of their bailed, bleached pulp comes from Rayonier, one of our West Coast plants. They had stockpiled 200 cords of excellent pine pulpwood from the Dalat area but recently the VC tax increased so much they were forced to abandon the 150 mile truck haul. The grinders are idle but mill expansion continues.

As I mentioned the Bien Hoa area is one of the so-called secure areas. It is reached by a modern 4-lane highway from Saigon. There was a tremendous mixture of traffic on the road, such as 15 yard dump trucks, troop carriers, armored personnel carriers, water and gasoline tankers, thousands of Jeeps, semi-trucks, cars and taxis. Helicopter gunships and medical evacuation ("Dustoff") choppers filled the sky. Both sides of the highway were lined with miles of coiled barbed wire, pill boxes and watch towers bristling with machine guns and thousands of armed guards. Thousands of acres are covered with military camps, storage areas piled high with all types of materials and a large hospital with red-cross marked helicopters coming and going with their cargos of dead and wounded soldiers. That 10 kilometer drive impressed on my mind the significance and magnitude of the war going on around me.

The two lanes in the middle of the 4-lane highway were

used by cars and trucks. Occasionally they can pass one another when the road ahead is clear. The two outside lanes are for bicycles, scooters, Lambretta carriers and wagons pulled by small horses or oxen. Traffic is fast and although it is illegal for cars and trucks to use the outside lane, they use it for passing on the wrong side. The old 1956 Chevrolet station wagon of the Directorate of Forest Affairs went weaving in and out of traffic. The lack of a muffler added much excitement and enthusiasm to our passing. We were stopped once by MPs for passing in the wrong lane. But our driver talked his way out of an arrest and a fine. This was vital to him since the fines are heavy and that driver, according to my interpreter, made only VN$3,800 per month. At the exchange rate of VN$118 to a US dollar, he has a tremendous problem in supporting a wife and 7 children. This is particularly true when a 100 kilo bag of rice costs more than VN$2,000. Until people can earn a decent wage, I fear there will not be a great deal of progress or reduction of corruption.

Interspersed along this route were a few small agriculture plots, where children tended water buffalo and farmers and their families worked the fields. The scorched appearance and clearing of some of fruit, other tree and brush cover provided evidence of the efforts to eliminate ambush cover. The farmers seemed oblivious of the helicopters looking down their necks and all the roaring traffic. I imagine they are keenly aware of all this activity and hope they will live to see better times.

Yesterday afternoon in the boiling sun I took three friends on a photo trip. There were many worthwhile subjects. The old lady "physician" was working on a swollen ankle of a well dressed lady. Fortune tellers were doing brisk business and beggars were stretched out for a rest. I enjoyed greeting people and bargaining in the market. I am becoming an expert and the other fellows have me make purchases for them. Prices start at about 10 times what they will finally accept. The price comes down quicker when the merchants learn you speak Vietnamese. I believe they enjoy dickering with customers and trying to take them for all they can. I bought you a set of small old fashioned looking scales. I bargained and obtained it for about MPC$2. I finished my purchases and can now mail you a box of interesting, different items.

Next week I will be on a Top Secret mission, the details of which must wait to be told in the future. I may end up in Sumatra and Java before it is over. It is nothing dangerous, just something extremely important to Vietnam and the US.

It is 0900 hours and time for breakfast. Later I plan to go for a walk and look for some new photo opportunities and subjects to write about. I will check on some of the apartments under construction. At the rate they are moving people into apartments I will probably be in this hotel room for the duration.

I have been involved in Top Secret work recently. It relates to rubber plantations and herbicides. I provided advice that resulted in more care being exercised in one phase of the military operations. For a while it appeared I might have to go to Malaya for information concerning the management of rubber plantations but I was able to obtain it here in Saigon from the Rubber Institute and other sources.

Yesterday I attended a Top Secret senior officers briefing session at the Embassy. I was invited because of my grade, so-called status and the program I administer. Please recall the lack of grade is what made the first candidate for the job back out, indeed he was correct. The meeting was held in a safe, sound proof room. In attendance were five civilians, including me and Ambassador Lodge, 10 Generals, 20 Colonels and five Navy Captains. Ambassador Lodge talked for some time. He was impressive and although he is due to leave shortly, he did not sound or act like a "lame duck" ambassador. General Westmoreland bought us up to date on his thinking. He appeared to be as interested in the USAID Pacification Program as he is in conducting military operations. He is a very dynamic person and an obvious leader. He is highly respected by all who come in contact with him. The General had been to Manila and back that morning. He is on the move at all times.

The staffs of the Ambassador and the General bought us up to date on the political situation, the status of the USAID Pacification Program and the war, including campaign plans for both. The only real Top Secret material we covered was related to future military operations. These were spelled out in precise detail, down to the moment of execution. I was impressed by what I learned and the importance of keeping it secret. The session was very informative and provided me with a good picture on current developments in South Vietnam.

Our ID was carefully checked by Marine guards as we entered and departed the briefing room. I noted when I arrived at the Embassy that the whole area swarmed with armed guards for several blocks around. This occurs wherever General Westmoreland is located. He is a prime target. The Embassy and USAID offices are significant targets and are guarded by the largest contingent of Marines stationed at any US Mission in the World.

At another session I attended this week an Army officer described the war. He said the US military has had much experience with brush wars around the world and has studied, in great detail, the wars guerrillas have waged in other countries. He talked about the British Army in Malaya, where 5,000 rebels were fighting the British. But in that war the British had a simple solution for ending the fighting. The rebels had to come to the villages for food or starve. There were no sources of food in the jungle, so the British just controlled the villages and finally captured or killed all of the rebels. It was also significant that the rebels were Chinese and different in appearance from the natives of Malaya.

In South Vietnam you can not tell one side from the other. The Vietnamese are a distinct people, but the Viet Cong, the North Vietnamese and South Vietnamese all have the same general features. The people from the North have a different accent and prefer different foods, just like in our country. But the South has over a million northern refugees who fled south in 1954 after the defeat of the French and Communist take over. All can trace their ancestry to one of a hundred families. In Malaya the British had 5,000 to route out. Some 5,000 VC are said to go through the Saigon zoo on a Sunday afternoon, just looking at the elephants. This may be an exaggeration, but it illustrates the problem. Also in Vietnam an army can live off the very fertile country. The forests and mountains contain many edible plants and animals. The Montagnards, or highland people, practice shifting agriculture and have plots of cultivated crops throughout the mountains. Even after a field is abandoned there is food to be found.

One of the success stories of this war has been publicized in the news. It is the Chieu Hoi (Chew Hoy) program, literally meaning "open arms". This is part of the PSYOPS Psychological Operations, i. e. propaganda) program. Under this provision VC and North Vietnamese regulars are given a safe conduct pass to surrender. Last month some four million of these passes were dropped and the payoff is large and growing. VC men and women can turn themselves in to any of the so called Free World Forces and are guaranteed safe conduct passes to a secure area. They are paid for the weapons they surrender. A heavy machine gun brings a large reward. These people are interrogated for security information and then taken to a camp for retraining and finally, after a period of time, if they are from one of the southern villages, they are returned to their family. While in a retraining camp they are provided food, clothes and a dry place to sleep and paid a small wage. As a matter of interest I have enclosed some of the safe conduct passes being dropped all over the South and in Cambodia and Laos. These passes are designed to be dropped from airplanes in special bundles which come apart in the air. The individual pieces fall rather rapidly to the ground and will float for long distances on water surfaces.

The temperature was 105 today. I took one of USAID's shuttle buses to work this morning. During the lunch break I went to the Embassy for a meeting. Since the buses were not operating I took a pedicab. On one of these vehicles you sit in a box-like seat in front and the operator sits behind and peddles. It gives you an exposed feeling to be out front and have traffic coming at you from all angles. Fortunately traffic was light and I had no problem in covering the 10 blocks. The cost was VN$40 (34 cents). The men that pump these affairs have firm legs of steel. It would be good exercise for you girls. That's all for tonight. Gosh it is hot!

Chapter XIV

War Zone Forestry

Travel was interesting and certainly different from what we normally experienced. We rode in cyclos (motorized rickshaws), pedicabs (foot-peddle-powered rickshaws), chauffeured cars, DC-3s, C-46s, C-47s, C-118s, C-123s, C-130s, turbo-prop Porters, Helio-couriers, helicopters and everything else with wings. Our pilots ranged from young Air Force officers, CIA's Air America pilots, and the standard commercial Vietnamese and American airline crews. All are overworked and stressed out. Collectively our forestry crew had many near misses, but we were lucky and just a step ahead or a step behind trouble.

Our "work horse" was the UH-1 or Huey helicopter. We flew countless hours in Army gunships and Air America's white, unarmed Hueys. The C-Model Hueys were gunships manned by two pilots and two door gunners, one of which was the crew chief. The pilots sat side by side, separated by a radio console. Behind them was a passenger-cargo compartment which was accessed by two large sliding doors on either side of the aircraft. Most of the time the doors were slid back along the fuselage and secured. Door gunners sat on either side of the aircraft at the rear of the passenger-cargo compartment where they could fire M-60 machine guns from open doors. Some D-Model Hueys had rocket pods and one supported a mini-gun (high speed Gatling-type gun). I can still recall the tortuous sounds and sense the ringing in my ears resulting from the many reconnaissance hours spent in those helicopters. The whomp, whomp, whomp of the blades, the high pitched scream of the jet turbine engines and the ear shattering blast of two machine guns took away some of my hearing. But that was the method we used to get from camp to camp, to size up the forest resources and determine the extent and magnitude of Ranch Hand's Agent Orange defoliation activities. We either flew at tree top level or at 5,000 feet. At the lower level we could come and be gone before anyone could get a bead on us and at the higher level we were out of range of heavy machine guns. We had complete confidence in the young military helicopter pilots. You learn not to take many things for granted when you are flying along at over 100 knots/hour, just a few feet above a solid canopy of 300 foot trees and goodness knows how many people you pass over that would take great pleasure in shooting down your aircraft. All it would have taken is one miscue, one bullet, or one slight malfunction and in the blink of an eye we would become distant memories in the minds of family and friends.

I recall being out along the Cambodian border in War Zone C near Tay Ninh City. We were examining the effects of defoliation in what once was a magnificent tropical hardwood forest. We received a radio message warning us to get out of the area. Our helicopter lifted off with a gut sinking swoop, in a swirl of dense red dust and its nose rotating into a heading away from the border. I sat with one of the door gunners and as usual had no ear protection. As we departed all hell broke lose below as one of the frequent military operations was kicked off. We could see the skirmishes between the ground forces and hear the explosions of mortars and the chatter of the machine guns over the scream of our Huey. Our machine guns added to unbearable noise level. When we returned from that trip we found small tree branches and bamboo caught in the helicopter's skids.

* * *

16 April 1967

It is Sunday morning and before it becomes unbearably hot I went to the roof top for a look at the city and countryside. The heat waves were shimmering in all directions. Delta-winged jets and helicopters have been going overhead on many missions this morning. The other evening we had a tremendous display of artillery fire just north and to the west of Saigon. It was just like sitting on the Washington Monument grounds watching fireworks on a 4th of July evening.

Tomorrow I take off on my first trip to the countryside. The purpose of this trip is to complete plans for setting up a sawmill and logging operation at a Montagnard Refugee Camp. The idea for this project originated with a Mr. Dwyer, who headed the Secretary's (Dwyer's) Forestry Study Mission. Dwyer made arrangements with OCO (Office of Civilian Operations) and the US military to set up a sawmill in the extreme northern (referred to as Region I, I Corps or "Eye" Corps) part of SVN near the DMZ (Demilitarized Zone between North and South Vietnam). Essentially this was a private business venture, since USAID was not in on the arrangements, nor was the Director of Forest Affairs. This is a portable mill, manufactured in Oregon and known as the Mighty Mite. The Vietnamese are unfamiliar with the installation, operation and maintenance of this type of mill. Mr. Dwyer has made arrangements for a technician from the factory to come and set it up and teach refugees how to maintain and operate it.

About two weeks ago we received a telegram from Manila from the technical representative of the company inquiring when to come and set up the mill. The shipment had no import permit

179

and USAID had no record, other than informal word it was to arrive in Danang during February. Considerable detective work was required on our part to find said sawmill. We sent a telegram to the company representative in Manila advising him we would check and let him know when we located it. The shipment was located in an OCO warehouse in Danang. About that time our telegram was returned with notification that there was no record of any such person at the Manila address. Such are the frustrations of operating over thousands of miles of water!

Last week we received word that the OCO people in Danang had made arrangements for use of the sawmill and to send the man to install it. We advised OCO that the man could not be located. The OCO response was to "send hundreds of axes with hangles (sic) and either the Philippine Civil Action Group (PHILCAG) or Cravens". Rather than be replaced by PHILCAG, I am going at OCO's request to support planning and implementation of what may prove to be an interesting operation.

I leave in the morning by Air America. *TIME* reports AA is the CIA's scheduled airline serving Vietnam, Laos and Thailand. I will go to Danang, which is two provinces south of the DMZ. From there I go to Quang Ngai (Quang Ni) and then to the resettlement camp at Tra Bong. You will find these places on your map just south of the 16 degree line. Danang is approximately at 16 degrees and the other places are about 75 miles south. I expect to be away for two or three days.

I have mentioned the children of Vietnam before in my letters and would like to add additional thoughts. The war, resettlement of tens of thousands of refugees, and the normal problems associated with a city trying to cope with 10 times as many people as it was designed to support is bound to create many homeless children. Many orphanages are operated by the Government, churches and other groups. Every week or so there are photos in the newspapers showing a Vietnamese official making a donation to an orphanage. I see many children I judge to be orphans, just living on the streets. None of them appear to be starving since there is plenty to eat. Some of the food does not look very appetizing, such as odds and ends of various parts of hogs, chickens (hatched and unhatched) and rice. Nothing seems to be wasted. I see children salvaging food at the rear doors of restaurants. Salvaged rice, from garbage cans, is spread out on a flat surface, such as the sidewalk, to dry. This keeps it from spoiling. Speaking of spoiling I have learned what is inside the small squares wrapped and neatly tied in banana leaves. It is chopped raw pork, mixed with salt and spices and allowed to ferment. This is considered to be a tasty "tidbit" and I see them for sale everywhere.

Back to the children . . . those with families seem to be loved and cared for to one degree or another. Some are neat and clean and others very dirty from lying on sidewalks, gutters and in back alleys. The babies are either naked or wear only a shirt or pants with a slit up the rear that opens at appropriate times. There is no sign of diaper rash . . . no diapers! I shudder when I look at them and then think about all the care you took with your babies to keep them clean. These toddlers sit in the dirt or in various kinds of filth and then suck their grubby little fingers or dirty pacifiers. I now understand why every Vietnamese has had hepatitis and recovered, or died or has it now. They do not receive shots of gamma globulin every five or six months like we do. Those surviving are tough, but many die of tuberculosis, liver degeneration, plague, etc.

Most of the children eat with chop sticks. It must be difficult teaching a child how to eat with sticks. The other day I saw some Vietnamese children squatting and eating rice from bowls. Nearby was an Indian child eating curried rice with his fingers. The Vietnamese children were sneering and making faces at the Indian, laughing and pretending to eat rice with their fingers.

Children of downtown Saigon are probably better off than others. The older ones shine shoes, do chores for a few piasters and purchase food from street vendors. Some vendors do not appear to charge children for food, but perhaps these are their children. At least they are fed out of "relatively clean" bowls. Some of the children were seen to gamble with an old lady who has a dirty bandage over a large lump on her jaw. She spins a wheel and picks up the children's one piaster coins. I never see the children win in this juvenile gambling operation.

Children living in refugee areas appear to lead an animal-like existence. I have observed them from the car. I do not have enough nerve to walk through these places. The children are naked or wear rags. Most are very dirty because water is a real luxury and has to be carried in cans of one kind or another. The children look as mangy as the mongrel dogs that frequent these areas. Markets and food vendors are filthy.

When I look about and see the health and sanitation problems I agree with you, Gwen. These people have a long way to go. They are trying to move up from 18th Century conditions to the 20th Century. I see hazards in going too far into a modern society, too quickly. The effects can have serious consequences for many generations. The ideal place for their society is somewhere between their present and our level of living. Our presence has created many problems for their society, some problems are very serious. The traumatic experience of failing to reach our level may set them back even further if they fail. My philosophy is to help them improve their machinery and their methods. We can introduce them to new techniques but I believe is impossible to take them into the "rocket generation". Setting goals to help them reach the level our country had attained at the beginning of this Century seems more appropriate and then let them develop from that point on . . . on their own.

One of the Vietnamese told me the other day they appreciated the help the Americans were providing but that "you are like a big elephant walking around in our garden, you break things". I thought this was a good observation.

Much is being accomplished here, to the credit of the local people, the Americans and the "free-world" efforts, take education for example. There are 1,800,000 children in school. Under the French control never more than a few hundred thousand were in school. The Vietnamese are quick to learn and grasp any opportunity for education. But the real challenge is not just building new school houses, it is providing them with competent teachers. Some villages have new schools that have never been used because there are no teachers. Some success has been achieved in expanding secondary and college-level facilities. But again there is a serious teacher shortage.

My program has inherited problems related to the French administration and education systems. The French made a deep impact on the functioning of this country which will be extremely difficult to overcome. For example, much of the GVN effort is strictly administrative in nature. A large bureaucracy tracks the inflow of money and supplies and another group follows the outflow. In this system there is great respect for the older experienced administrators. Under this system the new college graduate is placed in a very subordinate position. He does menial tasks and is seldom if ever permitted to utilize his technical knowledge. He is considered too young to teach the older administrators anything. As a result many modern techniques are never put into practice. Until I can break this barrier I will be reluctant to recommend candidates for training in the US or other countries. USAID encourages us, pressures us, to identify candidates for such educational opportunities. Candidates themselves pressure us to nominate them for training in the US. One of the most important prerequisites for training in the US is the ability to speak, read and understand English with a high degree of proficiency. Most of them are grossly inadequate. The group of Vietnamese foresters I work with have BS and Masters degrees from universities of Oregon, Georgia, Michigan State and Yale. They are very intelligent and have much to contribute to the management and development of their county's natural resources, but they are never given the opportunity. It certainly is frustrating for them and for me. Perhaps in time this is one of the historic barriers I can bring down and make real progress in forestry.

Another problem area is transportation. For example there are old broken-down buses, horse drawn carriages and handcarts to haul people and all sorts of cargo at a relatively slow pace. Add to that 1,000's of bicycles, scooters, pedicabs, cyclos and you have a congested mess. Into this insert longhaired teen-agers and a few hot-rod Americans on motorcycles and you have a safety hazard. Americans wear white crash helmets, the Vietnamese none. Add to this mixture 100's of Renault taxicabs, Citroen cars, Volkswagons, a number of new Japanese cars, a few large old American Buicks and every conceivable type of military vehicle and you have a deluxe problem. The problem becomes acute when some car and truck drivers speed through traffic, using as little brake as possible and "heavy on the horn".

If the price of cars ever comes within reach of the average citizen these people will not die from VC attacks, their own drivers will kill them off. I saw two serious multiple vehicle accidents on my way to work only this morning. In these crowded conditions the hit-and-run driver has little opportunity to escape. A notice in the BOQs urges military drivers to exercise caution and help reduce accidents. Statistics show that during a six months period last year that military vehicles in Saigon were involved in almost 400 accidents with 78 US fatalities. Numerous non-war related deaths result from traffic accidents, and horseplay with guns, grenades and knives.

I see many military convoys going through town, either to the docks or heading out loaded with supplies. Most of the drivers appear to be disciplined and vehicles are well spaced. Other drivers and pedestrians keep out of their way. GI's look hot and dirty. Usually the person "riding shotgun" on the seat beside the driver looks bored or is asleep as they roar through the crowded streets. Located in each convoy are many Jeeps containing armed guards. Most of the GI's wear flak jackets since it is well known that these vests save lives. GI's hate them and say they are about like wrapping up in a heating pad and turning the controls on high.

That's all for this letter. I need to study my language lesson and do some office work. Reports are due. I will write again as soon as I return from my trip to Danang.

22 April 1967

It was good to receive your letters when I returned last Thursday night from my first trip out of Saigon.

The trip was interesting. I was picked up at my Hotel at 0600 hours Monday morning and taken to USAID #2 where I transferred from a car to a small bus for the trip to Tan Son Nhut Airport. Traffic was very heavy at that early hour. At the airport gate there was a long line of workers having their ID, packages and bundles checked and their bodies frisked. This is a very strategic installation and one of the few places subject to a body search. Americans are not subject to such scrutiny.

Air America's terminal and the airline are not quite lik
United Airlines. It is more comparable to "Lil' Abner's Dogpatc
Airline", run by "Capt. Eddie Ricketyback". They have goo
equipment but informal operations. When I boarded the plane,
twin engine Volpar Turbo Beech, the pilot was calculating h
load on a round slide rule. He muttered he had to do everythir
from calculating loads to mechanic's work. As a retired airlir
pilot he objected. He estimated he had 1,000 pounds too muc
fuel and nine passengers for eight seats. He said with the fu
overload, nine passengers and their luggage the plane would t
200 pounds over allowable gross weight. He was still mutterir
when only five passengers showed up. We taxied out in a line
fighters, bombers, cargo ships and an assortment of planes
types I had never seen before. Before we could take off w
pulled out of line and shut down one engine while the pil
climbed out on the wing and fastened down an inspection plate
mechanic had neglected to secure. He told me it was not like th
when he flew for TWA and King Saud of Arabia. He was we
armed with an M-16 by his side and wearing a shoulder holst
supporting a .22 Hi-Standard semi-automatic pistol (standar
equipment for CIA agents). He said his flight plan called fc
flying at 11,000 feet to Nha Trang. We would be flying withov
oxygen and the pilot said he had flown 15 hours yesterday wit
only six hours sleep. He said management pushes them too har
These conditions would give the Forest Service fits since mar
basic safety rules were being violated.

Tan Son Nhut is the world's busiest airport. Literall
hundreds of huge cargo planes, both military and civilian a
landing and taking off at all hours. Some bring supplies in ar
others distribute them throughout the country. All military plan
have a coat of camouflage paint, while Air America is silv
colored. As we climbed out I could see planes coming and goir
in every direction. In one respect we have helped Saigon catc
up with our air quality, the air over the city was thick wit
smog.

Rivers appeared to be choked with silt. The fields looke
parched since the monsoon rains have just started in Nor
Vietnam and have not reached the south. I could see irrigate
areas near Saigon where there is an ample water supply. Son
fields were irrigated by gravity flow and others were watered t
people packing it in plastic buckets or pumping water wheels lit
a bicycle.

After all the pilot's muttering and filing a flight plan fc
11,000 feet we flew at 7,500. At this altitude I felt like th
"Cremation of Sam McGee" in reverse. Instead of longing for th
heat of the furnace, I enjoyed the first cool relief of my stay
Vietnam.

Above 1,200 feet a plane is out of range of small arms fi

and over 4,000 reasonably safe from heavy machine gun fire. We could see few signs of the war as we flew north towards Ban Me Thout ("Toot") and Nha Trang. If you read my letters with a map at hand you can follow the routes I travel. The defoliation has been described in *TIME* magazine and was evident over vast areas. It has killed or set back many of the large tall trees but everywhere I could see into the defoliated areas the understory was dense enough to hide several large armies. Forest fires were scattered here and there and will continue to burn until it rains. No one dares to go out and suppress forest fires. I saw many blackened areas in the tropical dry forests.

As we flew north I could see glimpses of the highways and the north-south railroad that once tied the two Vietnams together. None of these facilities can be used due to the lack of security. Here and there I could see a gap in bridges crossing streams and rivers. Bridges had been destroyed by the VC to disrupt communication lines. Near Ban Me Thout the scenery was beautiful. Mountains are covered with dense forest cover and contain white water streams and waterfalls. This is the famous hunting country of Vietnam. The French preserved this area as their private hunting ground. At one time the area supported many tigers and elephants. We were too high to see any game but this is in some of the relatively secure country where we will be working. Later I should be able to see this area on the ground.

Throughout the highlands I could see clearings, some were freshly cleared and still burning. Others are in crops and yet others have reverted to worthless jungle growth. (Note: the term, jungle, comes from the Hindi "jangal", meaning wasteland. Morris, 1975. Contrasted to this are the tropical rain forests and the tropical dry forests, both of which are productive.) Openings are created by the Montagnards and their shifting cultivation practices. It is difficult to imagine these hardy people fell huge trees with small, handmade, primitive axes. After a period of drying the dead material is burned creating a seed bed for dry land rice, corn or manioc. Some of these clearings are made in very steep county which I would describe as "steep as a goat's face". Much of the soil and fertility are soon lost due to erosion during heavy rain. Some areas at the higher elevations receive as much as 170 inches of rain during a six month period. For about two years these "moi" (Vietnamese slang term for savages) scratch the soil and raise meager but life sustaining crops. Within two years the nutrients created by the decaying forest vegetation and ash are completely gone and they move on, clear another patch and repeat the cycle. This is such a terrible waste of labor and natural resources . . . soil and trees.

We flew near a beautiful 7,800 foot mountain named Chu Yang Sin. The only signs of human existence on the mountain were a few small clearings. Here and there we could catch a quick view of the Montagnards' long houses. North of this mountain the country fell away to a heavily timbered valley, containing a clear, white water stream. South of the stream the timber had been sprayed with 2,4,5-T (so called "Agent Orange", derived from an orange colored band painted around the drums of herbicide for ready identification). The large trees appeared to be dead but the undercover was alive and dense. Here and there I could see smoke from agriculture clearing and the war. North of the stream were many Montagnard villages containing a number of long houses. I could see people moving around at one of them. In this area the timber was dense and untouched by defoliants. Here and there, on some of the nearby steep slopes were signs of clearings that must have been 3-4,000 acres in size. These may have resulted from escaped Montagnard fires. The stream below these cleared and/or burned areas had huge deposits of sand, evidence the clearing, burning and erosion had been excessive.

Passing east of Ban Me Thout we flew over Qui Nhon which is on the coast of the South China Sea. Through an opening in the clouds I could see the port facilities and a large air field. From this point on up the coast there were generally heavy clouds. Beautiful, huge cumulus clouds towered about us and extended east over the Sea. We navigated through these great white valleys in the sky. Through a break in the clouds north of Qui Nhon I could see vast areas of white sand dunes, some of which have drifted across small fields. I will see these later since we have been requested to start an afforestation project to help stabilize some of this sand dune county.

At one time while we were cruising through the valleys in the sky I saw the pilot shake his head as two F-4 jets came boring through the clouds in a steep climb. I am grateful they had radar to guide them around us. Our pilot probably was remembering all the sophisticated navigation equipment he had when he flew for TWA and the King.

Near a place called Tam Ky I saw many fresh bomb crater where B-52 bombers had dropped their strings of 750 pound bombs. It is difficult to visualize but each of these planes carry just over 100 of these bombs. They create large openings in the timber. I heard over the radio tonight a battle is taking place near Tam Ky.

North of Chu Lai I could see stretches of beautiful white sand beaches. The only occupants of the beaches were armored personnel carriers of the Marines. There has been heavy fighting in this area. Near Hoi An and at the mouth of the Song (River) Cai were hundreds of fishing junks. From the appearance of the magnitude of this heavy use I can fully appreciate what a monumental task the US Navy and Coast Guard have (referred to on the Sea as "Operation Market-Time" contrasted to the Saigon and Mekong Rivers where it is called "Operation Game Warden"). Their mission in each area is to stop the infiltration of VC supplies and troops. In this sector of the South China Sea the US Navy is stopping and searching 1,000 boats each day and the SVN Navy is searching around 700.

At 0930 hours we dropped through at least 1,000 feet of dense clouds and came out just over the Marine Airbase at Danang. Near the end of one runway is a large mountain. The airbase is a sprawling affair crowded with buildings, a few planes and helicopters. Most of the planes are over the DMZ and involved in battles with North Vietnamese Regulars. The area around Danang was beautiful. I could see islands and a long wooded point, containing some very beautiful beaches.

I was picked up at the airbase and driven to OCO headquarters. As I noted OCO is the civilian coordinating group operating the Pacification Program in the field. It is composed of USAID, Embassy, Army JUSPAO and two or three other civilian agencies. OCO headquarters is in a four story concrete and rock building. It is completely surrounded by a heavy wire mesh fence. I noticed on their roster of personnel that they have more guards than any other class of personnel. I learned later the home of each American in Danang has a 24-hour guard on duty. Outside the OCO office I saw one of the guards pick up an injured sparrow and put it in a safe shaded place. It seemed odd to watch this display of compassion from a heavily armed guard.

At Danang I began what I am certain will be one of the frustrating parts of my job. The man who had requested my assistance shook my hand and informed me as he went out the door that he was going on visitation to Manila to be with his family. That has been the story since I arrived in Vietnam . . . they are either going on visitation, home leave or R&R. At least he made arrangements for me to get to my destination the next day. I was to use his house and bed that night and then some one would take me down country to the next stop.

I spent the remainder of the day in Danang reading my file on the Mighty Mite sawmill and assisting one of the irrigation engineers edit a paper on irrigation. Not too productive a day I would say.

Danang is an interesting old French colonial-style city. It was a peaceful fishing village and had limited port facilities before we arrived. This is the first place the Marines landed when we became involved in the ground war in Vietnam. Danang has a fine harbor but there is a bar only 14 feet below the surface at low tide which limits shipping traffic. The harbor will eventually be dredged, when one of the dredges can be spared. The houses are of French architecture and have red tile roofs. The French Tri-color flies over the large French Embassy which is facing one of the waterways. Most of the side streets are loose sand and the main streets consist of chunks of asphalt between chuck holes.

The drivers of Danang must have received their training in Saigon. They drive as fast as the crowded roads permit and with horns blowing. I saw where an Army Jeep had run over a man on a bicycle near the Airport. From the looks of that bicycle I doubt the man survived.

If you wonder how a place like Danang which is protected by thousands of Marines, Army and Air Force can be attacked by the VC with such ease, I will tell you. To the north of Danang are high forested mountains, to the west are low forested hills and to the southwest and south there are rice fields, winding streams and waterways. The last attack on the huge air base came from the vicinity of the waterways and the southwest. The Marines have patrols out day and night, but it is a big country and again you are unable to tell the friendlies from the enemy. At night the Marines use dogs on their patrols and helicopter light ships. In spite of all this the enemy frequently slips in and causes heavy damage and casualties.

Danang is about like Kanab, Utah as far as lights are concerned. If you will remember that leaves much to be desired. The electrical system is still operated by a French company. It is terribly overloaded and without notice the power is cut off in one part of the city to reduce the load. That happened while I was there. Being on the inside of a strange house, which is as dark as the inside of a cow and in hostile country, is an interesting experience. But I could always see the shadow of someone going by outside my window. I hoped it was my guard. The Vietnamese guards are not armed and they are said to have a quick escape route lined out for themselves in case of trouble. Most of the civilians have an arsenal of their own. One of the men I visited had a submachine gun, an automatic rifle and several pistols. Just about like home. I hope you have not had to use yours on anyone since I left. When I come home I will be sure to whistle and sing out so you will know who it is!

The country and Danang reminded me of what a crazy "Alice in Wonderland" type of place this really is. You do not know who or what will pop into or out of a rabbit hole. During the black-outs the flares could be seen continuously over the marine and Air Force bases. I ate my meals at the Navy Officers mess. The food was excellent, inexpensive and clean. Fresh milk and vegetables are flown in daily from California. Sooner or later I will branch out to Vietnamese foods, but so far I am enjoying good health and not tempting fate.

I survived the night. Tuesday morning I was driven to the airport. The driver, like everyone else I have seen or ridden with, needed a good course in driver's training. Most of their vehicles are 4-wheel drive International Scouts. These are the right the Forest Service tested several years ago and rejected. Naturally that is what all civilian agencies in Vietnam purchased and they are literally falling apart. Here we have very poor maintenance facilities, extremely poor drivers and awful roads. For security purposes each vehicle is padded with sand bags which are placed on the front floor, under the seats and in the back. This added weight along with all the other problems is deadly on springs and shocks.

I waited about an hour at the Danang Air Force base for an Air America C-47 Gooney Bird transport plane to take me to my destination. While there I saw F-4 Phantom fighter jets and propeller driven dive bombers take off in group after group. Some were headed north to the Hanoi area and others were sent in support of ground troops fighting near Quang Tri and the DMZ. The faster planes made the round trip in about 30 minutes. Continental Airlines Golden Jets, Pan American cargo and passenger jets, and Flying Tiger cargo jets arrived and departed while I waited. I understand there is an average of a plane every 30 seconds either arriving or departing from this airport. Huge four engine Constellation radar picket ships were also seen departing. These are on station to guide our planes, identify crash sites and detect enemy sneak attacks from the North. One or more of these large planes are in the air around the clock.

The passengers and cargo I rode with to my next stop, Quang Ngai, were an odd assortment. One was 21 year old Wilson an automatic weapons specialist from Woodward, Iowa. He had been on a tour of Special Forces camps up north and provided them with advice on placement and maintenance of automatic weapons. A 35mm camera hung from his neck, a shot gun slung over his shoulder and a movie camera and a detective story were in his hands. He is interested in going to forestry school at Ames when he gets out of the Army this Fall. His next destination was Chu Lai. I heard over the radio tonight that Chu Lai is under attack.

One of the other passengers was a Vietnamese by the name of Tran Khiem, a CBS TV cameraman. He had just come from filming a battle near Quang Tri and land clearing being done in the 6km wide DMZ. He was on his way to Quang Ngai to film a special mode of transportation. In that town they have baskets fastened to either side of the rear wheel of motor scooters. These are used to transport people and just about anything else. New York had sent him a message that morning to go there and get some shots as soon as possible. Perhaps you will see his work on CBS news.

Two of the other passengers were Air Force Sergeants on their way to Tam Ky. One of them had just purchased a Canon zoom lens movie camera at the PX. He neglected to purchase film and batteries to drive the film mechanism. These young soldiers purchase high quality cameras at the PX and find they do not know which end of the camera to look into. The other passenger was a press photographer. He wheeled up to the airplane in a Scout. When he backed up to unload his gear he just about ran into the airplane. He unloaded two suitcases, several cardboard boxes, two tennis rackets, a brief case, attache case and a camera bag. He had a long-nosed camera and six rolls of extra film taped to the strap along with a supplemental lens. He was the type you immediately dislike. He charged up to the pilot and said "when do we go?" The pilot replied, "when I am good and ready". Mr. Long-nosed Camera announced he was going for lunch and to hold things for him.

In addition to the pilot we were accompanied by an American civilian who was the cargo checker. To escape the rain we waited for our departure under the wide wing of the plane. A crew arrived and unloaded the cargo of concertina 4-barbed wire. Then they loaded bulgur wheat, yellow cornmeal, sugar, dextrose in solution and plague vaccine. This was the beginning of many long waits at airports. This is a war of waiting . . . for the VC to attack, for transportation, for mail, for a signature and for supplies that may or not appear. After awhile, as expected boredom, sets in.

As soon as we were loaded the pilot put on a devilish grin, taxied and took off. Of course Mr. Long-nosed Camera was stranded. Presumably he could wait a little longer, on a full stomach, in Danang and hope to catch up with his gear somewhere in Indo-China. (Break in story.)

It is now 2200 hours on a lonely Saturday night. Writing these letters helps pass the time. I will take a break, take a look from the roof and go to the BOQ for a Coke. It is unbearably hot tonight. The air is stifling and it is trying unsuccessfully to rain.

From the roof I watched flares being shot into the air and dropped from planes near Cholon. That area is less than two miles from here and is attacked every night. We have many troops stationed there, as well as the large PX and commissary. While watching I could see tracer bullets fired from the ground at a helicopter overhead and then a steady return fire of tracers being shot to the ground. Other than in that area the traffic seemed to be flowing smoothly across the Bien Hoa Bridge, one of the few major bridges still standing.

I brought my Coke back to the hotel just as a nearby explosion rocked the city. The police on Tu Do Street started running in the direction of the sound and sirens were heard converging on the area. I went to the roof, but could see nothing except the flares being dropped over Cholon. Returning to my room I heard the American Forces Radio report on the explosion, but gave no indication of where or what was hit. Perhaps the VC were pepping up things for the next round of elections. I will sleep on it.

(continued)

Good Morning! It is bright and sunny and after a hot humid night it feels quite comfortable. After all the excitement the explosion turned out to be a loud sonic boom. Big deal, it was probably one of the super sonic courier jets that commute daily between Saigon and Washington. At breakfast most of the officers were laughing it off. Most of them thought one of the other BOQ's had been destroyed. After breakfast I took as short walk and now back to my story of last week's trip.

Flying south of Danang I could see were Casurina (Australian-pine) had been herbicided to reduce ambush cover. This tree species is highly sensitive to 2,4,5-T and was dead. With the loss of these trees the sand was starting to move and form new dunes. Vietnamese foresters had made good progress in stabilizing sand dunes with Casurina but much of it was destroyed during the war with the French and now we are finishing it off in this war. While it is illegal to cut these trees, dead or alive, local people cut it for firewood. The theory goes . . . all is fair in war and the battle for human survival.

After leaving the safe area over the Danang Marine base we headed out to sea. The flight to our next stop was made at about a mile offshore. Much of the coast is in the hands of the VC and is considered unfriendly. We saw many beautiful offshore islands and hundreds of small fishing boats. Then clouds closed in and it began to rain. The C-47 leaked and we moved from seat to seat to keep dry. Nylon bucket seats are uncomfortable so it was a relief to move around. The cargo was lashed down with large nylon straps and fortunately was not under any of leaky sections of the plane.

The plane came inland and amazingly the pilot came down through dense cloud cover and was over the airfield at Tam Ky. In landings and approaches in this country all of the pilots climb out in a spiral over the airfield and land in the same fashion, with sharp banks and turns to keep over the so-called safe area. There is no long approach in or long climb out of any airport in South Vietnam. As a reminder of the importance of following these rules we saw the remains of a C-47 beside the runway. It had been shot down two months earlier. Our pilot told me everyone got out safely but the plane was a complete loss.

At the airport the Vietnamese Chief of Police came running out with the troops. But the OCO/USAID man that was to meet the plane and receive the shipment did not show. The pilot was irritated and through an interpreter told the police to take the cargo. The Chief would not accept the responsibility. So the good old CIA Air America pilot started throwing sacks of bulgur wheat, corn meal and sugar out the cargo door. It landed in pools of water, broke open and spilled the contents on the wet taxi strip. As I watched these broken bags which were stenciled to show they were "Donated By the People of the United States of America" I decided the people of USA and me as a taxpayer did not go for this foolishness so I told the pilot in no uncertain words that if he threw off one more sack I would see that the people of the USA and the powers-that-be in Saigon heard about this waste of resources. That stopped him for the moment, but I could hear him muttering about the interfering civilians and the so and so Vietnamese. At this time the Chief agreed to bring his truck and put the remainder of the load under cover. He also picked up the broken sacks. I suppose these ended up being distributed to the police and the bulgur wheat fed to their hogs. The Vietnamese refuse to eat bulgur wheat but it is being distributed until the current supply is exhausted. I took some photos of this episode. I was surprised the CBS TV cameraman did not record the action. Walter Cronkite could have made a good story out of this episode.

The American checker on the plane showed real concern at the pilot's action and told me that the pilots do not respect these valuable supplies. Fortunately an Air Force truck arrived and the checker was able to get the Sergeant driver to sign for and accept the plague vaccine (marked "Urgent") and the dextrose before it was also thrown on the ground. I am told that what I witnessed at Tam Ky happens all too often. The pilots just will not take the time to see that supplies are delivered to the proper person. They deliver the cargo and get out as quickly as possible. They do have frustrations and problems, like the weather, long boring flights, overloaded planes and schedules, no-show passengers, no one to receive the supplies and the VC ready to shoot them down.

I suppose our shipment of forestry supplies that were shipped to Vietnam in January could have been lost in the same manner. They have never showed up but USAID records show someone signed for part of them. I shipped 200 axes to Quang Ngai last Wednesday. Records show they were delivered on 16 April but when I left Quang Ngai on 20 April we could find no trace of them. Perhaps they will show up someday, somewhere. I am making a report on these delivery problems and perhaps the handling of valuable supplies and receipt of shipments can be improved.

Now to continue on my travels. We roared down the runway, banked sharply, climbed into the clouds and went out to sea again. We dropped down to about 500 feet when we were a mile out from the coast and continued south. The clouds lifted and we flew into the Quang Ngai airport and landed on a paved runway which was flooded with about six inches of water. A recent cloudburst appeared to have flooded the entire countryside.

A civilian came running out to the plane and called out my name. He was Woody Stemple a missionary from the Christian & Missionary Alliance. He had a list of items to pick up. These belonged to Mr. Long-nose Camera. He had sent a radio message to have someone retrieve his gear. I felt like throwing them into the nearby river but I did not. Like a good Christian I helped unload the gear and load it into Stemple's VW bus. About that time a USAID warehouse man showed up to receive the balance of the cargo. As I departed the pilot gave me a "cheerful" look and gnashed his teeth.

I rode into Quang Ngai with Stemple since he was going to the Army base where I was to meet Captain Walter Carlton. On the way into this provincial capitol we drove in and out of water filled chuck holes. Sand bagged gun emplacements, tanks and various kinds of military equipment stood about in varying depths of water. Everything and everybody appeared to be half drowned. We passed a huge unfinished Catholic Church. Stemple told me this was started under President Diem's regime and never finished. The walls stood about a hundred feet high. There was no roof and the interior was filled with scaffolding. It was also filled and surrounded with pathetic refugee shelters.

At the perimeter entrance of the military post we were met by armed guards. Within the compound was a large ARVN (Army of Vietnam, rhymes with Marvin) base and a 200-man US Army advisory group. The entrance was over a small bridge and facing the approach was a fully manned tank, with the crew ready to pop in and fire at intruders.

I found Captain Carlton's quarters which he shared with an Air Force observation pilot by the name of Williams. They were listening to a tape and a ballad about Billy Sol Estes (person accused of swindling the USDA). Carlton, an infantry officer, was from west Texas and a distant relative of Billy Sol. He said his mother would not appreciate the ballad. Their room looked like typical bachelor's quarters. It was full of odds and ends of souvenirs, *Playboy* pinups, automatic rifles, carbines, pistols, knives, canteens, etc. The beds had mosquito netting because this area is bad malaria country.

We held a meeting in the local forestry office and discussed the proposed sawmill project. The local Cooperative Association is enthusiastic about the prospects, especially since they see this as a way to help reduce their VN$8 million debt. The Coop man spoke no English and talked to us through an interpreter. He impressed me as being a good business man. He could play the part of a Japanese Army officer. His father could have been in the Japanese forces which occupied this country during World War II.

We went to a lumber yard to check prices. The owner told us

all of his timbers were shipped in by boat. He had two sawyers who were cutting squared timbers into 1-inch boards. They were doing it the hard way and using a long whip saw. The owner said he could pay VN$18,000 per cubic meter for lumber delivered to his yard. This calculated out at US$152.54 for 227 board feet. That fantastic price illustrates a serious problem and an opportunity for us to help reduce lumber costs.

At quitting time I accompanied Carlton to the Officers' Club We had a Schlitz beer while listening to the officers describe what was wrong with the war and how it should be run. The place looked like one of the bars you see in war movies. Some of the men were FAC observers. They cover the area from two miles outside of Quang Ngai across the mountains to the Laos border. Their job is to try and spot VC and trap them into shooting at the small observation planes, then they call in the jets to finish them off. One of them described long foot bridges he had seen that day suspended on vines across deep ravines and the beautiful scenery of the highlands. Other men were advisors to ARVN and ROK (Republic of Korean) troops. ARVN troops are generally very ineffective. The ROK's on the other hand are highly effective and feared by the VC and the North Vietnam regulars. These officers have many frustrations and constantly live with danger.

After a good meal at the BOQ I went to the OCO guest house for the evening. And there was Mr. Long-nose Camera in the bunk next to mine. He was muttering about his new assignment which was working with refugees. In just a few hours he had succeeded in making everyone in Quang Ngai angry. He had left a similar job in Quang Tri when frequent VC attacks made his efforts ineffective. I just bet OCO decided to transfer him when they could not contend with both him and the VC.

The mosquitoes were hungry but I had a good supply of "Off" and avoided all but a few bites. We had no mosquito netting. The night was noisy, first it was crickets and frogs, then the artillery harassment fire started. ARVN troops spent most of the night shooting flares into the sky and periodically firing 105 howitzers, mortars and automatic weapons. But in spite of the noise it was cool and I was able to get some sleep.

Following breakfast Carlton and I were taken by Air America helicopter (Delta Model Huey), along with the Coop man, a local forestry official and a interpreter to Tra Bong (Cha Bong). Two OCO men went along on another assignment. En route we saw a flight of jets firing rockets into the foothills just outside of Quang Ngai. I could see fields which armored equipment had destroyed. The secure area at Quang Ngai ended at the edge of town. The foothills were covered with 100's of craters created by B-52 bombs. The area looked like craters on the moon. Some were full of water which could be used for fish ponds, livestock water holes and mosquito breeding. A 750 pound bomb dropped on a small rice paddy does not leave much for the farmer. Some of the rice fields had been sprayed with herbicides to deprive the VC of a food supply. I recall one of the spotter pilots commented on the fact the fields were harvested but they never could see anyone working them during the day. The VC did it at night. The poor farmers were so frightened they could do nothing, except hate everybody. You can see what a bind the farmer finds himself in, he is caught in a squeeze between ARVN troops, our modern war machines and the VC. What a tragedy, since it is the farmer we are all trying to serve and save!

When we had climbed to 5,000 feet for safety reasons we leveled off and headed for the Special Forces camp at Tra Bong, some 20 miles to the northwest. This Air America helicopter is the same type used by the Army and Marines for gunships and medical evacuation. We carried no mounted machine guns but each of the two pilots had an automatic rifle beside their seat. These jet helicopters are terribly noisy and since we had no helmets or ear protectors my ears were ringing.

Arriving over Tra Bong we could see the village and farms extending up and down the narrow valley and on the south side of the muddy Song (river) Tra Bong. Along the south edge of the village was a short dirt airstrip. Next to the fortified Green Beret fort and team house was a small dirt helipad. In addition to the resident Vietnamese there are 6,000 lowland Vietnamese and 4,000 Montagnard refugees in this narrow valley. We spiraled around to lose altitude while keeping away from the adjacent mountains and landed on the helipad within the outer barbed wire perimeter of the heavily armed Special Forces camp. This camp serves the dual purpose of providing protection for a major refugee settlement and blocking a major route used by VC and NVA to infiltrate from the mountains to coastal areas. Just saying the camp was armed is a gross understatement. It had miles and miles of barbed wire, some was fastened to steel fence posts and other in great coils of concertina barbed wire which completely surrounded the knoll of what was formerly an old French Foreign Legion fort. Attached to the coils of wire were many beer cans with rocks in them to serve as rattles that sound a warning if touched by infiltrators. Every few feet I could see a wire leading from reinforced, staggered bunker positions and trenches to claymore mines which were set to have an outgoing, overlapping field of fire toward the perimeter. I also learned the area was heavily mined and thick with "Bouncing Betties" and booby traps. The bunkers were all neatly sandbagged. The underground quarters was a cement building, set into the side of a flat-crested knoll, with a roof covered by several layers of sandbags. The top of the hill was occupied by a well padded tower which supported a .50 caliber heavy machine gun. Around the tower were mortar pits, machine guns, rockets, and artillery pieces. Six American Special Forces were on duty and two others were sleeping. I did not meet Captain James E. Callahan, the Commanding Officer, since he and two of his men were deep in the mountains on a listening-post patrol mission, living off the land and gathering intelligence information.

As soon as we arrived Carlton and I walked down through the village to the District Chief's headquarters. District Chief Captain Nguyen Huu Tung was a short well-built man in an ARVN officer's uniform. He carried a short bamboo swagger stick, but no weapon. Through an interpreter he welcomed me and presented me with a beautifully decorated Montagnard crossbow. I thanked him for receiving us and acknowledged the gift in a few words of Vietnamese which he understood. We explained the purpose of our visit and told him we looked forward to helping his people establish a logging and sawmill operation. He escorted us to a corner of headquarters and there surrounded by concertina barbed wire were three large wooden crates containing the Mighty Mite sawmill. These had been flown in a few days earlier by an Army C-118 Caribou transport plane. The Caribou is an old World War II glider with an engine and can operate on a very short runway. One of the crates had been dropped during transit and was broken open on one side. Nothing appeared to be damaged. I told him we would return when the manufacturer arrived and install it. I noticed next door to his fort was a Buddhist temple and a graveyard with a number of freshly dug graves.

As I looked at the nearby mountains I could see vast areas of slash and burn agricultural clearings and sparse stands of timber. This was not what I expected to find. The word brought back by the Dwyer Mission was that the site selected was in an area of unlimited timber resources. The country to the north and west of Tra Bong rose steeply into a wilderness of triple canopied forest, but that was economically inaccessible. I told the District Chief I would like to go into the forest and identify the species of trees and verify the quality. He said it was not safe for either him or me to go into the forest. Since he had survived six months in office as District Chief I followed his advice. I asked to visit the nearby Montagnard village and see one of the long houses. He tapped his swagger stick on the desk and armed troops appeared from nowhere and formed a circle around us and off we went. On a narrow log that served as a bridge, we crossed a small, clear, fast-moving stream where women were washing clothes. On the way to the village we passed a number of Vietnamese huts where rice was spread out to dry in large flat woven baskets. One woman was winnowing rice. At another hut a Vietnamese man used a small handmade axe to cut the frame for a bed. He was doing excellent work with high quality hardwood.

With guards leading the way and covering our rear we walked on a narrow path that curved through the rice paddies. Apparently a message had been sent ahead and we were met by the Montagnard village Chief who saluted us and led the way to one of the long houses.

Sitting in the narrow doorway of one of the communal long house were children and women. Both groups wore a minimum of clothes. Men that came to the doorway for a look at the visitors wore various odds and ends of dirty cloth shirts, some castoff army clothes and loin cloths. They were interesting looking people. This is a race separate from the Vietnamese and are thought to be a mixture of races from India, Indonesia and South China. They have no evidence of Oriental facial characteristics. They showed us some of their musical instruments, a flute, one stringed instrument and a variety of working crossbows. They probably wanted to sell them to me. I will look into that on a later trip.

The long houses provide shelter for a number of family groups. The structure was supported about four feet off the ground on poles. The roof was made of palm leaves. The houses are about 16 feet wide, 12 feet tall and 75 or more feet in length. The elevation is practical since it keeps deadly snakes, like cobras and mambas and varmints out. Debris can be swept or kicked through holes or gaps in the floor to the sway backed pigs rooting around in the accumulation of stinking refuse.

At this point we were much closer to the forest and I still wanted to get into it, but the District Chief and the Montagnard Chief vetoed the idea. I had to be content to look up into the timbered mountains. The timber was sparse and apparently had been subject to shifting cultivations for many years. To me, the remaining timber looked marginal, poor at the best, but the equipment was there and the people had been promised an industry. Both the Vietnamese and the Montagnards have been given many unfulfilled promises over the years, by the French, by the Japanese and Vietnamese officials. I decided if they were ready we would help them.

As I looked at the beautiful, high, rugged mountain peaks, river, crystal clear streams and waterfalls with naked Montagnard maidens splashing in the pools, I was impressed by the fact this could be made into a tourist paradise. But then the tourists would soon ruin it. It is so sad these primitive people do not have peace in their lives . . . as Sherman said, "war is hell!"

We returned to the District Chief's meeting hall. The place was crowded and we explained the proposed sawmill and logging program and what we would try to do to help the people. One or two in the audience asked some sarcastic questions which indicated they may belong the other (VC?) side. Most of the people are very interested in the project. They said they were suspicious because they had so many promises broken before. In a practical approach they wanted to know what share of the Cooperative's profit they would receive. The Coop man was very evasive and it was apparent to me that he wanted 90% of the profits for marketing the lumber and these people would receive only 10% for their back breaking labor. The division of income must be worked out through negotiations. These people are very poor and have no equipment, no oxen or water buffalo or elephants to move the logs from the forest to the mill site, they have only their strong backs. The US Army will transport the lumber from Tra Bong to Quang Ngai by truck or Caribou at no cost to the Cooperative.

I looked around at people at the meeting. Some Vietnamese sat in chairs or squatted on the floor. The Montagnards stood in a group and listened very intently. These interesting people have strong features, but if I met them on a trail at night I suspect I would jump out of my skin. They do not look especially clean, I guess the maidens are the only ones who bath. The men smoked some strong smelling stuff in small pipes fashioned out of wood or brass cartridge cases. One man wore ear rings made of curled aluminum, probably obtained from the wrecked airplane I saw at the nearby landing strip. That 350 foot long airstrip is used by small planes and the Caribou.

The District Chief concluded the meeting with a few remarks. We walked to the sawmill site to examine two logs that had been packed in while we were at the meeting. One log was 10 feet long and had been squared to about 12 inches. The other round log was six feet long and 19 inches in diameter. It required 20 men to move each log from an area being cleared for crops, down a mountainside as steep as a goat's face, across the river and up a short slope to the mill site. The logs were so heavy they had to be supported in the water. Much as I tried I could not make either log move a fraction of an inch. We were asked if the sawmill could be placed at the foot of the mountain. I told them if they could provide day and night security we would place the mill wherever it was most convenient. They just shook their heads. The cleared area was said to have many downed trees that could be moved to the mill and more are to be felled.

The District Chief and his guards escorted us to the Green Berets' camp and thanked us for coming. He probably never expects to see me again. At the Special Forces camp Lt. Francoe was in charge. He was a handsome young Italian-looking officer with short black hair and dark brown eyes. He wore no shirt and dog tags hung from his neck. He welcomed and fed us. These fellows have good quarters and they eat very well. While we were eating two troop carriers of Army Rangers landed. They unloaded their gear and joined us for lunch. They were on a routine patrol of the area between Tra Bong and Quang Ngai. It was nice to have them around. Sgt. Stevens, a huge Negro had been assigned as our escort while we were in the area. I had noted him in the background everywhere we had been that day. It was comforting to have him nearby. He had a radio and an automatic weapon. I noticed some different looking men serving as guards on the perimeter of the Special Forces camp. I learned these were Chinese Nungs who come from the border country between China and North Vietnam. These are some of the finest, fiercest fighters in the country. These paid mercenaries are loyal to the Green Berets. They did not respond to my smile or greeting. They looked mean and had a stare that would curl your hair! The Green Berets swear by them and say these are the kind of guys to have around when the going gets tough.

We met our helicopter at the scheduled time, climbed to the prescribed altitude over the Special Forces camp and headed across the mountains to another Special Forces unit at Ba To. This was a similar camp situated across a VC infiltration route. We just touched down and picked up some Vietnamese. Along with more people than we could possibly transport was a woman with a small baby. She and everyone else appeared to be frantic to get aboard. One of our crewmen held the others back and helped her aboard. You could see the relief flowing over her face as she settled down in a seat near me. She was probably the wife of one of the ARVN officers. Soldiers' families travel with them and live in the filth under hazardous conditions.

Back in Quang Ngai I head the powers to be were ready to turn Mr. Long-nose Camera over to the VC. I spent Wednesday night at Quang Ngai and went to Danang in an Air America Twin-Beech the next morning. The balance of my trip involved waiting . . . waiting to get to Danang and then waiting to return to Saigon. At Danang I waited with a young Vietnamese who had his right hand missing. He wore a black arm band to signify other losses. He was traveling with his wife and three small children. Vietnamese who have a security clearance can travel on Air America on a space-available basis providing they are personally approved by a US civilian. The Danang airport was busy with fighter bombers taking off every few minutes. I learned these planes were attacking a power plant near Haiphong, North Vietnam and supporting our troops who were involved in heavy fighting around Quang Tri and Hue.

Finally we flew south in an Air America C-47. En route I saw beauty spots where the mountains meet the sea. Between rocky outcrops I could see white sand beaches, all very beautiful, very insecure and full of VC. I saw the newly created harbor and port facilities at Qui Nhon, Tuy Hoa and Nha Trang. Each has a huge airbase. Near one base we passed low enough for me to see GI's swimming in the surf, under the watchful eyes of life guards sitting on towers behind machine guns. At one stop a large man, with a wide brim hat got on. He carried a guitar and a set of bar bells. This added to the "Alice in Wonderland" atmosphere.

I am glad our friend Gene Jensen warned me about eating oysters. I see many people collecting them near sewer outfalls. Hepatitis is so bad in some places, to enter or leave, we have to produce our immunization cards to prove we have current gamma globulin shots.

Arriving back in Saigon at 1830 hours our landing was rough and delayed because we came in behind a huge Air Force C-141 Star Master jet cargo plane. It is larger than the B-52 bomber.

I was pleased to receive your letters. Sounds like you have your hands full with the activities of our teen aged daughters. I regret you bear these burdens alone. I predicted to myself that my mother would be her own sweet self. She can really be mean, I should know, if anyone does. If she does not write to you or me, just ignore her. I will go by and see her when I return. All I really care to do is visit Dad's grave and the ranch he so loved and return to you.

I am glad we have the Volkswagon Beetle. You better have the fenders straightened and repainted before Cindy starts to drive. She can have Gene Jensen give her some additional lessons. I am glad the Chevy station wagon is being spared.

While I have written this very long letter it was rained about an inch. This will cool off things briefly. Now I am off to the Rex BOQ to barbecue a large steak.

Chapter XV

On the Move

30 April 1967
A most significant event involved my living quarters. You might call this a "moving" letter. This is an example of why it requires 10 times as much effort to accomplish anything in this country. Nothing is simple.

Wednesday one of the women from the USAID office asked if I wanted to move into her air conditioned hotel room. As she is a sweet young thing, on the surface it sounded like a splendid idea, besides the room was air conditioned. Before I get in trouble and you begin to think I have taken up diversions other than talking with and photographing the natives, she had moved out of her room and into an apartment vacated by one of the men who had gone home for two months home leave (granted when someone signs on for another tour of duty). He wanted someone in there to look after his possessions. She figured she would have her own apartment by the time he returned.

Her move set up a chain reaction. I moved into her room on the 5th floor of the Astor Hotel, just one block from the Hotel Oscar. Another man moved into my room at the Oscar. She was to report this to the Embassy housing office. We all thought we had improved our living quarters until Mrs. Yen, a pleasant lady at the Embassy, hit the roof. She said we could not do this without her approval. She needed the room for an Embassy VIP. I went on the offensive. I told her I was a VIP from the USDA, Washington, DC and had asked for no favors and been given none. I reminded her that she had given special treatment to USAID and Embassy people and those of us from USDA received second class housing. She denied this and said she would move me in with someone at the Astor until the VIP was gone. I told her I did not want to move in with anyone and besides the rooms at the Astor were too small. She studied her charts while apologizing for asking me to move. She said she had a nice room at the Park Hotel and would like for me to look at it. I agreed because the Park is just four blocks from my office and half a block from the Palace where Premier Ky meets all the dignitaries. It is in a much better neighborhood.

I had my driver take me to the Park. It was a very nice hotel and the room was twice as large as any of my previous ones. And it was air conditioned. It had three large closets and space to store my footlocker, suitcases, case of beer, case of Coke, empty boxes and all the gifts I am accumulating to bring home. I called Mrs. Yen and agreed to take it providing she did not move anyone in with me. I wanted this assurance since it had twin beds. She agreed and was probably glad to be rid of me. I also negotiated the right to select a person of my choice to move in when I moved out to permanent quarters. I figured I could maneuver one of my team into this room.

When I moved to the Astor Hotel my night floor man at the Oscar Hotel helped me move. He is a friendly, small, thin man by the name of Mr. Ty. He has helped me with my language pronunciation. He lost his wife last year and supports seven children on his hotel earnings. I had given him a few tips when I learned he was paid VN$50 (42 cents) for working the floor and taking care of 11 rooms from 1930 to 0730 hours. After he carried my heavy gear to the Astor I gave him VN$400 and he was overjoyed. Wages are a great problem. It costs the average family VN$20,000 each month for the bare essentials. You can see they do not live very well.

I moved again today to the Park Hotel. I arranged for a car and a young man from the Oscar Hotel to help me. He is 18 and goes by the name of Pham Van Kien. He is a high school student from a very poor family. I met him just last Saturday night when I returned to the Oscar to play poker. I saw him studying English and he asked if I would help him pronounce a few words. His language book contained a simple story about a dog. He was having trouble so I helped him rather than playing cards.

Kien helped me for about four hours. He packed and carried my gear to and from the car and helped me put things away. I gave him VN$400 for his work and he was very appreciative. He gave me his photograph and asked for one of mine. His is enclosed for you to see a Vietnamese teen-ager. He tries to name everything around him in English and then tells me the Vietnamese name. When he departed he asked if he could return to my hotel for an hour each Saturday and have me help him with his English. People believe they can improve themselves and find a better job if they speak English. I agreed, provided I am in town. I also agreed to help him when I return to the Oscar to visit friends. He borrowed my government pen to write down my name and accidentally put it in his pocket. I did not notice it was missing. Within an hour Kien came running back from the Oscar all covered with sweat. He discovered my pen in his pocket and returned it. Some of these people would steal you blind. It is refreshing to meet an honest one.

I enjoy the Hotel Park. The address is 35A, Nguyen Trung Truc (When Chung Chuck). My room is 12 by 20 feet. It has a good view to the northeast over the city and the central market. It has drapes, Venetian blinds and a pleasant balcony. The roof is a good place to sit in the sun and exercise. The air conditioner can cool a can of Coke in 10 minutes. Ice is available at the hotel, but I do not consider any ice safe. The air conditioner is a blessing because it is becoming hotter each day as we approach the monsoon season. I have slept well with an air conditioner.

Yesterday the Air Force flew me to their new F-105 jet fighter air base at Tuy Hoa (Tuey Wha). They asked for my technical assistance in solving a serious erosion problem. I flew 250 miles northeast to the coast in one of their large deluxe VIP transports. With me were Mr. Le Viet Du, a Vietnamese forester (Yale graduate), three civilians, the Major who had invited me and a handful of Colonels. The flight was over some beautiful mountains and was as smooth as glass. The flight required 45 minutes from Tan Son Nhut Airport. It was easy to see the problem. Engineers and construction people in their haste to construct the base had bulldozed everything level. In doing so they removed native bunch grass and brush. The coastal areas have very high winds which caused the loose white sand to come unraveled. Conditions were miserable for everyone. The jets had major maintenance problems, sand had drifted against and into barracks and warehouse areas, bomb storage areas were nearly filled with fine white sand. Bombs had to be dug out of the sand before they could be hauled to the planes. It looked bleak!

Earlier to remedy the situation the Air Force had brought in engineers from Hawaii to solve the problem. Their plan involved shipping in 100 miles of snow fence, tons of grass seed from the States and Australia, along with heavy reseeding equipment. I saw they had a real problem with huge stock piles of expensive seed. It had been stored in the open and the hot sun and rain had spoiled it. I cut into some of the sacks and found the seed hot from the sun and spoilage. It smelled fermented. The seeder was a 15 ton hydro-seeder which no one knew how to operate or could maneuver through the deep loose sand. I suggested they let the machine sit where it was. In regard to the grass seed they asked me how to plant it. I suggested they dig a trench six feet deep and bury it. That surprised the Major and he asked "will it grow?" I told them the seed was dead and would grow just about as well if they dumped it in the South China Sea. In most areas where the snow fence had been installed only about two or three inches of the top was visible. I was told the Air Force people were given a bad time when they sent a requisition to the Pentagon for 100 miles of snow fence! In Vietnam?!

While there I drew up a simple plan which prohibited dozers, vehicles and other heavy equipment from encroaching on the critical areas. I could see grass, vines and brush trying to sprout. I demonstrated how to cut section of various plants and vines and stick them in the sand. I advised Mr. Du to recommend that the Directorate of Forest Affairs supply them 60,000 Australian-pine trees. These seedlings are presently growing in a nearby forest nursery and can be planted in late September as wind breaks. The Commanding Officer of the Base and the Colonel in charge of such work in Saigon were pleased with my analysis and recommendations. The Colonel asked me to accompany him tomorrow to the huge Cam Ranh Bay Air Base and Port and come up with a solution to similar problems. I will be celebrating Vietnam's Labor Day at Cam Ranh Bay.

While at Tuy Hoa I was reminded there is a war going on. The base was bristling with guards. Watch towers were spaced all around the base behind a 12-foot high anti-personnel fence. During the seven hours we were on the base fully loaded F-105, called "Thuds", were landing and taking off in a steady stream. Each one took off "maxed", in Air Force jargon that means two 500 pound bombs, 24 rockets, two containers of napalm and fully loaded 20mm and .50 caliber machine guns. Even at a few feet above sea level the powerful planes required most of the 10,000 foot runway to takeoff. Ground crews worked feverishly to keep them loaded and flying. One battle was taking place just three miles away in the mountains. The VC are attempting to move into position to attack the air base. These planes were flying in support of Korean ground troops who were involved in the fighting. That place was like "Dante's Inferno" with jet's taking off and landing every few minutes. Bomb concussions added further insult to my poor ears.

The fighting continued as we departed and made a swing over the safety of the Sea. The coast and islands were unbelievably beautiful. That's all for now. I will write more following my trip to Cam Ranh Bay.

4 May 1967

On Monday, 1 May, I flew with the Air Force to Cam Ranh Bay. You will find it on the map 200 air miles northeast of Saigon and just below Nha Trang. We flew in a huge C-130 Hercules, a large 4-engine turbo cargo plane that could carry a semi-trailer. We carried 20 passengers, including the General in charge of the air cargo transport group, a crew of six and tons of mail. Cam Ranh Bay is located on a large, natural deep water bay that has become one of the Country's busiest ports. I could see 100's of ships of all sizes coming and going.

We came in over vast white sand dunes and landed at the Aerial Port Facility just in time for another General to dedicate a new passenger terminal. This facility handles 750 outbound passengers and 11,500 inbound ones each day. Many hospital carrier transports take off from here. All of the cargo ships are equipped to handle stretchers, but the long hauls are made by C-130 and huge C-141 4-engine jet Star Master. The latter carry 200 stretchers. Other cargo planes continuously haul cargo into and around the country.

The problem area I visited was at the air base. The peninsula that forms the ocean side of the bay is covered with small brush and sand dunes. Here again the engineers in their haste to construct this facility cleared and leveled the well stabilized dunes so completely that all of the brush and grass were destroyed. The problem was compounded by repeated passes of heavy equipment. I could see little or no vegetation. Considerably more area was cleared than needed for the facility and the sand dunes are actually galloping over the runways, filling bomb storage areas, creating terrific aircraft maintenance problems and causing serious morale problems for the troops. It was easy to identify the cause but solving the problems created will be as difficult an erosion control project as anyone can imagine. The sand is very fine, contains no organic material and appears to be sterile. In fact hundreds of tons of this material are hauled to Japan for use in their optical industry.

The wind and jet exhausts were seriously undermining the runways. Our inspection took us out between the parallel runways and we had to cover ears very tightly as the powerful jets turned on their afterburners and roared down the 10,000 foot runway. Snow fences had been installed, but were completely buried in a few days. The engineers welcomed my recommendations. The Commanding Officer agreed to provide the remaining live trees and brush around the perimeter as much protection from bulldozers as he is protecting the base from the VC. If this facility can be planted it will be the greatest success story of any reforestation project in the world. It is comparable to stabilizing the Sahara Desert!

I ate lunch in the officers' mess. The girls would have drooled over all the handsome young F-4 Phantom jet fighter pilots. There are 100's of pilots and ground crewmen stationed there. During the 12 hours I was at the base the planes took off every few minutes. These were fully loaded with bombs, rockets and a "pistol" The pistol was actually a Gatling-gun mounted directly under the fuselage. It spits out some 6,000 rounds of 20mm fire per minute. These powerful jets, loaded as they were, required about 8,500 feet of runway before they became airborne. Takeoff is a particular problem in view of the weight and high temperature. In the face of all of this massive fire power, sophisticated technology, death and destruction, the enemy troops continue to flow into the country. There was a fire fight in the mountains just across the bay from the air base. It lasted for the several hours I was there.

Our return flight was in a 2-engine C-123. The Provider is a smaller transport designed to take off in a short distance. We carried 20 passengers and a 4-man crew. The one hour flight to Saigon was uneventful. We were in dense clouds for most of the distance and landed in light rain.

Yesterday we had a localized tropical cloudburst just at quitting time. I was soaked as I ran to the bus, but my wash and wear outfit was not soiled. When I got off the bus there was no rain downtown. After dinner at the McCarthy BOQ I went to the Oscar Hotel to visit friends. I do not know a soul at my new hotel. After visiting with the old bunch I helped Kien with his English lesson until another downpour started. The rain drove in from the front and back of the building. There are no windows or solid walls and the hallways are open. Rain came in and was flowing into some of the rooms. Kien stopped his lessons and started mopping.

I borrowed an umbrella and took off for my hotel. The rain fell in solid sheets. I hailed a pedicab and bargained the operator down from 100 to 30 piasters and crawled in. When it rains these contraptions have top, front and side curtains. You feel trapped inside but I kept dry for the few minutes it took him to peddle me to my hotel. It was not raining a drop when I arrived. I have two problems at my new hotel. The air conditioner is either on or off and is about to freeze me. I now sleep in pajamas and under two blankets. At least it keeps my things dried out and beer cold.

I see many sweet and innocent Red Cross girls around Saigon and in my other ports of call. It brings back fond memories of one I found in Colorado two wars ago. They look and dress about the same. Skirts are perhaps shorter. They are busy at USO's and hospitals.

I had a meeting with Mr. Tan today. He had just returned from Australia. I briefed him on our progress and identified several areas where I need his support. He is cooperative and agreed to help me. He told me about an invitation he had received from Chief Cliff (written at my request) inviting him to the US this fall and to attend the Society of American Foresters October meeting in Montreal, Canada.

Mr. Tan invited me to accompany him to Dalat next week. Dalat is in the pine country of the Central Highlands. It may be found on the map due west of Cam Ranh Bay. He outlined plans to visit a forest nursery, a logging operation, sawmills, fire protection facilities, a hydroelectric project and participate in an evening hunt. He said the area has tigers, elephants and deer, with only an occasional VC. That should be an interesting trip. Mr. Tan's organization is also responsible for game management and enforcement of game laws. I am uncertain what that amounts to with a war in progress.

I was pleased to learn you had a good vacation in the Seneca area of West Virginia. Girls, please check with Gene Jensen on the VW's maintenance needs. It does require more than gasoline!

I am glad you received the box of gifts. The cloth I sent is not to wear! That is the Republic of Vietnam flag! Most

Vietnamese do not like it. Yellow stands for the yellow race and the three red bands for North (Tonkin), Central (Annam) and South (Cochinchina). Those were French administrative divisions the Vietnamese hated. The flag will probably be changed as soon as elections are held and a new government is in place.

Please greet all the neighbors. I am sending you a letter by Vietnamese air mail service to see how long it takes. You will find the stamps interesting.

Chapter XVI

Variety

I miss sharing your day-to-day experiences.

Mrs. Helen Black, my secretary on the Coconino, wrote about one of my former ranger's death. He was the alcoholic ranger at Beaver Creek. We were successful with his rehabilitation, but unfortunately he slipped back into old habits and died. I see many alcoholics in Vietnam. Booze is readily available and cheaper than Coke.

I enjoy receiving your news clippings. Coverage here leaves much to be desired. A friend gave me an April 23 issue of the *Washington Post*. I read it from cover to cover, including advertisements. Our news coverage is provided by two local Saigon English papers and the *Stars and Stripes*. Each office receives a supply of the well done *S&S*. However with the exception of Dick Tracy, it is censored for our benefit and morale.

Enough of my personal comments, now, shortly, I will start my report on the Dalat trip. That will take the remainder of the day. I worked at the office until 1600 hours today. Yesterday morning I briefed the forestry team. The forestry team has expanded! Martin Syverson, Forest Service from Portland and a sawmill technician arrived. The four hours work on Saturdays provides me eight additional hours of compensatory time per pay period, which can be used in lieu of annual leave. It's a good deal.

Happy Mother's Day! And now for a letter on last week's events.

My week started, for the lack of a better place at Hotel Splendid BOQ watching Dean Martin in a cowboy and Indian farce about "Two Flags for Texas" and ended with Elizabeth Taylor's demise in "Butterfield 8". Color movies are shown, rain or shine, after dark, in an open patio between the two wings of the building. The open side faces a noisy generator and a high fence. Guards outside try and keep missiles out. We sat in comfort in a warm rain and enjoyed ourselves. Cigars become soggy, but so what. I neglected to tell you I have taken up that vice again, but that is the only vice!

Some of the officers sit in the dry comfort of their balconies and watch the movie from at least six floors up. Clothes lines run on pulleys between the wings and fill the space with wet dripping clothes. Those watching from upstairs arrange the clothes so they can see the movies. That means the wet clothes are directly over our heads and add to the rain. Usually the rain is over before the end of the movie but clothes continue to drip.

On Monday morning, 8 May I was picked up at the Hotel. The drivers do well in meeting our transportation requests and are always on time. The routine is the same . . . car to USAID 2, car inspected underneath by using a mirror on a long handle for evidence of tampering or bombs, and transfer to a Ford Microbus for the trip to Tan Son Nhut airport. None of the cars or buses have seat belts. I perched on the edge of the seat ready to hit the deck in case of trouble. Buses have special passes permitting entry to the closely guarded airport; in addition our ID is checked. Security has tightened.

At the Air America terminal we checked in and were assigned to a C-47 transport. In addition to a crew of four there were six Americans and three Vietnamese. Cargo consisted of about a ton of flour, bulgur wheat, rice, wooden crates and a number of loose wheel rims for cars and scooters. Nothing was tied down! It was an accident looking for a place to happen! The only escape route I could detect, in case the load shifted, was out the cargo door. With only one visible parachute for 13 people, it would have been over loaded.

The weather was bright and clear. I shifted from the bucket seat to watch the activity. At the present this is rated as the busiest airport in the world. Just about every type of airplane made in the past 25 years can be seen. This ranges from slow Helio-couriers and high performance Porter Pilatus turbo-props which can land on 100 foot dirt strips, to the fastest and latest fighter-bombers, huge cargo jets and hospital planes. Helicopters operate from near the perimeter and they range from small one man observation ships, Huey gunships, to huge Chinook and Sikorsky flying cranes. We see the flying cranes transporting howitzers to battlefields and returning with wrecked airplanes or helicopters suspended at the end of a 60 foot line. Waiting for your turn to take off provides time for watching the activity. On Monday we waited 45 minutes for our turn, moving about 100 yards at a time. Air traffic control must be a nightmare. Crash trucks and cranes are on standby and fully manned at all hours.

The flight north to Dalat was very smooth. I am impressed by the Central Highlands. Forest covered plateaus and mountains are bisected by rivers and streams. The magnitude of the clear cut patches made by the Montagnards and the scope of the Air Force bombing and defoliation is striking. The higher mountains range up to 8,000 foot peaks in the north. Evidence of shifting cultivation and defoliation can be seen from the foot hills to the high mountains. En route to Dalat we could see extensive tea plantations. I have been enjoying the quality Dalat tea, now I will see where it is grown.

Approaching Dalat I could see extensive pine forests. While rare in tropical regions, Vietnam's highlands support five species of native pines. I could see evidence of reforestation and many blackened scars of forest fires. Another beautiful sight was the extensive vegetable farms. These are laid out in small plots of neatly arranged rows of many kinds of vegetables. In almost every field I could see workers in their typical pointed straw hats. Driving in from the airport we passed many fields and could see farmers at close range. Some fields had sprinkler irrigation, others were watered from a garden hose, however most fields were watered by workers carrying a yoke on their shoulders from which were suspended two 15 gallon sprinkling cans. The cans were filled from a sump dug in the field or at a nearby stream. Workers trotted to the fields and watered two rows, one on either side. The more prosperous and modern farmers use commercial fertilizers brought in and sold to the farmers under USAID's Commodity Import Program. Many farmers continue to use "night soil" and therefore vegetables are likely to be contaminated.

We were met at the airport by John Chitty. John is a handsome 6-foot-4 forester working for IVS (International Voluntary Services). He came to Vietnam from Algeria after USAID phased out his work in that country. He is a graduate of Oregon State University. He is dedicated to helping others. He and others in IVS work for $80 per month and are independent of the programs of others. USAID does offer them certain services and support but generally they live together and eat off of the local economy or what is given to them by the people they help. John was assigned to Dalat on my recommendation. He is an expert in forest nursery and reforestation work. He will also support our other programs by providing advice and identifying forest industries in need of technical assistance. John was promised a job in Laos as a provincial representative but someone else was given the position while he was en route to Vietnam from Algeria. He then signed on with IVS. He would like to join our team. We will consider him if the forestry program expands.

Other IVS members, men and women, teach agriculture and home economics to anyone who will listen to them. All are fluent in French and are becoming proficient in Vietnamese.

Dalat is a cool delightful city high in the pines. French colonialists built this city for themselves. Only their servants were permitted to enter the city. Guards around the city enforced that requirement. Homes are French-style with red tile roofs, very clean and beautiful. There are some Swiss chalet types. Homes are constructed on hills and are well spaced. There is generally no concentration of houses which is typical of other

Vietnamese villages and towns.

The air is cool and clear. Traffic in Dalat is slow and peaceful . . . a wonderful change from Saigon. There are few cars, mostly bicycles, motor scooters and an occasional military vehicle. Most of the military is ARVN. There are only a handful of US advisors and Signal Corps people in the area. The Dalat area is off limits to all US military, not stationed there, and is open only to US civilians on official business.

The central market is typical and has the normal colors and smells. Market stalls are concentrated in and around a large open structure the French built to meet their needs. This is where I purchased hand woven Montagnard cloth and a Cham blanket. This task required two days of bargaining to get the price down from VN$15,000 to 400, a little over US$3. They love to bargain and I enjoy doing it, all in Vietnamese. The variety of Montagnard bows, baskets and tools makes it difficult to decide what to buy. I am certain you will approve of my purchases from the very fascinating (dirty) Montagnards.

Following a meeting with the USAID Agriculture Advisor and the OCO Provincial Representative, John and I went on a tour. John had been there 10 days and had checked out which roads were safe to travel in the daytime. There had been a small attack on one of the police check points in Dalat the previous evening, but blood trails were all that could be found the next day. We went to some of the nearby pine plantations where three year old 3- needle pines were 4 inches in diameter and 12-feet tall. Since there is no frost, just a wet and a dry season, the pines grow most of the year. Some growth rings are produced and we are able to estimate the age. Experience and actual knowledge of the age of a tree is the best way to determine the correct age in the tropics. One beautiful plantation was seriously damaged by a recent fire. An ARVN military post is on a mountain dominating the area and soldiers periodically fire artillery and mortars at night to let the VC know they are on the job. ARVN remains in a fortified structure and never ventures out at night.

The remaining unburned plantations covered the rolling hills and made a beautiful sight. Tops of the hills and lower drainages had not been planted in order to provide livestock feed and wildlife. This was a good multiple-use decision because the soil at those hilltop sites is thin, unproductive and subject to erosion, but they did provide ample forage. The lower drainages were not planted because the workers did not like to work in these steep areas. I corrected that omission the next day when I recommended to Mr. Tan that these areas should be planted to reduce erosion and to take advantage of the increased moisture. I identified examples of natural stands in the draws which were much larger and taller. Mr Tan ordered his man to plant the areas during the next planting season. John and I remembered to turn around at the point where we had been advised that the road was considered unsafe. In that peaceful deserted part of the forest it was difficult to believe that trouble comes in many forms.

We visited species adaptation plots where the forestry research people (a different organization) had planted other varieties of pine, eucalyptus, etc. I must learn more about the research program and their organization since I detect little or no coordination or cooperation between the Directorate of Forest Affairs and the Forestry Research organization. My own personal "private intelligence organization" indicates this will make an interesting story.

While in the research area we came across a tiny man hiding behind a huge boulder. He was not a VC but rather a hard rock miner who was cutting cobblestones by hand. He had a lifetime's supply of raw material in the boulder. We talked to him and coaxed him to return to work. His stained teeth indicated to me that he might like some of my Beechnut chewing tobacco. He smiled when I gave it to him but at the first taste he spit it out and made an awful face. He had a few choice words that were not in my Vietnamese vocabulary. He was more discriminating than he first appeared to be. His ragged clothes matched the small dirty lean-to that served as his shelter. As we departed he was still pecking away at the rock with his mallet and rock drill.

On our return we stopped in an area where two men and a woman were clearing a quarter acre of pine poles. Their tools consisted of small handmade axes and hoes. The land was being prepared for planting upland rice. Stumps were about 2 feet tall and are typical of the heights I have observed in all cutting areas. This is poor utilization. Four children appeared from nowhere while we were taking photographs of the farmers at work. The children gathered twigs and piled them on a wire frame which I am certain their mother would carry home. We saw no indication of what was to become of the fine pine poles.

John and I walked through a nearby stand of old growth 3-needled pine. It supported a good understory of seedlings and saplings and it was time for the older and larger trees to be harvested. Under the dense overstory there was but little forage. The large trees were 24-30 inches in diameter and contained select grades of lumber.

As we continued our drive we came across on old Montagnard woman carrying a large basket of firewood. We stopped, talked to her and took her picture. The old woman understood enough Vietnamese to ask where we were from and what we were doing. We learned she had taken cloth to the market and had picked up a supply of fuel wood on the return to her village, still some 10km away. She said this was better firewood than that near her village. That frail little woman had a ubercular cough and a long way to go with her 80 pounds of food.

As the sun dropped behind a nearby hill we remembered the night belongs to "Charlie", so we scooted for the safety of ARVN check points. Back in Dalat the view from my hotel, the Dalat Palace, was spectacular. The setting sun illuminated a beautiful church across a nearby lake. Huge towering clouds created a spectacular background. As swallows wheeled in the sky it was difficult to believe the war was nearby.

My room in the huge Dalat Palace had 18 foot ceilings, heavily shuttered windows, massive furniture and a king-size bed with a canopy supporting a sweeping mosquito net which was gathered together 12 feet over the head of the bed. The ancient plumbing, with a 10-inch shower head, a pull chain reservoir over the toilet and a very dim light bulb suspended from the high ceiling reminded me the French had been there. But they must have enjoyed themselves. At that moment I could not help but recall my old friend Ranger Pat Murray's favorite expression when we were out in the mountains of Idaho and surrounded by fish and beauty, "I wonder what the common people are doing?"

I looked over the hotel rules and regulations and decided they had been put into affect after the French had departed. Among other things the rate of VN$1,065 included a meal. I decided to go to dinner at 2000 hours, the hour the dining room began serving. While I had a glass of cold bottled water I wondered what had happened to Director Tan.

Director Tan had met me when I arrived in Dalat. He was dressed in an immaculate Italian suit as he departed down the sweeping driveway with his family, in his black, shiny '47 Ford. I was surprised to see his car since there are hords of VC between Saigon and Dalat. I learned later the security people had checked out the "signs" and his driver had driven up that morning over mined roads and past VC check points.

At 1915 hours Mr. Tan and a man I had not seen before appeared at my door. The stranger was a GVN security agent. He bowed and Mr. Tan invited me to have dinner with him and his family at 2000 hours. Tan then explained he had known the security man since he was a boy and the agent had good eyes (Tan held his hands up to his eyes like binoculars), good ears (cupped his ears) and an alert mind (tapped his head). As I dressed for dinner I used a flashlight to see. The dim light bulb became dimmer as it grew darker. In the beam of my flashlight I could see mosquitoes and since I was in the highlands I sprayed the room with my bug bomb.

Before dinner we had drinks in the lobby bar and I met Tan's family. This included his charming wife, son, niece, mother-in-law, father-in-law and Mr. Le Viet Du, the Yale forestry graduate from the Saigon office. At dinner we were joined by two of the local foresters and their wives. Dinner consisted of many courses, including fish, very rare steak, nuoc mam (fermented fish sauce), hot red peppers, salad, bread, fresh pineapple and strawberries, cheese, red wine and strong, sweet, black coffee. My limited vocabulary was soon exhausted, since it was more useful to converse with woods workers, stone cutters and waitresses. Poor Mr. Du was assigned the thankless task of being my interpreter. It is no wonder he only weighs 90 pounds, since he has no opportunity to eat when the boss is around. The father-in-law was a fascinating old gentleman of about 80 years. He had been a judge and an inspector. Mr. Du would say "Mr. Inspector says . . . or Mrs. Inspector says . . . or Mr. Advisor says . . ." (that's me). Mrs. Inspector was an interesting old lady with stained teeth . . . "like they did in the olden days". It was an enjoyable evening.

The next morning I walked around the lake, along with the wealthy Chinese and Vietnamese. Gardeners were at work in their fields, others carried produce to market on their backs, heads, shoulders or in 3-wheeled Lambrettas. I had a light breakfast of dry toast and strong, sweet, black coffee.

After breakfast I glanced out the window of my room and saw the drivers washing and adjusting their cars. When Tan's driver looked up and saw me he clasped his hands, bowed and returned to work on the car. I had been with him on three other occasions and he recognized me.

We bid the ladies "adieu" and assembled for the trip. We had a game of musical chairs placing people in the proper car. IVS forester, John Chitty was with us for our three days of field trips. One of the Directorate of Forest Affairs Jeeps had only two cylinders working, the other had no brakes, so John drove an International 4-wheel drive Scout . . . later it lost all hydraulic brake fluid and limped in on what John could drain out of a hydraulic jack. Most of USAID's field vehicles are International Scouts. I recalled the Forest Service tested Scouts when they first came out and determined they were unsuitable for our work. USAID has bought them by the thousands and they are falling apart by the hundreds.

At our first stop we were met by 20 Montagnards who were fire crew and reforestation workers. They ranged from 4 to 6 feet tall and were an interesting group in their green head gear which consisted of real safety helmets, steel military helmets, helmet liners and pith hats. They were very dirty and ragged in appearance but very responsive in their bows when the Director stepped out of his car. A nearby fire tower was pointed out to us. It was a 20-feet tall structure, made of wood and located at the bottom of a hill in a deep draw. When I asked about the location and pointed out a logical high point, the answer was that it would make too good a target for the VC and ARVN. Also it was beside the road and easy to reach and could not withstand high winds. I could see our work was cut out for us and we had a long way to go to help the Vietnamese with their fire planning.

The plantations had a good network of roads. Some of the roads had wooden culverts which were deteriorating and resulting in serious erosion. The width of the roads would have met our SL-16 (16 feet wide) standard and had been constructed with primitive hand tools. If you had seen the heavy cuts and fills you could fully appreciate the magnitude of what a tremendous job it would have been. Firebreaks were properly located but were over grown with tall grass. The dense bunch grass creates a serious fire hazard. I believe prescribed burning during the rainy season and after the trees are about eight years old will reduce the fire hazard and facilitate fire suppression.

Most of the reforestation in the area northwest of Dalat was above Ankoret and a series of other reservoirs. While we were standing on a hill overlooking thousands of acres of plantations I noticed widely dispersed pieces of paper fluttering to the ground. These were Chieu Hoi passes that had been dropped from a high flying airplane. When I picked one up and identified it for the security man, he whisked everybody into cars and we sped off to a Montagnard village. Was the VC nearby? Who knows? This was the fire and reforestation camp and our lunch stop. The village contained the same type of interesting men, women and children I had seen at Tra Bong.

The square between the rows of Montagnard houses was swept clean, but inside the houses were filthy and smelled pretty strong from smoke and other human refuse. A lunch table was set under a thatch shelter and surrounded by recently cut pine saplings which were tied to upright poles. Mr. Tan was visibly disturbed when he saw the freshly cut pines, but I saved the local people from his wrath when I suggested these had logically been cut as part of a thinning exercise. All the stands I saw needed thinning since the trees are planted about 18 inches apart.

Our lunch was a typical, delicious Vietnamese meal. It consisted of soup, rice, several varieties of meat, including duck, chicken and liver, green beans, nuoc mam, red peppers, fruit and swarms of flies. As I thought about the inside of those houses and reflected on where the food had been prepared, and by whom, I devoured all the food with the exception of the fresh vegetables. I enjoyed the fresh strawberries, pineapple and bananas. The red peppers, nuoc mam and delicious Algerian wine served as a catalyst and I had no stomach problems. In Vietnam the host is expected to provide more food than the guests can possibly eat, to do otherwise is an insult. When we finished, much of the food remained. I hope the villagers were given the leftovers because they looked lean and hungry.

Following lunch and a short rest we visited the forest

nursery. The foresters do an excellent job in selecting and preparing the site. It is always located in the middle of the area to be planted and where they can dig for water or have a nearby lake or stream. The Montagnards move in, set up a small village and do the work. Seeds are planted in small plastic bags and watered twice each day during the dry season by workers walking between rows carrying sprinkling cans supported on poles over their shoulders. Within six months the seedlings are taken to the planting site. Each tree is planted in a small circular area that has been cleared of dense grass. The nursery operation can benefit from John's advice and suggestions. The trees need fertilizer and irrigation practices can be improved. They are doing quite well since the survival success of their plantations is better than 90%.

We visited a direct seeding area where they planted five seeds to a spot and competition among the surviving trees was serious. The foresters gasped and shuddered when I used my pocket knife to cut all but the dominant tree in each group. But it was needed. I suggested they should work out an agreement with the Vietnam Boy Scouts and train them to do the thinning and fight forest fires. They could earn merit badges for their efforts. Fire was a serious problem in this area.

The Vietnamese have an interesting jurisdiction problem. If an area has trees it is government property. If cleared and planted ownership passes to the farmer. So it is a continuous fight to see who wins. The Military Academy at Dalat, their West Point, also wants land so they have land use conflicts that are very similar to ours. I suggested the Boy Scouts and the 4-T (4 H) could be encouraged to develop an interest in the Directorate's forestry programs and thereby help the Directorate hold on to the land. Mr. Tan approved of my suggestion and ordered the local foresters to do it . . . so be it! How's that for action? Followup is something John can support.

Later we visited the market, had more excellent dinners, trips through VC areas and more interesting things that I can write about later. But I am running out of time. I must pack for my trip tomorrow and get some rest. It is now about midnight. Tomorrow we go to the Special Forces camp at Tra Bong in Quang Ngai Province of Region I (Eye Corps) to install the Mighty Mite sawmill. That should make an interesting trip and provide more photographs and stories. I will write the conclusion to the Dalat trip and tell you about Tra Bong when I return.

Chapter XVII

Projects

Detachment A-107 Company "C"
5th Special Forces (ABN) 1st SF
Republic of South Viet Nam
APO San Francisco 96337
21 May 1967

For the past six days I have been at the Tra Bong Special Forces Camp sharing their experiences and problems. The Green Berets are an interesting group and all have very strong, distinct personalities. All are specially trained in weapons and demolitions, but their mission here is to help the local people and refugees. They train the Vietnamese and Montagnard tribesmen ("Yards" in military jargon), keep track of the VC and NVA, and teach agriculture, medical, mechanics and other skills.

We arrived with two sawmill technicians and Mr. Thuong my interpreter. One of the SFs problems has been the defense of this old, former French Foreign Legion outpost. One day we received intelligence information that we would be hit by a mortar attack at 0300 hours. In preparation for the attack I received training on arming, loading and firing mortars, both 81mm and 4-duce (4.2-inch) heavy mortars, machine guns, grenade launchers, automatic rifles and claymore mines. During the alert my assigned post was the medical bunker where I had charge of a heavy machine gun. We apparently softened the opposition by firing over 100 high explosive mortar charges into their reported location. For insurance the CO requested standby support from the 175mm guns operated by an artillery outfit some 20 miles away. Since then we have had no reported contacts.

A number of VC have turned themselves in since my last trip. A platoon is expected to surrender tonight. We hear the forestry project is influencing these defections. The people see new hope, employment and a better life. We are known as heroes in civilian clothes to these people. The VC in the surrounding hills must respect us, since they shoot direct at other people, and we only have bullets zing into the rice paddies within 15-20 feet of us. Since a moving target is harder to hit, I keep moving!

Tonight we have a patrol out on a listening post to gather intelligence information on the VC and NVA that are reported to be near us. The CO, Lt. Katz and Sgt. "Doc" Jones, a medical specialist, are with a 150 Vietnamese Civilian Irregular Defense Force (mixture of Regional Forces, Popular Forces and ARVN known as Ruff-Puffs). I just came from the command bunker and heard their radio message. They have set their defense perimeter and are secure for the night. In the steep mountains, with 100+ degrees and 100% humidity, the three SFs each carry a 90 pound pack. They have had no contact with enemy forces today but expect contact tomorrow. The Vietnamese on this patrol have been a problem today, they are virtually useless, tire easily, make too much noise, and are probably sympathetic to the VC.

Today, Sunday, has been a hot one. The temperature reached 117 degrees. We operated the sawmill and scaled logs until 1100 hours. Then our crew of Vietnamese trainees quit, went to church and then home for rest.

The Vietnamese ARVN commander, at this outpost got miffed today when he was caught stealing food supplies from their own warehouse. So he had his men unload all the food he was about to take and stack it 50 feet in front of the Special Forces warehouse. Then 12 of us, including our crew, moved 300 cartons of rations, 50 bags of salt and 50 bags of rice into a very hot Special Forces metal warehouse. I worked inside the building (gosh, it was hot even for me!) and stacked 50 kilo sacks of salt and 60 kilo sacks of rice (60 x 2.2 = 132 pounds). The Green Berets are pleased with our support.

Tomorrow the District Chief wants to talk to me before I leave for Quang Ngai and Danang. Then I go direct to Dalat with one of the sawmill manufacturer's representatives before returning to Saigon by the weekend. It will be a while before I return to the typewriter and finish last week's Dalat story and this one.

Young Sgt. Stanley Bowen says "hello" to you. He is a graduate of Washington State University and plans to be Governor of the State of Washington. He will mail this letter with their outgoing mail.

27 May 1967

From the sound of all the murders in DC it appears you are in more danger than I am. So watch yourselves and keep the girls under cover. I am pleased to hear you are doing better with the finances. At the rate you are going you could purchase two $500 bonds. Or you could invest in one of the insured Savings and Loan companies in California and earn 5 1/2% interest. Income tax would have to be paid on the interest but we would still come out ahead. I do not think stocks are too good an idea unless sister Leela has a hot tip. Gene Jensen's stocks are so-so and you may lose in the stock market. Gamble if you like. It is your choice. My Mother always told me you will have it all some day My leave plans call for a 3 July departure, that should put me in Cedar Rapids by the 5th and I will be home by the 8th on sooner.

You have a great responsibility in taking care of the girls, the house and all the other chores. I owe you a great deal for doing it alone. I do believe I am doing some good for the people here. In a small way I can help improve conditions and have an opportunity to restore some of the war damage and make this a better place. In that respect I see progress and people respect my judgment and efforts. If you can bear with it for a while longer, I will appreciate it and make it up to you and the girls.

I regret you have had so much rain. I know how that eats on you. By the way some of the women who have come along with their husbands and live in safe haven posts in neighboring countries are not so enchanted. Those in Hong Kong are not happy . . . too much acid rock and Communists. In Bangkok they are tiring of the bridge circuit. Those in Manila are suffering from heat and humidity.

I am typing this letter on my hotel roof. It is cloudy and I can see several rain showers around the country. I will end these personal thoughts and after breakfast return to the office and finish my letter about the Dalat trip.

Back to Dalat . . . Please recall I mentioned in my last letter that special approval is required for all Americans to visit Dalat. This is one of the few places in Vietnam not overrun with GI's and other Americans. The cool climate is delightful and would attract thousands and quickly exceed the carrying capacity. Towns like Danang and Nha Trang were once small beautiful sea coast villages, now with overcrowding the streets and the entire infrastructure are falling apart and refuse is piled high . . . they are one big mess.

Lunch at the Montagnard village filled me and I retired early, without dinner. Chopsticks are provided for all meals. Melissa, you did well in teaching me how to use them. As you would guess the Dalat Palace offered fine silverware, no chopsticks. Before falling asleep I studied my Vietnamese lessons by flashlight in the soft bed and under the large canopy and mosquito netting. When I did turn out the light, the quiet was so deafening it was difficult to fall asleep. In Saigon I can hear noises of the city and the war even though I am in a closed room with the air conditioning going full blast.

The next morning I was up at 0600 hours and met George Watts, an irrigation engineer from the SCS in Phoenix. He is stationed at Nha Trang. George and another engineer were in Dalat on a one day assignment. After breakfast George and his associate took off for work and I still had two hours before Mr. Tan was to pick me up for the day's event. I walked around the lake and took some more pictures of farmers at work, women washing clothes in the spillway and Montagnards carrying various products to the market. The striking differences between Saigon and Dalat are the leisurely pace of life, absence of cars and the purity of the air. Dalat is delightful!

When I returned to the hotel our party had gathered in the lobby. Through an interpreter I talked with the Minister of Agriculture's wife. Her daughter was dressed in very tight blue jeans and this made me homesick for my own two lovely teenagers. She acted about like you two girls. I hope you have an opportunity to meet her when she is in the US next month with her parents.

We were assigned cars and took off for the second day's field trip. This time we headed south. After passing a police check point and barricade I noticed a fresh excavation on my side of the road. I was told this was where the VC detonated a claymore mine the previous week. The thought of it makes the hair on the back of your neck rise and you sit ever so lightly.

We drove through beautiful stands of Pinus khasya, a 3-needled pine. Our first stop was at Gougah Falls. The falls were spectacular and set in a beautiful stand of high quality pine and hardwood trees. This was near the entrance to the Dalat Zoo. We walked through the grounds and saw tigers and some mean looking leopards. On our walk through the Zoo the security man whipped out his pistol and fired a shot into the air. I was told this was just a warning to anyone interested and notification that he was around and armed. Before leaving the Zoo Mr. Tan insisted I take a ride on an elephant. I fed her some sugar cane and she behaved like a lady. You ride those critters by sitting on their head, grip your legs tight and hold on. It was a short interesting ride, since she was chained and could not go far.

From the Zoo we headed down the road. The only traffic was ARVN armored cars, trucks and Jeeps full of soldiers. All of the vehicles were armed with machines guns and I was told they were on patrol to keep the area secure for us. Our next stop was a sawmill which was cutting the 2-needled pine, Pinus merkusii. Logs were about 24 inches in diameter and the lumber produced was high quality select grades until the saw reached the 4-inch center core. I have never seen pine yield such beautiful, high quality lumber. The sawdust conveyor was two women who filled baskets and carried them out back to a sawdust pile. This was one of the few piles of sawdust I have seen at any mill. Most of it seems to be carried off and fully utilized. This mill had three head rigs and is one of the largest in the area. The owner told me he would appreciate help in upgrading his mill but he had other more serious problems to take care of first, and these involved the Army. The Army provides a permit to the sawmill operator to cut timber. This is done under a program called "road runner" and consists of removing the "scenic strip" or about 200 feet on each side of the road. This is done to reduce ambush cover. The sawmill owner had to pay loggers and log haulers, both of whom were taxed by the VC. Then he had the cost of manufacturing the lumber. After that, and here comes the "kicker", the Army has the right to take whatever lumber they want and at no cost. The owner wrung his hands as an ARVN truck pulled out with a load of select grade lumber.

The next stop was a pottery plant. We had met the owner the day before in Dalat and he had invited us to visit his plant. The operation was managed by a Japanese man who seemed startled to see us. He invited us into his office and we could see people scurrying off in all directions. When we went on a tour we were detoured around several areas. We saw kilns and stockpiles of firewood. We received hostile glances from the 100 or so workers. Many of the workers were young children, some barely in their teens. Many were performing very difficult, heavy and dangerous work. It reminded me of early industrial England and child labor exploitation. We walked out the back of the plant to see a small hydro-electric plant that provided power for the operation. I noticed a man standing behind a tree holding a sub-machine gun. Before leaving I picked up a couple of insulators for your collection. The manager was pleased to give them to me and even more pleased to have us leave. We learned later that this establishment is a suspected rice distribution point for the VC. I neglected to mention the security man left us when we pulled out of the Zoo. He would have had a fit if he had known we visited the pottery plant.

We continued south through a village which looked like one of our frontier towns of the 1880's. Houses had slab siding and over hanging porches. We visited another beautiful waterfall and inspected some Montagnard shifting cultivation areas. When I see these areas within the forest I think of the terrible waste of manpower and natural resources and loss of soil. This is due to too many people being in the same place at the same time and too many mouths to feed. But that is a worldwide problem.

We had lunch that day at a former spider house (o nhen lau xanh . . . this expression requires several accent marks to denote correct pronunciation). This structure was a three story, blue trimmed house that had been taken over by the GVN. It was deserted and made a peaceful place for a rest stop. What is a spider house? We once had one on a mining claim on the Tonto National Forest . . . it was, shall we say, a place for rest, relaxation and recreation. We were provided individual towels and a splash of shaving lotion to wash our hands and faces. We had a box lunch, consisting of chicken, ham, pork, shrimp, Spam, egg, bananas and a cup cake. I hesitated to break the egg shell for fear that I would disturb an unhatched chicken, but fortunately it was a genuine hard boiled egg. For liquids we were offered delicious Algerian red wine, rice wine (120% proof!), Scotch, tea and water. I passed on the water and tea because they are always suspect. Since I was on duty I took only moderate amounts of the other liquids, it was my duty, you know, in respect to my host!

Following lunch our folding chairs were taken to the edge of a nearby white water stream for, as Tan called it, rest and contemplation. Sitting there in the forest I told Tan, Chief Cliff and former Chief McArdle would have enjoyed being with us. I stressed the value of the forest as a place for rest and relaxation. At that moment the war was only a distant dream.

Our scheduled afternoon visit to the naval stores area was canceled for security reasons. We headed back up the mountain through the pines to Dalat. On the way the trusty old Ford heated up so we had to stop and let it blow steam in a magnificent stand of 3-needled pines. We saw a nearby trail which led to another waterfall. Tan, John, two of the young Vietnamese foresters and I headed down the trail. I started to go first but then I remember is was wise to let the chief go down the trail first. The pines and mixed hardwoods completely shutout the sky. Orchids and other wild flowers made a beautiful sight. We reached the roaring falls in a few minutes and sat there watching the cold, clear water tumble over the rocks. What a paradise this could be under other circumstances.

We returned to the hotel at 1700 hours and that was all for the day. I took a walk around the hotel grounds and watched children at play. They played hide and seek, hop scotch and ring-a-round the rosy. These were the poor little rich kids.

At 1800 hours I went to the bar for a beer. It was like a tomb because no one goes to the bar before 1900 hours. I sat there enjoying the view and watched the setting sun back light the huge towering clouds and shine on the red stone church across the lake. The rolling green, pine covered hills made the scene look like a beautiful painting.

That evening I had dinner by myself. The Tan's went elsewhere and the young foresters always disappeared at the end of our trips. There were about 20 waiters in the dining room, all dressed in white uniforms, ready to serve the guests, which consisted of a Chinese man, his wife and me. I greeted the other guests in Vietnamese. While the waiters were all Vietnamese they were programmed to speak only French. I said in English, "dinner please", and then the following started coming:

> Consomme au Tapioca
> Filet de Vielle a la Normande
> Poulet Roti
> Pommes Rissalees
> Salade du Pays
> Patissere
> Cafe ou The

It was delicious and I ate everything except the fresh,

luscious, green lettuce. One of the waiters brought me a fresh King Edward cigar and another one lighted it for me. The KE is a good 5 cent cigar. But in that setting it tasted and smelled like a $10 Havana! I sat for a while over my coffee and watched the other guests come streaming in at 2130 hours. Up to that time I had much personal attention and the services of about 10 waiters.

Following dinner I walked around the hotel. It was dark and very peaceful. In dimly lighted corners of the hotel I could see armed security guards. There were no cars, no blue exhaust smoke and only two scooters went by on a distant road. Lightning flashed in the distance. On a distant mountain I could see a fire, probably from a Montagnard clearing. It was a beautiful scene, but it was a lonely night when I thought of you, my loved ones, on the other side of the world. I wish you could have been there. With nothing else to do I went to bed.

I have a few more reflections about the Dalat Palace Hotel. I learned that most of the employees work for VN$50 per day or about 44 cents. They work long, hard hours for their tips. I hope that a portion of the assessed 15% service charge reaches their pockets. The hotel reminded me of some of the huge, old upstate New York hotels that were once a tourist attraction for city dwellers. As I noted before there were only three or four Americans around, the rest were rich Vietnamese and Chinese. Just getting to Dalat is expensive. For a matter of safety visitors must fly in on either Air America ("Dog Patch Airline") or Air Vietnam ("Air Nuoc Mam"). The hotel was substantial, but with abundant interior woodwork and poor wiring I kept my fire escape routes firmly in mind at all times. Being inside those thick rock walls made the place a veritable fire trap. Once Dalat had another large hotel, the Langbian Palace, but university students became upset over one of their many causes and burned it down for kicks.

The next morning I was up at daylight to take photos of the sun rise. It was a cool 65 degrees and my black nylon jacket felt comfortable. The grass sparkled with dew. Other early morning risers and walkers wore gray sweaters . . . that was the only color I saw the Vietnamese wear. Sweaters looked odd over the women's ao dais. The hotel grounds were very clean and a few water buffalo served as practical lawn mowers. I noticed a few scattered French and Belgium cigarette wrappers, but no Camels or Pall Malls. People appreciate being offered American cigarettes. When I travel I carry a carton for that purpose. I do not smoke any of them myself, but I do have the occasional cigar.

While waiting for our group to assemble I looked at the National Tourist Office's display of maps and curios of ivory and Montagnard handicrafts. A sign reminded me of some of Flagstaff, Arizona's Chamber of Commerce propaganda:

"*Welcome to Dalat*—Dalat offers you during all the season its spectacular waterfalls—dreaming pine covered hills—glittering lakes—flower gardens and its agreeable climate all year round. The National Tourist Office is at your service and complete information on the area in and around Dalat."

Following breakfast on Thursday 11 May I was greeted by Tan and his family. I practiced my Vietnamese with the family and they patiently helped me with the pronunciation. That is the difficult part of this language, since a single word can have as many as five different meanings depending on how you pronounce it. Mr. Inspector was particularly helpful and eager for me to use Vietnamese. Tan reminded me the French never once spoke Vietnamese and completely played the role of master with the servants. With much emphasis and gusto he said "today we proudly walk with your great Nation!"

Mrs. Inspector had a small basket on her lap and I asked if that was her lunch for their trip to Ankoret Reservoir. She spit into a small brass can and then took all the items out of her basket to show them to me. These were her chewing materials. The white nut which was cut in half was betel, large green leaves were from the plant and a jar contained ground limestone paste. In addition there were peanuts and a tough stringy fibrous material. (According to *American Heritage Dictionary* "betel is a climbing Asiatic plant, the leaves of which are chewed with the nut to induce both stimulating and narcotic effects". Morris, 1975) She gave me some of the fibers and peanuts to chew. I guess she did not want to stimulate Mr. Advisor with any of the betel. At that point my good friend, forester Du arrived and was asked to relate the story connected with the chewing materials. In a very polished style he related the following:

"Back in the olden times there were two brothers and a beautiful young princess. Both brothers were deeply in love with her. One brother finally married the princess and the other brother sadly went to the forest and was turned into a limestone rock. The married brother was sad at the disappearance of his brother and went out into the forest to search for him. He searched for many years and one day as he sat down by a limestone rock to rest he was turned into a nut bearing tree. Soon afterwards the princess went looking for her husband and his brother. She spent the remainder of her years looking through the deep forest for them. As she neared the end of her life she sat down by a limestone rock and under a nut bearing tree and was turned into a leafy vine which twined around the tree."

Du concluded, that to this time the nut, the leaf and the limestone are chewed by older women in remembrance of this story. These three items are used in today's marriage ceremonies. While she did not understand a word of English, Mrs. Inspector was visibly pleased to have the story told to me. She smiled at me, showing her stained teeth and then took one of the leaves, spread on the limestone paste, rolled it up and placed the mixture into her mouth along with one of the nuts and fiber. The combination makes a bright red juice which can be seen splattered all over the streets and sidewalks of Vietnam. When the French first came to Vietnam and observed this act they thought all Vietnamese women had tuberculosis.

We were ready for departure for our morning trip. Tan told me we could not travel on the scheduled route due to "security problems". For a short distance we repeated part of a previous route and then turned off and drove over a low mountain pass and entered an interesting valley. This area contained a dense mixture of pine and hardwoods. Large areas in the valley bottom had been cleared and primitive agriculture was being practiced. We stopped and I took photographs of wooden plows pulled by oxen. Irrigation was done by hand or by people peddling a bicycle-like arrangement to lift water from a stream, pond or sump to ditches in the field. We passed one small community containing a cemetery with many fresh graves. New ones were being dug and a military funeral was in progress at the church. A Vietnamese funeral is impressive with an abundance of flowers and an ornate hearse which is covered with bright colored carvings of dragons. Nearby a squad of soldiers fired rifles, machine guns and mortars into the hills.

In this same area there were many Montagnards carrying heavy baskets and bundles of fuel wood. Vietnamese hauled huge loads on small carts pulled by tiny horses. We could not stop for pictures due to security reasons.

Near Don Duong we visited the Bien Hoa paper company's timber concession. This area could supply excellent raw material, but for security reasons, it is inoperable. Their greatest problems are the long truck haul over poor, steep, winding, mountain roads and VC tax collectors. The operators refuse to pay VC taxes and have closed down their woods operation. You will recall I described how the paper plant uses bleached pulp shipped in from the US and bagasse (dry pulp remaining from sugar cane after the juice has been extracted) from Formosa.

Reforestation is required on all cutting areas. Successful regeneration of previous cutover areas was very evident. As insurance the operator had spared five seed trees per acre. Since these large seed trees had accomplished their purpose and were suppressing planted trees and natural regeneration, I suggested they should be harvested for sawtimber. Tan ordered this done on the spot and the local manager was delighted to have access to these trees. From the many fire scars it was apparent the local district forester had both fire prevention and suppression problems.

Our next stop was the Dran Hydro-electric Project. This

project supplied much of Saigon's power until VC sabotage made the transmission lines inoperable. As I was preparing to take a photograph of the forested watershed above the reservoir a nearby ARVN unit, just 100 feet above where we were standing, shot off several volleys of 105 howitzer shells. I just about jumped out of my skin! We learned later they had a patrol out on the opposite side of the lake and the support fire was to encourage the VC to leave the vicinity of the hydro-electric facility.

A lunch table was set up under a shelter overlooking the dam and reservoir. We sat in the cool shade and enjoyed a delicious meal of cold cuts, vegetables, spiced rabbit in a very hot sauce, pressed goose, bananas, apples and a few flies. For liquids we had our choice of Scotch, Algerian rose (Domaine de la Trappe, Vin Fin, Pres Alger) and Ruou de (rice wine). All were welcome substitutes for the polluted water.

After lunch we visited a monument a Japanese engineering firm had erected in memory of workers who lost their lives during construction of the dam. Heading the list was the name of the Japanese engineer who had dropped a tree on himself. This reminded me of a forestry student back in 1948, in a world far away, who felled a dead tree on a forest fire and was knocked down the side of a mountain.

On the return drive to Dalat, Tan announced security had improved and we could return by the route he had planned to use that morning. As we climbed the steep mountain road we passed several ARVN forts. These were fully manned and served as check points on the road parallel to the railroad. Surprisingly this is one of the few areas in Vietnam were the train still operates. I was told it carries a large contingent of guards when it rolls through the forest, down the mountain to the coast and Nha Trang.

In sight of the ARVN forts we ignored unauthorized cutting of fine pine poles. I sense there is a serious, unresolved conflict between the Army and the Directorate. We also passed crews standing on the road watching forest fires. They were watching the fires spread because it was unsafe to go into the forest in that particular area to control them. Nearby was an interesting fire prevention sign. Mr. Du translated it for me . . . "Please do not be so crude to set fire and burn the forest. The streams will dry up and the trees will die". Pretty sound advice!

We drove by a high point containing a huge, well guarded US communication center on our way to a 924 acre tea plantation. We saw the entire tea processing operation as we toured the premises with the Chinese manager. At the end of the tour we were presented with a package of choice Orange Peko tea. All of the production of this particular plantation is packed in waterproof wooden tea boxes and shipped to England. This choice brand of tea is not a favorite of the Vietnamese, who prefer light, green tea.

The next 10 miles of our trip were through a well known problem area. This was a VC crossing area, and while it was not signed like our deer crossings, there were numerous guards along the way to indicate it was dangerous. At times like that you can not help but think that the peace and quiet could be shattered and your life changed in the blink of an eye. It all depends on whether someone is crouching in the grass or trees waiting for the right vehicle to pass before touching two wires together to detonate a mine. But we passed the danger zone safely and through the check point into the security of Dalat. We were not to be the target that day.

Mr. Tan then turned me over to Mr. Du, along with his car and driver, to visit the Bao Dai and Diem Palace. This was a huge structure used in the past by the late President Diem and his cabinet. Royalty must have enjoyed themselves in the many large, well decorated rooms and spacious gardens. The gardens are not kept in very good condition but the house was spotless. We took off our shoes for a guided tour. Our next point of interest was Premier Ky's palace. It was locked and deserted except for a nearby ARVN outpost. We drove in and looked through the windows. The gardens contained colorful roses and orchids. On that particular day Ky announced his availability for the presidential election. He was due to arrive at this place the next day for meetings and rest. I was surprised we were permitted to drive in and look around, unescorted, the day before his arrival. At times security is surprising informal.

That evening we were invited to dinner at Mr. Ta Cau's home. He is the Division Forest Chief and a very interesting old gentleman. I noted he thoroughly enjoyed his drink of Scotch each day during our lunches. We had another game of musical chairs in deciding who would ride with whom to dinner. The two young foresters were changed in and out of several cars before we started. On the way we stopped at a lovely house and picked up the wife of the Minister of Agriculture, Madam Tre (Tree), her daughter and niece. By the time we arrived at the dinner party Mr. Du and the other young forester had been dropped by the wayside. I learned later that Mr. Tan decided he had invited too many guests for Mr. Ta Cau's dinner and someone had to go. Being low on the hierarchy totem pole, the young foresters were dismissed.

Mr. Ta Cau's home was neat, clean and modest. A family altar stood in one corner. Walls were covered with paintings and lovely lacquered gold inlaid plaques. The meal was served by the same four men who had served us so well at our noon luncheons. The wife, as expected, never once appeared, since the wife does all the work and supervises the kitchen.

The meal consisted of eight courses. Actually I lost count of the number of dishes. The first one was a delicious soup. It was thin and contained small bits of gelatin. I asked what it was and was told it was made from the sea swallow's nest. I had seen the nests on display at the market. After that I ate every delicious thing that was put before me, but I did not ask what it was. I did very well with ivory chop sticks and loved the rice which I flavored with nuoc mam and red hot peppers. We had duck, chicken, fish, preserved fruit and fresh fruit. Memories of the market I had visited did not dull my appetite one bit and it had been several days since I had survived lunch in the Montagnard village. The same Algerian wine was served along with some wine made by Mr. Ta Cau. It was made out of ginseng root and was good and powerful. Mr. Ta Cau loved his wine and delighted in seeing me enjoy it. He got a little higher than I did. Before leaving I paid my compliments to Mr. Ta Cau and asked him to express my appreciation to his wife. That night I had interesting dreams of dragons and Chinese mandarins chasing me around people frozen in limestone and trees. The after dinner cigars of strong Filipino tobacco added color to my dreams!

The next morning, Friday, Tan, John Chitty and the other local foresters took me to the airport to catch my Air America C-47 cargo plane. At the airport ARVN troops, brass and a number of US Generals and Colonels were all lined up at the arrival/departure gate. No, they were not there at Tan's request, but at that stage nothing would have surprised me. They were waiting for Prime Minister Ky's arrival. I was set to take his photograph as he started to emerge from his plane, but I was signaled to run to the Air America plane. Tan had departed to make arrangements for Minister Tre who was arriving in the Premier's plane.

Aboard my plane was the usual assortment of people, pallets and junk. I sat next to Brother Joseph My (Vietnamese) and Father Matthew Chung (Chinese), a Salesian brother and father respectively. As the plane took off they counted their beads. I thought I was in good hands after a delightful visit to Dalat. I had obtained enough information to plan and design a number of forestry projects for our crew.

Brother Joseph was an interesting young man of about 20 years. He was orphaned by the war in North Vietnam and taken in by the Salesian fathers and trained for education and missionary work. He spoke perfect English, three Chinese dialects, French, Italian and German. He was a remarkable person. He described the country to me as we flew. He knew the names of all the rivers, mountains and villages. His Catholic order was founded by an Italian priest by the name of Don Bosco Govap over two centuries ago. As best as I can recall one of Brother Joseph's quotable quotes goes something like . . . "Do not say all you know, but know what you say". That is good advice, and please relax, I have just one or two more items before I close this long letter.

We flew over many tea plantations and landed at Blao. This was a small dirt airstrip in the center of a vast tea plantation . . . producer of the best tea in all of Vietnam. About 20 people were waiting for a seat to Saigon. There is always someone waiting for a ride to somewhere, anywhere, it seems, will do. If Vietnamese have clearance from the local USAID official and the security police they can ride on a space available basis. We landed about noon and the USAID official had sent a letter with all of these people's names on it rather than bother to come himself. He was on his noon break. The pilot refused to take any of these people without an American to personally vouch for them. This airport was served by Air America on rare occasions. I learned some of these people had been waiting 10 days for a ride. There is always the danger of the VC hijacking an Air America plane. An OCO/USAID official sitting next to me and I were disgusted when that local AID official would not get off his rear and help these people get aboard. The disappointed and bitter looks on their faces were graphic evidence of how far we had to go to win the hearts and minds of the people. As we flew on to Saigon I had a bitter taste in my mouth as I looked around me in the plane at 20 empty seats.

The other item worthy of mentioning is the defoliation. We flew over vast areas where our Air Force spray planes, known as "Operation Ranch Hand", had defoliated the forest in strips that stretched as far as I could see. In some of these areas I could see evidence of bombing and fire scars. Forest cover had been removed in some areas but for the most part the areas treated with Agent Orange have such a dense canopy and understory that the combined VC, NVA and Chinese armies could hide out. I observed these conditions in a vast area north of Saigon. I could see no real accomplishment other than killing or setting back the growth of huge hardwood trees. I am supposed to be here to help the Vietnamese manage and utilize their timber resources and the spray planes and bombers are doing everything within their power to destroy a valuable resource. Time will tell and I must accumulate more experience and information before I dive into this can of worms.

We flew over the huge Bien Hoa Airport and staging center and landed at hot, muggy Tan Son Nhut Airport in puddles remaining from a recent rain. That night the VC hit the Bien Hoa Airport and inflicted considerable damage and casualties.

This concludes my Dalat trip report. We did not go on the scheduled tiger hunt because the security people told Tan that we might be the ones who were being hunted. He told me we would do another time when there was more security. When I learned

229

these magnificent beasts were hunted with a shotgun loaded with buckshot I was not so sure I wanted to participate in a tiger hunt.

My next letter will be about a most interesting trip and return to Tra Bong. Living with the men who wear the Green Beret was certainly different from any experience I have ever had before. Setting up a forestry and sawmill operation in the equivalent of the middle of a busy rifle range was not your usual run of the mill forestry projects. So goes life in Vietnam.

Note: Sometimes I refer to Mr. Tan as Tan. He has encouraged me to use this greeting as an expression of close friendship. However I do not address him as Tan when his men or other officials are present.

Chapter XVIII

Lost in Saigon

3 June 1967

This week I worked at the office, including Memorial Day. The war goes on seven days a week. Barry Flamm arrived on schedule and Mr. Long one of the Vietnamese Administrative assistants looked him over and said "he looks like the most dignified of your crew". Barry made a good impression on him. I am very fond of 55 year old Mr. Long. He came from Hanoi in the 1954 American airlift. He is one of many people who help me with my Vietnamese and he provides me good advice. While he does not work for forestry he will do anything I ask of him.

Barry was 30 minutes late this morning. He left his room at the Excelsior Hotel, fresh and clean and ready for work. But his elevator stalled between the 2nd and 3rd floors and he spent 30 minutes calling for help from that hot, dark elevator. A crew took 30 minutes to crank him down by hand. He was not clean and fresh when he emerged. He vows never again to set a foot in an elevator. I had a interesting experience recently with my French elevator at the Park Hotel. When it stops at a floor and people exit, the door closes and the light goes out unless a button is pushed to send it up or down. Several times I have gone to the elevator and pushed the button and waited and waited, often the elevator is there waiting for me. I feel silly when someone walks up, opens the door and walks in. The other day I left my room on the 5th floor (RC+5) and went to the elevator. I opened the door and started to step in. Fortunately I looked first, it was very dark, but that was because the elevator was on the ground floor. That would have been my last big step! As Ray Housley said in a recent letter he believes the elevators will get me before the VC.

It has been some six weeks since I lost my language teacher. He was a medical student and had no time for teaching after he was assigned additional duty at the hospital. Since then people around the office have been helping me. Last week I accompanied a friend to his language course at the VAA (Vietnamese American Association). After sitting in on a class we went to a handicraft show in the same building. While I purchased some Montagnard cloth for you, my friend departed. I planned to have dinner at the Hotel Splendid BOQ and then return to my hotel. Being a good woodsman, I figured I could leave by a different door than we entered, make a couple of easy turns and be at the BOQ in 15 minutes. After taking two turns and over an hour later I was in a totally strange part of Saigon. It was dark and I was lost! In the poorly lighted streets I could tell the area contained large homes. Later I was asked why I did not go up to the front door of one of those homes and ask for directions to my hotel. I did not do this because all the houses were behind locked gates and high fences with angry, mean sounding police dogs in each yard.

I stopped and tried to reconstruct what I had done. Gwen, I had pulled one of your tricks and taken off in entirely the opposite direction from where I intended to go. I then opened my brief case to pull out my map and locate myself. I had no map. I had left it at the office when I was pointing out locations to Barry Flamm. Occasionally a person would pass and look me over carefully before they crossed to the opposite side of the street. But no one bothered me. I finally found a street sign that read Duong Cong Ly—Duong Mot Chieu. I knew that meant Cong Ly Street—Oneway. I was relieved because this was a street that goes within one block of my hotel and in the opposite direction it goes to Tan Son Nhut Airport. I knew the oneway went from the airport to the city. So I started walking in what I thought was the direction to my hotel. This street was well lighted and there was some traffic, but no taxis. As I walked faster, because it was getting late, I noticed all the cars were coming towards me. I thought this was strange since the cars from the airport came to the center of town. I was puzzled until I finally located a policeman. During the day they are plentiful. You can find two on every corner directing traffic, holding hands and watching or creating traffic jams. But I remembered at night it was not safe for them to be out so I was lucky to find one. He could not speak one word of English. Fortunately I could ask him for directions to my hotel and he quickly pointed in the direction I had just come from. I felt very sheepish to find I had been walking towards the airport and directly away from where I wanted to go. If I had continued for another mile I would have been passing through an area that is reported to be very dangerous at night. I thanked him and turned around and in an hour reached my hotel. All BOQ's were closed so I had a can of Sego and sat down and studied a map and attempted to determine where I had been.

Last night at the BOQ I took a "Fantastic Journey" by seeing a movie of the same name. It was very peculiar. These three guys and a gal are reduced in size and they go on a mission inside a guy's circulatory system to his brain. Their mission is to destroy a tumor and return before the white corpuscles "whup them". Please do not bother to see it. It did not rain during the movie and we did not have to sit under dripping clothes. But a pipe in one of the upstairs room burst and water poured down on us just as we were passing through this guy's lymphatic system.

By this time you should have received the letter I mailed from the Special Forces camp at Tra Bong. I will provide background information before I pick up my story from there.

You recall the portable Mighty Mite sawmill was sent into the country illegally without approval of USAID, or the military or GVN. Arrangements had been made by an "Alexander Botts-type of salesman" . . . remember A. Botts, the Earthworm tractor salesman from the old *Saturday Evening Post* magazine series? He had talked the military and the Ambassador into selecting a suitable logging area. I will never forget Rufus Page's letter from Vietnam to Chief Ed Cliff, pleading for advice on what to do about this sawmill deal. I was given the letter while I was still in Washington and had wisely written to Rufus, for the Chief, and told him in our judgment . . . "we recognized this resulted in a problem but we should overlook problems and utilize opportunities." Now that government "hogwash" I had written and the chickens had come home to roost, on me, feathers and all! When I finally located the mill behind the barbed wire at Tra Bong I acquired a whole flock of chickens!

Two men showed up in Saigon to install the Mighty Mite portable dimension sawmill. One of them, Bill Bingham was a fine person. He had come from Portland and had gone through the proper procedures, had approvals and had made correct arrangements. But the other one, who I shall call Bob, came into the country unannounced and created many diplomatic problems. Bob was the manufacturer's vice president and general sales manager for SE Asia. He lives in Manila and had wired me in March that they were ready to come to Vietnam and install the mill as soon as we located a suitable site. The site had already been selected by the time we located the mill. We sent a wire to Bob in Manila which was undelivered because he had left for Taiwan shortly after he had wired us. That should have told us we had problems ahead.

We were all set to leave Saigon on Monday 15 June. The Tra Bong Special Forces sent word that loggers had produced 100s of logs. I will not go into all the gory details on how I worked all day Sunday to obtain Bob's travel authorization and reservations on Air America. Actually USAID does not function on Sunday! We were scheduled to depart at 0945 hours and had arranged for our transportation to the airport. I had two large peavies (log handling devices) stored in my room. A small, four foot bellboy carried these tools down to the street. The peavies have five foot handles and weigh about 25 pounds each. It was comical to see my helper struggle with them. These tools have a large hook and a sharp point and created much interest on the part of the pedicab and cyclo drivers in front of my hotel. I told them I used them on the VC. They looked on with awe at these secret weapons. They may have been VC spies. After all the town is full of them.

My driver had first gone to pick up Lewis Metcalf, our new forestry millwright and logging specialist from Verda, Kentucky and the two Mighty Mite men. The driver was unable to locate their hotel and passed my hotel without stopping. I yelled, whistled and waved my arms but he went right on by. I waited since I thought he was surely going to turn around and come back. After 10 minutes I called the dispatcher and learned the driver had been sent back for me. Thirty minutes later he arrived. I almost lost my temper, but then I realized why my Vietnamese teacher, Mr. Long and the others had refused to teach me any Vietnamese cuss words. I only knew the more refined phrases.

Finally we reached USAID #2, the airport shuttle bus point, where we picked up Mr. Thuong, our Administrative Assistant and interpreter. I made it clear to the dispatcher we could not wait for the next shuttle and we would accept the same "number 10" driver. (In the Vietnam vernacular people are rated numerically with No. 1 . . . "so mot", being tops and No. 10 . . . "so muoi", very bad or very poor.) En route to the airport we were detoured to a side street due to a street paving job and we ran into the worst traffic jam I have ever experienced. There were 1,000s of bicycles, cars, cyclos and trucks. A narrow two-lane street had increased to four solid lanes. I almost gave up at that point, particularly when we came upon an Air Force convoy. This consisted of 30 semi-trucks loaded with 500 pound bombs and many security vehicles full of armed guards. The guards are required to wear their flak jackets at all times. They looked ready to cook, as all of us were. At that moment in time it occurred to me that the only way to get out of this mess was to detonate one of those bombs. When we finally reached the Air America terminal we found our plane had departed 15 minutes earlier. I practiced counting to 10 forward and backward in Vietnamese; it helped relieve tension. There was nothing for us to do but return to town and arrange for new reservations through the Embassy travel section since nothing could be done at the airport. We struggled back through even worse traffic congestion and arranged a flight for the next day. That gave us additional time and I spent the day obtaining Bob's permits, ID, BOQ pass, etc.

The next morning we went through the whole routine again . . . hauled the peavies down to the street, told the pedicab drivers what they were and went to the airport. This time we made it. The plane was delayed for two hours and this gave us an opportunity to listen to Bob's tales of sales and other conquests.

On the plane I sat by a Special Forces Sgt. He was coming from a SFs camp in the Delta region and was going to Nha Trang for a few days of rest, which he said included the beach and a bottle. I told him where we were going and he said that is dangerous country. His outfit in the Delta had been hit hard by VC mortar and satchel charges during the five previous nights. He said they just keep coming in waves and it seemed like an eternity waiting 30 minutes for the "Dragon Ship" to arrive. That is a C-47 cargo gunship armed with a series of 20mm Gatling guns. These deadly guns can cover the length of a football field with a short burst of fire. That soldier was ready to unwind before he returned to the Delta.

We arrived at the airport at Quang Ngai on schedule and in spite of our advance notification, no one met us. The phone to OCO's headquarters was out and the Provencal Representative was having a siesta . . . his phone line had been torn down that morning and not repaired. We waited at the sun baked airport for two hours before we could promote a ride to town.

We rode in with an ARVN soldier in a Jeep. All five of us and our gear overloaded the wretched little vehicle. We stuck out and over all sides and the front. We made quite a sight with our peavies and axes Metcalf had brought along. We also had a good supply of tobacco, cigarettes and candy for the Montagnards. You will recall my description of the streets of Quang Ngai . . . a series of chuck holes and tank traps with intervening strips of broken asphalt. It was a rough, axle and tire busting ride. We went to the Provencal Representative's office. He was surprised to learn we had arrived. Oh, yes, he had received my telegram, but you know how it is . . .! After briefing him on our plans he loaded us and all our gear into a Scout and had his assistant haul us back to the airport. We stood around the airport for an hour in the boiling hot sun waiting for the promised aircraft. Storm clouds looked very heavy in the direction of Tra Bong.

Finally a plane, a single engine Helio-courier appeared. Bob, Thuong and I, along with our gear, went on the first trip. Cloud cover and rain kept us low and we flew about 500 feet off the ground. I had a good look at villages, fields and timber. I remembered on the first flight to Tra Bong the pilot said it was not safe to fly under 5,000 feet. We arrived over Tra Bong and I was pleased to see many logs decked at the sawmill site. The crew of Montagnard loggers had produced.

Some one popped a smoke bomb, which produced a trail of purple smoke to indicate wind direction. A Special Forces Sgt. drove a truck to the air strip to meet us and transport our gear. He said they were glad to see us. The District Chief was also waiting and said he was pleased by our return. These people have had many promises made and broken and find it difficult to believe anyone. The French bombed this village during the Viet Minh war and our forces bombed it a time or two. These people had gone all out to produce when they received word we were returning. They had brought in over 100 logs. Several Chieu Hoi had turned themselves in, surrendered their arms and were working on the logging operation.

Bill and Metcalf arrived an hour later. They had a near miss when they were put on a plane going to another destination but fortunately they rechecked the destination and waited for our plane to return. The Special Forces offered us bunks in the above ground team house or in the sandbagged bunkers. The latter are safer but the thought of rats playing tag all night on my mosquito netting was not too appealing. The bunkers have another drawback in that they never cool off from the daytime heat and frequently rats die in the closely packed logs and sandbags. The odor is not aesthetically pleasing. With these thoughts in mind I chose to sleep under the tin roof. During our stay at Tra Bong the daytime temperatures ranged from 110 to 125 degrees. At night it dropped to 100 degrees. Not your kind of weather, Gwen, real cozy and very humid! I must admit when it reaches 100 degrees and 100% humidity, it is hot!

In the absence of the Commanding Officer, who was on a patrol, Lt. Katz showed us around the fort and acquainted us with its features. Tra Bong was the site of an original French Foreign Legion Port. The present facilities were constructed and first manned by Australians. Several fields of barbed wire entanglements tied to steel posts surround the outer perimeter. Local Regional (RPs) and Popular Forces (PFs), or referred to derisively by the Special Forces advisors as "Ruff Puffs", were assigned the task of protecting Tra Bong and were located inside this outer barrier of wire. An inner barrier of dense coiled, concertina barbed wire with steel posts separated the Americans and the RPs. The inner zone could be reached only when the cold-eyed Nung mercenaries dragged back rolls of concertina barbed wire which sealed off the only entrance and exit. Placed underneath and around both sets of wire barriers were sharpened bamboo punji stakes (dipped in feces) pressure mines and trip wires which could set off additional explosives. At the high point on the knoll were the SFs facilities. The team house is a 50-foot long galvanized, sheet iron building, containing a kitchen, lounge and dinner table, several rooms for SFs who sleep above ground and an office. Leading away from the back door is a network of trenches which lead to bunkers, ammunition storage dumps, mortar pits and machine gun emplacements. The place of final retreat to is a tall concrete tower mounting a .50 caliber heavy machine gun which can be maneuvered to fire from slit openings on each side. Beneath the tower is the command post and communication center. Scattered between the various rows of barbed wire and along the outer rim of the trenches were 100s of claymore mines which could be detonated from a number of central locations by squeezing a hand operated generator or "clacker". These wicked, shaped charges were filled with large pellets and were highly respected by both sides.

We were each given assignments in case of an attack. My post was a M-60 machine gun pit and the hospital bunker. My assignment was to man the machine gun until someone was injured, then help "Doc" treat them and return to the gun. They gave me some on-the-job training in operating the M-60 and I burned up considerable taxpayers' money in the process. I was trained in how to load and fire the mortars. We fired several rounds of the large 4.2-inch mortar shells which have a range of up to 6,000 meters and the smaller 81mm mortar.

After that excitement we settled down and visited with the SFs crew. The nine Americans assigned to this camp were pleased to have extra hands. Their assignment is to train the local forces, protect the Tra Bong population and obtain intelligence information. They eat like "kings" on their $10 per person per day allowance. They have various methods for obtaining additional food from Army, Navy, Air Force and Marines. I suspect most of these methods would not have the stamp of approval of the General Accounting Office. They have a thriving industry which markets Montagnard crossbows, handmade bloody VC flags and captured weapons. This provides a good return to the Montagnards for their work and still provides a profit for marketing service. Air Force observation FACs and Caribou cargo pilots for many miles around come to Tra Bong to purchase its famous crossbows.

To accommodate frozen food and beer there are three refrigerators and two freezers. One refrigerator is devoted exclusively to beer. Beer costs less than 10 cents per can at the commissary and everyone pays 20 cents to a cash fund used for the purchase of more beer and other supplies. We all kept track of our "beer score" on a chart. The cold beer and a jug of ice cold green Kool Aid were welcome relief in that "hotter than Hades" tin building.

The latrine was near the firearms cache and could be located day or night by the odor. The odor became unbearable each day when the half-drum oil barrel was lifted out and the contents burned in an open pit.

Following an uninterrupted, good night's sleep we were ready to install the Mighty Mite. We met with the District Chief and agreed on a site for the mill. This site was immediately changed by ARVN Sgt. Bay who said he planned to erect a guard house at that exact place. To keep peace in the family we packed up and moved to a different site.

The local people were fascinated as the Mighty Mite emerged from wooden crates. In fact we had so many spectators, men, women and children, that we had to continually direct them to move back. Metcalf and I scaled logs and the others set up the mill. Thuong was indispensable. He explained to our six trainees how to assemble the mill, service and operate it. The SFs had trouble unloading the 1,000 pound crate containing the engine and parts. There was no problem removing it from the C-118 plane when it arrived, since the Caribou has deck rollers. The loading dock also had rollers. We placed pipes on the SF's truck to facilitate loading and unloading. The problem was lowering it to the ground from the Dodge Power Wagon. In spite of the number of helpers it was dropped. When we uncrated the materials we found the air blower and air filter on the Volkswagon industrial engine were slightly bent. We were able to straighten these items and by 1630 hours we were ready to start. The audience gathered around closer and when we did start it required all of us to keep men, women and children away from the rapidly spinning saw and occasional flying slabs. For the next 30 minutes we let the District Chief operate the controls. You could see pride showing on his face as his people cheered him on. At 1700 hours it was 115 degrees and we were glad to quit for the day.

Back at the SFs camp Captain James E. Callahan had returned from patrol with his two American enlisted men and 150 "Ruff Puffs". They had been out for five days on an intelligence gathering mission. Earlier that day they had almost trapped a small VC force but the local troops fired prematurely and scared them off. That happens all too frequently on these patrols and indicates the sympathy of the local friendlies. Information gathered by this patrol indicated the VC was preparing to hit Tra Bong. The CO verified this information with the District Chief who had his own intelligence sources. A FAC observer called in just before sundown and reported he spotted the VC cutting a path through elephant grass to a location that appeared suitable for launching a mortar attack against us. Yes, against us! While we waited for further instructions I helped "Doc" with the local sick call. We probed and cleaned out bullet holes, lanced boils, cleaned and dressed sores and treated cases of malaria and dysentery.

The Captain then briefed nine SFs, four civilians and three interpreters on the plan of action. This was nothing new for Thuong since he served the SFs near the DMZ before he had gone to work for USAID. I was assigned to the same post . . . hospital bunker and machine gun. The regular medic's alternate post was a 81mm mortar. He showed me his surgical supplies, demerol, morphine, plasma, bandages, splints, etc. I then pitched in and helped Sgt. Pascual and SP/4 Bowen adjust the 4-Duce (4.2-inch) mortar and the very substantial base supporting it. We prepared 200 rounds for the 4-Duce, including W/P projectiles (known as Willy Peters or white phosphorus marking shells), illumination flares and regular high explosive charges.

238

Sgt. Pascual was satisfied with the setup and then directed me to fire 20 shells into the suspected VC area. This was about two miles away on a ridge overlooking the valley. Mortars are fired by setting the projectile in the opening and holding on to it with one hand. The shells are quite heavy and you must have a firm hold. The next step is to look around, yell "on the way", let it drop into the tube, take a step away, bend over and hold your ears tightly. "Bam" and the projectile is on the way. The distance is determined by the number of attached powder increments. More bags of powder equals more distance. Instead of exploding out of the tube, one of the shells burned and left the tube in slow motion, going end-over-end. Everyone hit the deck as it landed about 100 feet away and just missed, by about two feet, falling into a 55 gallon trash barrel. Those crazy RPs came out of their bunkers and hit at it with a stick. The CO was heard to mutter something like, "I wish the damn thing would go off!" But it did not since it had not traveled far enough to fuse. Before I ceased fire we sent up some illumination flares. These are what I had been seeing around Saigon all these weeks. They hang in the air on parachutes for a minute or two and light up the area as bright as day.

We secured the area at 2100 hours and played poker. It was a 25 cent limit with unlimited raises. I won $28. With the exception of the guards we went to bed at midnight.

About 0230 hours a machine gun went off and we rushed to our assigned stations. I was in the trench near the hospital bunker. In front of me was an M-60 machine gun. Beside me were clackers ready to fire a number of claymore mines. The hair on the back of my neck stood on end and the adrenalin was flowing. It was a tense moment! (My thoughts at the time: (1) Recruiting in Iowa during World War II was never like this and (2) they never told me about this in forestry school). Someone had given the signal initiating a number of events. The mortars and the RPs 105mm artillery began firing at the suspected position. Then at a prearranged radio signal the 175mm howitzers located at Cu Chi, some 20 miles way started firing. Beyond the mountains to the northeast we could see the flash of those big guns, then hear the projectile fly overhead, then hear the sound of the gun firing, then see the flash of the explosion and almost immediately hear the blast and feel the rush of wind from the impact area. After 10 of these heavy rounds nothing was heard from the suspected VC position. Since the artillery shells were passing directly over our heads I asked one of the SFs what kept one of those shells from falling short. He shrugged his shoulders, looked up at the heavens, crossed himself and said "computers".

At 0400 hours all of us, with the exception of 6-foot-6 Sgt. Gray hit the sack. He remained with the other guards for the remainder of the night. The other guards, the Chinese Nungs never seem to sleep. There is another very effective guard. That is a large goose, which never sleeps and raises a loud fuss if anyone strange is detected in the area. Its effectiveness determines how long it stays out of the cooking pot.

At 0600 hours the bugler at the CIDF (Civilian Defense Force) gave out with his tune of the morning. It was not a Harry James' rendition but it suggested it was time to rise. On the American Forces radio the colorful Saigon DJ give out with a loud "G-O-O-D M-O-R-N-I-N-G, V-I-E-T-N-N-A-M" to start our day. In this far north area there is too much interference from Radio Peking to have good reception, therefore short wave is used.

On this day, Thursday 18 May, we conducted a training program for six refugees. Thuong, our faithful interpreter, accurately relayed instructions. We could not have carried out our program without him. The SFs provided taxi service to and from the mill. They also helped us scrounge miscellaneous parts and supplies. If they did not have it, or could not make it, or steal it, they could have it flown in within minutes. I gained high respect for those Green Beret warriors.

One of the trainees had caught my eye during my first visit to Tra Bong. His name was Phan Van Le, a refugee from the coastal area of Tuy Hoa. The only mechanical item he had ever operated was a water pump. He was keenly interested in the sawmill and listened to every word of instruction. You could tell by his expression that he appreciated the opportunity to learn and work.

At the mill that day we had over 300 spectators. It kept us busy just keeping them out of harm's way. Vietnamese farmers carrying plows on their shoulders stopped to look at the thing. Naked children and Montagnards dressed (or undressed) in their loin cloths watched in opened-mouth fascination. Children followed me everywhere I went, begging me to take their photographs and referring to me as "Number 1". A few Montagnard women and girls came by on a path near the mill. They were shy, giggly and difficult to catch on film, but I did catch a few. I must try and round up a Polaroid camera so I can give them pictures of themselves and their friends, however that might start a riot. You will be interested to know I have taken black and whites and color slides of everything I have described and more still to come.

I had a supply of cigarettes, pipe tobacco and candy. These kept the loggers and sawmill trainees in smokes and the children happy. I gave the Montagnards tobacco for their pipes, which were made out of cartridge brass and aluminum. These must have smoked terribly hot. Other pipes were made from curved hollowed-out roots and others were only a straight piece of wood or bamboo that was pierced and flared at the end. Whatever they used for a pipe they enjoyed my pipe tobacco. I offered several people some of my Beechnut chewing tobacco and showed them how I used it. Men and women alike would take the offered chew and immediately spit it out. They are discriminating people after all. I suspect some of the people I gave tobacco and cigarettes to must have been VC. At least I gave them a different perspective of an American. One small child followed in my shadow and had a small top that rested on the hollow base of a gourd. When he pulled the string it spun and he grinned at my interest.

About 1100 hours the heat became absolutely unbearable and I was happy for the local "pot time". I am uncertain if that means they sit on it or smoke it, but anyway I welcomed it. By the time I had returned to the camp I was weak and felt like I was going to be sick. I took two salt pills and drank a quart of water. That was the solution and I quickly recovered. From that time on I took two salt tablets in the morning, afternoon and evening and had no more problems.

Accompanying the excessive temperature was 100% humidity. I can tell you it was really cozy. I wore only work pants and a cruiser vest, which were soaked with sweat in a very short time. One day we had a light haze over the sun and that combination just about finished us off.

The District Chief told us he would place a fence around the work area to keep the people out. He had given strict orders that nothing was to be taken from the mill site or be disturbed. We had a difficult time convincing him the sawdust should be removed each noon and evening. We did this to reduce the fire hazard and to prevent someone from hiding explosives in the sawdust.

The only problem we had with the mill was a roller on the return off-bearer. It was knocked off when lumber handlers were too slow in guiding the lumber coming from the saw. It was nothing serious. We wired it back on and requested a welder, who flew in and repaired it.

That afternoon and evening was devoted to weapons. I had target practice with an M-1 30.06 equipped with a sniper scope. That highly accurate weapon would make a good deer rifle but the 12 pound weight would finish you off. I found the M-16, weighing only three pounds, quite accurate. That is the type of rifle Sgt. Brown, shown on the front cover of a recent issue of *TIME*, was carrying. That magazine contained an excellent article about Negroes in Vietnam. As long as these rifles are kept clean they operate without a problem. But let them get dirty and they are certain to jam. The one I used jammed occasionally when I used it on full automatic but the malfunction was easy to clear. That evening I fired off two belts of .50 caliber ammunition from the heavy machine gun on top of the tower. That is a very respected and accurate weapon. I could easily hit a target at 1,500 yards. My ears are ringing from all the noise, but the Green Beret insist on us being familiar with all their equipment.

That night we played poker and I came out even for the series. I was relieved to lose the $28 I won in an earlier game because I did not like taking their money back to Saigon. That night at 0300 hours the Sgt. on guard was cleaning his .45 caliber automatic pistol and it went off accidentally blasting a large chunk out of the concrete floor. I was sleeping about 20 feet away and did not hear a sound.

The next day was Friday, Ho Chi Minh's, the North Vietnamese leader's birthday. The CO expected problems that day and suggested I carry an automatic rifle. All day I carried an M-2 automatic carbine. It was very awkward to have the rifle hanging over my shoulder, in a ready to fire position, while I was scaling logs and helping around the sawmill. We had only a few visitors that day and we did not know whether it was due to Ho's birthday or because the District Chief had warned people to keep away.

The 1100 hours rest stop was welcome when the temperature hit 117. After returning to camp I had several glasses of Kool Aid. The cook kept a 5-gallon thermos of ice cold cherry or lime on tap at all hours. It certainly was a life saver on those hot days.

Needless to say I returned to Saigon safely. Gwen, it is Sunday evening and I have just about exhausted my supply of paper. I have more items to cover and will conclude the story of our activities at Tra Bong in the next letter. I believe we are helping these people. It has been encouraging to see their response to our training and their appreciation of being able to perform productive work.

Tonight Barry Flamm, Martin Syverson and I enjoyed delicious fillet mignon and all the trimmings at the REX BOQ. Afterwards we relaxed and listened to the bands. We always enjoy hearing both the one from Manila and the local talent. The latter is always very popular with the GI audience and played American songs, from hill billy, rock and roll to classic. They play two 1 1/2 hour sets with only one 10 minute break. It was quite remarkable. They had a lead electric guitar, drum, piano and two women singers who had excellent voices.

I worked in the office all last week and plan to be in for the next two weeks. In less than a month I will be coming home. I can hardly wait to see you all. While I keep very busy I miss you all very much. I reactivate my language lessons during lunch break tomorrow and can keep busy in my spare time studying. Be assured I do not take any more long walks around Saigon at night. That is all the news for now. Please keep your letters coming and I will let you know when my trip home is confirmed.

I received some sad news yesterday from Ray Housley and Frank Carroll. The Southwest Region has been having a severe fire season. A former associate of mine, Rolfe Hoyer, was killed on a forest fire on the Apache National Forest. We knew him and his wife when he worked on the Coconino. Rolfe apparently was searching for an Assistant Forest Supervisor who he believed was on the head of the fire. A crown fire exploded and trapped him. He inhaled enough hot gases to sear his lungs and died five hours later.

Ray Housley told me my old boss Director Byron Beattie was on the Coconino recently and tried to talk him into transferring to the Washington Office. Ray told him he had stopped following Cravens when he became Forest Supervisor of the Coconino and had no interest in going to Washington or following Cravens to Vietnam. It appears Ray has settled in for a while.

I planned to return to Tra Bong on the 19th to check the sawmill operation but I will send two of our men instead. I will be tied up with planning, paperwork and problems until I depart for the US.

I will finish my Tra Bong letter this weekend.

I am on the roof in the early Sunday morning semi-sunlight. It is relatively cool at this time. As I look across the lichen stained red tile roofs I can see where dozens of electrical generators are belching blue diesel smoke into the early morning air, already polluted from hundreds of smoky motorcycles, scooters, cyclos and taxis. All US billets are required to have generators and we are major contributors to air pollution. As I type this letter the hotel electrician is squatting on the floor beside me watching my, not too nimble fingers run across the key board. We have been having a limited conversation in Vietnamese. Most of these people are fascinated when they hear an American speaking their language. I enjoy the practice.

The successor to the U-2 spy plane, the SR-71, just passed overhead flying in the direction of Cambodia. The plane is a jet with huge wings in comparison to conventional jets. Other air traffic consists of the helicopter gunships which patrol the town and the environs continuously. There is no artillery fire at the moment but there has been a tremendous increase in that activity at night surrounding the city. We never hear what it is about, but it is probably to show and tell the VC we are still here.

For a month I have been without a language teacher. Now we have been assigned a young Saigon University student by the name of Kim Chi and we have returned to our lunchtime language sessions. I kept up my studies and consequently did quite well in the review. We have lost four of the original eight

243

students. The remainder of us will benefit from the smaller class. Our class consists of Roy Yamamato, Miss Hamilton a middle aged secretary with a strong British accent and a young college graduate assigned to USAID's personnel management program. I had to laugh when the young man told me about the excessive paperwork. I told him it went with the personnel territory and was typical of every government agency. He wonders if he is in the right field. I will not give him the facts, perhaps someday he can improve the system, if he stays with it.

I have added some class to the Forestry Branch with the addition of a new clerk. Our Division Administrative Officer gave me the task of interviewing a new applicant. Her name is Mrs. Hoang Minh Hai. She is a very pleasant person and has been taking typing and English classes on her own time while she searched for employment. She impressed me as a candidate with potential so I recommended the Agriculture Division hire her. While all the branches were arguing about who would get her, I put her to work in forestry. Thuong is training her on procedures. She is a so-so typist and limited in her ability to speak and understand English, but the other members of the Forestry Branch approved of her. The neighboring branches acknowledge she adds class to a room that is festooned with maps of Vietnam, pine cones from Dalat, exhibits from a match factory, sawmills, plywood plant, etc. Our walls also display framed National Forest color photographs, including Red Rocks Crossing on the Coconino. Mr. Long says "the Sedona picture is very beautiful, but where are the trees?" I told him Ray Housley is working on that!

Friday after work I was invited to go with a friend, Ray Russell, to the Cercle Sportif Saigonnais Club. Formerly this was an exclusive French club, where the Vietnamese were not members, only servants. The Club has 20 tennis, volley ball and badminton courts, an olympic size swimming pool, riding stables, track for running and a wonderful gymnasium with all kinds of exercise equipment, including foils and masks for fencing. There is a delightful veranda and inexpensive dining facilities.

I participated in an hour-long exercise class which was conducted by a fat, jolly Frenchman who counted out cadence and exercises in a continuous chatter of French. I did not understand a word, but I followed his actions and did well. Vietnamese men and women were at one end of the floor and Americans at the other end. I surprised Ray and the others by being able to keep up with the strenuous exercises. Remember on 1 December I weighed 212? I now weigh 83 kilos (182 pounds) pretty good for an old Iowa farm boy!

Yesterday Barry Flamm and I were invited to return to the Club. We spent Saturday afternoon enjoying the facilities. I played two hours of tennis, for the first time in almost 25 years, and I could still cut a mean ball! My partner was a pleasant turbaned Indian who was a member of the ICC (International Control Commission which is supposed to enforce the DMZ). He was an excellent tennis player and must spend all his time at it. Other members of the ICC are Polish, who are said to be well trained Iron Curtain spies. We have orders from the Embassy to report any contacts we have with this group. Once I rode in an elevator with two tough looking Poles. They have a cold look, never smile or say a word. I suspect they speak perfect English and probably speak and understand a dozen other languages.

The word "perfect" reminds me of a recent conversation with Mrs. Hai. One day she turned out one of her rare, but flawless letters and I told her it was perfect. She looked up at me, shook her head and said, "No, only Heaven perfect!" That was something to think about.

Now to return to our experiences at Tra Bong . . . the afternoon of Ho Chi Minh's Birthday was hot. At the Special Forces' camp the thermometer read 117. We had only five spectators, but the full mill crew was at work when we returned. The District Chief came by at regular intervals and we could see his troops watching over us from their watch tower and concrete bunkers. During the afternoon one of the FACs dropped in for a visit to the mill. He had been flying his "bird dog" plane over the Cambodian border. He reported no sign of activity. Those guys are a walking arsenal with all their ammunition, pistol and light weight version of the M-16 automatic rifle. In addition they wear a jungle survival pack to use in case they are stranded. With all that load, in such high temperatures, it was all his small L-19 plane could do to get airborne. I would like to have gone with him to take a look at the nearby forest areas but we are expressly ordered never to fly with these pilots for obvious reasons.

That afternoon we viewed a dazzling exhibition of acrobatics by a F-105 jet. This pilot is a friend of the local SFs and comes to Tra Bong for his R & R (Rest & Relaxation). He was less than 50 feet above the ground and doing 600 knots when he passed over the mill. He then climbed straight up performing rolls, loops and flew upside down. The SFs believe the presence of this pilot and his cohorts, some eight minutes away, impresses the local VC. We could see his bomb and rocket racks were empty. He reported in by radio to our camp saying he was just "stopping by" on his return from a ground support mission along the Cambodian border. I suspect these planes and the SFs make a number of unauthorized excursions into Cambodia.

After work I went with one of the SFs to the village laundry to pick up clean clothes. They had none of mine because I did not think my clothing could stand up to beating on rocks. The other civilians had sent theirs and when the clothes were returned they found several items missing and others were worn out from scrubbing on rocks. So much for the local laundry.

After picking up the laundry we drove by Jeep to the west end of the village and stopped at a gate. No one goes beyond this point without a full complement of well-armed men. The nearby tree line provides perfect ambush cover. This route took us through several Montagnard villages. The men and children came out to greet us. I tossed them some gum and candy. Some of the children ran along side and behind us yelling "So Mot, So Mot" ("Number 1, Number 1"). That was a good indication we were accepted. Up north near the DMZ, Quang Tri, Hue and in parts of Quang Ngai Province the children yell "So Muoi, So Muoi" ("Number 10, Number 10") at Americans. This indicates VC control over the local population and Americans are not welcome. Some of the Montagnard men waved and I recognized them as loggers and those I had either taken their photograph or given them pipe tobacco. We caught brief glimpses of the women peering from the doorways of the long houses. We saw other women bringing in baskets of fuel wood, cinnamon bark and others filling jugs at the well and bathing. They all turned their backs toward us as we came into sight. I took some photographs of their rice fields and the tiny grain storage huts perched atop stilts. I noted baffles on the stilts to keep the rodents out. My color slides of this scene at sundown are very good.

That evening I relaxed and listened to tape recorded music and worked on my notes. It was a peaceful night and the conclusion of "Uncle Ho's" birthday.

Early Saturday morning intelligence agents brought word the VC was going to hit the mill. Captain Callahan asked me to accompany him on a trip to evaluate possible alternative locations for the mill. We went fully armed and accompanied by the CO's interpreter "Henry". We looked at a site adjacent to the SF's camp. This could be covered by all the guns of the compound and was used by the helicopter gunships that bring in Rangers and other reinforcements as needed. We called this Site No. 1. We then went to an area within the densely populated village. This was a good location and readily accessible to loggers and others. However it was out of sight of the SF's camp and the District Chief's fort. We identified this Site No. 2.

While at Site No. 2 we were invited into Chief Interpreter Henry's new 3-room home for refreshments. Walls were plastered with cement and the inside mud walls were white washed. The place was very neat and clean. Located in a prominent corner was the family altar. Joss sticks and a candle were burning in remembrance of his ancestors. He turned his battery operated record player on good and loud, and we listened to several Vietnamese classics. The music was very high pitched. His wife, who we never did see, prepared a delicious coconut drink. I drank it with some reservations, but after all I had been drinking treated river water all this time and did not want to offend Henry. He served us small delicious bananas and mangoes. Next door at the school we could see children leaving morning classes. They clutched their American-supplied text books, notebooks and pencils and waved to us as they went home.

The Captain, Henry and I returned for a look at the existing sawmill site, designated as Site No. 3. We then went to the District Chief's office and discussed the problem and alternatives with him. We agreed Site No. 1 should be eliminated. At this location the VC could take the mill with a large attack force. True, they would suffer great losses but they could probably destroy the mill and demonstrate the Americans were not strong enough to defend the mill, or as a matter of fact, the village and the people. We also eliminated Site No. 2 on the basis that it could not be defended and its presence would favor one part of the village over others. We settled on Site No. 3, the present location. The District Chief believed in that location it belonged to the entire village and all of the villagers would help him and his forces defend it. This was exceptionally sound thinking.

I returned to the mill with the crew for the balance of the morning. We were only too happy to quit at 1100 hours when the temperature reached 115. I wore only my Filson cruiser vest, work pants, bush hat and boots. I was completely soaked with sweat when we quit. Salt pills kept me in good condition and I suffered no more heat exhaustion.

When we returned to work that afternoon the thermometer registered 120. There were no visitors. We did well with our lumber production, stacking, log scaling and other chores. As we left at the end of the day the logging crews came across the river with logs larger than any we had seen before. I was determined to help devise a wheeled trailer to help ease this back breaking labor. (Later in Da Nang we arranged with the local vocational school to make a man-powered logging trailer.) On the return to our camp we saw the District Chief in front of his headquarters admiring the decorations that had just been hung up in honor of Buddha's Birthday, next Tuesday. The numerous barber shops were busy as we drove by. The SFs patronize only one barber and when they go for a shave they take along an extra man who holds his rifle ready and pointed at the barber in case there is any tendency for the razor to slip!

After a dinner of liver (again) we drove through the western part of the village to the river. We stopped to look at the village's other sawmill. This is a hand-operated whipsaw operation which was the only previous source of lumber. These men glared at us as we watched them work. I made arrangements later to have the sawmill provide them with some timbers and slabs to saw. This helped restore more cordial working relations between the two competitors.

We then drove up the river bottom past Vietnamese hootches (poles covered with palm fronds). Vietnamese were bringing in the cattle and ducks for the night. Some were already in pens to protect them against marauding leopards, tigers and VC. The Montagnards were rounding up water buffalo. They do not use these for work, instead they are kept for sacrificial offerings on special occasions. We stopped a short distance below the waterfall and turned around as the setting sun made the sky and countryside very spectacular. We scooted for camp and waited while the Chinese Nungs pulled aside the barbed wire barricade to let us enter. I then went to the top of the watch tower to talk with two SFs and the interpreters. It felt cool in contrast to the recorded high temperature of 125. The valley looked very beautiful as the setting sun back lighted the towering cumulus clouds off toward Cambodia. Again we had another peaceful night.

Sunday we worked at the mill. I scaled logs while the others trained the crews, including a Special Forces mechanic and a Vietnamese named "Mo", to service and maintain the mill. Most of the Vietnamese working for the SFs were given nicknames. Interpreters were named "Playboy", "Philosopher", etc. Henry, of course, was the chief interpreter and SF's top scout.

Shortly after dark the moon dropped behind the clouds and vanished. An unexpected torrential rain came slamming down on the tin roof of the team house. The rain was deafening. Trenches and mortar pits became pools of red mud. The short rain was followed by attacks of hungry, relentless mosquitoes which loved the Green Berets "bug juice". With the temperature remaining over 100 degrees, it was not a memorable evening!

The next morning we detected a plan to discredit the District Chief. His ARVN CO apparently heard there was a good thing going and money to be made at Tra Bong and he came from Quang Ngai to get in on it. When the District Chief told him the mill profits were going solely to the people, the CO became enraged and called for an immediate muster of the District Chief's troops. The purpose of this roll call was to try and catch the District Chief padding the payroll with non-existent soldiers . . . an all too frequent ARVN scam. Fortunately all troops were accounted for and the Chief came out clean. Personally I believe the District Chief is one of those rare individuals who wants to help his people and make no personal gain. At the request of Special Forces Captain James E. Callahan I later brought this matter to the attention of top US military advisors in Da Nang and Quang Ngai. They promised to watch this situation closely and protect the District Chief.

That was the last day for me at Tra Bong. I departed with one of the Mighty Mite sawmill men and went to Da Nang and

Quang Ngai to report on our progress. Before leaving the District Chief presented each of us a large crossbow and select specimens of cinnamon bark that had been harvested in the nearby mountains. A Montagnard chief gave me his brass ceremonial bracelet. The District Chief had all his troops stand at attention as we climbed into a C-118 cargo plane and departed for Da Nang with a load of lumber.

At Da Nang we briefed the Regional people on the project. With some reservations I believe the project has been successful. The people have been given new hope. I neglected to mention the District Chief flew into Quang Ngai one day and purchased additional rice and fish sauce for the loggers and sawmill workers. The purchases were made with receipts of the timber cooperative. The people were visibly pleased with their reward. This was the first productive income that most of them had in many years. My reservations about the project concern the machinery. The mill is small and lightweight. I doubt it can hold up under continuous use in the hands of relatively unskilled workers. Time will tell. As far as the local people are concerned its value will increase with each day of use. This was an interesting and rewarding experience. I believe we did a great deal to raise the morale and income of the remote settlement of Tra Bong.

After meeting with the Regional people in Da Nang we crossed the inlet on a Navy ship to the Headquarters of "Eye" Corps and briefed Marine Corps Commanding Officer General Lew Walt and his staff on the results of our project. They agreed this was an excellent example of what could be done by the military, civilians, and local people working together.

The next day we flew by Air America in a Porter Pilatus airplane to Quang Ngai for another briefing, this time with the Provencal Representative. The Porter is a strange looking, ugly duckling airplane that can take off and land in about 100 feet. It climbs practically straight up and is ideal for short runways in populated areas or for those areas which have snipers around the airport. Most of the day was devoted to briefings and return to Saigon.

On the day we departed Tra Bong an entire platoon of VC attempted to come in (Chieu Hoi) and turn in their arms but other VC fired on them and spooked them back into the hills. The following day, on Buddha's Birthday, the VC leader and one of his men surrendered and came in to the government side. The others were said to be waiting in the mountains to see how their leader was treated and if he received work at the sawmill they would all come in.

I will advise you as soon as I learn what dates the Minister of Agriculture and his family will be in Washington. I suspect US teen-agers will make a major impact on their 17 year old daughter.

Tomorrow I will contact the travel section and obtain my flight numbers and verify my TDY in Washington and travel to the West. Then with knowledge of my flight numbers you and the girls can make your arrangements to fly west with me. Communications between Saigon and Washington are slow this week due to the Middle East War. It appears that the Israelis fought a "good, clean" war, but after all, they did not have President LBJ's interference. The Army officers I eat with at the BOQ talk about the Middle East War and long for a "good clean" war where there is a front and a rear, no jungles, no swamps, no mountains and most of all, a war in which you can tell who your enemy is.

18 June 1967

Happy Father's Day to me! I wish I was there to celebrate with you.

I am still waiting for clearance and approval of my TDY and travel authorization. The round trip communication between Saigon and Washington takes so much time. There should be an easier way to facilitate obvious plans and approvals.

Last week and up through today has been very busy for us. Everything seems to take twice as long to arrange. I worked with my men most of the week preparing work plans and firming up their field plans.

Last Wednesday I went with Tan to install a new supervisor at Phuc Cuong, northwest of Saigon. We departed by car at 0900 hours and traveled through some interesting farm country. The plantations supported coconut, betel (the nut women chew), manioc, rubber trees and a variety of garden vegetables. We experienced no problems since the road was well traveled by the Army, both US and ARVN.

The new supervisor was a former National Assembly representative, in fact he helped draft the constitution. He apparently had enough influence to be appointed or buy the job. This presents a problem I want to investigate before I leave here. The sordid details make these appointments much worse than our spoils system. The new man welcomed me in a carefully prepared speech. After the meeting and speeches we looked over his pathetic facilities and equipment. The motor pool was empty. Building maintenance was completely neglected. It is a mystery how such officials are able to carry out any program.

Following the official program we visited a lacquer ware factory. This particular shop has a reputation for producing the best lacquer ware in Asia. They had gorgeous chests, plaques, boxes and furniture. Gwen, wait until you see the three items Tan and the owner gave you. That should keep you guessing! I will try and bring those items plus the others I have collected when I come home next month.

Following the visit to the factory we went to lunch at a Vietnamese restaurant. Tan knows how much I enjoy the food. We had rice, noodles, pork, chicken, duck, soup, hot fish sauce, eggs, rice wine, and mangos. Tan was proud of me when I ate the eggs. Yes, you may have guessed, they had been incubated for 11 days and were actually delicious. On the 12th day the feathers form. When you break open the shell you see the head, a beak and a white opaque cooked eye staring at you. When you crunch into the skull it provides an odd sensation and a perceptible sound. When I described this delicious meal to the women at our office they had to leave the room. I expect you would also be squeamish to receive my first hand report.

Barry Flamm and I continue going to the club with member Ray Russell on Monday, Wednesday and Friday at 1830 hours. Sessions last for one hour and I have been doing very well with the exercises. Weight lifting, stomach and leg exercises are firming up my muscles and help keep my weight down. The well equipped gym is not air conditioned so we have a good steam bath at the same time. Following the sessions we have a cold shower, since there is no hot water. It is a very invigorating place! Later we have dinner on the patio porch. The French food is delicious, non-fattening and less expensive than the $1 Special at the BOQ. Barry and I have submitted applications for membership and we may be elected in two or three months. Meanwhile we go as Ray's guests. He has been trying for months to interest someone from the office to join him, but they are all too fat and lazy. Most USAID workers have their drivers take them home after work, where they have a drink, eat and watch TV. I believe my life style is much better. Sometimes after one of these sessions I go to the Hotel Splendid BOQ and have a large dish of vanilla ice cream and take in a movie.

Today I prepared a statement for the Director of USAID to use in his briefing session this week with Defense Secretary Robert MacNamara. At an Agriculture meeting all of the branch chiefs were briefed on what was needed. I had our forestry statement and graph done in an hour following the meeting. Other branch chiefs were still trying to figure out what to prepare when I left the office.

Flamm and Syverson leave tomorrow for Quang Ngai and Tra Bong to become familiar with the sawmill operation and help the people prepare a more effective marketing program. Metcalf and the representative of the sawmill manufacturing company have been in Tra Bong all this past week. The manufacturer was so interested in the project that he sent his representative from Portland back to Vietnam to make modifications which should help the operators. While my men are in the field I will stay behind and take care of the office and the crises as they develop.

This outfit seems to have an hourly crisis and demand instantaneous reports. The Forest Service provided me with good experience and training to handle this task. I was notified that Chief Cliff sent Secretary of Agriculture Freeman a copy of my report on the Tra Bong project.

You may recall Mr. Passey the Agriculture man from Boise. He was the oldest of the group from FHA that had dinner with us. I mentioned he appeared too old and not in very good condition for these conditions. He has had numerous sieges of dysentery, infected sinus, a kidney stone and finally an operation to remove a cyst from his kidney. Currently he is in the hospital and slowly recovering from the operation he had last Thursday. I see many elderly overweight Americans and they are suffering.

Life goes on about the same in Saigon and I have resumed my lunchtime language classes. I have not been lost recently since I take no more evening hikes. However I believe I am safer on the streets here than a person would be in southeast or northeast Washington, DC or Cincinnati. Sounds like things have been a mess all around the US with the riots and protest marches.

Thanks for sending me the clippings. Most everything printed in the articles about Saigon is true. The smell of rotten garbage which accumulates in several areas along my various routes to work is bad, as are the rats. The smell of fumes from vehicles and generators is terrible and the air is blue. Walking from the BOQ to work this morning I passed four American secretaries. It was refreshing to smell their perfume which momentarily mitigated the carbon monoxide, garbage and urine odors.

Take a deep breath of that fresh clean Bethesda air for me.

Entering mine field surrounding Quang Tri Sawmill near DMZ 196
(USAID/Forestry Photo)

Chapter XIX

Home Leave

24 June 1967

My request for leave (visitation, as it is called around here) and TDY (temporary duty) has been approved! I will be heading in your direction on 3 July. I miss you all very much. Being on the other side of the world from you is a strange and lonesome feeling.

Enclosed is a copy of my flight schedule. I leave Saigon at 1300 hours on 3 July, go to Hong Kong on Air Vietnam ("Air Nuoc Mam") and 1 1/2 hours after arrival I head across the vast blue Pacific to Guam, Hawaii and San Francisco, where I arrive at 1945 hours . . . the same day I left Saigon so to speak. It will be a long hard trip but I will spend a night in San Francisco before heading for Cedar Rapids. I will call you as soon as I contact my Mother, so expect a call about midnight your time on the 3rd.

Please make your reservations to Flagstaff and San Francisco and we can travel together. A full day in Flagstaff should provide me with the time I need to obtain information on the Forest Service's herbicide program. Then we can all go on to San Francisco and meet Director Tan. He is looking forward to the tour of the United States we arranged for him. I want to show him the redwoods at Muir Woods and some of the sights of San Francisco before he goes on to the Pacific Northwest. This will provide time for you to do some of the touristy things and for us to be together.

I went to Tan's for a very pleasant dinner party last Thursday night. His food was delicious . . . no eggs!

Weather has been typical of the monsoon season . . . rain each day and very high temperatures. Air conditioning at the office has been out for the past six weeks and we see no evidence it will be operable in the near future.

I will close now. Keep your fingers crossed for me and I will be with you before long.

* * *

Note: The flight to Hong Kong and San Francisco departed as scheduled on 3 July and I was miserable! I picked up a bug during my last two days in Saigon and had a severe case of dysentery, all across the Pacific. I was really wrung out and exhausted when I arrived at my hotel in San Francisco, about midnight on 3 July.

The trip to Cedar Rapids was little better. My Mother had adjusted well to being a widow and had things under control. She had sold the ranch, a better term would be "Gave it away" and moved into an apartment in the city. I paid my respects at my Father's graveside and departed for Washington.

The brief time with Gwen, girls and Myrl and Gene Jensen was very fine. TDY in Washington, DC gave me an opportunity to report to Chief Cliff of the Forest Service and Lester Brown in International Agriculture. While in Washington I prepared a story for *American Forests* about Tra Bong, "The Sawmill in Vietnam". (Cravens, 1967)

The trip west with the family gave me time to be with them and obtain additional information regarding Forest Service herbicide activities. I was particularly interested in operational and safety aspects. I met Director Tan at the San Francisco Airport and spent two days showing him Muir Woods and San Francisco, including some of the off Broadway attractions. I escorted him to the airport for his flight to the Pacific Northwest. I had a good day with Gwen and girls before we went to the San Francisco airport for a parting of our ways . . . they departed for Washington at 1300 and at 1400 I boarded Pan Am for the long flight back across the Pacific.

* * *

13 August 1967
I guess we all arrived safely at our respective destinations. Our days together rushed by too fast. Speaking of being fast, I calculated with me traveling west at 630 MPH and you going east at a more rapid rate, we separated at the rate of about 1,400 MPH. Every seat of my plane was filled on the flight to Hawaii. The movie on the plane, "Barefoot in the Park", helped pass the time. I arrived in Hawaii about the time you arrived in Bethesda. Rain had set in for the night and I caught a bus into Honolulu and took in a double feature "Spartacus" and "The Young Warriors". Those movies helped pass more time and at 2300 hours I returned to the airport, read and waited for my departure. I have the impression I am sentenced to spend the best years of my life waiting in airports.

The plane to Guam was completely filled with family members of military people. The ones I talked to were Air Force families related to the B-52 bomber crews that "service" Vietnam. From Guam on there were only 20 passengers on the big Boeing 707. We rattled around and had plenty of room to stretch out and try to sleep. I did not get much sleep, but I had time to think about the wonderful short weeks we had together.

254

You all seem much dearer to me and I fully realize how much all of you mean to me. Our time together in San Francisco was really nice. Everything worked out just right.

I arrived in Saigon on schedule. The only problem I experienced was a missing suitcase. One arrived but a second one that was Barry Flamm's suitcase did not appear. I had a difficult time explaining to a Vietnamese woman in Pan American that the bag had my name and Saigon address on one baggage tag and Barry Flamm's name and a Cody, Wyoming address on the second tag. With the claim made I was greeted by all the forestry crew. We had lunch together and then went to the office. Everyone greeted me like a long lost friend, particularly the Vietnamese at the office and the hotel. It made me feel good. I worked at the office until 1600 hours, and then "bam", jet fatigue caught up with me. I barely reached the hotel before I collapsed and slept for the next 14 hours.

The crew took good care of business while I was gone and there were no immediate crises so I checked on the missing suitcase. Pan Am at Guam and Honolulu could find no trace of it. I met the next Pan American flight and on the fifth and last luggage cart I could see the missing suitcase. Mrs. Tan's gifts, a telephoto lens and valuable forestry instruments were safe at last.

I have had time to rest and catch up with myself. Jet fatigue is a real factor on those long flights. At least on the return flight I did not have to contend with dysentery. This weekend I swam, played tennis, practiced at the pistol range and saw one of our favorite movies, the "Return of the Magnificent Seven".

Next week I go to Quang Tri to a dedication and graduation of a sawmill crew that has been trained by PHILCAG (Phillipine Civil Action Group). Quang Tri is near the DMZ and will be the closest I have come to North Vietnam. I enjoy visiting PHILCAG since they always have a good supply of San Miguel beer, reported to be the best in Asia, or elsewhere. After that I go to Tra Bong to check on progress at the sawmill.

It will be next weekend before I write again. Thanks to the Jensens for the hot cakes, Vermont maple syrup and lobsters. All were great!

20 August 1967

I was glad to hear you arrived home safely. It appears we all felt the same about the trip. It was wonderful.

I have learned how to iron shirts. The air conditioner helps dry the shirts. I can iron a short sleeved shirt in five minutes and the others in 10 minutes. I hope I do not get into the habit of doing this after I return home.

Last week USAID issued orders requiring all of the branches to be covered during the holidays. That will be no problem for us since I am the only one eligible for leave over the holidays.

The other fellows will be going on leave starting this month and running into November. We have instructions to attend a two weeks training session in Washington in connection with our leave. It appears I will be able to spend two weeks longer in Washington in January. I hope you all can put up with me for that long since I will need another rest by that time.

I plan to be in the office for the next several weeks and let the others travel. It requires a great deal of effort just to keep up with the office work. You know, the bureaucracy must be served. I plan on being in Saigon during the election, that will give me an opportunity to study my language lessons and keep out of harm's way.

Last week in addition to catching up with activities that occurred while I was on leave I had to take time to care for my clothes. In this tropical climate a closed, dank, dark room produces a healthy growth of mildew. My shoes, camera case, ties and non-synthetic clothing that were hanging in the closet were covered with the stuff. A light bulb left burning in the closet and air conditioning are supposed to prevent the growth of mildew. However the hotel management required the cleaning people to turn off all lights and appliances. I washed, scrubbed and treated leather goods with neatsfoot oil. Now everything seems to be restored to normal and was not damaged.

Monsoon rains continue in the Saigon area, most of the Central Highlands and the Delta. Today the rains were veritable downpours. The temperature is in the 90's and low hanging clouds keep rolling by. I worked for a short time in the office today. But with no air conditioning the place was so hot and muggy that I packed up my papers and worked on them in my room.

Last Wednesday I went to Quang Tri with PHILCAG (Philippine Civil Action Group). These are mainly foresters and forest engineers. I traveled to "Eye" Corps on an Air America twin Beechcraft with three PHILCAG officers (a Major and two Captains), a US Navy Lt. Commander and a civilian from CORDS (Civil Operations & Revolutionary Development Support . . . can be further shortened to CIA). Most of the country around Saigon and over the Central Highlands was completely obscured by heavy monsoon clouds. As we neared the coastal plains near Danang the weather cleared. We witnessed several aerial and ground battles. Fighter planes out of Danang where bombing and strafing in support of ground troops. We could see bombs strike and canon and machine guns pound away at targets on the ground. We witnessed ground tracer fire pouring into areas being hit by the planes. We saw what appeared to be mortar fire landing in the strike areas and among the friendly troops. In all we saw four separate actions in the foothills and along coastal areas. From our elevation of 5,000 feet it was quite a sight to see the F-4 jet fighter bombers diving and firing on their targets.

High winds almost aborted our landing at the Quang Tri dirt air strip, but just before we landed the winds dropped in strength and blew down the length of the runway instead of across it. We were met by an armed escort of Philippine and US military personnel. I asked one of the local civilians in charge of the airport when the VC last hit Quang Tri. He said they hit in or around the city every night. Quang Tri is a small city or more correctly a large village. There are forts, a prison and evidence of military presence throughout the area. The prison was hit by the VC and NVA two months ago and 250 prisoners released. The prison was surrounded by a high wall and covered by several guard towers. Attackers blasted holes in the walls and after a heavy fire fight prisoners were released.

Homes are built right up to the streets. They have bamboo, palm thatch roofs and mud or cement plastered walls. Near the houses are trees, banana plants and hedges. The place has a rather hostile atmosphere and contains many suspected of being VC. In this town the children do not come running out, yelling "OK" and "So mot" like in other areas. Here they shake their fists and yell "So moui" (No. 10) and that "ain't a good sign". This is the environment which PHILCAG Captain Gemoto and eight enlisted men moved into last May, set up a sawmill and trained refugees in the use of it.

This sawmill sat idle for almost three years, defying attempts by the Vietnamese to make it run. It was an Army surplus circular sawmill shipped in from Germany. There are half a dozen of these things in the country and all have been unsuccessful, since the Vietnamese just can not operate and maintain them. They are much more complicated than their own mills. This one had originally been set up by a former USAID advisor, who knew absolutely nothing about sawmills. Logs were brought in and village officials were present when it was started. As soon as the circular saw hit the end of the log it came to an abrupt halt. These particular circular saws have insert teeth that fit into a gullet. The problem was they had no teeth and the operators had attempted to sharpen the gullets. That of course did not work. Our millwright discovered the problem at this particular mill and had ordered a supply of insert teeth from Tennessee. The mill was then turned over to PHILCAG to train operators.

PHILCAG had constructed a shelter over the mill, a classroom and an office. They had designed a training program complete with excellent visual aids and charts. One round of classes had just been completed. They had practiced cutting some old rotten logs and were awaiting a shipment of fresh logs. The logs came from across the river in an area controlled by the VC so there is some uncertainty as to when the sawmill can operate. The mill layout looked neat and clean. Workers were busy around the mill on various construction and painting projects.

The difference between the set up at this mill and a similar one in the US was striking. There were a number of armed guards and all of us were armed. I was provided with a Thompson sub-machine gun. The mill area was surrounded by a double row of concertina barbed wire entanglements. Mines were laid in the intervening area. We carefully followed a guide and were instructed to step where he stepped. In the background traffic inched across a newly constructed bridge. The old bridge had been destroyed a few months ago and the remains just barely protruded from the river.

Captain Gemoto explained his training program and showed us around the mill. He did not discuss the management problems or the lack of coordination that had plagued the operation from the beginning. In three years this mill had cut less than 1,000 board feet of lumber. From the mill we went to the Province Chief's office. We convinced him he should take an active interest in the mill and help get it into production. He agreed and requested the PHILCAG team to remain at Quang Tri. I offered to provide technical support for the operation after PHILCAG finished their assignment.

After our meeting with the Province Chief we had lunch at the US Army camp. Lunch consisted of sliced onions and cucumbers, onions and beets, onions in a salad, onion soup and liver smothered in onions. Burp . . . pardon me! This camp had a mixture of Army, Special Forces, Air Force, Marines and Navy advisors. Everything was heavily sand bagged and surrounded with enough barbed wire to fence all the cattle grazing allotments in Arizona. Through one maze of barbed wire I saw Negroes and white GIs playing tennis and basketball. In Vietnam I have observed no racial conflicts or problems involving our troops.

Following lunch we visited PHILCAG living quarters. They were provided a large two room house. It was surrounded with sandbags and contained an arsenal of machine guns, riot shotguns, rifles, pistols and grenades. Wires led out from the house to hedge rows filled with claymore mines. This house and most of the other buildings in town had bullet and shrapnel holes. I gained the impression this was not a friendly town and I was ready to go.

We went to the airport at the appointed time and learned our plane could not land. We were taken to Dong Ha near the DMZ in a smaller plane. We were told this frequently happens. You may have read about the snarl up when the SVN presidential candidates flew to Quang Tri recently. The reception committee was waiting for them at the Quang Tri airport but due to high cross winds they were unable to land and were taken to the bleak, heavily fortified Dong Ha airport. When no one showed up to receive them, except for two US Marine trucks, they left in a huff and accused the Ky regime of sabotaging them. That is the true story of what happened. It was the wind and weather that resulted in the party being diverted to another airport. However the press did not report that to be the reason.

As we spiraled out of the Dong Ha airport we were only 10 miles from the DMZ. The hills of North Vietnam were plainly visible. There was much military activity in the vicinity. I was glad we do not have a sawmill to service in that hateful area. All of the artillery and bombs used in the area must impact most of the trees with metal fragments.

I was dropped off in Danang where I met with the Regional people. They are pleased with our program and the service we offer. They are now asking us to set up three more sawmills in their Region. They have not followed up on their agreement to help the people at Tra Bong with their political problems or marketing. Follow up and follow through on commitments are not well developed skills in the USAID organization.

After spending the night in Danang, I caught a ride on the morning courier plane to Quang Ngai. When in the air I talked the pilot into dropping me off at Tra Bong. That saved me hours of waiting at the Quang Ngai airport. Arriving at Tra Bong I could see the sawmill was idle and the log supply had dwindled down to six logs. The District Chief met and invited me to his office. Without an interpreter I had to rely on my knowledge and ability to handle the language. I learned the workers had not been paid for lumber they had produced for the Special Forces and the Province Chief. They needed the money to pay loggers and mill workers. The USAID/CORDS people had failed in their commitments. They had not placed orders for lumber and had not supplied roofing materials, cement and gasoline they had agreed to provide. There was a further problem concerning an operating permit for their cooperative. The cooperative had added cinnamon, honey and lumber to their list of products. That created a serious problem since the harvesting of cinnamon and honey is currently illegal. I asked why? It seems the VC levies heavy taxes on those harvesting operations.

After identifying the problems we walked to the mill. The machine appeared to be in good condition. Two good stacks of lumber were stored and available for the Special Forces. All they had to do was pay for it and take it away. According to our previous suggestion the mill workers had erected a shelter over the mill to protect it from the rain and the workers from the intense sun. The District Chief stated they would operate the mill the next day with the six logs they had on hand. I spent the remainder of the day talking to the Special Forces. They agreed to pay for the lumber and haul it to their camp. They also agreed to supply roofing, cement and gasoline if USAID did not come through.

The afternoon brought us some excitement. Intelligence reported the presence of a large VC contingent in the hills just above Tra Bong. The Special Forces ordered a large Chinook helicopter and air lifted 150 Vietnamese troops into the nearby mountains. The airlift was proceeded by a hammering of the area by jet bombers and helicopter gunships. All this took place within 1/2 mile of us and I recorded the action on film. It difficult to hold steady when bombs are dropped so close. It was all very exciting to watch.

The makeup of Americans in the Special Forces group had changed since I had last been there. Several had been wounded. One had picked up four Purple Hearts since my last visit. Others had become ill and been rotated out. They showed me a large display of captured weapons, predominately rusty old Chinese bolt action rifles. Since the sawmill was shut down I helped the weapons officer repair and clean a number of their weapons. In turn I was given a new Remington M-1 carbine and a new Colt .45 semi-automatic pistol and holster, plus magazines and 100 rounds of ammunition for each weapon. For souvenirs they gave me a captured VC flag, a knife, a Green Beret cap and a pair of black pajamas, the type worn by VC and peasants alike.

That night we had an alert. We fired mortar rounds, machine guns and artillery into the nearby hills. A check was made to determine how many of the remaining local forces were at their defensive positions. Remember, 150 had been air lifted into the hills. Out of the 200 that were supposed to be on duty we could account for only 79. That is typical of local civilian defense forces, when they get tired, homesick or afraid they take off. Next morning the Special Forces discovered where they had slipped out through the outer mine field. As I was leaving the SFs were making plans to change the spacing and location of mines in an effort to keep troops on the job.

Before leaving I returned to the sawmill with a SF's interpreter. I learned the District Chief's CO was coming in that day, therefore the sawmill would not be operating. The interpreter was able to confirm the information I had picked up the previous day was correct. I agreed to go to Quang Ngai to help solve some of the cooperative's problems. I had some of the District Chief's famous tea. I was told on my first visit to watch out for the tea, since it was guaranteed to cause dysentery. I must be acclimated since it caused me no problem. The Chief gave me more cinnamon bark and invited me to return and help them.

A Montagnard Chief gave me his sacrificial bracelet. That makes two I have been given. It is made of brass and is a tight fit. It will be difficult to remove. It has 60 lines etched into it, with each line signifying a sacrifice of one water buffalo. According to the Special Forces to be given one of these bracelets you are supposed to drink a cup of either buffalo or goat blood. Fortunately I missed that treat, and equally fortunate, I was not offered a nose or ear ring.

I flew out on an Air America C-118 Caribou cargo plane to Quang Ngai. I was glad to leave since the temperature was 120 degrees and expected to increase. The timber chief from Tra Bong came with me to collect payment for their first shipment of lumber, which had been taken out earlier by a motor convoy, the first one in years. The road is again closed.

That afternoon and the next morning I worked with local US civilians and military personnel and local Vietnamese officials to unscramble some of the cooperative's problems. It becomes very frustrating to have to follow up again and again on everything that is done or left undone. I believe some of the Vietnamese officials in Quang Ngai are "dipping in the till". I have no proof but bribery, graft and corruption exist in most all Vietnamese efforts. As you may have noted in a recent issue of *TIME*, it is prevalent throughout Asia. Vietnamese tell me it is done because of the low wages and as long as you take only enough for your family it is not really serious. But to us it is a real problem and very frustrating.

Conditions were about the same in Quang Ngai as I observed during my previous visits. The place is an armed camp and everyone carries a gun, reminds me of Dodge City, Kansas back in the olden days. Election posters are draped across streets and posted all over the country. Everyone believes the VC will disrupt the election and discourage voters from going to the polls on 3 September.

The large unfinished Catholic Church is slowly taking shape. In advance of the election the Vietnamese government is putting up additional construction funds to speed up the process. Catholic refugees have moved into more respectable quarters near the church. I Corps has gigantic refugee problems. Military operations go into an area, move or force people out, burn villages, and spray herbicides on crops. The VC and NVA are moved out for a while, but these efforts do not "win the hearts and minds" of the people. All this reminds me of Sherman's march through Georgia to the sea. In the long run it may accomplish something. I certainly hope so.

I completed my business in time to catch Air America's morning cargo-passenger flight to Saigon. As usual it was late and I waited with all my gear in the boiling sun. There was not a cloud in the sky since this is the dry season for the northern coastal area. I had a good view of the country as we cruised north towards Danang, then towards Laos, to Pleiku and Kontum, then down the coast to Nha Trang and finally across to

Saigon. It was a long day and a long return flight. I returned to my room at 1800 hours.

On the flight south "Dog Patch Airline" was its usual interesting scene. Passengers included US and SVN soldiers, civilians and clergy. The cargo door was not properly latched so the cargo handler fiddled with the door while we were in the air. It is frightening to see a person, with no parachute or safety harness, open the cargo door and close it while we were 5,000 feet in the air. With 25 passengers, their gear and miscellaneous cargo we had a problem taking off in the high temperature. But I figured if we experienced trouble it would be easy for the cargo handler to push out loose anvils, wooden crates, heavy truck tires on rims, etc. Nothing was tied down. Four Korean mechanics came aboard at one stop and their gear consisted of four large storage batteries, two large, uncrated bottles of sulfuric acid, four large tool boxes and two truck axles. It makes you shake your head when you recall how fussy airline stewardesses are about requiring you to place a light handbag in overhead storage compartments.

The high country was green and we passed through many rain storms. The air was extremely rough and before we returned to Saigon all the air sick bags had been used. How shall I say it? It was rather "close" when we landed at some of those hot landing strips on the return trip. I picked up a *TIME* magazine along the way and tried to overlook the surroundings and odor. I was not air sick, it must have been due to the onion and salami sandwiches I purchased at one of the many stops.

One other item . . . one of the Tra Bong sawmill operators at Tra Bong was fired from the job for stealing two boards. I believe it may have been the one I wrote about in my story to be published in *American Forests*. That is the way it goes . . . the little man takes two boards and gets sacked, while the officials are dipping into the money bag up to their shoulders and get by with it. So goes another week in Vietnam. Please do not get the impression we are not making progress . . . we are! War damaged timber is being salvaged in a number of areas, new mills are being started, old ones are being reactivated, and reforestation is taking place. I look at it this way . . . if there were no problems there would be no need for any of us to be here. Tomorrow will be another day of challenges.

Chapter XX

The Election

24 August 1967

I hope you are over being annoyed with me and my letters are now reaching you. We have had difficulty getting our letters to jibe ever since I first arrived here. I received two of your letters and enjoyed reading about events at home, from the "real World". I find your letters very interesting and look forward to reading about the latest boy-girl affairs. Hearing about GIs falling asleep on the couch brings back pleasant memories . . . sounds like history repeating itself.

I had your San Francisco slides processed. The hippies at Sasalito and Haight-Asbury look good. Some of the slides were overexposed or perhaps it was the fog. I will prepare them for mailing this weekend and shop at the PX for a Canon camera for you.

We have experienced much terrorist activity and are advised to watch ourselves and stay close in. Most problems have been in Cholon, Saigon's Chinatown. Unfortunately that is where the largest and best PX and Commissary are located. I may not be able to shop for the camera until after the election.

The weather has been hot and muggy. It rains some every day. I am pleased to learn you have had a pleasant stretch of weather for a change.

My traveling has put me behind in language training. It is becoming more complicated each week and I must devote more time to my study. I took a break last evening and drove to Tan's to present our gifts to his wife. She was not there so I left them with their son. Today I went to the airport to greet Tan on his return to Saigon. After leaving the US and Canada he went around the world, through Europe. I visited with Mrs. Tan before he arrived and she was delighted by the gifts and expressed her appreciation to you. Some 75 members of Tan's staff were there to meet him. He arrived bubbling over with enthusiasm, so I guess he really enjoyed his trip. He told me he was very proud of his improved ability to use English. It was good for him to be away and see our forestry practices. I will learn more about his experiences when I talk to him later.

I have had a miserable cold for the last couple of days. It has given me a "blah" feeling. It is very difficult to shake a cold in this place. Many of the people at the office have colds. With no hot water to wash your hands, the office is an ideal place to

catch one. It is 2100 hours and my throat feels scratchy so I will
go to bed. I will write more this weekend. I am still on deck and
there are no stewardesses, etc.

27 August 1967

It is Sunday evening and time for a letter. I spent most of
the day studying Vietnamese and trying to catch up on the
lessons I missed while I was on leave and traveling. I spent 1 1/2
hours yesterday and the same today at the home of my teacher
being tutored. The rest of the time I studied and repeated the
lessons to myself. The secret of learning the language is repeating
and repeating . . . they say it takes 20,000 repetitions to become
really proficient. I have a long way to go, but I am trying.

My teacher is a young college student who lives with her
parents, brother, two sisters and a house full of relatives. They
sat in the background while we studied at a massive dining room
table. The wood in the 4 by 8-foot table is hardwood. The top is
one solid piece which is five inches thick. When I tried to lift it
by one corner it was so heavy I could not budge it. The other
furniture is also massive and displays good workmanship. Most of
the solid furniture I have seen has large cracks, but not this
furniture. Teacher's father was once a sawmill operator. He cut
timber, sawed and seasoned lumber for many years. In the end
he lost everything to the VC. They ran him out of the forest,
stole logging equipment and destroyed his mill. That family has
no love for the VC.

I am enclosing a few news items concerning terrorist activity
taking place in Saigon. It appears to increase each day as we
approach next Sunday's election. Several people, both Americans
and Vietnamese have been shot . . . some were killed and others
wounded. A number of people have been wounded by grenades
tossed by children and women. That adds another hazard in
addition to the traffic and smog. As noted in the clippings the
US Embassy and Palace are considered prime targets. My hotel is
a short block from the Palace. As I look over there I can see the
place bristling with Vietnamese Army and police.

Guards have been doubled at most American establishments
and streets bristle with US and SVN MPs, police and soldiers. We
have been advised to keep off the streets after sundown. As
noted in the attached letter from the Ambassador all Americans
are confined to quarters from Saturday 0900 hours on 2
September until 0400 hours on 4 September. The objective is to
keep Americans away from any possible disturbances during the
election on 3 September. This means many Americans living in
hotels without restaurants will have to eat out of a tin can. My
hotel has a restaurant but it is not very good. I plan to stick to
the tins.

As tension builds you just keep your eyes on everybody,

watch where you step and keep an escape route in mind at all times . . . it is just like fire fighting. Adding to the tension is a continuous artillery barrage all around the city. I was on the roof a short time ago and the sky is continually lighted with flares. Helicopter gunships and "Spooky", the flying Gatling gun, are in the air continuously. I could see flashes from explosions of artillery, some originated at ground level and others were air bursts, just above the surface. As I type I can hear the artillery and planes over the sound of the typewriter and my radio. This is similar to sitting in a fort under siege and waiting for something to happen. The next few days should be interesting.

Travel to the provinces has been restricted since we are expecting stepped up enemy activity in all parts of the country. Most of the forestry group will remain in the office and try to catch up on reports and planning. I put Lew Metcalf on a plane yesterday. He will be on leave in Kentucky for the next 20 days. His knowledge of sawmills makes him a real asset to the program. Evidence of the election can be seen everywhere. Every available wall space has been plastered and replastered with election posters. Cruising vehicles equipped with loud speakers are extolling the virtues of candidates on most all of the busy street. "Busy streets" is a gross understatement since traffic is becoming impossible. As numbers of civilian and military vehicles increase we are nearing the "choke point" or "grid lock". Driving to my teacher's home today was nearly impossible, but mysteriously the traffic periodically opens up and somehow you get through. My teacher lives near the edge of the city adjacent to a police check point. I decided not to go out there anymore. I will settle for the hour of class work each day during my lunch period.

Added to my other frustrations are increased power failures. When power fails it is usually raining and my surroundings become as dark as the inside of a cow. I would not care to be caught out on the streets at such times, but none of us venture out at night. My hotel has an emergency generator which comes on and provides feeble light, but no air conditioning. At least we can see who is stranded in the elevator. I have not used an elevator for weeks.

Director Tan met with me on Saturday. He is still elated over his trip. He is very grateful to Chief Cliff who made the trip such a great success. He was especially pleased with the gifts we presented to his wife. He again reminds me when we are not in public the forestry team should call him Tan and he will call me Jay and the others by their first names. Few other advisors have that kind of rapport with their Vietnamese counterparts. I have you and the girls to thank for the time you spent with him. Tan's men are well aware he is the boss and they treat him with respect.

That's all for now.

Today is election day in Vietnam and the expected VC attacks and peak of terrorism failed to materialize. It is now 1300 hours and I am on the roof soaking up the sun and sights. The sun feels good for a change after all the torrential downpours. Activity in the sky is interesting at this moment. Planes are rotating in a clockwise pattern and are stacked at different altitudes, ready to pounce on any trouble spots. Arranged around me on the roof were two groups of men playing cribbage. Stretched out on pads are a few American secretaries catching a few rays. That is probably all this group will ever catch.

Down on the street local people are coming and going in all directions as usual. I see no Americans among the pedestrians. We are all grounded and confined to quarters. Our special curfew started last night at 2100 hours and we are not permitted to stick our noses out of our quarters until 0400 hours on the 4th. Those living in hotels with no eating facilities are provided special curfew passes for limited trips to nearby BOQs.

In many respects the past two weeks of the political campaign have resembled ours. Vietnamese are experts at mud slinging and digging up dirt on one another. Platforms of the different parties are remarkably the same . . . all favor peace, prosperity, amnesty, anti-corruption, higher wages and equal treatment for all people. These slogans have been expressed in speeches from the rice paddies and villages, to the streets of every large city. Supporters of all factions have put up posters. Some posters carry the symbol of a dove, others depict the sower of seeds, a rice plant, a torch and various other symbols representing that particular group. Every accessible square inch of wall space displays a poster. Some posters have photographs of the "bets", as the candidates are called. Platforms are printed in bold type for all to read. People have spent a great deal of time reading these posters, listening to the radio, reading newspapers and watching TV. They are probably just as confused as most of us are in the US during an election.

Many posters are now hanging in shreds as a result of last night's heavy rain. People have either voted, are going to vote or have been frightened out of voting by threats and terrorist activities. Some of the people have a problem related to the required voting cards. I talked to one of the men in our office who had such a problem. His application was on file but had not been processed. When he went to the police station to get a card he was told it was too late and he could not be issued one. Others are apparently schooled in the art of New Mexico-type politics and have two or three voting cards. I understand all of the occupants of the graveyards are registered and voted today.

Throughout the country many are not voting today. Hundreds have been killed by terrorists, many kidnapped and others frightened away from the polls. Many had their voting cards taken from them by terrorists. I presume that accomplishes two purposes . . . keeping the owner of the card from voting and the thief can use the card illegally to vote for the party of his choice.

The measures to keep Saigon secure this week have been impressive. On several evenings we could see tracers pouring into the ground from helicopter gunships. This action took place at the edge of the city which is about five miles from here. Guards on all US military and civilian establishments have been doubled. Last night in anticipation of today's confinement I obtained a special pass and went to the Hotel Splendid BOQ and had a steak, baked potato, vanilla ice cream and tapioca pudding. I sat in on the movie "The Great Race". At 2020 hours a monsoon rain hit us, the movie was stopped while we moved under overhanging balconies to get away from the dripping clothes hanging from many clothes lines. At 2025 hours the movie started again. At 2030 hours the movie was shutdown and we were ordered to leave unless we were billeted there. I walked two blocks through a cloud burst to the Hotel Park. I did not dare to run because in that short distance I counted over 500 guards standing in the rain with rifles and machine guns ready. This morning I could still see hundreds of guards. The Palace and BOQ appeared to be undisturbed.

We received a report that the city of Quang Ngai, where we have been working, was hit by the VC. They achieved their principal objective and released 1,200 prisoners, of which 400 were recaptured. They hit the MACV (Military Assist. Command Vietnam) compound with mortars and caused many casualties. That is where we eat and use the PX. It is well that we stayed out of the north during this pre-and post-election period. We have no reports from Tra Bong.

Yesterday a special notice was posted on all bulletin boards and voiced over the radio. The message stated:

"All US personnel are not to buy any of the following items on the local economy during the immediate election period:

Cigarette Lighters
Fountain Pens
Oil Cans
Alarm Clocks
Candy Bars
Bottled Aspirin

The VC have fused many such items to explode when opened or used."

That reminds me Tan gave me a new sterling silver Parker "75" pen. I hope he purchased it in the US. All and all this has been an interesting election campaign.

Each week I have been watching a couple of hours of American football on TV in the hotel bar. For the price of VN$50 I have a bottle of Ba Muoi Ba (No. 33 is the brand of the local beer) and see the previous Sunday's game. Film is flown in each week. I have been planning to purchase a small TV set but the PX has none for sale. They were snapped up in advance of the football season.

You probably read about the black market. I have written about it in my previous letters. It appears to be with us to stay. Some of the big items have disappeared because the PX has stopped bringing in fans, refrigerators and other large appliances. These items flowed freely into the black market. Many soldiers have been caught in this practice. Their excuse? They were just storing it in someone's house. All of the bars have a good stock of PX liquor, Salem, Camel, Pall Mall and Kool cigarettes. None of these items are legally imported for local consumption.

In my limited travels this week I saw several sights of interest. I notice the number of amputees and soldier beggars has greatly increased. The SVN government makes little or no provision for disabled veterans, they must shift for themselves. One election promise was to develop a program for war veterans. Many of the amputees lost their limb(s) as the result of land mines and the tendency of Vietnamese surgeons to cut and remove rather than repair. This results in terrible mutilation, just like in our Civil War times. I understand Vietnamese hospitals are full of soldiers with recent amputations, many of them suffering multiple removals of arms and legs.

Women beggars with babies frequent areas near American hotels and apartments. The babies are pathetic, frail, dirty little things. Many of them have never experienced the benefits derived from soap. Their mothers nurse them and I suppose that gives them a certain amount of immunity and an opportunity to live. Infant mortality is extremely high and is obviously caused by lack of hygiene and medical care. Near some of the downtown hotels I see many children of all ages laying, sleeping and playing on the sidewalks and streets. Many may be orphans. Some appear to have been fathered by white and Black Americans. I have seen some of the babies' mothers, who are maids and waitresses, come out of the hotels to nurse and feed them during the day. Many of these women appear to be pregnant.

USAID has done a great deal to help improve the city. Streets, water and sewer systems and garbage collection have been upgraded. However problems persist. Garbage and refuse are dumped on sidewalks and gutters for scavengers, including people, dogs and rats to work over. There are only one or two public toilets in the city and people are generally left to their own choice of resources. The toilets must be awful since it is very unpleasant to be down wind from them. Main water lines have been improved to the point where increased pressure continually blows out the old French distribution system. The resulting floods create a real mess and undoubtedly contribute to contamination of the water supply. I see people gathering around leaks, it does not matter whether it is a water or sewer break, to bath and gather water for cooking or home use. Bringing a primitive people into a large overcrowded city just does not work.

This must be the puppy season. I see many baskets of small pups for sale. Some will be used for pets but it is well known that the North Vietnamese favor dog meat over any other meat. There are many adult dogs running wild on the street. This is a source of great concern since vaccination for rabies is unknown.

Barry Flamm and I received notice our applications for membership in the tennis and swimming club have come up for action. Tomorrow after work (a Labor Day work day) we are invited to appear before the board for an interview. We both enjoy the gym, tennis, pool and inexpensive French food. When we are in the city we go there at least three times each week and frequently more often. Occasionally I take a chance and order something from the menu without asking what the French words mean. The other evening I had a delicious casserole of vegetables and meat. I wish I had not asked what it was, because it tasted better before I learned it was tripe. Strange people those French.

The other day we had another of those interesting staff meetings. The air conditioning was off as usual and all doors had to be open. The outside sounds were very disturbing. There were generators roaring, carpenters pounding, jet helicopters going over at tree top level, fighter jets scrambling with their after burners shattering the air, motor cycles with and without mufflers, convoys blasting their way through traffic with air horns, all accompanied by the shrill sound of police whistles and sirens. It was difficult to make any sense out of the staff meeting.

I had lunch with one of the men from the USAID office. He has been trying to get me to move into one of the apartments in his building. Frankly after seeing it I was not impressed. Access to the building is though two narrow, dirty alleys. While it is closer to the office than my hotel, it is in a very tough looking neighborhood. His apartment reminded me of your Cousin Ella's . . . books and boxes occupied most all of the available floor space. He has a part-time maid who is supposed to clean and do his laundry. None of the rooms or his clothes looked like they had been cleaned for some time. Furniture was falling apart, bathroom fixtures were broken and leaking. He has no stove, just a hot plate. This man has lived under these conditions for three years. He believes it is good setup. The more I think about it, I believe I am better off in a nice clean hotel. I do not have to pay a housekeeper or purchase groceries. I can go to the club and BOQs for meals and relaxation. Most USAID supervisors have very nice, clean apartments but most of the other places leave much to be desired.

The last time I checked I was No. 87 on the list of those eligible for housing. Even if I was interested in moving into an apartment there is not much of a chance. The only thing I would appreciate would be to have some comfortable furniture and a good place to sit and read. I really do not need an apartment since I spend a great deal of time at the office, traveling, going to the club and BOQs. If I had such a place I am certain I would eat more snacks and that does not sound too appealing.

On Tuesday I have been asked to go on an aerial reconnaissance and evaluate some military projects near the city. At times I get a little paranoid. Most of our work is to help forest industries, improve forest management, reforestation and economic development. Then I go out on a military operation which may have just the opposite result. Such is the life of a forester in Vietnam.

I note an interesting, positive switch in my Mother's attitude. She is really unpredictable. One last item for this letter . . . the patent medicine worked on that wart which was on one of my fingers. It is gone! (the wart that is!)

9 September 1967

It is Saturday night and it is raining, again, yet! Power is off all over the city. Fortunately my hotel has an emergency generator and I can see to type this letter and listen to the radio.

After working all day today on work programs and the Fiscal Year 1968/69 budget I went to the Hotel Splendid BOQ for a good dinner. I missed lunch due to work and a heavy rain. After dinner I thought I would take in the evening movie, but alas, the title was "Mother Goose A-Go-Go" . . . yuck! That did not sound like my kind of movie so I returned to my hotel to watch TV in the lobby. The power was off and there was no TV. I thought about all the really good movies we had seen together as a family. I enjoyed reading about Sandy Dennis in *TIME* last week. I liked the part about her having 21 stray cats and the stacks of "kitty litter" that keeps guests away. She can not resist a homeless cat. That reminded me of Gwen and girls and home!

Last week we received a report of the meeting the Saigon "brass" had with "brass" from Washington. Priorities have changed and the Agriculture Program is to concentrate on rice, pork and fish. The remaining activities will be subordinated to support the favorites. We were told that Grant, a new Assistant Administrator of USAID, considered the forestry program to be a "boondoggle". He mentioned that during his first couple of days in office he spent most of his time on the forestry program at the insistence of Secretary Freeman (our supporter!) Now at least Grant knows we are here, to stay, we hope. Being No. 2 like AVIS (or "So hai" in Vietnamese) we will try harder. I am confident our program will make more progress. We have more than our share of attention and we are receiving more requests for assistance from all over the country. Some are routine requests and others are of an emergency nature. I was out just last week on a classified emergency military project. The military calls on us frequently. They respect our evaluations and recommendations; in turn they provide us with extremely valuable transportation, maps, aerial photographs, supplies and security.

I just walked up to the roof to see what the increase in helicopter traffic was all about. There are six helicopter gunships hovering overhead between my hotel and the river. That particular area and much of the city is illuminated by flares. We are having a very heavy rain and the view was quite a sight. It is not a good night to fly and the crews must be up there just in case they are needed.

Since the air conditioner does not work off the generator I have my door open. A few insects come in but that is no problem. I open the door frequently at night to attract a new supply of food for my pet gecko. He/she is a small lizard living on the ceiling near one of my light fixtures. His/her toes have adhesive pads that enable him/her to climb on vertical surfaces. He/she cleans up all the mosquitoes and other insects that are attracted to the light. These critters can be seen in all Vietnamese homes. They are considered to be good luck charms and very beneficial.

Tuesday Barry Flamm and I were invited to a reception at the Cercle Sportif Saigonnais Club. The purpose was to interview us as prospective members. You recall we have been going there and using the gym, pool and tennis courts for three months as guests of Ray Russell pending consideration of our application for membership. Guests are not supposed to use the facilities more than once during the application period. As guests we signed in each time using a different name. History and their records will show that Ed Cliff, Art Greeley, Byron Beattie, Red Nelson, Frank Carroll, Ken Cook, Norman Johnson, Billy Swapp, Gene Jensen and goodness knows who else were guests. We had just about exhausted our list of Forest Service acquaintances and looked forward to using our own names once again. We were called by our own names and went together for the interview. We greeted the Vietnamese board members in formal Vietnamese and carried on a brief conversation in the same language. They were visibly pleased and welcomed us as members without, fortunately, asking us a single question. So again the language ability opened doors and paid dividends. Please recall this was a private French club for many years. Never once did the French make an attempt to speak Vietnamese. Today most of the French are gone but the staff and waiters at the club speak only French. Strange place isn't it? Barry also enjoys learning Vietnamese and teaches English two nights each week at a Buddhist orphanage and one other school. I caution him about going there every week, on the same days and at the same time. This sets up a pattern and could be hazardous to his health.

All of my activities, such as going to the club and the BOQs keep my spare hours occupied and help pass the time. Still it is a very lonesome feeling to be so far away from all of you. That is the major sacrifice connected with this job. The rest of it is interesting, rewarding, exciting, humorous, and frustrating. It pays to retain a sense of humor and somehow things mysteriously fall into place and are worked out.

In our USAID/Washington training we learned about clandestine activities and intrigue. Remind me when I return home to tell you about a very fascinating episode Barry and I were involved in recently. This helps keep us on our toes and on guard. Really there is no danger just a warning to be cautious. This instance is typical of the local political situation and the Vietnamese way of doing things.

We have good coverage in the *Stars & Stripes* and radio regarding the dangerous and destructive forest fires in the western US. Those fires must have disrupted Forest Service activities from coast to coast.

Tomorrow Barry and I go to the pistol range to practice and then to the office for work on the program and budget. We now have good running "wheels" for our use around Saigon. The vehicle is a World War II heavy duty Arctic Jeep (in the tropics?) . . . but this is USAID!

Barry goes north to Danang and Hue next week. I will do one of two things. I may go to a camp along the Cambodian border for a special meeting with PHILCAG and others. Or I may go to Dalat with Tan and the Minister of Agriculture for a dedication ceremony and a tiger, deer and/or boar hunt. Then I am scheduled to go to the Danang area for a week. While there I will try to look up old friend Pete Stilley, from Flagstaff. Trying to locate someone in this country is like looking for a needle in a hay stack, but I will try.

It is time to sign off. I am unable to read by the poor lights in my room. I finally borrowed a Sears catalog and ordered a suitable light from Los Angeles.

Note: The following letter was received by Gwen and a copy to Saigon:

United States Department of Agriculture
Forest Service
Washington, DC 20250

IN REPLY REFER TO
1600
September 20, 1967

Mrs. Jay Cravens
5624 Green Tree Road
Bethesda, Maryland 20034

Dear Mrs. Cravens:

I'm sure you know how much we in the Forest Service appreciate the sacrifices you and your family are making so that Jay can head up the important USAID-USDA forestry mission to the Republic of Vietnam.

We think very highly of Jay. He has an exceptional record of accomplishment in the Forest Service and his ability is already producing results in Vietnam.

We all enjoyed seeing Jay briefly on his recent visit. We are very proud of the job that your husband is doing.

In an alcove in the second wing of the third floor of the South Agriculture Building we have special displays of important Forest Service activities. You will be interested to know that currently being displayed, is a 29x32-inch poster featuring photos of Jay and his team. We are sending Jay photos of the poster, but I was sure that you and your family would like to have one also.

Should you, at any time, visit the South Agriculture Building, I would be honored to have you call on me.

Sincerely yours,
/s/ Edward P. Cliff
EDWARD P. CLIFF, Chief

Enclosure

23 September 1967

I returned from Dalat this morning after spending three days with the Minister of Agriculture and Tan. The purpose of the trip was to participate in a Directorate of Forest Affairs ceremony recognizing the gift of forest fire control equipment from the Republic of Germany.

Tan, accompanied by Mr. Du and a Mr. Thin (security agent) picked me up at my hotel at 1300 on Wednesday. All air bases have strengthened their security measures recently and I expected a delay in going through the gate in a private car with four Vietnamese, including the driver. Guards have been stopping our Jeep and other USAID motor pool cars for inspection of the vehicle and our ID cards. I was horrified as we approached the gate and check point without reducing speed. I hunkered down fully expecting warning shots or a chase. But the 10 US and Vietnamese guards made no attempt to stop us. I looked over at Tan and he grinned and said it is good to have the security people take care of all necessary arrangements. We drove on to the VIP lounge past more guards. Lew Metcalf was there to meet us and help us with the complex arrangements. The procedures we went through for this trip are examples of what we go through to get a job done.

Our Embassy gave me the task of arranging a special VIP flight for the Minister and the German Ambassador. While this did not sound too difficult, it was a typical Vietnam affair. With these high level people involved it did not take much to get things started, but by the time it was done it became very complex. There were to be ten in the party and a suitable aircraft was arranged early in the week, but then the complications started. The Minister changed the members and the number in his group. The Ambassador did the same. The names and numbers changed each day. We finally traveled in two separate planes, on different days and returned with a different number, on different days.

The day before our scheduled departure I received word from our travel people that the German Ambassador's plane had been canceled. Sorry about that! I got on the "horn" to Air America's manager, gave him my name and State Department rank and suggested they make the necessary arrangements for us or I would have Ambassador Bunker personally take the initiative. Needless to say AA made the arrangements.

Lew Metcalf was at the VIP lounge to help with the luggage. AA procedures require luggage to be taken from the VIP lounge by our driver, to the AA terminal to be weighed. Then an American is required to ride in the plane as it taxied to the VIP lounge one mile away. The VIP lounge is a very plush waiting room the size of a large ballroom. It is beautifully furnished, thanks to you nice US taxpayers, and comforts the waiting brass. The fact is the principals never use the facility because their plane is always waiting. Parking slots for aircraft have a certain degree of reverence. No. 1 at the front entrance is used only by the Prime Minister or the new President. No. 2 is for commissioners and ambassadors. We used No. 2 since the Minister is a commissioner in the war-time cabinet.

The Minister is an interesting person. He is my age but looks 10 years older. His two years in office has taken its toll. He was educated in a French school in Vietnam and speaks flawless French. During the past two years he has learned English "from studying Webster's dictionary" and he speaks it without an accent. He opened up and became very cordial when I spoke to him in Vietnamese. From that time on I was asked to ride in his car and sit in the honored seat on his right.

The flight to Dalat was not without incident. Twice the gasoline tanks became air locked when the pilot switched from the main tank to the auxiliary tanks and both engines stopped momentarily. The Minister and I looked at one another and shrugged . . . both of us paying our "respects" to good old Air America. I noticed Tan and Mr. Du turned several shades of green as they looked out at the stalled engines and the VC infested forest some 5,000 feet below. The pilot knew what to do and quickly corrected the problem. We probably lost only 50 feet of altitude.

When we climbed above the clouds in our twin Beech we entered a veritable paradise of towering cumulus clouds. We wound our way around the clouds, just like driving down a winding, crooked road. It was beautiful and bright. We were controlled all the way by radar and we were given a good wide berth by other planes. Occasionally we could see other planes 1,000 feet above us or 1,000 feet below us, winding their way through the canyon in the clouds. Our approach into the Dalat area was by GCA (ground control approach) radar. We came out of the clouds exactly at the airport and made a smooth landing. I give much credit to the Vietnamese GCA operators. They learned their lessons well from their American counterparts.

All during our stay in Dalat we were surrounded by troops and security agents. On the ride into Dalat from the airport we were escorted by the police who ran everyone off the road. I was more fearful of our driver than the VC. He drove with much horn and little brake. I was happy to arrive intact at the Dalat Palace Hotel.

I learned plans were under way for me to go on a tiger and deer hunt that evening. But it turned out there has been VC activity in the happy hunting grounds and the hunt was called off. With the hunt called off arrangements were quickly made for a Chinese dinner. First we had drinks in the spacious bar on the veranda of the hotel. The Minister had some of France's finest cognac. He told me it gave him a bad liver but it was still his favorite. I had a glass of very good, light Vietnamese beer. The conversation ranged from hunting, to agriculture, to forestry, to politics, all covered in a bewildering mixture of French, Vietnamese and English. The Minister has a keen interest in the forestry program and believes with peace and good administration it can contribute much to his country.

Nha Hang Shanghai was one of the best Chinese restaurants in Dalat. I had just learned about "nha hang" in my last Vietnamese lesson so I was able to comment about it in Vietnamese. That impressed the audience! The meal was one of those delicious, multiple course affairs. I enjoyed this meal thoroughly because I did not ask what each dish was and how it was made. I operated under the assumption of "what you do not know will not hurt you". The only things I will not eat are rare meat and fresh vegetables, unless they are prepared in one of the Army-operated BOQs. There the meat served comes from the US or Australia and the vegetables are carefully prepared and washed in chlorine and iodine solutions. I have seen a Vietnamese slaughter house and I would not want to write about what I witnessed.

The 10 Vietnamese at our dinner were astonished by my use of red peppers in soy sauce. These are hot, but really no match for the hot Mexican peppers. I find them very satisfying and not too hot. Vietnamese use small quantities of the peppers with all of their meals. Peppers impart a particularly good flavor to the Vietnamese fish sauce nuoc mam. The final course consisted of delicious lichees a very fine tasting Chinese fruit. I must bring a couple of cans of those with me when I come home for Christmas.

Following the meal we returned to the hotel. Security guards were present on the approaches and surrounded the hotel. I later learned many of the additional guards were on duty to protect a meeting of 30 US and Thai Air Force pilots.

Friday morning I was up early and took a pleasant walk around the lake. People can be seen fishing at all hours. Fishermen catch nothing of any size but they are persistent. At 0630 hours vegetable merchants were on their way to the central market and airport. Dalat provides US military installations fresh vegetables. All our vegetables are flown in by plane since the highway to Saigon is insecure. Within the past two weeks two railroad bridges have been destroyed. Last month the train was derailed twice.

I had been advised by the Embassy to go armed while in the Dalat area. The Marines at the Embassy issued me a .38 revolver and two boxes of ammunition, but obtaining a permit was another story. I went to the USAID department which handles gun permits. I was given an application to complete and was told there was no charge, but it would take at least six months to obtain one from the GVN. I completed the application and discretely enclosed VN$1,000 (US$8.47). My gun permit was delivered to me in less than two hours! In carrying the revolver I had to be discrete so I stuck it under the waistband of my Levis and wore a black nylon jacket. Back at the hotel I had a delicious breakfast of bacon and eggs on the veranda. Sitting in large soft chairs on the veranda was uncomfortable with the revolver jammed under my belt.

After breakfast I walked to the market and enjoyed observing the activity. Smells were a delicate mixture flowers, fruit (oranges and bananas) and fish. The fish and red meat looked fresh, but at that early hour they were speckled with flies. The price of red meat was about 90 cents per pound. Large numbers of dogs scavenge around the meat stalls. I get nervous when I see them since Vietnamese dogs have never felt the sting of a rabies shot. The meat merchants displayed a large number of puppies and older dogs. There are many North Vietnamese refugees in the Dalat area and their favorite meat is dog.

I did some bargaining to check prices. Later I came back with forester John Chitty and had him bargain for the same items in French. I could do just as well in Vietnamese and purchased tea, Montagnard cloth and baskets.

At the hotel I found a company of soldiers lined up in a formation. I learned the Minister had organized a hunt and these soldiers were to provide security. I could now tell you about hunting deer, tigers and elephants. It would make a good story but it would not be true. This is really the way it happened. We drove 10 miles to the city zoo. One of the Minister's body guards handed me a loaded .38 S&W revolver and told me to stick it under my belt just in case. Being a 2-gun hombre made my tight Levis even tighter! The guards fanned out and we stood between the elephants and tiger cages. We shot birds, about the size of sparrows! This was supposed to be a dove hunt, but fortunately for the doves, they were not in the zoo grounds. The Minister had a beautiful .22 automatic rifle with a 4-power scope. He shot several times and missed each time. He handed me the rifle and I connected with a couple of lucky long shots. He exclaimed what a good shot I was and then I remembered the cardinal sin is to embarrass the leader so I deliberately missed the next few shots. The Minister then had better success. I became a little suspicious when the Minister's men retrieved all of the 15 poor little birds that had been shot. The Minister said these birds were delicious to eat and we would have them for dinner. After our hunt we relaxed and ate pounds of small apple bananas, which are absolutely delicious. The Minister peeled off a large number of bills from a huge roll of piasters and paid off zoo security forces. Those birds must have cost VN$500 each. I suspect part of the payment went to pay the other side (VC) for cooperating and thereby perpetuating this strange game.

Following lunch Chitty and I went to the National Geographic Society of Vietnam (no connection with our NGS) to look at maps. They have some beautiful ones and I was prepared with a request on a US Embassy letterhead to purchase one set for each member of the forestry team. The procedures we went through provided a real education in the GVN bureaucracy. I went up through channels to obtain a Colonel's approval and then back down through channels, receiving a stamp at each station, until I reached the clerk in charge of the map room and purchased the maps. Apparently these controls are used to keep these very detailed maps out of the wrong hands.

The hour for the dedication of the German fire fighting equipment arrived. The German Ambassador arrived and I had an opportunity to retrieve some of my almost forgotten college German. The dedication was held at the forestry compound. A stand was decorated with palm fronds, Vietnamese flags and bunting. There were no freshly cut pine trees. The crews must have remembered Tan's outburst of anger when they decorated our lunch stop with small pine trees. While the speeches were short, it took time to repeat them in German, French, Vietnamese and English. Fortunately I had reviewed the draft speeches. They were guaranteed to make everyone angry. Oversights and slips in protocol can create serious problems. In my redraft of the draft speeches I recognized the contributions of the Republic of Germany to South Vietnam, South Vietnam's to the Republic of Germany and even slipped in a very discrete mention of the USAID forestry team's contribution. Our role was to provide clearance and assistance in bringing the equipment to Dalat by sea and air. I hope we do not have to go through such an international orgy again in the near future.

Next came a demonstration of the Mercedes-Benz Unimog truck, fire plow, pumper and tanker. The fire plow was so rusted from the sea journey that it could not be hooked up to the truck and demonstrated. John Chitty had trained the Vietnamese crew how to start and operate the pumper. However the crew did not hold the nozzle securely and it was very exciting when the hose and nozzle lashed around and soaked and scattered the spectators. The 4-wheel drive truck has a top speed of 20km per hour when pulling the large 4-wheel water trailer. At that speed it will take forever for the truck to reach a forest fire. No equipment can be stored in outlying work camps since there is no security. Another problem will be involved when they reach a fire. It will take about 40 acres of space to turn this unmanageable rig around. All of this is rather academic because they will be unable to do much fire fighting until peace arrives. I can not blame the Vietnamese for not rushing to every forest fire because that would certainly be an ideal way to set up an ambush. Such is fire suppression in present-day Vietnam.

Following the demonstration the visitors and officials also got wet on the inside. There were lots of goodies. I drank warm Coke because I had seen where the ice was stored, uncovered and on the ground in a heavily used chicken coop. I enjoyed talking with the local people. They were surprised to hear an American speak Vietnamese.

That evening a state dinner was held in the grand ballroom of the hotel to honor the German Ambassador. There were eight courses of delicious, well prepared and beautifully served French

278

food. Fortunately my fillet was well done. Also fortunately the baby birds harvested by the mighty hunters must have been served to someone else. It was interesting to watch the Vietnamese use utensils. The educated ones had learned their lessons well. They also cover their mouth with one hand while using a tooth pick. The waste of food at one of those affairs is disheartening, but in their society it is impolite to clean up your plate. This is an insult to your host and shows he did not provide enough food. I left delicious looking fresh lettuce as my concession to good manners. Most of the table conversation was in French. When I found a break in the conversation I said in Vietnamese, "I regret I do not speak French". That brought us back to English and then to Vietnamese. It was an interesting way to end a very interesting day.

The next day dawned cool, bright and clear. I had an early morning walk before breakfast. I was accompanied by one of the security agents. Our secretary Mrs. Hai had asked me to bring her a "pensee" from Dalat for her scrap book. I finally understood she meant pansy. The security agent said he knew where some were growing. On the far side of the lake we stopped at a beautiful private flower garden. It was early and no one was around so he just hopped over a barbed wire fence and returned shortly with a bouquet of pansies.

Back at the hotel the Minister was having breakfast on the veranda. He invited me to join him for "croque monsieur". This was a delicious concoction of Canadian bacon and jack cheese placed between two slices of bread and fried in butter. It was very good!

Flight arrangements called for departure at 0830 hours. We departed for the airport with the same escort that had greeted us on arrival. I rode with the minister and again I feared his driver more than the VC as he careened around the curves. But we made it! I returned the .38 revolver to the security guard and his eyebrows raised when he saw I had my own revolver tucked under my belt. The security guards and forestry representatives, with the exception of Tan, were dismissed and directed to shop for flowers and vegetables to take to Mrs. Minister in Saigon. They were to return on a later Air Vietnam flight. Their seats were taken by the Minister's daughter and two teen-age friends. The pilot had no objections to taking them as long as we filled the seats and were ready to go to Saigon. The teen-agers chattered just like our teen-agers, first in French, then Vietnamese and then English for my benefit. One of the girls was thrilled to sit in the empty co-pilot's seat. The return flight was uneventful. For most of the way we were in or just above the clouds. For the last 100 miles until we touched down at Tan Son Nhut Airport we were controlled by GCA. That device is accurate and provides peace of mind in those crowded skies around Saigon.

That was a good letter and photo you received from Chief Cliff. It is my pleasure to serve him. That's all for now.

Chapter XXI

Vietnamese Bureaucracy

25 September 1967

It is 0525 hours Monday morning and I will leave soon for Danang and other points north. The purpose of this trip is to unravel and reduce some of the GVN red tape associated with lumber production. Marines and Special Forces have a critical need for large timbers to reinforce their bunkers. I will not be near the DMZ or current battle zones, so do not be concerned. Metcalf and Thuong will travel with me.

I spent yesterday afternoon and late into the evening working at the office and was unable to include these items in my last letter. Martin Syverson has a hernia. He developed it while I was on leave and decided to do nothing about it. Now it is giving him trouble. I worked with him and prepared the necessary report and memos on the subject. He leaves for Portland tomorrow and will have the hernia surgically repaired by his own surgeon. He will be gone for up to two months. Sy is a strong member of the team and we shall miss him.

I close and jog to breakfast and return here by 0600 hours.

1 October 1967

I returned to Saigon late yesterday afternoon after spending the week up north. Come to think about it, I am covering many air miles. A round trip to Danang with all the stops made by Air America covers almost 1,000 miles.

When I returned to Tan Son Nhut Airport yesterday it was raining very hard. On the drive into the city I saw many troops on the move. I later learned the Buddhists and students were demonstrating for their causes and against the election results. I was advised to stay off the streets until morning.

This morning en route to pick up Barry Flamm for the drive to the airport I found the entire downtown section of the city closed off by riot troops. They were armed with a variety of weapons . . . rifles with fixed bayonets, machine guns, riot gas, gas masks and wicker shields to ward off rocks and bottles. It took me 45 minutes to size up the situation and drive within a block of Barry's hotel. I whisked him aboard and took off for the airport. He was scheduled to depart for Cody, Wyoming and TDY in California. He will return on 1 November.

I dropped Barry at the airport and high-tailed it for my hotel. Troops were on the move all over the city. I have been in my room all day. I hear sirens in the distance. I neglected to take my camera with me on the trip to the airport but I am not about to go out on a photo expedition. We have no information on what is taking place and the radio has nothing to announce. It is just like the day terrorists blew up the Chinese Embassy. There was much artillery action that day and another dull boom was scarcely noticed. All I know about that episode is what I read in the newspaper. Contrary to warnings some people rush to the scene of disturbances. I have learned that a moving target is harder to hit and I keep moving and go in the opposite direction. So far I have been in the right place at the right time to keep out of trouble.

Last Monday Lew Metcalf and I were at the Air America terminal at 0630 hours for our scheduled departure to Danang at 0700 hours. Thuong was to accompany us but at the time of departure there was no Thuong and no airplane. Something was wrong with the first plane and by the time another plane could be brought in Thuong arrived. He had been tied up in one of those monumental traffic jams. When we were ready to depart I checked and found our baggage was on the first plane. After another delay luggage, people and airplane were ready to go.

The flight north was through smooth, clear skies. We flew over some of the battlefields north of Saigon where the forest is severely damaged by defoliation, artillery and bombing. Fortunately not all of it is in that condition. Last week Syverson was near the Cambodian border on a log salvage operation. He visited a sawmill and the owner gave him some samples of the plague of the forest industries, that is shrapnel. Some of the case hardened pieces are eight inches long and weigh almost a pound. These sharp, wicked fragments are often buried deep in the wood and are not detectable until the saw hits them. Many saws are seriously damaged or destroyed. This problem will be around for a long time. We plan to introduce the mills to metal detectors and thereby save saws.

Buddhists were on the rampage in Danang also. We were rushed into town to conduct our business. Tension is high in that northern city. While in Danang we accompanied local forestry people to a forest nursery. We had to pass through a Vietnamese Army Ranger camp. The locals had no entry problem but they did not want to let me pass. I talked to the guards in Vietnamese, called the Lieutenant a Captain, made everybody happy and I was permitted to pass. The nursery covered two hectares and supported a healthy crop of eucalyptus and Australian-pine. The nursery and their techniques appeared to be primitive but successful. The district forest ranger asked me to help him obtain more funds to accelerate his nursery and reforestation program. The Saigon office makes few visits to inspect or train field people. Consequently there is little understanding of the field organization's needs and problems. Again the excuse is the lack of security. The Saigon people are afraid to go out in the field and the field people are afraid to go to Saigon. Each group finds themselves more comfortable with their own terrorists or VC.

The main purpose of my trip to I Corps was to help stimulate the production of heavy timbers for the Marines. Some of them are holed up along the DMZ and in other areas in open trenches. They need the timbers for constructing bunkers with overhead protection. The eight sawmills in the Corps area all suffer from the same problem, a shortage of logs. I learned logs were available, could be cut and transported under military protection. But at present the Army, Marines and ARVN are fighting battles with the North Vietnamese and VC and can not be spared for escort duty. Tra Bong remained the sole source of timbers. They have logs and are operating under the protection of the Special Forces. However to my frustration I learned even that operation has a problem with the Vietnamese bureaucrats. They had been shut down because the Directorate of Forest Affairs had neglected to process cooperative and sawmill authorization permits through the Ministry of Agriculture. While Tan and the Minister had endorsed and approved the cooperative and the sawmill operation, the local officials had been left out of the chain of command and had shut down the mill. You could say their noses were out of joint. Now I had an interesting situation and I had to devise a means to cut through an unbelievable amount of red tape. It is one frustration after another!

Metcalf went to another Special Forces camp to size up the opportunity for establishing a sawmill in a secure area. Frankly I do not believe anything is secure in the entire I Corps area. Thuong and I flew to Quang Ngai. I was glad he was with me because my limited ability in Vietnamese could not possibly cope with these local bureaucrats. We visited the local forestry office to determine why the Tra Bong mill was shut down. I learned the decree authorizing the timber cooperative had not been received from Saigon. Before leaving Saigon I had checked and learned the cooperative papers had been approved and signed. But no notification or papers had been received in Province headquarters in Quang Ngai. After checking with the Cooperative Chief, Economics Chief and Province Chief (many chiefs and few Indians) I learned their hands were tied until they received approval from Saigon. I sent a telegram to USAID Saigon asking them to run an immediate tracer on the approvals and to report back to me.

Meanwhile I asked the local officials to take care of the necessary paper work, charts, maps, records or whatever was needed for the authorization and operation of the cooperative and sawmill. I believe the only thing these people learned from the French was how to make the Vietnamese bureaucracy even more complicated. Following that liberal education in local red tape Thuong and I flew to Tra Bong for a meeting with local cooperative people. With the exception of minor repairs to the sawmill, we found they were ready to resume operations but afraid of the possible consequences if they operated without official authorization. I told them this was their decision. After all my years in the US Forest Service I could not in good faith tell them to operate without a permit. I did tell them they could fill a large order from the Marines and if they wanted the business they had better make the right decision.

In my meeting with local officials I inquired as to why they had discharged Pham Van Le, one of the original operators we trained in May. They said he had stolen lumber, loaned out some tools that were lost and had lost a few vital parts from the mill. I asked to meet with him to verify these charges. They agreed to try and contact him and set up a meeting in a day or two. I inspected the mill and discovered we could not resume operations until Metcalf arrived to repair the damage.

The Special Forces were pleased to see me and happy to have Thuong as another interpreter. They showed us our beds and provided us with some of the tools of their trade . . . M-16 rifles, pajamas. The crew was down to five Americans. Two were on a patrol to protect the rice harvest, one was at Cam Ranh Bay recuperating from malaria and one had been wounded and was in the hospital in Danang. The Commanding Officer assigned me to the communications bunker and gave me a machine gun to operate. The only enemy activity that occurred while we were there involved the patrol. It was hit by a small VC force and the local artillery unit tossed in a few rounds and sent them scurrying into the hills. I helped the medic patch a few bullet holes in the local defense forces.

Wednesday was the day I was scheduled to meet with Pham Van Le. This little episode illustrates the complexities of working in Vietnam. This is the man I referred to in my article "The Sawmill in Vietnam" in *American Forests* magazine. He was our best sawmill operator, the one who expressed his gratitude for his first job in 10 years. He turned out to be the VC Lieutenant in charge of the Tra Bong District. Apparently the real reason for his being discharged was the security risk. The Special Forces were elated when they learned I had requested a meeting with him. They had been trying to find him for many weeks. The meeting in the timber chief's office was interesting. The officials and two of the mill operators were there, but no Pham Van Le. The Special Forces and the District Chief's intelligence people had agents hidden all around the office, in trees, bushes, banana groves and rice paddies. Everyone was very surprised when Pham Van Le arrived. I talked with him and found that he indeed had

284

aken the lumber, had lost the tools and the sawmill parts. He expressed his regrets and offered his promise to improve, if he could just regain his job. He expressed appreciation to me for giving him the training and a job. Unless he really had a change of heart he was a very effective liar. Time will tell what will happen to him because he was carefully shadowed as he left. The Special Forces expected he would lead them to VC in the nearby mountains.

I spent the remainder of the week checking records of the yet-unapproved timber cooperative. They had good records, and if true they are making money from lumber sales. We have been concerned about the waste resulting from the mill operations. I calculated it to be about 50%. There had been some suspicion the Montagnard loggers were being cheated, but the records showed they were being paid a fair rate. I verified this by talking to Montagnard leaders. I would have had great difficulty in conducting all of this business without Thuong. I know just enough Vietnamese to follow a conversation and be sure my words are used and the response is conveyed accurately. Too many times interpreters think they know what you want to say and what you expect as an answer and they put in their own thoughts. Thuong did his work very accurately. He was well liked and trusted by the Special Forces since he had once served them.

Thuong, three Montagnard laborers and two Special Forces interpreters were the only non-Americans permitted within the inner perimeter of the camp. Nungs carefully guarded the gate. There is a valid reason for not allowing local Vietnamese into Special Forces compounds. A number of SF's camps have been overrun and everyone killed. Subsequent investigations showed in every case that spies had worked within the compound and had carefully mapped the strategic locations of communications, weapons, mine fields and ammunition supplies. Attacks focused on these key locations and outposts were easily overrun.

After forestry work was done I became an expert gunsmith. I helped piece together carbines for local recruits. Ten of them had no weapons. We used pieces and parts of damaged and captured guns, all junk left over from World War II and Korea. The VC have a good supply of US arms they have picked up in one way or another. We managed to assemble 10 smooth working carbines. I learned how to convert an M-1 semi-automatic carbine into a fully automatic M-2, and vice versa. It was simple. Local troops are provided only M-1s since they burn up ammunition too fast with an M-2. We test fired the weapons and threw a number of hand grenades. All of the noise and activity must have helped because we had no problems with the VC up to the time I departed from Tra Bong at noon Friday. Metcalf arrived that morning and I briefed him on the situation. He and Thuong remained to make the necessary repairs, provide training to reduce waste and supervise the construction of a lumber shed.

I departed for Quang Ngai City on an Army Caribou C-118 twin engine cargo plane via Chu Lai. The plane delivered a load of rice, dried fish and fish sauce for local troops. Needless to say the odor was not quite like that experienced on Pan Am's planes. Chu Lai is a large Marine Corps air station on the coast. While we waited for fuel and our plane to be loaded I saw continuous flights, one right after the other, of Marine fighter bombers taking off on missions. A number of Navy carrier planes came in for refueling. These were identified as from the aircraft carriers *Intrepid* and *Constellation*. Most of the planes from I Corps bases and Navy planes refuel, before going into North Vietnam on a mission from aerial tankers and again on return flights. This extends flight range and enables them to have a lighter load when they are over target areas. This provides better maneuverability in case of MIG attacks. That day there must have been more missions than the tanker planes could accommodate and Navy planes were diverted to Chu Lai for refueling.

I could see vast stockpiles of supplies at Chu Lai. There were hundreds of trucks of all sizes moving about or waiting to be loaded. Boredom was written all over the faces of the young GI drivers and guards. They do the same thing, day after day. Many drink heavily and resort to drugs to relieve boredom. Occasionally they have some excitement when VC probe the base's defenses. The only two events GIs have to look forward to are the day they can go on R&R and the end of their year in Vietnam. At that time they return to the states or elsewhere. The one year limit on tours of duty helps keep them more or less sane but it results in a huge turnover of personnel and inefficiency.

Our load for Quang Ngai turned out to be a Jeep. It filled the cargo compartment of the plane. I sat behind it on a nylon bucket seat with two Vietnamese soldiers who had been traveling with us to Quang Ngai. The Jeep was chained securely to the deck. But it was a little overpowering as we took off in a very steep climb to look up at that Jeep perched above me. There was no way to get around it and the only exit for me to reach would have been the rear cargo door. But the crew chief knew his job well and carefully checked all tie downs. En route we flew over the Sea. It was comforting to see a number of "Jolly Green Giant" rescue helicopters patrolling the area. We made it to Quang Ngai safely.

At the Quang Ngai airport I had a unique opportunity to solve my normal transportation problem. I signed for the Jeep and kept it until I departed the next day. I gave a young GI a ride and on the way into town we were detoured through the outskirts of the city. I did not know this area but I soon found myself boxed in by a convoy of armored personnel carriers so we stayed with them. I had to drive faster than I prefer but it was better to be bounced around by chuck holes and bumps than to be run over by one of those large tracked vehicles. I would have preferred to travel by myself rather than in a convoy of attractive targets. But those rigs bristled with .30 and .50 caliber machine guns, and provided a certain peace of mind. We made it through the detour and into the city.

Quang Ngai was an armed camp, more so than usual. It had been reinforced since the election eve attack. Local people were digging in and when we see that being done you can expect the worst. Word was out they are due for more trouble. The compound where I normally eat was hit with many mortar and rocket shells during the last raid. Many US locals went to the Quaker House and the Red Cross Headquarters just before that raid. Some of the explosives came close to them but missed. They told me on the night of the attack the VC were roving all over the town and each group of civilians and military had to look out for themselves. I was the only guest sleeping at the Provencal Representative's guest house. The only problem I had was with mosquitoes and the terrible heat. I drenched myself with repellent and survived. I had few sweet dreams as I lay on a lumpy cot with a .45 automatic under my pillow and a carbine by my side.

The next morning, Saturday, I drove my Jeep to the BOQ for breakfast. As I was concentrating on my driving and the other drivers I did a double take and there stood a half naked woman beside the road waving to troops as they went by. Traffic was heavy as people went to work, or wherever they were going on foot, bicycles, in baskets on motor scooters, and in carts of one kind or another. Many people just plodded along aimlessly, dirty, ragged and with no sign of hope on their faces. Most of these were refugees that have swarmed into the city for food and protection. It is a wonder that the fast driving Vietnamese Army drivers don't kill many of the pedestrians and bicycle and scooter riders. They drive down the middle of the road as fast as they can go with their horns blowing. At the last possible moment people step out of the way. As I watched these poor people I could not help but think that as soon as we leave they will have very few jobs and their vehicles will fall apart completely. But perhaps they will return to their farms and rice paddies and be much better off. It is sad to realize they have had no peace for over twenty years and with a life expectancy of 35 years they do not have much to look forward to. They are caught in a vicious squeeze. They frequently are targets of our sophisticated technology and war machines, the VC is after them and trying to win them over, local officials and business people are out to take them for all they can get through graft and outright theft and the VC fully exploits the graft and corruption angle. These people lead a terrible life and have to battle all the way through life just to survive.

I talked to a very discouraged, obviously fatigued medical doctor at breakfast. He told me he was rapidly becoming the world's specialist on black plague. He said he was handling hundreds of cases daily at his hospital. Frequently he finds a wounded soldier or civilian is forced to share a bed with a plague or cholera victim. On my way to my meeting the half naked woman was still there, sitting on the curb and waving. Others paid her no attention as they pushed their heavy loads or passed up and down the street. This is certainly a strange, cruel and crazy world.

I held a short meeting with the Chiefs of Forestry, Cooperatives and Economics. In my best Vietnamese, since I had no interpreter, I told them that if they wanted forestry assistance and new industries they had to cooperate and promptly get the Tra Bong mill back into operation. I told them my time was too short and there was much to be done all over South Vietnam. They acknowledged my message and said there would be no local delays. As my final shot I informed them we would be checking records and if there was any evidence of graft and corruption, the sawmill and our support would immediately be removed. It will be interesting to see what happens now.

I drove out of that terrible city to the airport where I gave the keys of my Jeep to the local Air America official. He did not know what to do with it either. This man had neglected to put my name on the manifest for departure at 1045 hours but the fact that I had given him a Jeep and had travel orders giving me the highest priority, I was able to board a not-so-luxurious C-47 and sit on hard metal bucket seats for the next 600 miles. We flew over vast forest areas and made three scheduled stops. Forests were spotted with clearings made by shifting agricultural practices of the Montagnards. Most fields and paddies in remote areas of the highlands are idle and have been sprayed with Agent Blue, a crop herbicide containing a trivalent form of an arsenic compound. Most of these areas have no security and the people have become refugees in the lowlands. In other areas I observed swaths of trees killed or damaged by applications of Agent Orange and/or Agent White. Some of the defoliation flight paths appeared to stretch beyond the horizon. As a result of the defoliation and intensive shifting cultivation I could see streams running thick with silt. High intensity monsoon rains carry off vast amounts of soil. In some of the more unstable areas I could see many large slips and land slides. In the untouched forest areas streams ran clear. The white water streams in such areas make a magnificent sight as they cascade over waterfalls and carry on through spectacular rapids.

I could see pieces of trail throughout the mountains. It is up and down this trail network refugees, VC and North Vietnamese move freely across Vietnam, Laos and Cambodia. In spite of our modern war machines and huge stockpiles of supplies, they keep coming. Their persistence, dedication and leadership sometimes makes me think we may be on the wrong side. They carry supplies on their backs or on heavily laden bicycles pushed along these trails. They certainly have the stamina and total commitment to a cause that is not always apparent in many of the people we are trying to help. It will require fancy footwork on the part of US government leaders to figure a way out of this quagmire. Please do not misunderstand me, much progress is being made, but the situation is so complex and obstacles are tremendous. It is a challenging opportunity to be here and have a minor part in this moment in history. I will do my best to try and help these people overcome some of their problems.

We stopped for lunch at Nha Trang and I bought a hungry looking young GI and myself a couple of sandwiches and root beers. He was a good looking young lad and until recently he must have been a Pfc. Marks on his uniform indicated his stripes had recently been removed, for one reason or another. Like many of the GIs he had an expensive Nikon camera and did not how to operate it. I gave him a short course in how to use a camera. He makes about the 24th one that I have helped understand their new PX cameras.

I just went up to the roof to see what was going on around Saigon. One of the men up there said the town was bottled up tight by GVN troops. Students and Buddhists are milling around again. Troops near us were in position to keep national assemblymen from reaching the Assembly Building to complete their work on the election results. By law their work must be completed by 2 October. So who knows what's going on? The troops are certain to be either keeping someone in or others out. Tune in later and perhaps we will learn more about it. Meanwhile back in my room I will finish this letter and listen to the Boston Red Sox defeat the Minnesota Twins. I wish I had more time to improve these letters but I am doing the best I can. I hope you find them interesting.

Chapter XXII

Life in Saigon

6 October 1967

It is Friday night and I have returned to my room early. The day went something like this . . . I worked at the office on reports until 1535 hours and then had a meeting at Tan's office. I arrived there just as the bottom fell out of the sky and a wall of water came down. I met with Tan until 1745 hours and then drove back to the office in a driving rain. I did not know it could rain so hard! If I was Chicken Little I really would have believed the sky was falling. I quit work at 1800 hours and walked out through a crowd of USAID people who were holed up on the first two floors waiting for the rain to subside. I drove to the club and had a good workout doing situps, stretches and weight lifting. Then I had a cold shower. Then it was off to the Hotel Splendid BOQ for dinner and a glimpse of Martha Raye and her USO entertainers. My, but she really does have a big mouth! After that I went to the courtyard to watch a Henry Fonda western. We sat along the sides under the overhangs as the continuing heavy rain and waterfalls came off the balconies and the clothes on the lines. It was difficult to hear the sound over the rain. About halfway through the movie the last projection bulb burned out, we all grumbled and filed out to our respective quarters. Rain was still coming down in buckets and as they say down South, it was a "frog strangler" or a "chunk floater". Our Jeep has a good canvas top, but with no side curtains, half of you can get very wet. I bless wash and wear clothes. Back at the hotel I have just washed out my clothes, lighted my pipe and sat down to write this letter.

I thought about you today as you were taking off for Vermont to see the fall colors. We have a fall color photo of the White Mountain National Forest hanging in our office to remind us what it is like out there in the real world, so very far away.

I enjoy reading your letters concerning the coming and going of the guys and dolls. Sounds like you are busy with jobs and school. I hope Melissa is enjoying college better than she did at first. I recall I had problems until I fell into a study routine. I did not have it quite as easy as Melissa's schedule. That certainly should provide time for study. So get with it, Lis!

From what I see, hear and read it appears everyone is growing tired of this war. We are still very much aware that it is with us. Artillery fire has been continuous around the city tonight. It must be programmed to prevent infiltration during the heavy rain. I keep well away from all the trouble spots like the embassies. However the Palace is just a block from me and that is where some of the demonstrations were taking place last week while I was up north. I have driven by it this week and an old monk and a few of his caretakers are still sitting in the park. I could see them tonight, sitting in the pouring rain, as I drove from the club to the BOQ. The club is adjacent to the Palace. Students have been threatened with arrest if they cause any disturbance so the government is trying to keep things calm. I hope there will be a break in hostilities, but it is doubtful it will happen in the near future. I would like to think Melissa's Bob might find Vietnam a more peaceful place if he has to come here, but it remains about the same as always.

I am about to fall asleep. Staying awake to listen to the World's Series broadcast starting at 0045 hours (12:45AM) is tiring. I will sign off and mail this in the morning.

8   October   1967

It is Sunday morning and not raining, for a change! The sun is trying to burn through thin hazy clouds. I grow weary of looking out my hotel window every morning and see pouring rain or rain just about to start. Today is welcome relief from the hot dreary weather.

I just returned from the BOQ after having a good breakfast of steak, eggs, potatoes, purple plums, toast, coffee and two fresh oranges. In view of spreading demonstrations, radio announcements advise us to keep off the streets. Activity seems to be building in the downtown area where I first lived and near the Palace which is a block from where I now live. This morning I could see newly erected barbed wire barricades and police and troops assembling.

In a park just two blocks from me and in front of the Palace the old Buddhist monk has maintained his fruitless vigil. He sat there all week in the heat and monsoon rains with his head bowed in silent prayer. A few of his followers surround him and hold an umbrella over him. He started a hunger strike on Thursday. He is the same one who fasted for 100 days a few months ago. That hunger strike failed to achieve the desired results and he abandoned it. The police have a barricade around him to hold back any possible congregation of supporters. I have driven by the place several times during the week in the course of my work. I have taken no photographs since we have been warned by the Embassy to refrain from photographing the event. You may have read that correspondents and CBS TV people were roughed up by the police. Actually they were severely beaten. Some of the protesting students have been beaten, jailed and others warned to keep away.

A young nun in one of the western Delta provinces is reported to have set herself on fire in protest. We are told they are fighting for recognition of their militant Buddhist sect. The government does not recognize this particular group, but the sect is determined to gain recognition.

In view of all of this activity I will heed the warning and stay in today to write letters, wash and iron clothes and start Christmas cards. Does that please you? I will box up the cards and send them to you for addressing and mailing. My address book has been stolen, probably by some low level VC intelligence group. Perhaps some of my friends will receive greetings from the VC.

The week has been marked by continually shelling of Con Thien, a Marine outpost up north. In the last day or two there has been a slight let up in pressure, but shelling continues. I feel bad because we have not been able to get the mills in the northern area into full production cutting badly needed timbers for their bunkers and lumber for refugee housing. All of the mills, with the exception of Tra Bong, are unable to operate due to a log shortage. In most cases logs are less than 10km from the mills. The military is tied down and unable to provide security for loggers and log haulers. Tra Bong continues to suffer from GVN red tape. I have used "scissors" liberally this week and made slow headway. The Tra Bong Mill is shut down until the GVN completes a security clearance of all workers and members of the cooperative, issues a cooperative decree and a sawmill authorization permit. On my last trip I told the local people that if there are VC in their organization they had better weed them out themselves if they want continued technical assistance and support. I have not heard anything about the discharged operator. When last seen Pham Van Le was heading for the hills with Green Berets and local forces dogging his heels. That situation may be taken care of by now.

Please read the 25 September 1967 issue of *NEWSWEEK*. Saigon Bureau Chief Martin wrote a straight forward editorial concerning his views of the Vietnam situation. It placed major emphasis on the Vietnamese to wake up and get with it if they wanted to solve this thing. The article has resulted in repercussions among the Vietnamese so the shoe must have pinched. In today's *Saigon Sunday News* Martin's letter to the editor is entitled "VN: Last Chance". Copies of both articles are enclosed for your study of the situation.

There have been purges this week in an attempt to eliminate corrupt practices. Participants are spread throughout the Vietnamese military and civilian organizations. Cleaning up corruption and "kick backers" will take some doing, especially since the practice according to *TIME* is widespread throughout Asia. It is a way of life with many of these people. I understand hey believe there is nothing wrong with the practice as long as ne is doing it for his family. From what I understand, some Generals and Colonels in the Vietnamese Army must have very arge families.

Lew Metcalf and I have been in the office this week catching up with budgets and work plans, reports and working with US and Vietnamese military forces. USAID's budget is all up in the air and we receive very little advice, while we revise he revised version of the revised budget.

After work yesterday I drove out to the MACV (Military Assistance Command Vietnam) theater to see "Hello, Dolly". The heater has just been completed and it required a search to locate t as it was tucked away in the huge Pentagon-West complex. I asked a civilian how to get into the theater. He took me in the back door and I was able to bypass several hundred GIs lined up at the front door. This gave me the choice of seats in the front ow. I hope my photographs of the show turn out well. In spite of Martha Raye's collapse from heat and humidity yesterday at a nearby base she was in fine form. The audience of GIs, hospital patients and a few civilians loved her show. Martha has made many trips over here and has endeared herself to the troops. After one particular active dancing episode she was puffing as she and her troops acknowledged the wild applause. When the applause subsided she puffed some more and said, "I'm getting too damned old to do this". (Some days that applies to me too!) Her jokes and antics brought roars of laughter. Two hours of "Hello, Dolly" were enjoyed by all of us. At the end of the show General Westmoreland hugged and saluted her and said what a great trooper she was and how happy they all were that she had recovered from yesterday and was in such good form. At that remark she turned aside to the GIs, threw out her ample chest, raised one edge of her long skirt and said "Pretty good for such an old gal". That concluding remark brought cheers, hoots, whistles, clapping and just about brought down the roof of the new theater.

After the show I drove the short distance to Tan Son Nhut Airport terminal and bid farewell to Director Tan's son. He departed for University in Switzerland. The Tan's, his brother-in-law the General, dozens of relatives and most of the Directorate's staff were there to see the young man off. It was quite a sight to see hundreds of people gathered there to see sons and daughters off to school in Europe. Many of the young men appeared to be very suitable candidates for the Army. But, as in the good old US of A, there are also ways to avoid the draft in this country.

My next stop was the home of my Vietnamese language teacher. She was off to France for five years of study. Her many friends and relatives were there to see her off. Three of her students, including me, were there to take her and the family to the airport. At the home I was entertained by her two youngest sisters, Tau (age 11) and Anh (age 9). They gave me an intensive course in Vietnamese. They could speak enough English to correct me and I learned a great deal as they chattered away on my lap like magpies. We loaded five pieces of luggage and family members into four Jeeps. With a prearranged pass and some fancy talking I was able to get everyone past guards and into the Tan Son Nhut commercial airport terminal.

Our teacher also showed us how things are done in Vietnam. There was a huge line at the Thai Airline counter. I could not see how she could be processed before departure time but she had a friend who worked for the airline and was taken to the head of the line. Papers were processed and she was quickly passed through emigration. I asked to look at her passport and visa papers. These authorized her to go from Bangkok to school in Geneva. I started to ask her about Paris, but she "shushed" me and took me aside. I learned Vietnamese students are not permitted to go to France. Instead they all go to school in Switzerland. Once they arrive in Geneva or Zurich a mysterious transformation takes place and they are off to school in France for four to five years. I suppose Tan's son will also end up in France. I noticed all students going on this long flight had vast amounts of food with them. Some must have had a good supply of nuc mam because the airport terminal smelled pretty fishy. I have always wondered what a flight would be like if a jug of nuc mam broke in the cabin. In the pure form nuc mam is overwhelming, but in the diluted form with hot red peppers it tastes delightful, at least after you get used to it or eat it in self defense.

There were many tears and the prolonged wait for their departure was touching. I do not believe an injection of tear gas could have produced more tears than were flowing that day. Vietnamese have very close ties to their family and departure for four to five years is a major step. Everyone in teacher's family shed tears, except the two bright eyed little sisters who kept up a continuous chatter. Before leaving teacher asked if any of us had any green dollars to exchange for piasters, but none of us did. We all were impressed by the severe penalties imposed for dealing in black market dollar exchanges. Each Vietnamese passenger is very limited in the amount of funds they can take out of the country. None of teacher's baggage was inspected, courtesy of her Pan Am friend, so she may have "made out". I held her purse for a moment and the weight indicated it could have held a gold bar or two.

The return trip from the airport was made in a hurry at 2030 hours. Back alleys and streets were clear of traffic and we were anxious to return to our sanctuaries.

My next stop was a reception for Vernon Johnson, a USAID/Washington official. Vern is a pleasant Negro who is in charge of Vietnam Agriculture. I had a good visit with him in July in his Washington office. He greeted me like a long lost friend. By 2100 hours the crowd was "well oiled". I had a delightful time visiting and eating fried chicken and Vietnamese goodies. The chicken was particularly good and I made up for the food I had missed during that busy day. For desert I had a Coke and a cigar. Some of the guests showed the results of their continuous partying . . . red faces, round, full stomachs and baggy red eyes. I am glad I am not on that continuous social circuit, which for some people is six or seven evenings each week. This is an occupational hazard experienced by professional overseas workers.

As I type this letter I can hear a continuous roar from the helicopters coming and going overhead. I will relax for a while and go up on the roof to absorb some sun and the sights.

I have returned and everything appears to be peaceful for the moment. I hope you all had a good trip to New England and the new camera worked well.

I received a long letter from my Mother a couple of weeks ago. She seems to be settled, satisfied and apparently wants to hear from you. Sounds like she has had a change of heart. She is difficult to understand. I enclose her letter from 2030 1st Avenue Northeast, Apartment 305, Cedar Rapids, Iowa 52402. I plan to drop off and see her for a day or two on my next home visit.

I have not been paid for the expenses associated with my trip to Washington and the West. The expenses have been approved, but since Congress has not passed the FY 1968 appropriation bill it has not been paid. According to my calculations I will have $1,000 in my checking account by Christmas. You should plan for a trip following Christmas. I plan to leave here on 15 December and following a brief stop in Iowa I will be home by the 20th. I have completed most of my Christmas cards which I will mail to you today in a package for addressing and mailing. I will request a five day official detail in the Washington office and return to Saigon on 17 January. Quebec would be a nice place to visit and will be fine with me. I will probably freeze to death and would favor the Caribbean. However it is your choice. I suggest you make reservations early and obtain a packaged deal for transportation and hotel, but no group tours since we prefer to do things on our own.

Cindy, I am pleased you have the job at the veterinarian's. That should be interesting. Please keep up with your French lessons and you can be our interpreter when we go to Europe next June. Gwen, you can do our interpreting in Wales and England. Melissa, you can take care of Scotland, since you are so thrifty. I will make plans to meet you in Paris or London. We should be able to have three weeks before going home. I can arrange a week or ten days of official time in Washington to extend my stay. So much for planning. I hope this will help make up for some of the family trips we have missed due to me being overseas.

I enjoy hearing about all your activities. Sounds like the boys' band is well organized. My ears are still ringing from that session they held in our house. Drums and electric guitars are rather piercing. They are good and I hope they are successful in landing a number of jobs.

Melissa, do you have any information concerning Bob Humphrey's next Army assignment? I hope it is not to Vietnam. If it is be sure and give him my address which is the Park Hotel, 35A Nguyen Trung Truc Street. I am just at the rear of the Embassy Hotel and all taxi drivers and pedicab men know where it is located. My office is at USAID No. 1 at 85 Le Van Duyet. My office is 6D in the Agriculture Division. We may be moved to another building but that move has been on and off for several months. Things move slowly in Vietnam.

Melissa, I am glad you took one of Ray's drums as security on what he owes you. I hope to heaven he pays off before you become a drummer!

The radio has just given one of its frequent messages on malaria . . ."take pill once a week, keep skin covered, sleep under mosquito net, etc." I told you one of the Special Forces men came down with malaria at Tra Bong. They shipped him out to Cam Ranh Bay where patients recuperate by spending their days filling sandbags. Most of them are said to recover quickly since there is an unlimited amount of sand at Cam Ranh Bay.

A Frenchman who lives across the way from me is presently yelling his head off. I can not tell if he is praising or berating someone. He is wildly waving a paper in the air. Perhaps he is rehearsing a Charles De Gaulle-type speech. Interesting people, the French! Bye for now I am about out of paper. It is 1830 hours and I will be off to mail this letter, eat and take in a movie at the BOQ.

15 October 1967

Your fall weather sounds delightful. With all the boys and girls around the place there should be sufficient help to rake the leaves. Aren't you pleased I split the fireplace wood before I left home? Gwen, you may have lost your knack of splitting wood. I recall you were good at it when we lived in the Arizona mountains at Spring Valley on the Kaibab.

Our weather is the same day in and day out. It is cloudy and rains very heavily sometime during the day or night. Saturday there was a little sun and I spent the afternoon at the club swimming and sitting by the pool, writing Christmas cards. I should be able to complete all of mine next weekend and send them to you for addressing and mailing. We will have another of those long curfews due to Lower House elections.

I hope you received word regarding Aunt Madeline Kester's death and took care of the flowers. Her death was a blessing since she suffered so from Parkinson's. She must have suffered a great deal as her muscles and joints "froze". She bore her pain in silence and I never once heard her complain. I know you liked her. She was the only one in the family who treated you decently. She was a good person. I shall always have good memories of her when she was young, beautiful and jolly.

The USDA has advised me I owe $481 in overpayment on my salary while I was on leave the last time. That was the amount paid to cover the 25% differential for hazard pay. USDA said it should have stopped when I left Vietnam. I told them politely to "drop dead"! I quoted USAID regulations which state differential does not stop when on visitation. I said in the event it turns out the other 3,000 USAID employees were also overpaid I would follow instructions and advise them how I wanted to pay it back in convenient amounts. I suggested I would consider paying it back at the rate of $2.07 per pay period and spread it out over the next 221 checks. That should just about carry me to retirement. Relax! Don't worry! I do not plan on retiring that soon. The USAID office here assures me the USDA people are incorrect in their interpretation. I hope I am successful in quieting the uproar they created here among other USDA employees. Seems they also received letters notifying them about the overpayment. They look to me to fight this battle for all concerned, as if there aren't enough battles going on in Vietnam!

In regard to my travel plans I plan to leave here on 15 December and return to Saigon on 15 January. That will enable us to make the trip to Quebec. I could do with some cold weather for a change. I am agreeable to going anywhere as long as we all can be together. That will give me 24 calendar days of leave and the balance in TDY at the Washington office. The TDY will cover (1) program review with Forest Service and other USDA administrators, (2) personnel appraisals with Forest Service Personnel Management and (3) timber management conference for development of Vietnam timber sales procedures. The purpose of the latter is to discuss Vietnamese timber disposal procedures with Forest Service timber sales experts and develop a system to help eliminate some of the leakage and seepage of funds.

I believe most people around here share your opinion of President Lyndon B. Johnson. LBJ certainly has all of us in a real mess!

For the past 10 days we have had trouble with our ancient World War II Jeep. It is a tired old relic, which according to a plate on the dash board, was shipped in from Antwerp. It has a canvas top that pools water to a certain depth, which then leaks through to soak either the driver or passenger. Wipers go slowly back and forth and touch only about 10% of the windshield. The instrument panel keeps falling out and no one is able to fasten it into place. It is held on by its umbilical cords from the speedometer cable, oil gauge and generator. None of the other instruments work but their wires help hold things together. When we had it washed recently we could see the road through the rusted-out floor. As to be expected the ancient tires result in many flats. So far all of these have been experienced by Barry Flamm. We have no tools, jack or lug wrench since they would be stolen as soon as we turned our backs. A flat results in a call to the shop and produces a mechanic, two young boys and a spare tire.

We lock the Jeep by using a large padlock to fasten a log chain, which is welded to the rusty floor, through the steering wheel. Flamm recently had the padlock stolen. Apparently the lock was more valuable since the thief did not take the Jeep. Barry has trouble whenever he parks the Jeep at his hotel. Usually some of the local boys show up and indicate they will watch the Jeep for a few piasters. Frequently Barry has found the air let out of the tires. On one occasion someone used the driver's seat for a toilet. Frequently the boys will say "VC work on Jeep", just to get a job of watching it. We look under the hood each time before we start it. When we drive into the USAID compound the guards use a long handled mirror to look underneath for concealed explosives.

Starting the Jeep has been my greatest problem. It floods easily and recently I ran down the 24 volt system made for arctic conditions and operated on two very large batteries. I called the shop to come and get me started. One morning following a very heavy all night rain I turned the key, punched the starter and after a flash of light and the sound of an electrical discharge from behind the fire wall the battery went dead. Mechanics came and merely recharged the battery. Metcalf picked it up and said it was now fixed. Driving to my hotel that evening a tremendous, loud electrical discharge and a bright flash of light came blasting up between my legs. I thought surely the VC finally got me. But when I recovered and looked under the hood I located the short and was able to separate it from metal and creep to the hotel. I borrowed some plastic tape and tools and solved the problem. The battery cable and some other wires had corroded through and were in contact with metal. The old Jeep doesn't go very fast, in fact it is difficult to outrun some of the bicycles and horse carts, but it gets us around Saigon and over rough roads to the local sawmills. Its size, weight and the indestructible appearance results in other drivers and pedestrians giving us a wide berth.

Filling the Jeep with gasoline at USAID No. 2 is another experience. I know most of the Vietnamese mechanics and they enjoy hearing my Vietnamese. They try to be helpful and swarm all over the old Jeep checking things I am certain they do not understand. After all this expert servicing, really all I ever need is gasoline, I sign the issue sheet and they always insist I fill in the remarks column. I do not know what good comes from the remarks but I humor them with by writing in such comments as "thanks for the service, hello, isn't it a nice day, it is going to rain, it is raining or this Jeep is no damned good." This satisfies the service men but I am certain it must confuse the American supervisor if he ever reads it.

After receiving a wire about Aunt Madeline's death from Parkinson's disease, I went to the Buu Dien and Dien Tin Ngoai-Quoc (post office and international telegraph office) to send you a sympathy message to forward to my Mother. That was another experience. The French constructed this huge building about 40 years ago. On the side facing the street a crew of three men were erecting a scaffold made out of three-inch bamboo. The men were working 100 feet above the ground and were still adding more twenty foot lengths of bamboo and cross supports. It gives me the creeps to see them running around on those flimsy structures in the absence of common sense and all safety rules. The front steps are lined with food peddlers, envelope merchants and black market currency dealers. It is necessary to weave your way through the scaffolding and the peddlers who have their junk spread out on the sidewalk and steps.

Once inside the high ceiling building the next task was to work my way through tremendous crowds to the international telegraph counter. People are always jammed together in this area. Most seem to be edging up to the international postal and registry windows to mail letters or packages. I suppose many of them are sending the earnings they have made or skimmed off of the good old American taxpayers to a numbered account in some Swiss bank. So much for that speculation.

At the telegraph window there must have been a system, but since no one spoke a word of English and all the signs and notices were in French and Vietnamese, it was a real chore for me to learn the procedures. I laid my telegram down at the end of a long line of telegrams spread out on the counter and found I was on the wrong end of the line. I worked my telegram into the proper sequence and waited my turn. While waiting I looked around the building. The French must feel sad when they see the deplorable condition of their structures. I assume when the French were in charge their Vietnamese servants kept things well maintained. Walls were cracked and dirty. On the walls huge maps of Saigon and Vietnam were peeling and covered with filth.

299

Overhead fans were covered with dust and dirt and only a few worked. Those that operated wobbled slowly in an unbalanced way and did little good in stirring up the very hot fetid air. The ornate grillwork over the counters were covered with spider webs and dirt. Blue, white and black floor tiles were dirty, broken, covered with gum, dirt and the bright red spittle of the betel chewing old women. Behind the counters clerks sat at small wooden desks and worked slowly at their chores in very poor light. Behind them 16 foot high doors led to a office where I could see more clerks working at similar desks in less light. The only concession to the 20th century was a few flickering florescent lights which did little to break up the gloom. When I reached the moment of truth, to pay the charges I did not have enough piasters! But fortunately there was one of the black market currency dealers at my side and he took care of my problem. The rate of exchange for MPCs was not very favorable but by that time I would have given anything to complete my transaction and get out of that place. I will be surprised if you receive my cable. The pace at that place seemed to go much slower than the hands of the huge clock which seemed to spin. That little transaction took place after work and required over two hours.

After the telegraph office episode I was off in my Jeep to the BOQ for dinner. That evening it appeared most of the waitresses were pregnant, some very much so. The very pregnant ones have a difficult time weaving in and out among the tables. The other day I saw one get stuck between a pillar and a customer's chair. They apparently work right up to the time their baby is due. Recently they carried one out as her labor pains started. And believe it or not they return in just two days. One day I saw one of them go outside to pick up her baby from one of her children and nurse it standing beside the MP guarding the BOQ.

Frequently I see one of the waitresses or other women on the street or at the office with a row of vertical, parallel red marks on their necks. At first I thought perhaps their husbands or boy friends were responsible. I inquired and learned that when they have a sore throat they pinch each side of their throat to get rid of demons. Interesting custom! Another strange sight is to see the Vietnamese pick their teeth. They learned that procedure from the French. They cup one hand around the hand that is doing the picking to cover what they are doing. Then, by gosh, I have seen them throw down the tooth pick and pick their noses. I wonder if the French have that repulsive habit and pick their noses?

The waitresses work eight hour shifts and are extremely busy at meal time. The very pregnant ones carry heavy plates and the way they lean backwards to balance their load makes me think their backs must be ready to break when they finish work. The present crew at the BOQ is a good one, including the cooks and dishwashers. Three mess Sergeants do a good job of training and supervising the work force. One of their most frustrating training jobs is to teach the kitchen workers and waitresses to stop picking their noses. The present crew appears to be "nose broke", so to speak. The mess Sergeants are huge characters, all are over six feet tall, and they tower over their Vietnamese employees. They do a wonderful job of keeping the place clean and everything in order. It must require much training and supervision to be certain the fresh vegetables and hands are properly disinfected and washed. Meat comes frozen from Australia and the United States. Other food items come in cans from the US. The vanilla ice cream has gone through a transition and really tastes great. When I first arrived it tasted like reconstituted ice cream. Now the local Foremost Dairy is working on a blend of coconut oil and it stacks up to the best of ours at home. You would be surprised how something like that makes war less like Hell.

After dinner the usual crowd of officers and civilians watched a bloody South Pacific World War II Marine movie called "First to Fight". I arrived later than usual and lost out on getting a choice seat under the overhanging balcony. True to form, while the Marines in the movie were sitting in the rain in their foxholes, we experienced a cloud burst. No one left, we all just sat there in the rain and soaked in the realism. It is strange but the war movies draw the largest crowds, more so than those featuring girls. Rain ceased after a while and we finished watching the movie in the wet steamy patio. Actually rain did not stop because the GI drawers, shirts, socks and trousers hanging on the roll-in clothes line kept dripping to the very end.

Today, Sunday, there are no scheduled, or unscheduled, demonstrations. Buddhists have quit their sit down in the park. They returned to their pagoda to lick their wounds. Students calmed down when they were warned that all demonstrators would immediately be whisked off to the Army. That cooled their demonstration ardor. If they cleared Saigon and sent all young eligible males to the Army, it would be relatively easy to drive all VC and North Vietnamese out of the South. It is amazing how many eligible young civilian men we see racing around Saigon on brand new Honda motorcycles and in new Honda cars. You will recall my description of students leaving for university in Europe. Many of them are eligible for the Army but they either know someone of importance or belong to someone with influence or money. When I look at all of our young GIs that could be in US colleges rather than being over here I feel quite ill in the pit of my stomach.

After breakfast today I loaded my brief case with my armament and took off for the pistol range. We are encouraged to go there for practice every other week. I drove through the usual crowded streets and tried to avoid breathing too much of the dense blue exhaust haze. There is no place for the faint hearted when it comes to driving in Saigon. All my driving life I have learned and practiced defensive driving. Not so here! To do so means you would be bluffed out continuously. I take an offensive role and put the old Jeep in 2nd gear and take off through the crowd. Miraculously it opens up at the last second and I am able to make progress. I do watch out for the large Army 6 by 6s whether they are driven by GIs or ARVN soldiers.

One of the places I drive by is the very large fish market. It is the fishiest, fish market you can ever imagine. You can smell it a half mile away. It stinks of fresh and rotten fish. Dogs and cats can be seen feasting on the remains or running though the stalls. At this point the street joins five other streets and traffic slows to go around a circle. In addition to the smell there are groups of children who stand in the middle of the circle waiting for some unsuspecting driver to slow down. Then they try and reach into the vehicle and steal whatever they can get their hands on. At this place recently Barry lost a sack of supplies he had just purchased at the PX. The kids have a system and work in gangs. First a group comes up on one side of a vehicle to attract the driver's attention and then a culprit on the other side reaches in and lifts whatever is in sight. Police are on duty at this point and I believe they share in the booty. I had no problem and the kids scattered when I put the Jeep in low gear and roared through the crowd.

The next scenic attraction was a community garbage dump which was piled high along each side of the street. It smelled about as high as the fish market, only more so. People who live in this part of town throw their garbage on these piles. There are always a few scavengers sorting through it. These people look like the dregs of the earth. Today USAID had donated garbage trucks to haul the stuff away. Men with pitch forks were tossing the wet, dripping, stinking mess onto a truck.

Proceeding on down the street, through heavy traffic, I turned into a poor, back-of-beyond neighborhood. There was no let up in the heavy traffic as I drove down a very narrow street. Some chuck holes were 20 feet across and filled with muddy water up to 12 inches deep. I dodged scooters, pedicabs, cyclos, push carts, taxis, a few cars and hundreds of pedestrians. I tried to be careful not to splash any pedestrians, but I noticed the other drivers take great pride in splattering people with filthy water. At one place the road passed through a roadside market and the same water splashed on the vegetables. I suppose that keeps them fresh. The first time Barry and I tried to locate the pistol range we missed a turn in this bewildering maze. We ended up at a police check point. We asked for directions to the range and even took out our pistols to demonstrate we wanted to shoot at bull's eye targets. The police laughed and pointed to the other side of the check point and said "VC" and pointed back the direction we had just come from and said "bull's eye place".

The range is a standard pistol range and is located in the middle of a thickly populated area. It has an ammunition store house, a police watch tower, covered shooting lines, and a 40 foot high mound of earth serving as a backstop. Along each side is a wall about 50 feet high and 20 feet wide. These devices are supposed to keep the bullets inside. The range serves local police and USAID Public Safety people. Sunday the facility is open to American civilians who want to practice. Supervision and training are provided by two local policemen.

We are given an unlimited amount of .38, .45 and .30 caliber ammunition. The .38 ammunition, called wad cutters, cut a neat clean hole in paper targets. The other is regular ball ammunition. Today I fired 200 rounds through my .45 Colt pistol and 200 rounds through my .38 revolver. We were given training in how to shoot from behind a barricade and a wall. In this fashion you can support the weapon with your free hand while exposing very little of your body. After that we were taken though a combat shooting exercise which consisted of rapid fire shooting at a man's silhouette. I placed 30 out of 30 shots on the paper and the instructor said all but three of the hits would have been "stopping" shots.

We were given instruction in how to fire fully automatic weapons, such as a Thompson sub-machine gun and a grease gun. These burn ammunition at a rapid rate. The sound is deafening when several of these weapons are fired at the same time. We use ear plugs which provide some protection. After about three hours of this it was time to quit. I gave the policemen 20 piasters each to pay two small boys to clean up the brass. I drove back to my hotel through the same crowded scenic places. Then I spent two hours cleaning my guns. I keep them carefully locked up when not in use and give no hint to anyone that I have any weapons. When I have completed the job I wrap oily rags and cleaning patches in newspaper and drop them in a garbage can at the BOQ.

This afternoon I watched a two weeks old Dallas Cowboys vs. the Los Angeles Rams football game. It was an exciting game and helped pass the time. The cost of watching the game in the hotel bar was one bottle of beer. That TV works very well. But after paying MPC$130 for a PX TV set and MPC$30 to a local TV man to install and adjust the set, my sad neighbor sits and watches only "snow". Our rooms are on the opposite side from the local TV station and apparently the concrete and steel building blocks TV signals.

I ate dinner at the BOQ with an acquaintance of mine from

303

the US Navy, Lt. Senior Grade Hagerty and an Ensign friend of his. He had helped me deliver a small shipment of trees to a nearby Navy base. Hagerty reminds me of the stage character "Mr. Roberts" and his friend was a splitting image for "Ensign Pulver" from the same play. These two were characters in their own right. They had been waiting for me and were desperate for help to solve a serious US Navy problem. What was their problem? Their Admiral has some sick palm trees. At first I thought they were putting me on (. . . Mr. Roberts . . .). But they were deadly serious. Seems like the Admiral's palm trees at the Nha Bay Naval Station on the Saigon River were failing. The Admiral had some steer manure flown in from the US. But when it arrived the Admiral thought it was spoiled and threw it away. I guess it did not smell as rich as the local stuff. The Admiral gave these two officers orders to collect some local manure or else! They were frantic. The poorly fed water buffalo and the few cattle around here scatter their droppings and the local people pick up the chips before this naval task force could gather any. They were desperate and could see their careers going down the river. I couldn't resist saying "no—!" They did not appreciate my humor or the present situation. They pleaded with me, since I was in the Agriculture Department and had access to commercial fertilizers. I told them I would supply them with a suitable mixture of what they needed. All they had to do was convince the Admiral that this was what his palm trees needed. The two officers relaxed when this logistic problem was solved and headed for the bar to celebrate.

So much for the war in Vietnam today. In my next letter I will tell you about my drive to Bien Hoa, a nearby town, and some observations about attending sick call at the 17th Field Hospital. I only went for a flu shot.

22 October 1967

Typing makes my fingers tired so after a week's rest I am again ready to peck away. I still enjoy three workouts during the week at the Club after work. On weekends I try to get in some reading and relaxation at the Club pool. I missed this weekend due to meeting Tan at the airport and being confined to quarters due to some unrest among the natives.

I have completed my 70 Christmas cards and they are finally in the mail to you for any additions you care to make, addressing and stamping. I have added a Vietnamese stamp on the back as a touch of reality (or something different). That should lighten one of your many chores. Let me know what your plans are for our Christmas trip. I will have over $900 in my checking account for Christmas presents and our trip. I am counting the days, only 54 of them before I will be on my way for an eagerly anticipated vacation with all of you.

Now about my trip to the 17th Field Hospital. Military hospitals outside of the US are called field hospitals. This one is located on Tran Hung Dao Street which is an extremely busy street between Saigon and its sister city of Cholon (meaning market). Most of the Cholon population is ethnic Chinese. Cholon has more than its share of terrorists attacks and I am not very fond of the place. The only business I ever have in Cholon is an occasional trip to the huge PX and Commissary. Most of my purchases can be made at the Brink's PX which is near my hotel or one at Tan Son Nhut Airport. Tran Hung Dao Street is filled, at all hours, with a grid-lock-jam of scooters, bicycles, carts, cyclos, cars, buses and military convoys. For added interest and creating a real traffic jam there usually are a few ambulances screaming to and from the Army's 17th and local Vietnamese hospitals. I just put the old Jeep in low gear and flow with the traffic. It gives me the creeps the way the scooter riders squeeze in on each side and try to crowd me out but as long as I do nothing unpredictable most everybody seems to make it through unscathed. At one stop scooters were practically inside the Jeep. One scooter operator kept edging closer and closer. When I started up he gave a howl of agony. Seems like his toes, covered only by sandals, were under one of my tires. When we started up he scowled at me, but I noticed everybody gave me a wide passageway.

The hospital is located in a converted multi-storied apartment building. Out front is the usual contingent of armed guards sitting in a sand bagged bunker, looking very bored and waiting for something that never happens. Inside the building the first stop is the record desk. The place is normally filled with walking wounded. GIs and civilians alike are found here. Most of the injuries handled at this hospital result from non-hostile action. A number of the casualties come from confrontations made on Tu Do Street . . . street of bars and bar girls. When I was here two weeks ago I saw GIs and civilians with their arms in slings and casts, legs in casts and a number of bandaged heads. Some of the GIs looked like they were burning up with fever. Many looked very ill, gray and worried.

At the record desk a Sergeant in change asked in a not too comforting tone "what's a matter mit youse?" The complaints I heard were stomach ache, diarrhea, fever and chills, dressing needs changing, here for an x-ray, runny noses and other parts. Those with the latter complaint were told to come back on Monday. I heard a few protests come from the line of patients, but the not-so comforting Sergeant's reply was "youse shoulda taght of dat 'fore youse went out with the co" (co means girl in Vietnamese). Some hung their heads, others shrugged their shoulders and departed when they were told to return Monday.

At the far end of the waiting room a bored looking Army

Captain MD listens to complaints and self diagnosis. He sends patients to the laboratory for tests, x-rays, examinations, prescriptions or into the hospital. Civilians go down another corridor and are treated by a USAID doctor. I only needed a flu shot but I had to wait for a nurse or corpsman to show up. That gave me the opportunity to observe the workings of an Army hospital. I was not impressed! Everyone, GI and civilian, is responsible for keeping his/her inoculation records up to date, no one in this place cares or has any interest in your health or welfare. The authorities publicize the required intervals for shots in bulletins, newsletters, radio and TV. Anyone missing the required inoculations either catches one of the terrible Southeast Asian diseases or gets hung up at the airport when trying to leave the country. Anyone who does not have the required inoculations is not permitted to leave the country. When your turn comes you tell them what you need and a bored corpsman or nurse plays darts with arms and bottoms. They do not check to be certain they shoot you with the proper medication. To date I have had 18 inoculations of one kind or another just to survive in this part of the world.

Now I would like to describe a recent trip to Bien Hoa, headquarters of CORDS (Coordinating Office of Revolutionary Development Support). I was advised to drive and was told the road was considered safe, during daylight hours. Since I did not consider the Jeep dependable, I signed up for a Scout from the motor pool. The Scout had only 1,000km on the odometer and was in good condition. I had to remove the seat belt from its original wrapping since it had never been used. General Services Administration, the US Government purchasing agency, has argued for a long time that it is a waste of money to install seat belts in motor vehicles because people just will not use them. Except for the Forest Service, that is true. I am more comfortable wearing a seat belt because travel around Saigon is always hazardous and would be a must if the traffic ever permitted high speed. The route to Bien Hoa is over a high speed 4-lane highway, the only one in the country. The road is crowded with a mixture of military convoys and heavy civilian traffic. Drivers of slow vehicles insist on using inside lanes, as a result passing is on the right side through a maze of bicycles, water buffalo, horse carts, duck herders, etc. This so-called diesel route has a continuous cloud of blue smoke. On some of my flights over this area I have noticed the dense exhaust smoke that almost obscures the road. There are many busy truck stops along the highway. I suspect they are used by the black market and drug dealers. One side of the highway contains a continuous buildup of military bases. The side of the highway that has no developments is controlled by ARVN and they refuse to release the land for refugee housing. This would be an ideal location and would help relocate the refugees away from the filth associated with the crowded resettlement areas, but ARVN refuses to relinquish the area.

Along this stretch of highway the US Army has one of their many Medivac Hospitals. These are portable hospitals that can be air lifted by helicopter and set up in a very short time. I noticed a steady stream of "Dustoff" helicopters, marked by large red crosses, arriving and departing from this facility. The helicopters are dispatched to pick up battlefield casualties. These crews do a remarkable job of rescue. It is good to know they are usually not over 15 minutes away in any part of the country, at any hour.

Meanwhile back on the highway . . . on the other side among the military establishments are huge warehouses and stockpiles of supplies used by the various contractors. It is unbelievable to see the amount of supplies we have brought over here for war or peaceful purposes. There is also a large new water treatment facility that delivers clean water to the local water distribution system where it is immediately contaminated and unsafe for anyone to drink except the Vietnamese. They seem to be able to live continuously with gastric illness of one kind or another.

At another place is a monument of a weary looking soldier sitting beside the highway. At this point a side road leads to a very large new military cemetery on a hill overlooking the countryside. Once this site was a VC stronghold, but the timber and brush were cleared and now it is the final resting place for thousands of Vietnamese soldiers. The country is gently rolling and in low areas there are productive rice paddies and farmers at work. I saw water buffalo working in some of the fields. Ducks were being tended by children. In some of the fish ponds I could see fishermen using seines of woven bamboo.

I turned off the highway onto a narrow winding road through the outskirts of Bien Hoa. Traffic is halted at four bridges which alternately become oneway passages. At these stops numerous young vendors swarm up to the stopped cars and buses to sell peanuts, grapefruit, oranges, fermented fresh pork wrapped in banana leaves, short stalks of sugar cane and a number of food items and fruit I could not identify. I passed the time by talking to the children but did not purchase any food or fruit. The vendors do well at these stops although it is very uncomfortable to sit and work under the boiling sun all day. When it is cloudy it is still very hot.

The bridges in this area are very old and rickety and are guarded by more bored looking guards. I am surprised these bridges have not been blown up. Most of the bridges in the country have been destroyed and replaced several times. Good sized rivers flow under two of the bridges. These rivers carry large, floating clumps of aquatic vegetation. A favorite stunt of the VC is to have their frogmen float explosives down the river and fasten them to bridge supports. I was thinking about this just as a truck stalled ahead of me and brought traffic to a sudden halt in the middle of the bridge. It was not a pleasant feeling when I looked down and saw large clumps of vegetation hung up on bridge supports and I could see more floating down the river towards the bridge. But this was not the day to blow the bridge and our group of vehicles finally crept across. At one crossing a speed boat with a crazy American on water skis went racing up the river. I noticed there was heavy brush and trees along the river which could have contained snipers.

On into Bien Hoa . . . the place is really run down. You could call it the pits. The streets are covered with very large water-filled chuck holes separated by narrow strips of asphalt. But the same masses of local people and GIs from nearby Army and Air Force facilities fill the streets and sidewalks.

My meeting with the Regional Director was very successful. We had prepared a good forestry plan for this region, complete with supporting visual aids and maps. The director enthusiastically endorsed our proposal and agreed to support it.

With that chore done I went to lunch at the former French Bien Hoa Air Force Base with local USAID people. While at the base I had an opportunity to observe the home base for the Ranch Hand operation. That is the group that delivers the herbicides and is trying to convert the forests to grassland. From what I have observed in many parts of Vietnam they really are not doing well in that conversion, but they are killing vast numbers of large trees and releasing dense understory. The dense undergrowth provides even more ambush cover and greater perils for the foot soldier. Herbicides are delivered by twin engine C-123 aircraft. These large transports fly at a height of 500 feet above the ground and are a prime target for enemy soldiers. Most of the planes were patched from end to end. I do not plan to ride on any of these missions. Our forestry plan involves salvaging some of the timber they are trying to eliminate. I will have more on this subject later.

Following lunch I received instructions from the USAID people on which route to take back to Saigon. The return trip was over two of the same bridges but then veered off through different villages and farmland. The area contained vegetable farms, rubber and coconut plantations, citrus groves, betel and other crops I could not recognize. Traffic was relatively light, except for some farm-to-market traffic and an occasional military convoy. Convoys usually drive in the center of the road or enter bridges without waiting for their turn. This requires other drivers to take evasive, defensive action or be run over. ARVN military has bad drivers. Fortunately I met only a few of them on my return route.

The countryside of Vietnam presents a beautiful, interesting pastoral scene. It would be a fine trip if there was peace. I would enjoy prowling through the country and visiting tiny picturesque villages. Naked children played beside the road and splashed about, either in reasonably clean ponds or in putrid water filled ditches. It sets my teeth on edge to see these children playing in water that appears and smells as foul as a sewer. But as I look around the country I can see many children are surviving and more will be born soon.

My route led me back into the crowded traffic of Saigon. Driving this route was comparable to going down a river, everything flowed in one direction, towards the city center. It is always easier going down a river. I saw many interesting side streets but I was not tempted to take any of them for fear I would end up in a dangerous neighborhood.

Last Tuesday, the week following my trip to Bien Hoa, I scheduled a flight to evaluate the forest resources of III Corps or Region 3 as it is also called. It contains a dozen provinces surrounding Saigon. I drove to the Air America Terminal at Tan Son Nhut Airport and boarded a Dornier, which is a two engine plane that can fly as slow as 50 miles per hour or well over 100 Our flight plan cleared us for a 2,000 foot elevation. These low level flights require two pilots. We flew to Bien Hoa to pick up the regional agriculture advisor. The flight to Bien Hoa took me over the same route I had driven the previous week. I saw diesel smoke billowing up from the 4-lane highway and congestion at the four bridges.

As I type this letter in my hotel room I just glanced out the window and saw two Vietnamese propeller dive bombers go racing by just above the roof tops. They were carrying their normal load of 10 bombs. It probably was a show of force on this House of Representatives election day. We are restricted to quarters today, just to keep us out of trouble. Off in the distance I heard two explosions which sounded like grenades. There may be trouble at voting places in Cholon today.

We landed and picked up the agriculture advisor. Then we waited 30 minutes for clearance to take off. The delay was caused by a scramble of jet and propeller fighter bombers who were off on a ground support mission.

We flew over several areas where huge Army Rome plows had cleared scrub jungle growth near military bases, villages and hamlets. I suggested that these areas were well prepared for cultivation and should be planted before sprouts take over and occupy the site. Some of the areas were near refugee centers and it appeared to be an excellent opportunity to put the land to good use and people to work doing something worthwhile. The agriculture advisor agreed this was an excellent idea.

A few days earlier I had provided the Commanding General of III Corps with information concerning the Ho Bo Woods area north of Cu Chi. Historically this area was a trouble spot for French forces long before our present involvement. I directed the pilots to fly over the area in order for me to verify my recommendations. As we neared the north end of this vast defoliated area I could see it contained dense low (20 to 30 feet in height) vegetation that had not been killed. In fact, as I had described the likely consequences of defoliation to the General, the understory of this two and three story forest had exploded in growth when it was released by the application of Agent Orange. When this herbicide was applied the large overstory trees were defoliated or killed and the understory was released. Repeated applications of herbicides only succeeded in releasing the understory which in the case of bamboo is highly resistant to the chemicals, in fact it flourishes.

As I was looking over the area from an elevation much lower than our assigned altitude of 2,000 feet a battle erupted less than a mile ahead of us. We had to bank and climb quickly to get out of the way of a swarm of helicopter gunships, dive bombers and artillery fire called in to support our ground forces. We had just observed an ambush of two companies of the Army's 1st Division, the "Big Red One", by a Viet Cong regiment estimated to contain a force of 2,500 men. We saw armored support, including tanks and armored personnel carriers racing to the rescue. The battle continued for the remainder of the day. We could look back and see the fierce activity after we were miles away. The next day the newspaper and radio reported 56 US troops were killed and 63 wounded in this Ho Bo woods encounter. I recommended they clear and convert this troublesome area into a very large soy bean field.

We flew over an area where loggers were at work salvaging timber that had been defoliated and cleared. I was pleased to see the progress in this area where we had recommended increased security for the loggers as a means of opening up the area for work and development. I would like to visit the area on the ground and see logging and utilization up close, however I am content to provide advice and observe these dangerous and remote areas from the air.

This brought us to the Cambodian border and we could see several miles into that country. Cambodia looks much the same as Vietnam. It has the same rice fields, rubber plantations, swamps and dense forest. The big difference is the forests are green and untouched by Agent Orange and bomb craters. We could see major trails spanning the border. These are routes enemy soldiers take to and from Vietnam. Cambodia provides a convenient sanctuary for Viet Cong and North Vietnamese soldiers.

We saw several other battles off in the distance during our six hour flight which required one landing for fuel at a military outpost. Many refugees have been moved out of trouble spots in

Region 3. We could see many thousands of acres of abandoned rice fields. The only areas being farmed were near villages and towns where there was military defense. Much of the timber to be salvaged can not be reached due to bridges that have been destroyed or to the lack of security for loggers and log haulers.

We followed the railroad for several miles and saw vast stands of very accessible standing timber. Not all of the forest in this area had been touched by Agent Orange and should be available for harvesting when peace comes. At this time we are not certain that all timber in defoliated areas will die. Some of the larger trees appear to be recovering and putting out new growth. Some areas have been defoliated two or three times in an effort to stop the regrowth and kill all the vegetation. Another depressing fact is that Operation Ranch Hand applies chemicals, for each treatment, at the rate of 28 pounds of acid equivalent per acre. In the Forest Service we use only 1 1/2 pounds AE/acre or less. Time and further observations may provide information on the future of these treated areas. At one point we could see where a complete train had been abandoned when the bridges on either side of it were destroyed.

On this flight I carried my survival kit and wore jungle boots given to me by the Special Forces at Tra Bong. The small kit contains waterproof matches, map, flares, flashlight, fishing kit, two rations, canteen, halazone water treatment pills, first aid supplies including malaria and lomatil pills, mosquito repellent, knife, and a .45 automatic pistol with extra clips of ammunition. It would have been a long walk out that day. It was most reassuring to see patrolling helicopters crisscrossing the country below and around us. Again the plane delivered me safely back to Tan Son Nhut at 1700 hours.

Following my thorough coverage of Region 3, I met on Friday with all of their agriculture advisors and presented my observations and recommendations. I enjoy these briefing sessions. We have good maps and visual aids to stimulate the interest of our audiences. I whip out a retractable pointer, just like the Colonels and Generals use in their briefings, and point out areas and items of interest. The advisors appeared to appreciate our information and advice. We expect excellent support for our forestry program in this region.

I was interested to learn about the agriculture advisors' other problems. One of their major areas of concern is the fertilizer program. That makes an interesting story. Much progress is made in the agriculture sector through organizing farmers' cooperatives and associations to distribute and train farmers to use improved seeds and fertilizers. One of the many major problems challenging this country is a half finished, inoperable fertilizer plant. After spending untold millions of US dollars, experts discovered the plant would lose over US$I million per year during the amortization period! This is the very reason we resist USAID's continuous pressure on us to develop proposals and plans to import machinery for large sophisticated forest industries. We believe the timber supply, marketing, labor force, transportation and many other factors should be carefully studied and evaluated before jumping in with huge expenditures for facilities that may fail for one reason or another. We continue to make good progress through training the Vietnamese to improve alignment and operation of their small labor intensive mills. These mills can meet the country's needs and put thousands of refugees and others to work. If something breaks in the mill they can make another part rather than ordering parts and awaiting delivery from Tennessee or Oregon.

I have mentioned our maps before. In USAID we are recognized as the people with the best maps. Our maps have been accumulated through our Vietnamese contacts, the National Geographic Society at Dalat and CICV (Combined Intelligence Vietnam) military groups. We have provided advice to the CICV on top secret military operations and through these contacts we can obtain special order maps and excellent aerial photographs meeting our specifications in a very short time.

Other observations made during the past week:

In addition to crazy water skiers, we see Americans practicing for cross country or long distance running. They select the rush hour for their practice and weave in and out of traffic. As they run in this polluted air, at only a few feet above sea level, they huff and puff desperately for oxygen. They are crazy! But come to think about it, I have never seen a happy or smart jogger anywhere! The high temperature and thick clouds of exhaust fumes must be doing a splendid job of preparing their lungs for cancer.

For lunch Saturday I went to the Brinks BOQ and had a delicious barbecue rib meal. I complimented the cook in Vietnamese as he was preparing ribs over charcoal. This rare compliment must have pleased him because he just about finished me off with a large extra serving of ribs, baked beans and potato salad.

Following lunch I went to the PX to purchase a few food items for my confinement to quarters from 2100 hours on Saturday until 0400 hours on Monday. I picked up my processed color slides and a few other odds and ends. Waiting in line ahead of me at the cash register was a handsome Negro GI who was apparently preparing for a big weekend. He had two bottles of vermouth and two bottles of rum. Does that combination sound good? No gin? When he arrived at the cashier he did not have enough credit on his liquor ration card to make the purchase of four liters of booze. He was pleased when I handed him my card and said I would buy it on my card. He gave me the money for the purchase. Then I checked out my few purchases and found I was 30 cents short. While I searched through all my pockets and billfold for money that was not there, a Korean officer just behind me told the cashier to put the charge on his ticket. That is how we help one another in Vietnam. Big finance!

After shopping and dropping off my purchases at the hotel, I went for a walk and drive to take photos of election activity. Election posters are plastered on every available tree and wall. Draped over most of the road intersections are banners carrying slogans in Vietnamese and Chinese which praise the candidates or announce the forthcoming inauguration of the President. Viewing stands are being erected behind the cathedral. Barricades of barbed wire and steel posts are being installed near the stands and the Palace. I suppose we will also be confined to quarters for those events. I note you are having your share of problems in DC this weekend. The world is really screwed up, isn't it?

Following my short tour I drove to the Airport to see Tan and his wife off to Rome. Tan has been appointed by the Presidency to represent South Vietnam at a United Nations FAO (Food & Agriculture Organization) Conference in Rome. He is becoming a worldwide traveler.

As I drove into the airport a couple of car loads of ICC (International Control Commission) Indians came rolling into the parking area. I carefully avoided them since I did not want another confrontation. I may not have described that episode for you. A few weeks ago I drove into the same airport parking area and was directed to an open parking spot by one of the Vietnamese policemen. I shouted a thanks to him in Vietnamese. Before I could reach my assigned area a turbaned Sikh burned rubber, sped around me and drove his car into my space. This made me angry and I put the heavy old Jeep, with its massive front bumper, in low gear and rammed into his rear with much force. That was the first time I had ever seen an Indian Sikh without a turban. His went flying when I hit his car. And was he mad! However the policeman had witnessed all this fracas and with a big grin on his face he ordered the Indian to pull out of the parking lot and directed me to the parking place. The policeman was still grinning when he bowed, saluted and directed me to the parking spot. So much for the worthless ICC . . . all they do is play tennis and drink at the Club. They would not dare go near the DMZ they are supposed to be supervising, it would be too dangerous.

Porters at the airport are a hungry, industrious looking bunch of vultures. They wait at the entrance to the airport and pounce on arriving cars. They apparently have no mutual understanding or agreement among one another. Every arrival is fair game and whoever gets the best hold wins the privilege of earning a tip. It is amusing to watch them work. Many of the passengers prefer to carry their own bags. But to do so they must have a firm hold to retain possession. I watched the porters swoop down on one likely looking car. Expectantly they jerked open the trunk and you could see their shoulders droop in disappointment when there were no bags.

A number of GIs were in the airport waiting room. Most of them looked relieved or filled with disbelief about leaving Vietnam. Many of these were young, but old looking, Warrant Officers. Most of the men with this rating are helicopter pilots with a long string of missions behind them. I have flown with many of them and I do not envy them their kind of flying.

A group of Indians from the ICC came into the lobby to board an Air France plane to India. I see these people coming and going every time I am at the airport. You have only one guess as to which country finances this worthless ICC operation. You are correct! It is the good old US of A!

The lobby began to fill with many affluent Vietnamese who were off to Europe. Family and friends make quite a ritual out of these departures and arrivals. They dress in the finest suits and ao dais.

At 1600 hours Tan's two security men made a sweep through the lobby. They did not spot me since I was standing off in one corner to observe the activity. Next some of Tan's staff and relatives began to arrive. All stationed themselves according to rank. The General (brother-in-law) was at the head and Tan's receptionist was at the low end of the "totem pole". I approached the body guards after a few minutes and asked them if he had arrived. The response was, "no, he will arrive at 1630 hours". We had some more conversation, all in Vietnamese, before they excused themselves and went scurrying off on some urgent security matter.

Mr. Linh, the German TV cameraman came over to greet me. He had been at the airport filming the departure of Japanese Prime Minister Sato who had taken off minutes before my arrival. This is the same TV man who accompanied the German delegation to Dalat for the fire equipment dedication. I had also visited with him at the "Hello, Dolly" show. He said his coverage of Martha Raye and company was good.

Vietnamese continued to arrive. Most of them show up in bright, shiny, new Mercedes Benz sedans which they either own or rent for the occasion. It is quite a show to watch this activity. I understand the Vietnamese are pressuring the US Mission to rebuild the commercial airport terminal facilities to make arrivals and departures more convenient. I have no comment that is printable at this time!

Right on time, at 1630 hours Tan's bright and shiny 1947 Ford wheels up and out steps Tan and his wife. There is a great amount of hand shaking and conscious effort on the part of all to be sure he sees they have come to see him off. After they made their presence known, a number took off after they had touched base with him. Come to think about it, that was the reason for my being there. I did enjoy meeting and talking with the ladies, relatives and members of the departure delegation. I find those older ladies and their chewing to be fascinating. I noticed they do not spit very often in such surroundings.

Tan's plane departed on time and I went to the nearby new MACV, Pentagon West snack bar. I have the proper license plates on the rusty (not trusty) old Jeep to get into that select parking lot. I was not permitted to park in the inner sanctum parking area which is reserved for staff cars of colonels, generals and admirals. Their cars are quite impressive, all parked in a line according to rank with the names and symbols of rank prominently displayed. Each vehicle has its own driver standing by. At that time the Americanized Vietnamese drivers were gathered around for a good old American crap game. There were few piasters changing hands, most of the currency being wagered was MPCs along with stacks of green dollars. Both of the latter are strictly illegal in the hands of Vietnamese. War certainly is hell!

In regard to money that reminds me I forgot to tell you about Tan's experience in Portland, Oregon. As he was leaving his hotel he went to the cashier to pay the bill. He had been entertained by the Mighty Mite sawmill manufacturer, Mr. Dwyer and Tan naturally thought when the cashier told him the bill was taken care of that it had been handled by Dwyer. The hotel believed that since the Forest Service had gone to great pains to reserve the room for a visiting dignitary that they would pay the bill. Recently the hotel checked with the Forest Service in Portland about the bill and were referred to the Forest Service and State Department in Washington. Unknown to me this information had come down through State Department channels to Saigon. The training office here, which Tan hates with passion due to their unfortunate treatment of him prior to his recent departure to the US, had called him on Saturday morning and informed him that he had run out on a hotel bill in Oregon. Tan told me about this unfortunate episode at the airport. He promptly peeled off three green US$10 bills and asked me to take care of the bill. I fully expected to be whisked off to jail when this exchange of green money took place in the middle of a busy airport lobby. But to date I am still at liberty. It really spooked me when he handed me the dollars. I am impressed with the consequences of illegal money transactions. But I will take care of his bill.

After a snack of chile and a thick milk shake at the Pentagon West, I visited the bright and shiny restroom. It is so clean because the cleaning ladies come in and check it frequently

315

and keep it spotless. I find it embarrassing to have them come and go so freely when I am using the facilities, but that's the way it is in Saigon.

I will be in Saigon all week. Our program has attracted the attention of some of the head people in CORDS. General "What's-his-name" and a couple of others want to push the Mighty Mite sawmill project at Tra Bong. They are having two elephants, yes, elephants, air lifted to Tra Bong to help with the back breaking chore of logging. I can reliably predict those Tra Bong people will soon be having elephant-burgers! Meanwhile I am trying to identify and eliminate corruption and red tape that has stopped the operation. Somehow I hope to mysteriously get everything worked out and the mill back into operation. Perhaps we will be successful in getting all the pieces of this crazy quilt in place.

Chapter XXIII

The Spooks

24 October 1967

How goes it? As I listen to weather reports for your area I feel for you. I know how much you hate cloudy, rainy weather. Here it is always hot, rainy and exhaust smoke from the vehicles is beginning to make the climate unbearable even for me. I look forward to our Christmas vacation and breathing some cold fresh air. If you wish we could drive to Quebec through New England or better still take the train. It will be good for all of us to forget about Vietnam, the war, State Department characters, Mother, boy friends, housekeeping and the Forest Service. You better plan to take along an electric blanket or something suitable because I will need a lot of warming in that cold country. But I am excited about the trip.

Our Secretary Mrs. Hai missed work one afternoon and the next morning. Her eldest son, age 11, was suffering from a severe asthma attack. When she returned to work she was in tears and told me she had been giving him a tonic she obtained from a Chinese herb doctor. It was probably ground lizard tails, buffalo dung and bat's wings. It provided him absolutely no relief. I gave her a supply of my 100mg pyrabenxamine antihistamine pills. I cut them into quarters and advised her to give him one every six hours. That did the job and the boy had his first good night's sleep in weeks. I am sympathetic with the child since I suffered from asthma for many years. This hot, humid weather and terribly polluted air are a hellish combination for asthma sufferers. Both Mrs. Hai and her husband are grateful for relief I provided for her son.

Now I am off to dinner and afterwards watch a five year old movie.

28 October 1967

I worked at the office all day. I will be glad when Barry returns next Wednesday and Syverson the following week. However Barry will not be here for long because he goes off to the Philippines to do some work for us on 15 November and will not return until after I leave for home. The work load is increasing. Keeping up with the office and field work is more than I can handle. Metcalf is an excellent sawmill and logging technician, but when it comes to reports he requires guidance.

I have had no time for swimming or relaxing today. I will go to the BOQ for dinner and a movie as soon as I finish this letter.

We had no rain today but it is hot and threatening. I wil find time tomorrow for a longer letter with more news.

I received two of your letters today! You are doing admirably with your letter writing and they mean so much to me I save every letter and read them two or three times. It helps me visualize your busy days and gives me the feeling there is a rea world out there.

I am enclosing my travel schedule and will be so happy to see you all at Dulles Airport. The Forest Service and USAID approved my TDY in Washington from 4 to 12 January. Tha will provide a few more days at home.

I listened to Hubert H. Humphrey spout off during his field trip in Vietnam today. The Nation should try hard to preserve President Lyndon B. Johnson. Heaven help us, if HHH should ever take over for LBJ. HHH says, "most countries have had their wars of liberation first and then taken on the job of developing their nation, but this country is doing both". What other country has had someone else foot all the bills? And I do mean *all of the bills*. All the talk about this wonderful help being extended by the Free World Forces is being financed by us good old US of A taxpayers. Korea, the Philippine Islands Thailand and all the other cooperators are growing fat and rich by sending their troops and civilian workers to Vietnam.

As I reflect on some of your beliefs, I am in complete agreement with you. We have made a mess of civil rights Democrats have certainly fixed us up good in that respect. Amer on the National Guard and please add the reserves to make your criticism complete. When I think of all the time and money we have invested since the end of World War II in training those "weekend warriors" (many that I know and have worked with over the years), it makes me ill. Now when they could be used in this fracas and replace some of these 18 year old kids whose minds will never be unscrambled, they are not called up. The situation is all screwed up and getting worse. Sorry but I find my patience running thin at times. Perhaps I am working too hard.

I went to the pistol range Sunday morning and enjoyed three hours of shooting. I am doing very well. Following that break I returned to the office and worked until 1900 hours on our forestry budget and work program. At times I do not see how this USAID outfit can behave so stupidly. They provide little guidance, have no instructions or pattern for handling their multi-million dollar budgets. My presentation was the only one in USAID Agriculture that has been supported and survived. I developed our forestry program in the same manner as we did in the Forest Service and USAID could not refute the logic. While it s a relatively small program I am pleased it stood the test. Today had to redo some tables to reflect changes I had suggested in he first place. Then I went to the BOQ for dinner and saw "The Blue Max", another war movie that has been shown before.

It is about midnight and time to sign off for now.

5 November 1967

This has been a busy weekend. I returned to Saigon after a short trip to Tay Ninh which is about 70 miles north of here. I hen spent Saturday afternoon and most of the evening studying my language lessons. I have been scheduled to take a test on Thursday. I do not believe I can pass the test because it is normally given after 15 lessons are completed. I have had only five formal lessons. But the training people were trying to fill out the schedule for the examiner so I will make an attempt to pass it.

I wish I could be at home to help repair faucets, haul in the fireplace wood, and spend time with you, Gwen, from the sound of your letters you have your hands full. My being here is a hardship and puts real pressure on you. I will be home in just over one month from now and I am really looking forward to my return.

For some reason your last letter came in a government envelope marked "Monitored". But I guess it got by. Watch it! Government envelopes are for official use and LBJ needs the money!

I will be interested to hear if my next few pay checks come through on time. USAID's appropriation has not passed and the outfit is out of funds. So let me know.

I am writing this letter in the BOQ TV room. The TV is blaring in one ear and a "live" Elvis is singing about a "hound dog". The "music" is coming in through the open door. I will look in on him for a while, although I have never been an Elvis fan.

I will sign off for now. The mail will be picked up in a few minutes.

10 November 1967

Now I can relax for a while. I completed my language examination today. It was a demanding test and included dialogue and translation. A man from the Washington Foreign Service Institute was in charge. The Vietnamese lady, who I had loaned my office to for classes during the lunch hour several months ago, conducted the examination. She spoke clearly and slowly for my benefit. The examination was conducted entirely in Vietnamese. She asked a series of questions and I had to answer them in Vietnamese . . . Where do you live in Saigon? How do you get there from here? When did you arrive in Vietnam? When do you leave Vietnam? Do you have a family? Where are they? How many children do you have? What are they and how old are they? What is your work in Vietnam? etc., etc. My answers had to be in complete sentences. Then I had to describe what I had for breakfast, where I would eat supper, what I would order, where is the restaurant, how much is the meal, and where do I pay? Then I described a trip to a hotel and rented a room for a week, described the size of the room and the facilities and when I paid for the room. Next the man told me what to do in English in regard to a taxi trip to an area where the taxi driver did not know the location. Then I had to describe this trip to the lady in Vietnamese and translate her responses to the man in English. Then I read a hand written page of Vietnamese and translated it into English. I had a series of items and expressions to translate. This was a demanding examination but I believe I did pretty well. I missed some of the questions. Time will tell if I passed. If I pass, so what? Well it was good experience and practice. If successful, I will be given a raise in pay of almost $1,000 per year, less tax. Big Deal!

I spent much time studying for the examination. My two Vietnamese employees provided me with good practice. I had dropped the lessons and formal class about one month ago when I was on my third teacher. The class was changed from 1300 to 1400 hours. That provided me with a 30 minute break before work resumed. I studied at night and on weekends. It is a relief to drop that demanding task. I know enough of the language and it has helped a great deal in my work with the non-English speaking Vietnamese. I am able to pick up an expression or two each day on my own and I shall continue to make progress.

Cindy, I enjoyed reading your long, interesting letter and seeing your good school report. Yes, I stayed away from the inauguration, parades on 1 November (Revolution Day) and from HHH's speech for the Americans. The mortar attack you referred to hit the Palace which is about as far from my hotel as the length of your football field. The explosion was much louder than the normal artillery fire heard around the city. So far I have kept out of the enemy's way and I intend to look out for myself and my men. I was pleased to hear you are a driver instructor. I always knew you would be an excellent, safe driver. I am sympathetic for you concerning the "all American souls" at high school. Those "pseudo hippies" make me ill! UGGGh! Yes, Cindy, I will be prepared to take you all on a trip following Christmas, please make your plans. I do not mind going into cold weather as long as you all guarantee to keep me warm with your antics. I would also like to plan on a trip to Paris, in June after you are out of school, and then to England, Wales and Scotland. Italy, Spain and Portugal can wait until I complete my assignment here next November. At that time we can take a month, if you can spare it from school. It would be nice and warm in that part of the world at that time. We can also include Monaco and Capri, if you like.

Melissa, it sounds like your small room is a joy. I hope your college classes are going well. I was sorry to hear about Bob's trouble with the Army. I would have liked to have been in the back seat to have watched your confrontation with the US Army. Let me see if I have it straight. Late one night you and your Mom took Bob back to his barracks. On the way out of the military base you were stopped by the MPs . . . you had no purse or driver's license for ID and your hair was put up using beer cans (?) as curlers. Did they think you were a pair of Russian spies or the Viet Cong? You called Bob to come and identify you and your Mom. He put on all of his uniform, except he wore house slippers instead of regulation shoes, and the OD about had a fit to see him out of uniform. His punishment consisted of being confined to base for a month and extra duty. It sounds like he has a good strong backbone and the courage of his convictions. Tell him the extra duty will be good experience, although his restriction will be difficult for both of you. But stick with it and it will all work out okay. That's all for now I am about out of typing paper.

* * *

All of us working in Vietnam carried a Top Secret security clearance. True we had access to sensitive political and military information, but at times much of the secretiveness seemed over-done and paranoid. In some instances I believe security classification was used to cover up incompetency. However we were literally under the gun at all times. Marine guards in our USAID building checked our office at any hour of the day or night for careless handling of classified materials. Brief cases were checked each time we entered or exited the building. Classified materials were kept in files secured with combination locks. We set the combination but we were advised not to use numbers such as our age, birth dates, street addresses, telephone numbers or social security numbers. I suspect our living quarters were checked during our absence. As expected security is a fetish around any diplomatic or government operation in a foreign country. Vietnam was no exception and had a well-developed intelligence network.

Spies were all around us at all levels and ranged from the wood cutter who was paid a few piasters for reporting a visit by a Viet Cong tax collector, to the CIA and Viet Cong types and Eastern European members of the International Control Commission (ICC). Every US mission overseas has a guessing game of identifying the spooks. My first CIA contact was not the glorified spy types I had seen in movies and TV or read about in spy stories, rather this one was a female agent posing as a USAID librarian. CIA agents (spooks is a good description) ranged from local Revolution Development workers found in most every secure village and hamlet throughout the country, to low level province representatives, agents in the field with a wide variety of credentials, the USAID librarian who came to our office at regular intervals to check our work, to William Colby (later Director of CIA/Langley, Virginia) the head of the civilian pacification program CORDS and a strong supporter of the forestry program, to economists and political officers in the Embassy and my neighbor General Edward Lansdale who had gone from the WW II OSS to CIA Vietnam, where he handled "other duties as assigned". Then there were the hundreds of Air America pilots flying throughout Southeast Asia in their unarmed C-47s, C-118s, twin Beeches, Porters and Hueys. Those pilots carried M-16 automatic rifles and silencer equipped .22 Hi-Standard automatic pistols, probably for shooting doves. At times rivalry among intelligence groups led to suspicion and distrust and made it difficult for the process to function as planned. Nice people, the spooks!

Repeated warnings came to me from the CIA when we reported some irregularity we detected or when we came too close to identifying sources of bribery and other forms of rampant corruption that permeated all sections of the country. I was told this was none of my business. Some of the people I reported were CIA cooperators. I believe the CIA and others believed corruption equals custom, and was a way of life in Southeast Asia. But it concerned me. Corruption in many forms and payoffs from the forest to the market were responsible for the very high costs of production of lumber and other forest products vitally needed for refugee housing and other domestic purposes.

As 95% of the forest land was owned by the government the disposal of government timber involved miles of red tape. Logging operators needed cutting permits. Log haulers, sawmills and lumber trucks required permits and licenses. Processing permits and licenses involved very complex, multi-channel processing that could be resolved only through bribes and substantial pay offs. I was horrified when one of the foresters diagrammed for me the detailed steps required to legally set up a sawmill. Many timber businesses had no formal or official approval and depended on payoffs to keep functioning. In the woods, on the roads and at the mills VC tax collectors promised and provided protection. The VC were less greedy than other sources. Many times those other sources were the police who regularly stopped boats and trucks for tolls. High level Vietnamese military officers and province officials extracted monthly tributes for protection. And some South Vietnamese government officials who were cooperative with the CIA received periodic payments. Corruption links reached through most all levels of Vietnamese society. The openness of official dishonesty could be seen in luxurious villas, many expensive cars and a striking rise in the living standards of large extended families. Wives, brothers and cousins were all in the business. During the last few months of my stay in Vietnam we made great progress working through the Directorate in eliminating corruption. When we identified a corruption problem and the forest industry people reported it, we had good support from Director Nguyen Van Tan and his successor Director Le Van Muoi.

Of all my exposures to the untold dangers in Vietnam, the identification of those responsible for corrupt practices was probably the greatest. We identified some significant sources of corruption and impacted some very lucrative (mega-green bucks) operations. We interfered with some very powerful operators. On one project we learned the Montagnard cinnamon gatherers loaded the bottoms of their pack baskets with cocaine fresh out of Laos for delivery in Vietnam and points in the western world. When we reported this activity we were again told by the CIA this was none of our business and to back off. At times I felt I had a prominent bull's eye target painted on my back. That was one of the reasons I kept moving, for I learned long ago that a moving target was harder to hit.

Another interesting set of spies were the French. A surprising number of Frenchmen remained in Vietnam after defeat of France in 1954. Perhaps they were double agents, working both sides of the street, for what they could get out of it. These smooth French/Vietnamese/English speaking businessmen and planters could come and go where ever or when ever they pleased, throughout Vietnam, North and South. Involved in this intriguing aspect of the Vietnam scene were the rubber plantations, run by the French and their Vietnamese managers and workers. These plantations were strictly (politically) off limits as far as military operations were concerned. "Lay off", was the warning we received through channels. These plantations were excluded from free-fire zones, harassment and interdiction fire and defoliation.

Director of Forest Affairs Tan suggested at the time of his ouster that VC spies in his organization were responsible for his removal from the Directorate. He accused his Assistant Director Nguyen Van Hiep as being involved. We had no way of verifying the facts but we avoided antagonizing either party and our working relations with Mr. Hiep were cordial and productive.

Unknown to us, and probably to Forest Service Chief Edward P. Cliff, the Forest Service had a fire research person in

Vietnam, before and until shortly after our arrival, devising top secret means of destroying the tropical hardwood forest (that damn jungle). This researcher was working with a think-tank outfit to utilize fire to destroy the forest and deprive the enemy of protective cover. They lit fires on the ground, had napalm and other ingenious devices dropped from high altitudes. They used every sophisticated and unsophisticated means they could dream up to create a fire storm in the forest. They tried in virgin timber, cutover and defoliated areas, but it did not work! Even a dirt forester, like me, could have told them it would not work. Why not? The combination of very high moisture content of the twigs and leaves and the 100% humidity caused most of the fires to fizzle out. Heavens knows how many millions of US dollars were spent on that ill conceived, aborted project. Fortunately for the future of the forests and people of Vietnam it was unsuccessful. The Forest Service researcher would not discuss the project with us because it was so secret. I am grateful our forestry team had nothing to do with that project.

There were valid instances when utmost secrecy was required. For example, as a senior level executive I participated in Top Secret briefings for the Ambassador and General Westmoreland. At these sessions we were given precise details about planned military operations, down to the exact day, hour and minute they would kick off. These operations were planned months in advance and involved thousands of American troops. In these instances a loose lip could have resulted in serious losses of American lives. We even helped design some of these projects by reporting on forest conditions in such infamous places as Ho Bo Woods and a beleaguered Marine base located on a worthless piece of real estate called Khe Sanh. It was frightening at times to think about all the facts we did know. In those instances, and all others, we fully respected security requirements.

* * *

12 November 1967
For obvious reasons as you will see, I made no copies of this letter. Today was the day for Tan's return from Rome. Barry and I spent some time at the pistol range and then attended a fine reception in honor of D.A. Williams, head of the Soil Conservation Service. We then went to the airport to meet Tan. At the airport we saw a number of Tan's friends and associates. One of the young foresters, Mr. Qui, told us Mr. Tan was no longer Director. He had been replaced yesterday by order of the new Minister of Agriculture. When Tan arrived and greeted us, he told me the VC had pulled off a coup d'etat. This news was very disturbing, but we had expected something to happen with the change of leadership in the government. Tan said he would

contact me soon and provide details. He informed me several weeks ago he had "confidence in my presence". At the time I suspected he feared the new government might bring about changes. I asked him when he was preparing to go to Rome if it was in his interest to leave the country at the time of major leadership change. He said he had no fear because his government had selected him to represent them at this United Nations FAO conference.

While I was at the BOQ for dinner and a movie, Tan came to my hotel. He left no name or message. However he returned again at 2200 hours and provided me with the details of his dismissal. He claims one of his assistants Mr. Nguyen Van Hiep had pulled this off. Earlier Tan warned me that Mr. Hiep had VC connections. I considered this to be unproven information. As proof Tan told me Mr. Hiep had once proposed cutting timber to prepare an area for reforestation. This scheme was said to involve a payoff of VM\$2 million. Tan and the previous Minister would not approve the project. Tan now says the new Minister and Director are involved in this large payoff. Tan asked for my assistance in reporting this fraud. I suggested he report this to the new Minister, but he said they would believe he was making up the story just to retain his job. I also asked if he could report this to the President or to Vice President Ky (who Tan had once said was his friend). Tan said he did not think he could do this and asked if we would do it for him. He told me he had proof, but I doubt he can produce it. Again I suggested he report this fraud to the new Minister and through his own security people to the National Police. Tan said he wanted me to have the facts and would do as I suggested. Tan said he prefers to be placed in the Ministry of Foreign Affairs. With Tan's departure three key men in the Directorate of Forest Affairs will be out. Mr. Hiep will be placed in the coveted position of controlling the Directorate's finances.

Now I have a delicate situation on my hands. Such is the intrigue and plotting going on all about us. This episode is not unique. I have read about similar situations in my study of the history and government of Vietnam. It will be interesting to watch the developments. Tan says if he can hold on to his job he will get rid of Mr. Hiep permanently, whatever that means. We have maintained good working relations with everyone in the Directorate, including Mr. Hiep. I do not believe this will create any problems for our program. Until this is settled, it may complicate our work program.

At the moment USAID has serious budget problems. You may have read about it in the news. We have not been paid our SMA (Separate Maintenance Allowance). Barry is scheduled to leave for an important meeting in the Philippines on Tuesday, but there are no travel funds. His trip will be canceled or at least delayed. Arrangements for his trip required some complicated arrangements with the US, Philippine, Vietnam and German governments. Nothing is simple around here. We will wait and see what develops.

Yesterday I worked until 1230 hours and then I had a scheduled meeting with D.A. Williams. Chief Cliff had asked Mr. Williams to check with me to see how the forestry program is progressing. He was delayed for over two hours. Working in Vietnam requires much waiting and more patience. Mr. Williams was delayed due to a Jeep breakdown while he was inspecting a project in the back country. He was relieved when he reached an airport and got out! We met and talked for two hours.

I am anxious to hear the results of your physical examination. I hope you are in good physical condition. From Cindy's last letter it sounds like you are all busy preparing for Christmas. I sent you a supply of photos of me for including in the Christmas cards. Please use them as you wish. They are a black and white of me at one of our sawmill sites near the DMZ. I am in my Forest Service uniform, holding a sub-machine gun, standing in a mine field with a destroyed bridge in the background. It is different than my earlier Forest Service photos.

Assuming USAID has some funding by December, I will be on my way home in about one month. From here it is difficult to believe winter is arriving back home. I see by the weather reports you are having freezing temperatures. The high temperatures here are in the high 80's and 90's. We have heavy rain most every day. However the sun is beginning to come out a little more each day which is an indication the monsoon season may be tapering off. Typhoon Frieda hit land north of here and apparently did little damage. All military aircraft were moved out of the typhoon danger zone to Tan Son Nhut Airport, where it was crowded with more airplanes than usual. We had more rain than normal out of the storm, but no wind.

It is now past midnight . . . whups, it has crept up to 0100 hours and I will conclude this letter and try to dream up solutions to help solve my current problems. At least it is something different and there are few dull moments!

16 November 1967

We received our checks today, but Barry still does not have clearance or funds to pay for his flight to the Philippines. Hopefully Congress will pass the appropriation bill in time for my departure which is less than a month from today.

Gwen, I am sorry to hear you stepped on a nail. That can be painful and you have so much to do, with helping Steve move and the million other items. When I am out in areas where there is danger of stepping on something sharp I wear the boots the Special Forces gave me. These jungle boots have nylon layers to protect your feet from nails, punji stakes and small mines.

I am glad to hear Bob has served his time for being out of uniform, in his bedroom slippers! What is his schedule now? I thought he would be in Vietnam before winter. From what I see of the young GIs around me he will be well off to forego this experience. It sounds like winter is well on its way to Bethesda and the northeast. I am all set for the Quebec trip. Just for kicks, please check on train connections to Quebec City. We should consider that as an alternative to driving. You will note my checking account is solvent and I know all of you enjoy train travel. Either way I will be ready. Please remind Gene I will need to borrow a winter coat.

No coats are needed here. It was 90 today. The rains are fading and the weather is generally fair, but hot. Barry and I have our lunch at the club pool each noon. Then we have a swim and sit in the sun and relax. Please do not publicize our lunch breaks to our friends, since they imagine us wading through swamps and jungles. Exercises at the club keep me in good shape and my weight down. My arm strength has improved. I will now be able to take on all of you pale faces.

Today I had an appointment to meet the new Director. His name is Le Van Muoi. He is 50ish, overwhelmed by the nature of his job and seems to appreciate our presence. Perhaps the change of leadership in the Directorate will work out after all. We still have no word on the outcome of Tan's ouster. I will try to find out and let you know. A number of Tan's assistants went out with him. The new group is interested in "keeping out corruption". I told Mr. Muoi we would limit our activities to technical matters and let them handle the administrative matters. That seemed to please him.

Today I was notified by the housing people to look at an apartment. They are making a strong effort to close out some of their leases on hotel rooms and place people in apartments. They have a new housing policy which provides you with the choice of three apartments and you are required to take one of the three or make your own arrangements. Conceivably this means, make a decision, or pay for your own quarters. Just before this new policy became effective Barry and a friend, who want to share an apartment, were given three places to inspect. All were in a very tough neighborhood, near one of the waterways and Tan Son Nhut runways, far from work, and generally undesirable.

I had an appointment to inspect one of the apartments. It was a new apartment building, about as large as my hotel. It has a high fence around it and guards on duty. I looked at some of the apartments. The rooms are small and completely furnished with new Danish modern. Tomorrow I will look at the others and then make a decision. People are starting to move in so I will have to move fast or the units will be filled. One item influencing my decision is the possibility civilian use at the BOQ will be terminated. So many civilians eat there that many times officers can not find a place to sit. I will write as soon as I make my decision.

19 November 1967

I am writing this letter on Sunday evening on my new teak desk. Surrounding it is my new apartment. When I sized up the housing and BOQ dining situation on Friday I decided to take this one. There are 58 apartment units in this new building leased by USAID. It is much like a hotel and has two guards and a generator operator on duty at all times. As soon as all units are filled it will have a concierge at the front desk 24 hours per day. I selected a unit on the cooler, shady side of the building, away from the main street and a noisy generator. There were larger units but I selected this one because it looked like I could take care of it myself. Now let me describe it.

My apartment is No. 217 and is located at 192 Cong Ly Street. This is a main one-way street leading into town from the airport. The location is what sold me. It is close to work, four blocks from the club and four blocks to the BOQ. Many of the men are selecting places miles out from the center of the city and in some bad neighborhoods. It is good to have a comfortable place of my own rather than share it with someone. Many of the people sharing apartments have problems getting along with one another. There is now a rule, once moved into a facility, no further moves are permitted. We are given three choices, all my choices were in this building. I had 25 units from which to make my selection. I am on the second floor, only one flight up and convenient to the parking area which is located under the building. I find it convenient to drive through the gate, park the Jeep by the steps, climb one short flight of steps and be in the apartment in a jiffy. I am grateful there are no French elevators to threaten or frustrate me. I have a balcony overlooking a lovely French-style mansion. If I had a long pole with a hook and a basket on it I could reach out and collect fresh coconuts from a line of trees located along a high wall. On the other side and across the street are military facilities with armed guards. I believe this is in a secure location. My hotel was subject to brown-outs every evening, lights were restricted and there was no air conditioning after 2200 hours. Here we have two generators, one provides power and the other is on standby.

On entering the apartment a short hall leads to the kitchen, which is on one end of the living room. The kitchen has an apartment-size gas stove, new Frigidaire, sink with drain boards and lots of cupboards. On the open side of the kitchen is a counter with suspended overhead cupboards. The counter opens on a combination dining-living room. We are given a hospitality kit which contains complete kitchen equipment, bedding, blankets, mattress pad, pillow, dish and wash cloths, towels and wooden clothes hangers. The furniture is teak, nice and plain. There are two green upholstered armchairs and two lamps, one each for the living room and bedroom. There are two ottomans. The couch is long and covered with material that is olive green with small orange stripes. There is a coffee table in front of the couch. The dining table can seat two when placed against the wall or ten when extended. Chairs for the dining table are teak with cane bottoms. There is a large buffet, telephone stand, two end tables, and three nested tables in the living room. The combined living room and dining room area is 15 by 15 feet. That is a lot of furniture for a small space but it is nicely arranged. The living room is lighted by a large florescent light concealed behind a valance above grilled windows. This will be a nice place to sit, read and relax. I ordered drapes for the living room and bedroom along with a bed spread when I moved in, at no extra charge, thank you. This is how Americans rough it in foreign posts!

The bedroom is 20 by 20 feet. It has a large 8-drawer dresser, bedside stand, chair, ottoman, 3-shelf book case, large desk with six drawers, lamp, twin size, comfortable bed with box spring and a firm foam mattress. Off the bedroom is a utility room with an electric water heater and storage shelves. The small bathroom has a basin, toilet, shower and hooks for a clothes line. Floors in the bath and kitchen are covered with white tiles and the other rooms have alternate red and white tiles. I only had to purchase a broom and a squeegee mop. Everything else was provided. For cooling I have an air conditioner in the bedroom and an overhead fan in the living room. I swept out today, cleaned and put things away. The place now looks neat and clean.

I purchased a small supply of groceries, including tea, coffee, chile, enchiladas, tamales, kipper snacks, soups and juices for use when I eat here. I will eat at the club and the BOQ as long as they permit us to use the facilities.

One of the other agriculture branch chiefs lives in this building. I have noticed several of the men I know from work. We are generally the ones who went through training together and have been living for the past several months in down town hotels. I have seen no women tenants in this building; they must be placed elsewhere.

Barry helped me move from the hotel to the apartment yesterday. We then went to the airport to meet Syverson who has been away for 53 days. He has recovered from his hernia operation and glad to return to work. Our fifth man will be reporting on or about 2 December. With the full crew I will spend most of my time in Saigon supporting the field operations of the others. As yet we do not have approval for Barry's trip to the Philippines. Congress is having difficulty approving USAID's budget. I have been fully occupied recently making presentations of our budget and revising the revised revision of the revised budget. Things are murky concerning most of USAID's program, planning and budget activities. We received a reduction in our proposed budget but that is what we expected and wanted in the first place. Tuesday we have a budget meeting with the Director of USAID and he will either approve our budget or wipe us out and send us home.

After doing some work in the apartment today I went to the pistol range with Barry. After that I worked at the office for four hours, had dinner and sat in on a movie at the BOQ. It is now 2300 hours and I will sign off and try out the new bed.

26 November 1967

I have accomplished a great deal this busy weekend. Saturday after work Metcalf, Syverson and I went to the Brinks BOQ for BQ spare ribs. They were delicious. In spite of notices that our BOQ cards must be surrendered this has not been done. We will continue to use the facilities as long as they are available. Later we went shopping at the PX. Then I returned to the apartment and cleaned, scrubbed, polished, washed dishes and clothes. I can wash clothes in the kitchen sink or bathroom basin and dry them on a line in the bathroom. The air conditioner dries things quite rapidly. Now I have a spotless apartment. I take my shoes off at the door. That is a splendid local custom and prevents tracking in crud off the streets.

As I look around at my furniture and what I have seen in other apartments, USAID could save a great deal of money by eliminating half the furniture. It is much more than any of us need. As I look closer at this leased apartment building I am appalled by the shoddy workmanship. There are leaks in the plumbing and nothing is level or square. The wood trim in this reinforced concrete building is poorly done. Much of the new construction work is being done by women and children. The better workmen are drafted into the Army or engaged in more profitable work.

We have helped the Navy on a number of projects. The Admiral's palm trees prospered and we now have an open invitation from the Navy to call on them when we need any supplies. So far we have requisitioned a heavy hoist and a large electric generator for our demonstration sawmill. I met a Navy civilian a few months ago while we were assisting the Navy on erosion control and reforestation projects. I had also met him at the Vietnamese-American language classes. He was having trouble learning Vietnamese so he got himself one of those "longhaired Vietnamese Dictionaries" to whisper words in his ear. Apparently he got in deeper than he expected. This was the day I was invited to his wedding reception and dinner.

I well remember my wedding day and all the relatives. But this guy had twice that many relatives and none of them spoke a word of English. He had many rituals and customs to perform. Relatives were all North Vietnamese. The older men and women were most interesting people. They dressed in their typical clothing which is quite different than the South Vietnamese wear. All their clothing was black, as were their head bands. All men and women over 40 had stained black teeth from chewing betel nut. The old ladies were very interesting and pleased when I talked with them. Their accent was much sharper and crisper than the soft South Vietnamese accent. They called me the "man from Hue", since my accent more nearly matches that spoken in the central part of the country. The old men sat at one table and after drinking too much they really whooped it up.

Poor bridegroom Vern had so many rituals to go through that he had little to eat or drink. He looked exhausted after three hours of this. Prior to today's session he had previously gone through three different marriage ceremonies. It will take him weeks to recuperate. It will take me considerable less. I had only three Scotch and sodas and a 12 course meal. I may have lost count of a couple of courses. The food was served at a fancy Chinese restaurant. I did not try to identify all the foods, however I did not eat any of the fresh vegetables or the meat. The North Vietnamese have a very strong preference for dog meat.

Barry has been bragging about his ability to eat anything and drink the tap water. I have cautioned all the men about eating fresh vegetables outside of the BOQs and warned them never to drink the tap water without treating it with halazone. Now Barry has severe dysentery, fever and chills. He appears to be a little better today and perhaps wiser.

I have spent the afternoon stretched out on my comfortable couch reading and relaxing. It is now 2000 hours and I have recovered from the wedding feast. I skipped dinner this evening. As soon as I finish this letter I have a few office chores to complete.

I worked all day on Thanksgiving, as I did on Veterans Day. In the evening I had Thanksgiving dinner with the bosses. It consisted of turkey and all the trimmings. Here we are entitled to take all the US holidays plus the Vietnamese ones. Although I was authorized to work on these two holidays I donated my time. I would have been paid $109.00 for each day. Isn't that ridiculous? It was against my principles to claim this pay. But many Americans working here did claim the time! After you are here for a while it is easy to understand the rifts that develop between the Vietnamese and the so called rich Americans.

Gwen, we should run you for President! I note more of the world is beginning to agree with your ideas. We can begin by reducing the number of Federal employees by 50%, eliminate many of the poverty programs, censor the press to eliminate references to Jackie's love affairs and trips, Lynda Bird's love affairs, Baby Lynn's (uggh!) gurgling, the Pope's prostate and by all means drown the man in Phoenix who just produced his 27th child. I note Professor Davis (*TIME*) at the University of California advocates tax relief for childless couples and increased taxes for each additional child and places limitations on the size of available housing, etc.

We still have no funds to operate USAID. I hope you are receiving the regular pay checks. Permanent USAID people are not being paid and we have not received our SMA checks. I limit my spending and I am keeping my Bank of Bethesda checking account balanced properly. I have not written any checks on your 1st National Bank of Arizona account.

It will soon be time for me to leave and I count the days. I hope the appropriation is passed this week so things can settle down to normal (whatever that may be) and I can obtain my travel clearance. All out-of-country trips have been canceled. We have no money and therefore Barry has not made his important trip to the Philippines. We are authorized local travel only. I have stayed in to take care of the budget and all the interminable revisions. The other men have been servicing our various projects. I have been out on two Top Secret operations, but these have been short, one day trips to nearby areas. I may have to make another short trip or two before I come home.

As you read and see on TV, the war goes on. Some of the battles lately have been big and bloody. So far our forestry group has either been just ahead of them or just behind them. Tomorrow Barry goes north to check on reforestation progress at two air bases. Syverson and Metcalf go north to Danang, Hue and Quang Tri. That's as far north as we can safely travel at this time. In fact there is an interesting story on that subject. Our Assistant Mr. Thuong has been talking about getting married for two years. When I told him I wanted him to go north for two weeks with Syverson, he said he had to check on his mother's health. He checked and she was doing well. As a last alternative he took the plunge, asked the girl to marry him and then requested two weeks off for a honeymoon. I am convinced we helped him make the decision. The other reason for getting

332

married is that he needs someone to take care of his mother and her house. He needs help too because he is so frail. That young woman has her work cut out for her.

I am sending a small package to the girls for Christmas, so do not let them open it or shake it. Hide it when it arrives. I do not know what to get you for Christmas. How about me? Will I do? How are plans coming for the Quebec trip? I hope you find my long underwear, fur hat, overshoes and L. L. Bean shoepacs (rubber bottoms and leather tops). My blood is thin from being boiled out in the tropics and I will probably freeze to death. However I understand the French in Quebec have good wine and other spirits to warm me.

That's all for now.

30 November 1967

Good news! We received word today Congress passed a resolution to pay our salaries, issue SMA checks and cover any obligations made prior to 2 December. The travel people called me today to report they are working on my airline tickets for the Christmas Holiday travel. This means they will issue my tickets and I can be on my way home in 15 days! That is none too soon as far as I am concerned. I will leave here on 15 December via Pan Am and arrive in San Francisco the same date at 1730 hours. I will spend the night there to catch up with jet lag and then take United to Dulles, arriving about 1715 hours. The reservations are made but I do not have confirmation of the flight numbers. I have TDY at the office from 4-12 January 1968. The return flight will depart on 13 January and I am required to return to Saigon on 15th, essentially this means flying straight (25+ hours) through from Dulles to Saigon. That will be a killer, in more ways than one!

I received a letter from my Mother today. She said she is expecting to see me. She will receive my Christmas card but no visit until next year. My plans are made for the holidays. She plans to spend the winter in Florida with her sister Mabel. She says she needs help in figuring her income tax. I will advise her that is what her attorney is supposed to do. Her tax report would be much too complicated for me to calculate. She is moaning about how much it cost her to take care of her sister Madeline's funeral, the grave stone and shipment of the body to Worthington, Indiana. I guess she figures she can take all her money with her when she goes. I told her she should be happy she has the money to cover some of her family's expenses.

I just finished my laundry. I take off all my clothes and go in the bathroom and have at it with Tide and elbow grease. That is quite a semi-automatic operation. I am becoming an expert at washing a sheet in a wash basin. It comes out quite well. When I finish I take a sponge and swab up the surplus water. Then I hang the sheet and other items on a line in the shower stall until the water stops dripping. Then I transfer the clothes to a line in the bathroom. The air conditioner dries my clothes quite rapidly. Each morning I make the bed, sweep the tile floors and put everything away. That takes about 15 minutes and looks good afterwards.

I returned to the apartment today after a light lunch to pick up my revolver, shoulder holster and camera for a trip to the Regional Headquarters at Bien Hoa. Before I could leave the apartment I had my first woman caller. She was from the drapery store and had come to make the measurements. She did not speak a word of English and this gave me an opportunity for more language practice. She will send a man to install traverse rods and drapes at 1500 hours next Monday. She asked me to call the Embassy and approve the expenditures. Being able to speak the language is a real advantage. You can imagine the delays and misunderstandings that develop in transactions when neither party can understand the other.

Tomorrow Barry and I will go on a short hop with some military people to look at a project. We leave at 0630 hours and are scheduled to return by 1200 hours. The military and civilian people are calling on us more and more as technical experts and arbitrators to help settle disputes between the two groups. It is very interesting.

Saturday I accompany Director Muoi, Mr. Hiep and a forestry expert from the Lillenthal group to do some post-war planning for Dalat. They are considering the possibility of installing a pulpmill in that area. We are scheduled to return on Sunday. If anyone thinks planning for a multi-million dollar pulpmill can be done in less than two days they are crazy. A feasibility study must consider the mill site, equipment, sources of water, wood supply, transportation, markets, labor, air and water pollution abatement and a multitude of other factors. The fall weather in Dalat is predicted to be clear and cool. We will travel by Air Vietnam.

It resembles fall around here since the rains have tapered off. We had only an inch of rain tonight after supper, and of course it came during the movie. Tree leaves are falling and people sweep them into piles and burn them in the streets. As leaf smoke mixes with the terrific traffic exhaust I am convinced we have succeeded in bringing these people up to our level of progress, at least in air pollution. On some days it is difficult to see or breathe with the blue haze hanging thick over the city. Add to that the odor of Agent Orange and other herbicides drifting in from the Ranch Hand operations and you have an impact that will manifest itself in future lung problems for the Vietnamese and all of us.

You may recall Mr. Passey as the older USDA man from

Boise who had dinner with us one evening. He lives across the hall from me and sometimes rides back and forth to work with me. I seldom see any of the men who lived at the Oscar and Park Hotels while I was there.

Please tell Myrl and Gene to stock up on buckwheat flour and maple syrup and ask Mrs. Jensen to bake up a double batch of her famous oatmeal cookies! In addition to those items I will need a gallon of real, un-reconstituted milk to wash it down. I expect breakfast to be ready for me in a little over 20 days. Please inform Gene I would like a martini for desert!

As a former bomber pilot Gene will appreciate the fact that Vietnam pilots have a rough time. They fly mission after mission with little rest in between. Weather is always poor somewhere along their flight paths and they are fired on by very sophisticated missiles. I doubt our actual losses are being reported. I have talked to fighter-bomber pilots who cover North Vietnam and they say the flak is unbelievable. Our forestry group has been called on to help refine rescue techniques used by the Jolly Green Giant helicopter pilots. Conducting rescue operations in the 3-story tropical forest is a challenge. Those pilots are producing miracles and some of them are in the sky 24 hours each day. It is good to know they can be reached in a moment's notice when I am up with the CIA's Air America or in a military aircraft on one of my many missions over remote areas. Gene would be interested in seeing the size of a LZ (landing zone) created by a 10,000 pound bomb. Yes, 10,000 pounds! He would be interested in seeing what a flight of five or six B-52s can do to a dense forest. It is unbelievable. Each of these planes, with 108 bombs, pack much more wallop than one of Gene's WW II B-17G bomb groups. The B-52s fly out of Guam and are so high we never see or hear them until they drop their bombs. But we can certainly hear and feel their "whump, whump, whumps" when they unload. When I return to the Forest Service I can tell them precisely how to construct large livestock tanks and fish ponds in a short time!

It is getting late and I will end this letter and read until I fall asleep. I now have a good bedside light to read by. All things considered I am enjoying the comfort and solitude of my apartment. I am reading *Treblinka* a story about German concentration camps and the only uprising of the Jews against their murderers. It is a sad story. We have access to a fairly good library and that helps ease the boredom of being away from home. Good night, dears.

3 December 1967

This is just a quick note to let you know I returned after my two trips. I had a good trip to Dalat with the new Director. I believe we will get along well. Mr. Muoi has more trouble speaking and understanding English than Tan. This makes it more difficult to talk with him.

I read you had a foot of the "white stuff" last week. I am looking forward to seeing snow and being with you.

It is now 1830 hours and I missed lunch at Dalat. I came straight to the apartment, cleaned it and myself and washed some clothes. Now I am starved and it's off to the BOQ for dinner and a movie.

6 December 1967

Happy Birthday, again! It was good to talk you to briefly. It was 2030 hours when my call finally went through. I waited at the USO for two hours for my turn to get a line. It was too bad the radio-delay signal faded out, but your voices came through loud and clear at the beginning of our conversation.

I will send you my airline schedules tomorrow or the next day. I miss you all.

8 December 1967

Just a week from today I should be packed and ready to take off on Pan Am #842 at 1100 hours. I will arrive in San Francisco at 1717 hours the same date, 15 December. I am scheduled to leave San Francisco on United 0050 and arrive at Dulles International Airport at 1505 hours.

I have much work to complete at the office before I depart. What I do not complete Barry can take handle.

Your dear little $4 gold watch finally stopped running. It did well and I liked it very much. Last night I went to the PX and purchased a new one. Now I am back on time.

I mailed the last of my cards yesterday morning so that job is finished. I am glad I could help you with those.

It will seem strange to be back with all you bleached pale faces. But I will love them. I look forward to our time together. This will be my last letter until I return here in January 1968.

* * *

All went well during Christmas home leave and temporary duty in Washington. We had a delightful train trip to Quebec City. Accompanying us were Susan Maynard, the girls' friend and practically our third daughter and Melissa's Bob Humphrey (he wore real shoes for the trip!). I did find that at age 46, my tolerance to cold weather was greatly diminished. Walking on the promenade at the Chateau Frontenac in Quebec City and riding toboggans down the chute was frigid. I just about froze to death. But I had my dear Gwen, the girls, Bob and French spirits to keep me warm.

The Forest Service, Secretary of Agriculture, involved USDA and USAID offices expressed their appreciation and complimented the forestry team for our performance. It was a good feeling to have their support. The *Dear Gwen & Girls* letters have been widely circulated around the Chief's office and created more interest in our Vietnam tour of duty.

Chapter XXIV

Calm Before the Storm

Bombing and chemical warfare affected our forestry project. Before we left Vietnam Barry Flamm and I gathered information and photographs depicting the impacts the war made on the significant forest resources. Later we prepared articles which were published in the *Journal of Forestry* and the *Congressional Record*. (Flamm & Cravens, 1968) Our job was to help the Directorate of Forest Affairs inventory, manage and utilize the timber resources. But bombing, the defoliation done by Operation Ranch Hand, and other destructive schemes, were designed to destroy forest cover that interfered with military security and provided ambush cover for the enemy. Very early in our involvement we observed the magnitude of the defoliation, by far the greatest impact, which covered millions of acres of forest in all four Corps Areas, from the DMZ to the Delta.

Early in the war the South Vietnamese with the support of their American advisors had declared free-fire bombing zones. (Sheehan, 1988) In these selective areas of guerrilla dominance, suspected structures or anything that moved or was observed by Forward Air Controllers or other sources, could be killed or destroyed. These areas could be hit at anytime, day or night, by artillery and mortar fire, aerial bombs or by helicopter gunships. While South Vietnamese peasants and loggers were warned to move out and keep out of these areas, the impacts on innocent wood cutters and the forest resources were tremendous.

My earliest involvement in the herbicide activity was a request by Lt. General Mearns to advise the III Corps military command what to expect as a result of defoliation and Rome plow land clearing projects and what might be done to facilitate military operations. At that same meeting the Commanding General, William Westmoreland, thanked me for helping them and told me he was not a strong supporter of chemical defoliation as a military tool or tactic. He said the Army's Chemical Warfare people had promoted the program. I told General Westmoreland they could expect repeated defoliation to eventually kill large overstory hardwood trees but it would eventually release highly resistant species. Bamboo would explode in growth and eventually make it much worse and more dangerous for the foot soldier. Later enemy troops moved down for the Tet offensive along the Ho Chi Minh trail from North Vietnam, Laos and Cambodia. They passed undetected through

War Zones A, B & C and other areas which been treated repeatedly by Ranch Hand. In fact, the enemy constructed roads and trails through defoliated areas under the dense cover of bamboo and other resistant species . . . leading direct to Bien Hoa and Saigon. Repeated air and ground patrols failed to discover these access routes or troop movements prior to Tet. Multiple applications of 2,4,5-T & 2,4-D were applied at the excessive rate of more than 28 pounds of acid equivalent (AE) per acre each time. Some areas were treated more than once with these massive doses of herbicides. In the Forest Service we applied herbicides at the rate of less than 1.5 pounds per acre AE, with satisfactory results.

My next involvement (all highly classified, Top Secret, and extremely sensitive at the time) in the defoliation activity came late in 1967 and early 1968. The Embassy began receiving an increasing number of claims for herbicide damages. These came from vegetable, fruit and rubber producers. I was designated as USAID/Agriculture's defoliation claims investigator. Damages to vegetable and fruit producers in areas surrounding Saigon and to crop production areas near specific defoliation target areas were indeed great. In one province alone a highly successful and profitable conversion to soy bean production was completely wiped out. Huge tomato fields and significant fruit tree resources were destroyed. A serious side effect of this destruction was the embitterment of farmers who were loyal to the South Vietnam government. Many became supporters of the Viet Cong. The VC were experts in utilizing such mishaps to their advantage. There were many valid claims and these were paid by the United States Government.

What caused this damage? It was 2,4,5-T and 2,4-D, known locally as Agent Orange, in reference to an orange band painted around the barrels and used by the non-English speaking or reading Vietnamese to identify the chemical content. And then there was Agent White (white band) a desiccant, and Agent Blue (blue band) a trivalent form of arsenic for upland crop destruction. Some damage resulted from drift of the spray, which could be expected when applying herbicides at 300 to 500 feet above the ground in temperatures ranging up to 90 degrees with high winds. In some cases the herbicide was released at much higher altitudes due to the intense enemy fire directed at the spray planes. Much damage occurred near Ranch Hand's base of operation in Bien Hoa and resulted from sloppy operations. I examined C-123 spray planes which were fully loaded with herbicide, sitting on the tarmac and waiting for dispatch. Herbicide was not dripping from the hoses, nozzles and tanks, it was streaming and sometimes collected in overflowing buckets. Great quantities of herbicide were spilled during loading and maintenance operations. Some barrels of herbicide, Orange, White and Blue, were piled so high for storage that bottom barrels ruptured and the contents ran off into irrigation and drainage ditches. Empty contaminated barrels were picked up by enterprising Vietnamese junk dealers and sold to farmers for irrigation and domestic water storage. It does not take much imagination to visualize the disastrous effects of these sloppy practices, as well as planes taking off and flying low over adjacent croplands. The Air Force had excellent aircraft mechanics and technical manuals which covered the proper care and maintenance of spray equipment, but the neoprene washers and other fittings were poorly maintained and leakage and seepage occurred. Instructions covering the proper disposal of used herbicide containers were generally ignored.

We received claims for damages to banana and rubber plantations. I found banana plants to be highly resistant to herbicides. It is doubtful if a banana plant could be killed if it was drenched in Agent Orange and holes drilled in the stems and a bucket full of Agent White poured in. But the plants are affected by age, insects and disease. The owners were trying to saddle the US with the cost of replacing a deteriorating resource. Rubber was another story. Agent Orange did have an impact on rubber trees. My study and investigation of Southeast Asia rubber culture demonstrated that heavy doses of 2,4,5-T could kill rubber trees. But light amounts of 2,4,5-T were used in Indonesia to stimulate latex production, without damaging the trees. Claims were filed for defoliation damages on many rubber plantations. Investigating these claims was not one of my fun jobs. Rubber plantations were off limits for the Allied military . . . no hot pursuit of enemy forces, no bombing, no artillery fire and no defoliation. These were dark and spooky places, full of VC and sympathizers. With the exception of high noon, 1200 hours, the trees kept the area in dark, deep shade, ideal for concealment and movement of enemy troops that utilized some of these areas for base camps, hospitals and transportation routes. I found only one valid example of defoliation damage. A C-123 spray plane experienced an engine failure as it was passing over a rubber plantation and the crew used the emergency release to unload the entire tank. In that amount Agent Orange killed all rubber trees on the area hit by the chemical. Based on my studies of rubber tree culture and management I found most of the claims were invalid. The owners of the plantations were trying to write off old, decadent, insect and disease ridden trees that should have been replaced years earlier, on good old Uncle Sam. Claims were disallowed, and I got out as fast as an armored personnel carrier (APC) and a company of the US Army's 1st Infantry Division, the Big Red One, troops could escort me to my helicopter gunship. I was relieved to be out of those hateful places.

Herbicide use was intensive around many Special Forces, Marine, Army, Navy and Air Force bases to eliminate concealment cover. Some chemicals were applied around these facilities from the air by Ranch Hand, but hundreds of barrels of the undiluted chemicals were applied by inexperienced GIs using fire fighters' hand spray pumps and truck mounted tanks and spray rigs. The rate of application was tremendous and repeated frequently. The applicators were normally drenched in chemicals that drifted onto them or more frequently deliberately sprayed on one another during a friendly GI herbicide spray fight. The stench of herbicides could be detected in the countryside, at isolated outposts, air bases and in Saigon. In some of these areas, where the vegetation was removed and prevented from becoming reestablished, the red laterite soils underwent a process called laterization and produced the equivalent of red concrete. We saw no evidence of laterization in the general forest stands, in fact some of the areas receiving one treatment by Ranch Hand appeared to recover and produce new foliage. We had not been informed about the dioxin problem. I learned much more about defoliation before I left Vietnam. From evidence collected we convinced USAID, the Ambassador and General Westmoreland that Ranch Hand and herbicide use was a poor military tactic and was making a very serious impact on an extremely valuable renewable resource. Our efforts and formidable opposition in the United States resulted in a full scale review of the defoliation program and eventual cessation of the practice. Barry Flamm and Dr. Fred Tshirley of USDA's Agriculture Research Service made the extremely dangerous field examinations of the defoliated areas along the Cambodian border where they collected plants and soil samples. I worked on the review in Saigon, since I was a "short timer". But the damage was done and the herbicides made very significant impacts on many valuable species of trees, including the very sensitive mangrove. The final verdict is not yet in on the significance of dioxin contamination of human resources. The body tissues of those of us who were in South Vietnam during this period may be loaded with the contaminates. In the late '70s, following retirement from the Forest Service I served for three years as an adviser to the legal counsel of Dow Chemical in Agent Orange litigation. I briefed a series of attorneys on my knowledge of both the Forest Service's herbicide program and the Vietnam operations.

The war in Vietnam was very real and very bloody. At first I could see real progress in the war, but as the months passed I was convinced it was a deadly stalemate. While my story does not dwell on this aspect, we were in the war zone every time we left Saigon and during Tet the war came to Saigon and to our front door. Hundreds of books and a million square feet or so of columns in newspapers and news magazines have been published about the war in Vietnam. Countless hours of TV coverage of the death and destruction were displayed on screens throughout the world. Our forestry team made every effort to avoid the conflicts and fortunately we were just ahead or just behind the battles.

Most battles were probably accurately described by the ubiquitous media. However many facts were deliberately covered up, presumably to deprive giving aid and comfort to the enemy and to cover gross negligence on the part of some military unit or commander. Others like the My Lai massacre in the coastal village of Quang Ngai Province resulted in the slaughter of 347 Vietnamese by a US infantry company on 16 March 1968. (Sheehan, 1988). From our work in Quang Ngai Province we were well aware of the hatred held by the Vietnamese toward Americans. Efforts were made to hide the needless My Lai deaths. Fortunately the press ferreted out this devastating story. There were others which should remain buried with the dead. Atrocities were plentiful on both sides. Give a teen-ager an automatic weapon, scramble his mind a bit with the pitfalls of Vietnam and then let him lose a buddy or two in a sneak attack and he will kill anything, water buffalo, pig, dog, chicken, duck, old Vietnamese men, women or children. Some of these tragedies can be traced to keeping score, like a basketball or baseball game, and calculating the death ratio. We were winning if we could count more of their bodies (any bodies counted) than our bodies. Conducting a war by body count was an immoral, shameful way to run a war. But after all, all wars are immoral.

Another interesting facet of our involvement in Vietnam was our advisory role. Everybody had an advisor. In forestry we were there to advise, not dictate, but to suggest methods for improvement. We were there as invited guests. Hog wash! We invited ourselves in! America slipped into this war without fully realizing what we were getting into! In the Vietnam war the US military can never be accused of approving targets to be bombed or selecting an area to be treated with Agent Orange. The Vietnamese approved the free-fire zones and the targets for bombs and Agents Orange, White and Blue. Actually it was the American advisors who identified those targets and maneuvered the Vietnamese into making those decisions.

Again the Tet offensive of February 1968 kicked off by the enemy from the DMZ to Phu Quoc Island in the Gulf of Siam has been thoroughly covered by countless writers and cameramen. My coverage of Tet in "*Dear Gwen & Girls*" is extensive, perhaps exhaustive. But I have a few more points to emphasize. The Tet offensive was an intelligence failure of major proportions, pure and simple. It can be directly equated to Pearl Harbor. We had warning. We knew when and where the attacks would come. But rivalry among the cloak and dagger agencies, conflicting intelligence information that could not be put together, plus the boredom and monotony of the war dragging on and on, allowed us to be caught with our guard down. The thud of bombs and the thump of artillery being fired into buildings and the rattle of automatic weapons just one or two blocks from where I lived, reverberated down my street and left deep unforgettable scars in my mind . . . scars that carry memories of the many brave young MPs that rushed into machine guns and into the arms of death in the early hours of the Tet offensive and the countless numbers of other deaths and the destruction of Saigon, no longer or never again to be the Pearl of the Orient. Then there was the tragedy of creating a million new refugees as we destroyed portions of Saigon and other cities to save them. The entire refugee relocation effort was irresponsible and a heartless decision. In herding bewildered families into vast ramshackle slums we devastated the lives of hundreds of thousands of people. We did not win many hearts and minds of the Vietnamese people from that point on. My final thought on Tet concerns the young Vietnamese forester who invited me to spend Tet with him and his family in the ancient imperial capitol of Hue. He and his family disappeared off of the face of the earth. There but for the grace of God go I!

* * *

16 January 1968

Guess what? I am writing this first letter of 1968 from Hong Kong. Ten of us from USAID/Vietnam were first stranded in Honolulu and now in Hong Kong. Pan Am made an error in our reservations and has been covering our expenses across the wide Pacific. We have been staying in the finest hotels where the food is great.

We sent a telegram through the American Embassy in Hong Kong to Saigon informing them about our delay. Now they know where we are and why we will not return as scheduled. This will not count against our annual leave or visitation privileges. There is not much we can do except see the sights, shop and wait for Pan Am to make arrangements to book us on another airline. As of today all flights out of Hong Kong are overbooked. I will keep you posted on my progress or the lack thereof.

I am grateful I was not delayed on my journey home. My time with you was wonderful. Wish you were here.

21 January 1968

It is Sunday evening in Saigon and I am more or less rested up from the trip. This letter will describe my delay in more detail and inform you about what I have been doing since returning.

The flight to San Francisco from Dulles was uneventful and

on schedule. The movie was "Gigi", one of our favorites. I had time to relax and reflect on our wonderful vacation. I still shiver when I think about Quebec, but it was great. The movie en route to Honolulu was the so-so "Last Safari". We arrived in Hawaii at 1730 hours. I went to the international departure desk of Pan American Airlines and asked when I should return to check in for Flight #841 for the 0130 hours flight to Saigon. I was told to return at midnight. I caught a bus into the city, had dinner and took in another movie to help pass the time.

When I returned to the airport just before midnight to check in I was informed there was no Pan Am flight #841 to Saigon. I was told I should have been at the airport to take flight #819 which departed at 2145 hours. I pointed out the desk man who checked my tickets and told me to return at midnight. It seems Pan American's communications broke down. My flight had been canceled three weeks earlier! The Pan Am people were very apologetic and took care of my transportation, room and meals at the Waikikian Hotel. I had a good night's sleep. It was much more comfortable than sitting up and keeping awake all the way to Manila that night.

I was up early Sunday morning and had a good walk along the beach. The surf was up and I watched surfers ride some huge waves. I had breakfast, went shopping, purchased a Hawaiian shirt and swimming trunks, took another walk and snapped some photos. After lunch I packed my small carry-on bag and returned to the airport where I found 12 of my USAID/Agriculture friends from Saigon, victims of the same foul-up. We departed on Pan Am Flight #001, their east-west, around-the-world flight. Due to head winds, the flight to Tokyo required 9 1/2 hours. It was freezing in Japan as we walked from the plane to the airport terminal building. We had just a hour to browse around the airport shops.

Four and a half hours after leaving Tokyo we landed in Hong Kong. After checking through customs and immigration we proceeded to Cathy Pacific for our next flight. We had no confirmed reservations but were told to return the next day at 0800 hours and "perhaps" our flight to Saigon could be arranged. Pan Am again provided meals and rooms at the Park Hotel, on Chatham Road, Kowloon, Hong Kong (Telephone 661371). We had another good night's sleep. The next morning after breakfast we returned to the airport where we learned only two seats were available to Saigon. These seats went to two of our group who had been stranded earlier in San Francisco by the cancellation of Flight #841. They had their request in before the rest of us reached Honolulu. Again Pan Am took care of us and asked us to return the next morning. Cathy Pacific is notorious for overbooking flights. As a result many passengers are frequently stranded by their sloppy operations. We had nothing to do but enjoy our enforced vacation at Pan Am's expense.

We spent the day sightseeing. I walked around with Mr. Passey of Boise and a Mr. Anderson. The weather was clear and warm. Shops were open and many, many people were on the move. There were none of the normal riots or protest marches. Rickshaw runners trotted up and down the streets with their passengers. The double-deck buses, double-deck street cars and left-hand flow of traffic made it seem much like England. The airport and our hotel were on Kowloon, which is the mainland. We traveled by Star Ferry to Hong Kong Island. The water was speckled with sunshine and large ocean liners and junks streamed by on both sides of us.

Arriving at the Island we took a taxi to the China Fleet Club for shopping. Anderson had been there before and told us about the excellent bargains. The Club has two floors of shops and is operated under a US Navy contract. The place would drive you out of your mind. They have clothes, suede jackets, dresses, cloth, suits, 100's of different kinds of cameras and electronic gear, cosmetics and wigs. Guess what I bought for you, Gwen? Give up? It is a beautiful, hand-tied, 100% human hair wig. The color is as near your natural color as I could remember. You once told me you would like to have one. I will mail it to you as soon as the Hong Kong Government processes and sends me the Certificate of Origin. You will probably be required to pay a small amount of duty. It cost only $26.80. I purchased ivory earrings for the girls at 27 cents each.

For myself I purchased a Sharp 4-band, battery operated portable radio for $31.70. It has a good tone and I can reach Manila and Hanoi. For $75 I bought a 9-inch Sony TV which works very well in the apartment. I watched American football today. On the next two Mondays I can watch the last two episodes of the "Fugitive".

Following our visit to the Fleet Club we window shopped and returned to Kowloon on the Ferry to deposit our purchases at the hotel and do some more sightseeing. There is so much to see and do in this popular tourist spot. We had good meals at the Park Hotel, on Pan Am's bill of course. We saw many GIs from Vietnam on their R&R trips. They are entitled to one Rest and Recreation trip to Hong Kong, Bangkok, Penang, Sydney or Honolulu during their one year tour of duty. And they really celebrate their momentary release from Vietnam and go through money in a hurry. I talked to one 18 year old Marine who had blown $500 in one afternoon and night. He was broke and still had four days to go. I gave him enough for a few beers. The US military and the hotels wisely require GIs to pay for their room and meals in advance. At least they can eat, sightsee and sleep (alone) during their stay.

The next day we packed up again and caught a Pan Am van

back to the airport. Cathy Pacific was full again but Pan Am told us they were trying to obtain seats for all of us on Air Vietnam. After a very long wait, and listening patiently to on-and-off again arrangements, we were finally accommodated. Mr. Passey was the last to be called and he was happy to pay $26 for a seat in the First Class section. Air Vietnam operates Caravelle jets on overseas flights. It was a smooth, uneventful flight of 2 1/2 hours. En route we could see Macao and the Peoples Republic of China off in the distance. There were Communist patrol boats all along the way. We arrived in Saigon at 1600 hours and I went direct to the apartment. Mine was cool, clean and in good shape. I had left the air conditioner on and Barry Flamm had been in once each week to check the place. Mr. Passey has a maid and in spite of his detailed instructions she had turned off the air conditioner and his place was hot, damp, musty and his shoes and clothes were covered with mildew. Many tenants are not satisfied with their undependable, inefficient maids.

Back at work the next day everyone seemed pleased to have me return. Several were envious of our enforced, all-expense-paid, stay in Honolulu and Hong Kong. Everyone in the office knew about the delay since we sent a telegram via the Hong Kong and Saigon American embassies. I had no jet lag following the trip.

During my absence many of the BOQs and PXs announced definite closings. Troops are being moved out of the city as rapidly as new outlying quarters are available. Moving into the apartment was a wise decision. Syverson has moved into an apartment at the edge of the city near Tan Son Nhut Airport. He is miles from work and the center of the City. I am pleased to have a place so convenient and within easy walking distance of the office, the club and other facilities I use.

Saturday after work Syverson and I went to the PX and Commissary at Cholon. I am always amazed at the amount of money spent in those facilities by GIs and civilians. I purchased a few things at the PX and picked up a supply of groceries to supplement what I brought from home. The Commissary is larger than Bethesda's largest grocery store. Items are stored in stacks of open cartons. Selection is fairly good and inexpensive. Lettuce from California is 26 cents for a large firm head, margarine is 20 cents per pound, a cut up frozen chicken is 65 cents, cans of tuna fish are 25 cents, one-half gallon of vanilla ice cream is 50 cents, bread is 15 cents a loaf and fresh milk is 17 cents per quart.

Concerning the food I brought from home . . . I checked my bag all the way from Dulles to Saigon. It was routed on the missing Flight #841 and arrived long before I did and was impounded by Vietnamese Customs. The bag was locked and no one could pick it up for me until it was opened and inspected. I went to the airport and after quite a delay my bag came out of storage. When I unlocked it the inspectors went over the contents in great detail. When they saw all the food it contained they asked how long I intended to stay. I told them a year. That must have been a good answer because it was acceptable. They were suspicious of the foil wrapped box of cookies and directed me to open it. I offered each of the inspectors a cookie but fortunately they did not take any before they closed and returned my bag.

I spent Saturday evening reading and relaxing. Sunday I swept, mopped, dusted and put groceries away. I went to the lower level where the maids wash clothes. Some of them tried to get me to hire them to do my washing but I declined. They were surprised to see an American washing clothes and even more surprised to have me talk to them in Vietnamese. I discovered it is much easier to wash sheets in a tub rather than a wash basin. I appreciate the contour sheets when it comes to making the bed. Please mail me some plastic clothes pins, the wooden ones mold and stain the clothes.

For lunch I fried a chicken, prepared rice, a salad and had ice cream and cookies for desert. Tell Myrl and Mrs. Jensen I enjoy the cookies. After eating I watched a two weeks old football game on TV and read. Then I went to the BOQ for a movie.

Tomorrow I leave early with some people from the Directorate and fly to Tay Ninh to present permits to sawmill operators. This will make the owners happy, since it will legalize their operations and free them of some, but not all, of the huge payoffs and graft associated with the timber business. The timing of this, just before Tet, should have a dramatic affect.

Gwendel, thanks for your letter. It was good to hear from you. I hope Melissa got her ring repaired. Please do not forget to mail the insurance premiums I placed on the dish shelf. Why should I bring up such routine matters? After all you are doing a splendid job of looking after all of our affairs and the girls. Sorry to hear about all your rain. It is bone dry in Saigon, but still raining up north. I will watch out for "falling elephants". Good night, dears. It is midnight and I must be up and ready to go at 0500 hours.

24 January 1968

I regret to hear you have to go through the car repair routine again. The girls must be more selective in who they let drive the car. They should check IDs and not let any of their friends under 25 drive the car. It is best for the girls to do the driving. I have much confidence in their driving.

In regard to the car repairs I believe you did the right thing and I hope the other party pays off. I would have recommended two things . . . replacement of the bumper and fender with new parts and the fender should be under-coated to protect it from salt. Those Volkswagon Beetles do not have very thick skins. Again I regret I am not around to take care of such things, but you are doing well.

My trip to Tay Ninh was successful and we returned the same day. We met with sawmill operators and local forestry people. We accomplished a great deal and eliminated two sources of corruption and payoffs. The Province Chief, a Vietnamese Army Colonel, did not look especially pleased by the turn of events. I suppose we eliminated one of his many sources of funds.

I will be working in Saigon this week. Everything on the Vietnamese side is closing down for Tet, their big annual family holiday. Tet lasts for four days, officially it runs from next Monday through Thursday. But for practical purposes the celebrations start early and run later. Our Vietnamese employees have four days off. They are also paid a Tet bonus, which means their annual salary is divided into 13 parts and they receive the additional pay just before Tet. The money is used to purchase new clothes, food and flowers for the family.

One of the Vietnamese foresters at the Directorate is going north to visit his family in the old Imperial Capitol of Hue which is between Danang and Quang Tri. He invited me to go with him. It would have been a good experience to live with a Vietnamese family during Tet, but I declined his invitation. I decided to stay in Saigon during this holiday season of Tet. I have office work and other chores, like sitting around the pool at the club, to do. I must leave for work now.

Chapter XXV

Tet

The past few days in Saigon have been different and rather interesting. I would like to describe Tet. This is the Vietnamese "Festival of the First Day of the Year". Tet is actually the contraction of "Nguyen Dan Tiet". A few years ago in peace time, (please remember a Vietnamese must be at least 25 years of age to have known peace in this troubled land) Tet lasted for three months. It filled the time between the end of the harvest season and the beginning of rice planting in March. During that period the farmer had little work to do and much time for celebrating. The Vietnamese have an old song which sums up their way of celebrating Tet, it goes like this:

We enjoy Tet at home during the first month,
The second month is spent in gambling,
And the third in parade and festivities.

In Saigon Tet is celebrated for four days. Our secretary, Mrs. Hai, asked for Saturday off to prepare for Tet. She said she had much food to prepare for the many guests they expect to call. She will have quite a workout because the family is poor and they have only one girl servant. After one of these family sessions she is usually worn out from all of the preparations, but she has always returns to work with a smile on her beautiful face (refer to the photograph I sent you).

Mr. Thuong, our Administrative Assistant, gave each of us a Tet greeting card. I have sent you Tet cards in a separate envelope. The purpose of these beautiful cards, with their colorful drawings, is to welcome the Spring and the "Year of the Monkey". The Vietnamese people are very superstitious and have mixed emotions about this particular year. The "Year of the Monkey" is not a good year and everything must go right to make the year a success. Enclosed in the greeting card is a small red envelope covered with gold Chinese markings. Inside is a lottery ticket. It is customary to give a small gift and it is particularly important to give the children an envelope containing new money. One is attached for you.

I gave cards to a number of Vietnamese friends. When I visited the Directorate to deliver cards I decided to give one of the people, selected at random, a card for each of his children. I selected the doorman who always bows, clasps his hands together and escorts me into the office. Well wouldn't you know it, when

I asked him in Vietnamese how many children he had, he said 10! I gave him ten small red envelopes. He was overjoyed. I just about ran out of envelopes with one presentation!

For the remainder of Saturday afternoon I drove around the city observing the activity. I was grateful I did not have a large, bright, shiny new car because the traffic was terrific. Streets were filled with shoppers. Tet requires much food, many gifts and flowers. Streets were full of flowers for sale and were being carried home by every means of conveyance. The area in front of the hotel where I first lived was devoted to flowers. Actually there were thousands of pots and vases of flowers of every kind and color, including flowering fruit trees, kumquat, peach, etc. Red flowers are required for the family altar. Yellow chrysanthemum, white narcissus, yellow daisies, or cuttings from peach trees or other flowering trees were in the bud stage and expected to blossom on the first day of Tet.

I found a man taking photographs of his beautiful wife standing among a mass of flowers. He gave me permission to take her photograph. I have a brand new Miranda Sensorex camera and took one shot before the darned thing jammed. I worked on it, cursing the "monkey" or whatever was responsible for my missing pictures of that beautiful doll. I retreated to the shade and worked on the camera for about five minutes before I unjammed it. I took off with a fresh roll of color film to find said doll. Luckily I found her and the camera performed well, I hope!

The transport of flowers was a fascinating sight. Huge loads were hauled into Saigon by bus, truck, Lambretta, horse and ox carts. Relays of people unloaded flowers and as fast as they hit the ground people were buying and carting them off on foot, scooter, bicycle, pedicab, cyclo and car. There was literally a river of flowers flowing up and down every street in Saigon. Before Tet I had noticed a few fields of flowers being tended near Tan Son Nhut Airport but I did not realize so many flowers were produced. Prices for the flowers ranged from the equivalent of US$1.50 for a small plant to US$15 for a large pot of narcissus. Even the poor people seemed to have money for flowers. Tet requires flowers for the deceased relatives' altars and flowers for the living family. I noticed the very oldest and poorest cyclo and pedicab drivers had a sprig of marigold fastened to their vehicles. Except for a few beggars everyone wore new clothes except a few beggars. I noticed one pedicab driver in a western style, cowboy hat, a white nylon shirt and black silk trousers. That same driver had a sprig of wilted marigold fastened to his cab and he let me take his photograph.

The traffic was wild. People were whipping in and out and around, especially the young, longhaired Honda motorcycle and scooter operators. American drivers are being criticized in the local press, but actually they are no worse than the Vietnamese. The worst drivers are the ARVN military drivers who wheel through crowded traffic at high speed with air horns blowing full blast. Emergency vehicles travel the streets at all hours. In spite of laws similar to ours, no one ever yields to ambulances, police or fire vehicles. Somehow the emergency vehicles squeeze through without causing more injuries.

Later I drove to the Cholon PX and Commissary. These areas were busier than I have ever seen them. Hordes of Vietnamese and Chinese merchants had set up makeshift stalls alongside the streets and unused railroad tracks. GIs and civilians streamed in and out of the big US compound laden with groceries, liquor, tape recorders, cameras, TVs, radios, fans, refrigerators, cases of beer, soda pop and clothing of every description. Large items and heavy loads are carried by Vietnamese bag boys who earn piasters from American shoppers. In this area there is a mixture of GI trucks, Jeeps, half-tracks, semis and every other kind of GI vehicle parked by the troops as they take a PX break. There are always many USAID and other US vehicles parked in this informal parking area. Drivers park on the sidewalks, railway tracks and some even park in the street.

Small Vietnamese boys, six to twelve years old watch for you to drive up and then run along ahead and mysteriously always find a place for you to park. Some carry dirty rags and a small can of water and offer to wash your car or watch it. We never get the Jeep washed because the dirt is the only thing holding it together. If you do not want these kids to take care of your vehicle they are prepared to offer you some of the most ugly pornographic photographs of Vietnamese or Americans for US/MPC$10 (no piasters please). Usually at least two of us go to the PX so that one can remain with the vehicle. Often standing beside these kids are Vietnamese police, referred to as the "White Mice" by both Vietnamese and the Americans. The place is literally crawling with police. Why are they there? They do not prevent thefts, or prevent the kids from selling dirty pictures, or even attempt to control the traffic or crowds. What do they do? You can expect to be approached by three or four policemen with a request to purchase something for them in the PX. They hand you a prepared list of items they want and ask you in English to buy the items listed, even to the model number. They offer piasters, MPCs or green dollars. These poorly paid (less than US$30 per month) police try to hustle up a deal to obtain items for the black market. Every item for sale in the PX can be found in the flourishing black market, often at double or triple PX prices. While these illegal PX purchases can result in Americans losing their PX privileges, I see such transactions taking place everytime I am near the PX or Commissary.

To enter the PX or Commissary you must produce ID and

PX cards for three MPs to examine. Inside the large double Quonset-type structures are found most of the items a GI needs or does not need. He can purchase three 40 oz. bottles of good US whiskey and three quarts of fine California wine on his ration card each month at the cost of less than US$10 for the six bottles. The incentives are high to get the GIs and us civilians to part with our money in a US operated facility rather than spend it on the local economy. In the PX you can purchase a top quality, measured to order Hong Kong suit, cameras, electronic equipment, watches and jewelry. Fords, GMC trucks, Chrysler and American Motors cars can be purchased for State-side delivery. The latest magazines are for sale. *Playboy*, is a best seller. The favorite reading materials for GIs in this war, as in the past two, are comic books. I could go on for many more pages listing items for sale.

I shopped at the Commissary that day, rather than at the PX. Usually there are 10 long lines, eight waiting to check out and two waiting for baskets. This little ritual is not one of my fun things, and it usually requires a couple of hours of searching and standing in one line or the other. This is one of the hazards associated with housekeeping. Prices are much cheaper than in the US but stocks are depleted and restocked by Vietnamese workers with different items and brands, sometimes several times during one shopping trip. There may be some system for their display arrangements but I find the best way to shop for hard to find items is to watch baskets going by and when you see something you want you ask the guy where he found it. Most of the shoppers are guys because this seems to be a man's world and women of the US variety are few and far between. No Vietnamese shoppers are permitted in these facilities, after all they are waiting outside.

The challenge is ahead when you learn where a hard to get item is located. You rush through the crowd and try and find and grab the item before someone else does. You may or not be successful in your treasure hunt. After you find what you came for or give up in frustration you must stand in a check out line with 25 or 30 people ahead of you. About this time, in this non-air conditioned tin building, I have forgotten about frigid Quebec, only to recall it is even hotter outside on the street. For 20 to 40 piasters you can get one of the bag boys to pack purchases back to your vehicle. The same policemen and kids are standing by for hand outs. You must watch carefully for busy little hands that try to rip off your purchases. Then ahead of you is the return trip to Saigon through the never ending heavy traffic. It is quite a trick to get frozen food and ice cream to your freezer before it is completely thawed. But it can be done and is worth the effort. Doesn't that sound like fun?

Sunday morning, the 28th, I was up early, cleaned the

apartment and baked brownies for the Vietnamese guests I expected that afternoon. Using the brownie mix and following the directions was no problem, however the oven was another matter. The thermostat cutout when the temperature reached the proper level but it failed to come back on when the temperature dropped. I had to manipulate the controls manually. Instead of taking the prescribed 25 minutes, the baking required an hour. This reminded me of some of the stubborn Forest Service stoves we had in the past. This stove had the same cleverly engineered sharp edges, corners and traps to catch and cut your fingers. Surprisingly the brownies were fine in spite of the prolonged baking.

My Tet guests consisted of Mrs. Hai, one of her sons, brother, brother-in-law and Mr. and Mrs. Thuong. All were well-dressed in new Tet clothes. They enjoyed a glass of wine, nuts and brownies. I gave the small boy a glass of grape juice which he promptly spilled on the tile floor. Fortunately the purple stain wiped right up. Mrs. Hai brought me two heavy, neat, little packages wrapped in banana leaf and tied with fibers. Mr. Thuong's new wife was pretty but looked very tired. He told me his mother was getting old and he needed a wife to look after her and take care of the house. It appeared taking care of mother, the house, a new husband and preparing for Tet was taking a toll on the young woman. At the conclusion of our talking, mostly in Vietnamese, I was invited to accompany Mrs. Hai and her family to visit her father.

We took off in the brother-in-law's smooth running 1954 Buick Super. He apologized for having such an old car. He felt better when I told him my car was a 1953 Chevrolet station wagon. He could not believe Americans drove a car that old.

The traffic was very heavy with pre-Tet travelers. Some Vietnamese live in very strange locations. We parked on a main street and walked 100 meters or so down a narrow, muddy lane between homes and a well-lighted US compound of some kind or another. The US facility was surrounded by a 20 foot high wall and encased in coils of concertina barbed wire. Understandably the Vietnamese do not feel safe living near one of these installations.

The 53 year old father of Mrs. Hai was named Hoang Giap. He was surprised to hear a "round eye" speak Vietnamese. He is typical North Vietnamese. His accent is as distinct as someone in our country from Boston. He was dressed in a white silk pajama outfit. His name, Giap, is pronounced "Jap". Vietnamese originate from 100 families and this family also produced the famous North Vietnamese General, Vo Nguyen Giap, who defeated the French and now is trying to do the same to the Americans. The Vietnamese have justifiably been suspicious of foreigners for over 3,000 years.

Mrs. Hai's 83 year old grandfather was dressed in the same clothing as his son. The old man was deaf but could see to read lips and was very cordial in his greeting to me.

Hoang Giap talked about Haiphong and their better days in North Vietnam. These refugees from the north have had a sad life since they came south in 1954 to escape Communism. In the north Mr. Giap and Mrs. Hai's husband were large land owners. They had a good income and valuable possessions. A US Navy ship brought them south with only the clothes on their backs and what they could carry in a small bundle. That was a significant change in their life and lifestyle. The culture shock would be comparable for an upper class Bostonian moving to Rolling Fork, Mississippi. Their greatest sorrow seems to be leaving the family home and the graves of ancestors and friends. They have heard nothing from relatives or friends they left behind in 1954. They speak frankly about our American bombers and their concern for people in the North. All North Vietnamese I have met express this same concern.

Mr. Giap and I carried on a conversation within my limited vocabulary. Mrs. Hai's brother filled in as an interpreter as needed. After a short time Mr. Giap produced at large bottle of Black Label Scotch and some fine Cuban cigars (pretty rich items for a poor family). I have not smoked a Cuban cigar for many years and, not intending to be disloyal to US interests, I thoroughly enjoyed it! Preserved fruits and candied melon appeared from the rear of the house and added to the pleasure of the visit. I did not see Mrs. Giap, since in the traditional Vietnamese family home the wife seldom appears when the husband entertains guests. Some of the young nieces and nephews of Mrs. Hai's edged their way into the living room where we were sitting to watch a Vietnamese TV program.

The large branch of a peach tree stood in a jar in one corner. One blossom was about the burst open. Mr. Giap told me the secret of having flowers open on the first day of Tet was to add sugar water and hot water at the proper time.

At 2100 hours Mr. Giap asked me if I had eaten. I told him no and that I was hungry (I perfected those Vietnamese phrases during my first lesson). With that he announced the family was going out for a pre-Tet dinner and I was invited to go with them. He looked very handsome and about 30 years of age after he had changed into a new white shirt, tie and black trousers. Fortunately I had worn a suit and more fortunately, I did not wear a shoulder holster and carry my .38 revolver. Frequently when I go out to evening meetings I have been advised to carry those accessories. I felt secure since Tet is a family time and both sides in the current conflict agreed to a cease fire.

Mrs. Giap and the grandfather did not accompany us. Mr. Giap, the group that had brought me to the Giap home and I crowded into the Buick. We drove for the next hour looking for a restaurant that would let us in. Most were crowded and out of food. It was obvious to me that the dinner arrangements were made on the spur of the moment and there were no reservations. We finally ended up in a rough section of Cholon back of the PX. We had a six course meal prepared by men cooks and waiters dressed only in black shorts. They just kept bringing food and with the help of chop sticks it disappeared. Long ago I stopped trying to identify the various foods. That evening I operated on the assumption that what I did not know, would not hurt me. I had a problem though when it came to the oysters. My knowledge of Gene Jensen's Public Health Service shellfish program is firmly fixed in my mind. I know oysters thrive in the deadly chemical and bacteriological environment of polluted estuaries. I know that this country has surpassed our level of progress in both water and air pollution. But the oysters had been steamed until the shells opened and at Mr. Giap's insistence I joined him in gobbling down these delicacies. He seemed to be pleased with my ability to keep up with him in eating. But as expected I always quit before a bowl or dish was empty. If you eat everything and leave your plate "as clean as a whistle" it offends your host because they have lost face by not providing enough food for you. Accompanying all this eating were the customary, and expected, "slurps" and "burps". I can assure you I do not follow this custom, because if I retained this habit on my return to the "Land of the Big PX" (as the US is referred to over here), I fear I would be clobbered.

As we left the restaurant, all around us, I could hear the sound of firecrackers and see fireworks exploding in the sky. These displays are a distinct part of Tet and have been increasing all week as Tet approaches. I read in the newspaper 30 children were arrested and fined for shooting off fire crackers before the officially designated starting time of midnight Monday night. I noticed there were more offenders than policemen in evidence that night. The purpose of lighting the fireworks at the beginning of the New Year is to frighten off bad spirits. There must have been many evil spirits abroad as we drove out of Cholon that night.

During the first three days of Tet firecrackers are set off on several occasions. They are used to signal such events as the first visitor of the New Year, important visitors and the first trip taken by the family. That explains the series of firecrackers I heard when I was at the Giap home. Each home has a supply of these extremely noisy items. The wealthier families shoot off the most fireworks. I noticed fireworks stands appeared all over the city in the 10 days preceding Tet. Firecrackers are packaged in cans, boxes or fastened together in long chains. There is very little shooting of firecrackers one at a time like I used to do.

They touch off whole strands at a time. Interspersed in these four to five foot strands of 2-inch firecrackers are giant six-inchers that sound about like a hand grenade exploding. If you don't think that doesn't add interest in a tense city like Saigon, then you have another thought coming. The children enjoy throwing them under the feet of older people, particularly Americans and MP guards sitting in concrete, sandbagged bunkers. Reactions of spectators are varied and when I heard one of these series of explosions it did not bother me. I associated the noise with firecrackers and not something associated with war.

Frankly I was glad to be dropped off safely back at the apartment that Sunday evening just before the 2300 hours curfew (for Americans). The midnight curfew for the Vietnamese was waived by government decree. As I entered the apartment I noticed the Vietnamese were streaming by my place in every means of conveyance. Overloaded buses with people hanging on the sides, rear and top continued to roll in from the outlying areas. In my judgment, I thought this relaxation of regulations by the Vietnamese government was a tactical mistake by the GVN and their American advisors. But after all, the Tet season promises every opportunity and hope for a better life.

The US Mission designated Monday to be on a Sunday Schedule, meaning for us to take the day off. Vietnamese workers had been given Monday through Thursday off. I spent most of the day working in the apartment, reading, writing a few letters and paying bills. In the afternoon I decided to go out on a photo excursion and capture some of the local scenes associated with Tet. As I took my dark glasses out of the case a crack in the frame allowed one lens to fall and shatter on the tile floor. That gave me another task to take care of. After taking photos of the flower buyers and huge piles of garbage, I drove to the Tan Son Nhut Airbase. En route I stopped to take photos of the so-called River Styx which is one of the many tributaries of the Saigon River reaching into the City. People have constructed shacks resting on mangrove poles out over the water. The condition of the water at low tide and at the hour of the incoming tide resembles and smells like the inflow to a sewage plant. I had a horrible thought, this is probably where those oysters come from!

The reason for going to Tan Son Nhut was to take my glasses to the only optical department. According to the American Forces Radio it was the only PX open. Security at the air base was normal and my rusty Jeep and I were waved into the parking area. I could see most of the helicopters quietly perched on their pads at the heliport. Everything was quiet, especially the PX. It was closed! A printed sign on the door announced "This facility will be closed in celebration of Tet on 30 January 1968". A hand printed sign provided the truth when it stated the PX was closed because the Vietnamese employees did not show up for work. Disgruntled GIs added other hand printed comments indicating in our letter words the Vietnamese workers should be fired or otherwise accommodated. This meant putting off repairs until "ngay mai" or tomorrow. GIs must have been particularly angry since they were confined to quarters from 1000 on Sunday to 0400 hours on Friday. This was done to keep them away from the celebrating Vietnamese and to keep them from utilizing public transportation. When I returned to the apartment I found Barry had come and gone. He left some New Year cards for me to deliver. His note said he had walked to my place because taxi rates were sky high.

On Monday evening, Tet-eve, so to speak, I had an invitation to have a meal with Mrs. Hai's family. I knew it would be a special meal because she had been working on it since Saturday. Tet is the only time of the year everything should be taken in abundance. There should be much to eat and drink. Many Vietnamese people speak of "eating Tet" rather than "enjoying Tet". I also had an invitation to go to Mr. Thuong's home on Wednesday evening and to his wife's family home in Cholon on Thursday. I looked forward to these meals with mixed emotions, but I did not look forward to the trip to the various homes after dark. Although the military had been confined to quarters, we civilians were permitted to be out until our normal curfew hour of 2300 hours. I could not understand why the civilians did not have the same restrictions as the military. But then I figured it out. The Embassy and department heads did not want a restriction limiting their attendance to the many social affairs.

I was due at Mrs. Hai's home at 1900 hours. I had to remember her husband's name was Lu Van Ry and at home she was Mrs. Ry. While this is confusing I knew I should not refer to him as Mr. Hai, that would have been insulting. Her home was in the vicinity of Tan Son Nhut and I parked the Jeep on a busy well lighted street and started looking for 466/7. This is a French-designed system in which the first series of numbers in an address is the address on the main street and the last is the number of the house which in this case was down a dark narrow lane or path. I found two 466's and selected the second one. I wove in and around chicken pens, shacks and Hondas to a row of houses. In the doorway of one house leaned a "stoned" GI dressed in black VC pajama trousers. I could see his GI shirt with Staff Sergeant stripes hanging on the wall. Behind him I could see an equally relaxed Vietnamese woman lying on a sagging bed. The GI starred into space and was totally unconcerned about being confined to quarters and could care less about where I could find 466/7. Wandering around in this dark alley was a hair-raising experience. I went to another house and was told by the lady of the house that 466/7 was nearby and just down the street. I had learned to carry on that kind of conversation in my second language lesson. I wove my way out to the main street and found another narrow alleyway, narrower than the previous one and tried again. This time a young girl guided me back out to the main street and to a third 466. Those French administrators certainly fouled up this city in more ways than one. The numbering system has absolutely no logic and makes no sense!

My third attempt was successful and I found Mrs. Hai's No. 2 son, the one who came to my place, waiting for me in the street. He ran ahead and announced my arrival. Mr. Ry met me at the door, introduced himself and welcomed me. I presented my gifts to the adults and gave the small red envelopes to the children. I wished him all the good things I could think of and could express. He did not speak a word of English, but spoke excellent French. I recommend that anyone coming to Vietnam to work should study French. A knowledge of French would help elsewhere in the world and be an asset in the Foreign Service. But I find there is no substitute for being able to speak Vietnamese, it opens many doors and the hearts and minds of the people. It enables me to better understand the people and their customs. Mr. Ry and I were able to carry on a conversation in Vietnamese. We talked about his life and wealth he left behind in North Vietnam. He was a good personal friend of General Giap before they split on idealogical and political beliefs.

The Ry home was very plain. The room we sat in was long and narrow. I did not go beyond this room. The room contained a plastic and wood couch and a solid wooden bed. These thin people sleep on a woven mat laid on top of the bed, which is more practical and cooler in the tropics. It would be comparable to sleeping on the floor. You can imagine what mattresses could pick up in the way of bug, vermin and mold. That was well illustrated back where I saw the "stoned" GI. Mr. Ry and I sat on two wooden chairs with a small table between us. I remembered not to cross my legs and not to point my toes at him, which is considered to be an insult. We talked and drank small glasses of Black Label Scotch. I was glad my gift for him had been a bottle of the same brand of Scotch.

The most interesting item in the living room was the family altar. It was 4 1/2 feet high, two feet deep and five feet wide. The top held the small family chest, brass candlesticks, incense burners of polished brass, a brass cup holding joss sticks and a small brass box about the size of a match box. The small priceless items were carried with them when they fled to South Vietnam. There was also a huge brass bowl on a tall pedestal holding a magnificent display of fresh apples, grapes, oranges, bananas and other fruits that I did not recognize. This fruit display would have cost the equivalent of US$25. On the altar beside the fruit bowl were a number of cellophane wrapped boxes. The fruit and boxes were gifts for the ancestors. The eldest son is responsible for these arrangements. Around the wall were photographs of Mr. Ry's grandfather, father and mother. They had been farm people in the North and were dressed in traditional North Vietnamese clothes and wore the black head band. I understand clothing of this type was worn by the ancient Chinese. Many of the northerners, men and women, who now live in Saigon, wear black head bands.

Mr. Ry invited me to eat with him and I accepted. Please recall the original invitation was made by his wife and his invitation formalized the arrangements. A servant girl brought in a huge tray loaded with about 20 dishes of different foods. One held a greenish-looking rice dish made into a block and contained some unidentified substance in the center. Another looked like head sausage. Some of the other dishes looked familiar but others were completely strange to me. For the next 90 minutes we ate and drank. It was only when I ran out of nuoc mam (fish sauce containing terribly red hot peppers) that Mrs. Hai appeared. She smiled and bowed to me. She was dressed in a obviously new white blouse and black silk trousers. This appears to be the outfit worn by the servants and by the wife at a husband's dinner party. At work Mrs. Hai wears a different beautiful ao dais every day. She did not remain in the room with us until after we had eaten and had our strong green tea. Then she and the children came in to be introduced. The oldest boy was 11 and very handsome. His father remarked that he looked like Japanese. The boy beamed at me. He is the one who had the serious asthma attack and was provided welcome relief by my hay fever pills. He looked like he was my friend for life. The supply of pills I ordered, through you and my Forest Service Secretary Anne Warner, arrived while I was on home leave and Barry gave them to Mrs. Hai. Those should take care of him for a long time.

We talked some more about the olden days in the North and how poor the family is now. He works for the Saigon Electrical Company in a clerical capacity. Mrs. Hai's (Mrs. Ry's) annual USAID salary is the equivalent of less than one pay period of my salary. Inflation has hit the Vietnamese people and this element in itself is responsible for much of the graft and corruption.

At 2200 hours I excused myself for I knew they had preparations to make for Tet which was only two hours away. I noticed the boys looking at the short string of firecrackers and knew they were looking forward to that occasion. I wanted to get out of that dark neighborhood fast. The two oldest boys walked down the lane with me. At one street the youngest one turned back but the oldest one accompanied me to the main street. He told me good night in perfect English and asked me to call on them again. That must have taken much coaching and practice. He turned back and I crossed the street to the Jeep. As I was unlocking the chain some children threw a five-inch firecracker under the Jeep. Outwardly I ignored it and drove off. They were chattering in a disappointed way because I had not jumped out of my skin. My ears are still ringing. The trip back to the apartment was uneventful in light traffic. Most of the people must have been at home preparing for the New Year.

Just before midnight I went up on the roof of the apartment and joined other men who were there. The sound of firecrackers built up gradually to the most awful roar of explosions you ever heard. The noise from the direction of Cholon was just unbelievable. An hour later, when you would have thought all the fireworks in the world had been fired off, the din continued unabated. I was told some rich Chinese merchants in Cholon hang long strings of firecrackers from tall buildings. Some strings are said to be as long as a five story building. Across the street from the apartment we watched a man rig a 20 foot string of firecrackers. At each foot he tied in an eight-inch one. When that was touched off it made a tremendous racket. All over the city we could see rockets and Roman candles being shot into the sky. Some enterprising characters used rockets to carry long strings of firecrackers high into the sky where they sounded like machine gun fire. I noticed quite a number of tracer bullets being fired into the sky. These are easily distinguished from conventional fireworks. We could hear large explosions all over the city. The equivalent of US$1 million goes up in smoke over Tet. What a terrible waste of money! The resulting fires and injuries number in the thousands. It makes me shudder to see children run in and around the exploding firecrackers. In some way this must be associated with the many one-eyed and blind Vietnamese I see around the city.

In my judgment the Viet Cong seriously miscalculated the timing of their Tet attack on Saigon. I shall cover that later, in a very long letter. If they had selected the 90 minutes after midnight (when all the firecrackers in the world seemed to be exploding) for their attack, they could have easily taken Saigon and no one would have noticed it or known the difference.

USAID had told us tomorrow would be on a Saturday Schedule. This was to be another half day. I worked hard my last day at work to complete a special Top Secret report on defoliation. It was due in the Embassy at 1800 hours on Monday but I missed the deadline due to a slow American typist. Only American secretaries can type classified documents. I called my contact in the Embassy and was told it would be satisfactory to deliver the report to him at 0800 hours at the Embassy just before our meeting on the same subject.

After watching the fireworks displays for a short time I returned to my apartment to listen to the late news report. I learned the war was building up around the country. As I continued listening it sounded like almost every province capital, places such as Nha Trang, Ban Me Thout, Pleiku, Kontum and Danang, had been hit and overrun in the past 24 hours. I decided this would be a good time to clean my carbine, revolver, pistol and riot shotgun. I was glad to have on hand a thousand or so rounds of ammunition for the various weapons. With that job done I went to bed a little after 0100 hours, on Tuesday, 30 January 1968, not knowing this was to be the first day of the siege of Saigon.

At 0300 hours I was jolted awake by a terrific explosion that was followed by automatic weapons fire. It sounded like it was coming from a few blocks to the east, in the vicinity of the American Embassy. The firing kept up continuously and at about 0500 hours the sky above me and to the east was filled with helicopter gunships. The siege of Saigon was under way and it had started on Tet, the Lunar New Year. The declared truce had been meaningless.

I am certain your news reports have provided much more comprehensive information than we have been given over the American Forces Radio and TV, but you may be interested in my observations from this little corner of the war at 192 Cong Ly Street.

At 0600 hours Embassy employees were warned by radio to "stay in your billets". About that time several helicopters were swooping over our apartment and descending to land on the roof of the Embassy, unloading troops, one after another. I could hear automatic weapons firing continuously from that location. Several hours later we learned about the attack on the American Embassy. I decided it would be inappropriate for me to go to my office, pick up my Top Secret report and deliver it to the Embassy by 0800 hours.

At 0630 hours USAID and other civilian employees, contractors and Tan Son Nhut military personnel were warned to stay in their billets. At 0930 hours the Embassy was declared clear. General Westmoreland came on the air with a report concerning the VC attack on Saigon. He called the enemy "very deceitful and they had taken advantage of Tet". A few hours later Embassy workers were ordered back to work. I am certain this was done for propaganda purposes to show we had not been bothered. At that time we were being attacked all over the city. Machine gun fire could be heard in every direction from our apartment, some of it less than two blocks away.

We were locked in our apartment grounds and protected by a single, unarmed USAID guard. It was not very comforting. However across the street in the Army intelligence headquarters

Nung guards were squinting across their machine guns with their fingers on the triggers. At the old MACV headquarters, about 100 yards away I could see five Green Berets sitting behind machine guns or holding weapons equipped with sniper scopes.

Mr. Passey, my USDA friend from FHA, had not turned on his radio that morning. He proceeded to work as usual at 0715 hours. As he walked past the Palace he saw a group of ARVN and US MPs herding along, not too gently, two nearly naked Vietnamese who were blindfolded and had their hands tied behind their backs. The American MPs who have supervision over us said nothing to Passey. However one of the ARVN soldiers told him to "Di Di". Although Passey does not understand a word of Vietnamese he got the meaning and returned to the apartment as fast as he could. "Di Di" is an order to "Go" and is frequently used to get rid of someone.

Fighting continued around us all day. Automatic weapons and bazooka fire poured into the Palace area. We could see bodies lying in the street between us and the Palace which is just two blocks away. American and ARVN forces were trying to blast the VC out of an apartment building where they had taken refuge after trying to storm the heavily guarded Palace. We could see down the street, but trees blocked our view of the Palace and the apartment building. We could see dive bombers operating in the vicinity of Tan Son Nhut Airport and MACV'S Pentagon West Headquarters. We learned later both Westmoreland's and the GVN's staff offices were under attack.

A friend of mine was due to depart today for R&R in Singapore. He and I went through training together in USAID/Washington. He has had no leave since arriving last March. It was a sad blow when he learned all civilian and military flights in and out of Tan Son Nhut had been canceled. To emphasize that point we could see huge columns of smoke coming from the airport, Bien Hoa Airbase, GVN Naval Base, Race Track Training area, ARVN training center, Cholon and other parts of the city. We had ringside seats and watched jets, propeller dive bombers and the new Cobra helicopters dive and fire rockets near the MACV Headquarters. Cobras dove in with their turbine engines going full power and fired rockets. This is the first time I have seen them in action. They look lean and mean. I noticed as I walked around the roof that there were several bullet holes in the walls of our apartment.

Truck loads of ARVN and US troops rolled past our building. Ambulances came roaring back periodically, going the wrongway on our one-way street. I could see one ambulance had a stream of blood pouring out the rear door. EOD (bomb disposal squads) Jeeps roared through the street with their sirens going full blast. Cars full of US military personnel roared up and down the one-way street. Our part of the world had gone crazy! With rifles sticking out of the doors and windows of cars and Jeeps I couldn't help think it looked like the Kaibab National Forest on opening day of deer season.

We kept a radio on all times. About 0815 hours our power went off. We ran out of diesel fuel for our generator and it was doubtful any could be delivered until Tet or the siege was over. I had two chickens defrosting in the freezing compartment. I fried them and fed two friends and myself. I helped one friend eat a gallon of vanilla ice cream that was melting. Beer was warming but that was not too serious. War may be hell, but at 192 Cong Ly Street our bellies were full!

At 1700 hours six of us on the roof viewed a sky filled with helicopters. I counted 75 in the air at one time. Jets were wheeling in and out of the helicopter traffic as SVN propeller dive bombers and Cobra rocket ships crisscrossed the sky. I noticed one helicopter coming in low just above the trees near us. About one block away be stopped and hovered over a Navy Headquarters building. He started down, got behind the trees and then I saw his tail rotor fly off into space. He lifted only momentarily but with no tail control the helicopter started rotating wildly and fell off to one side and crashed. Two us ran downstairs and alerted MPs across the street and they took off in the direction of the crash. There was no fire so perhaps the men got out safely. We could not tell how many people were inside the helicopter.

We used a battery operated radio but details were sketchy. We were "reassured" when we heard President LBJ had "considered the siege of Saigon a pretty serious incident". It sounded that way to us, especially with all the automatic weapons and sniper fire, mortars and bombing going on all around us.

We received reports five US marines and paratroopers of the 101st Airborne had been killed at the US Embassy and four wounded. Nineteen VC were killed in that attack. We received word we were confined to quarters until further notice. Later we were told over the radio USAID would work on Thursday; again the notice was canceled. The President of the Republic of Vietnam declared martial law. We were kept off the street but for some unexplained reason the crazy Vietnamese kept racing up and down our street on their Hondas and in cars. Of course some of these must have been VC messengers and spies. I will never understand why our troops did not clear the streets of all, except military traffic. We noticed one accomplishment. The air was relatively clear of exhaust smoke.

VC hit the Hotel Splendid BOQ where I ate and watched movies under the dripping laundry. That is four blocks from here. Action at the Palace went on all day and into the night. Still we received no word as to what was going on in that battle. We speculated that the VC were inside, then we heard they were on the outside trying to storm the place. Perhaps you know more about what is going on than we do.

We heard the curfew for the military had been extended until further notice. That GI back in the alley at the 466 address must be enjoying his confinement. All Tan Son Nhut GIs living off the base were told to remain in their quarters until further notice.

By nightfall all types of firing continued in and around the city, some of it within a few blocks of us. Flares were dropped continuously over Tan Son Nhut, our area and all around the city. It was as light as day. Helicopters whirled overhead with only small warning lights blinking from the top. There are no lights on the bottom of helicopters because that makes them a better target from the ground. Dive bombers and Cobras continued to work over an area near Tan Son Nhut. Puff, the Magic Dragon ships were all over the sky. We saw one unload not too far away. A Gatling gun firing at 6,000 rounds per minute produced a solid stream of tracers, a veritable waterfall of fire poured into the ground.

I did not stay on the roof too long at one time and I kept moving. I learned a long time ago that a moving target is harder to hit. Some of the heavy drinkers sat under a light on the roof and others and their girl friends barbecued steaks. Fuel oil was delivered shortly after dark and that made us feel like we were not forgotten. It helped give us a false sense of security. I left the roof at about 2300 hours. About midnight I turned in and before I went to sleep I heard the crack of a carbine from a couple of houses away. This was followed by a high velocity heavy rifle shot and a splat against the wall of our building, just above me. Artillery fire took over when other firing tapered off. That was all the action until I heard three whumps nearby. I have fired enough heavy and light mortars since I have been here to recognize the sound of outgoing mortar fire. I can assure you incoming sounds entirely different. A short time later I heard mortar shells exploding some distance away. A VC mortar crew is operating in our vicinity.

(continued)   1 February 1968

The battle for Saigon has been intense today (Thursday). We hear 639 VC have been killed and 27 detained. Some 197 ARVN troops were killed and 212 wounded. We have lost 46 Americans killed and 93 wounded. We have been told little or nothing about what is going on in other sections of the country, except that over 5,000 of the enemy have been killed and almost 300 US. This is a peculiar war where success is measured by body count ratio. We can tell by the number of cargo, paratroop and fighter airplanes heading west that there must be considerable action in the Delta IV Corps area.

Two separate, large columns of armored personnel carriers (APC), tanks and self-propelled 155mm artillery tanks have rolled by the apartment in the past hour. These troops were heading for the Palace, where automatic and heavy weapon firing has been heard off and on all day.

The shot which hit above my apartment last night was fired by a nearby GI. He heard the carbine shot which hit near him. He looked around and saw someone on our roof and snapped off a shot. Fortunately it just missed a friend of mine. Today that friend is keeping well-back from the edge. The wall around the roof area appears to be solid concrete but it is just plaster over hollow tile. The bullet made a neat little hole where it entered and came out producing an 8-inch hole. The bullet then apparently traveled across the roof, hit the opposite wall and produced the same results. Obviously the walls offer little protection. Today everyone on the roof keeps moving and away from the wall.

This afternoon two SVN propeller dive bombers made several low level passes over our roof. They looked large and quite imposing with their fully loaded bomb and rocket racks. They looked us over and apparently judged us to be friendly folks. We smiled and waved. Helicopters patrol the area continuously, apparently looking for snipers. All day while I have been in my apartment typing this letter I have heard sniper fire. That is probably the reason we are confined to quarters until the Provost Marshal issues an all clear signal.

I can not forget the GIs who were told stay in their off-base quarters and wait further orders. These were support personnel, not combat troops. Who knows the way things are going in this screwed up world we may be forgotten? But food will not hold out indefinitely. We already know of food shortages in our building. Speaking of food we hear over the radio there is heavy fighting in and around the Cholon PX, Commissary and nearby Buddhist Pagoda. VC are reported to have established headquarters in the Pagoda and are flying VC and NVN flags. There are said to be 100 enemy soldiers at that site. After hearing those reports I went up on the roof and witnessed the beginning of repeated bomb runs by the SVN Air Force over the Cholon area. We hope they miss our PX and Commissary. While we can not identify any specific targets being hit at that distance, we have a good view of the general area.

I just took a meal break and fixed my favorite meal, a Kraft Dinner. My food supply is still adequate. I am certain you would not agree that rice, kipper snacks, Kraft Dinner, grits and WW II "C" Rations are adequate. While my other food items are in short supply, I do not expect to starve out soon. After stuffing myself with my delicious meal, I contacted one of the Public Safety Officers who lives in our apartment and suggested we should provide our own security. He agreed and encouraged us to proceed. We called everyone in the building together to take an inventory of weapons and ammunition. In the building we have seven carbines, several pistols and revolvers, two sub-machine guns, one .375 rifle with scope (a highly illegal weapon in Vietnam!), a case of hand grenades and three satchel charges. The latter are enough to level this building. Taking inventory I had a .30 caliber semi-automatic M-1 carbine with 800 rounds of ammunition, a .45 semi-automatic pistol with 400 rounds, a .38 revolver with 600 rounds and several boxes of 12 gauge 00 shells for my shotgun. Other than a CIA agent, who did not participate in the inventory, I seem to have the most to add to our weapons cache. I was voted in as chief of the guard detail. This required preparing a list of instructions and a duty schedule.

Some of the volunteer guards were ineffective and we weeded out those who did not know how to put a clip in a carbine or how to operate the safety. I was more concerned about the danger of trigger happy guards than the VC. Along with three other men, I took a 0200 to 0400 hours shift. We heard several rifle shots during our shift. This activity was within a block of us, but we could see nothing in the dimly lighted street. Within a radius of a few blocks there were several heavy fire fights and a number of loud explosions. Two loads of men in police Jeeps went by about 0330 hours. Along with the guards across the street we kept our guns trained on the Jeeps, but they kept moving. The police have mostly changed to civilian clothes since the fighting began, because so many of them in uniform were killed in the first few hours. We have been notified the VC have captured several police Jeeps and are reported to be driving around the Palace area firing at random targets.

(continued)   2 February 1968

I turned in at 0630 hours for an hour of rest. The radio provided sketchy information. We were instructed to remain in our quarters. To date no one from USAID or the Embassy has come near us and we have received no instructions on what we should be doing at 192 Cong Ly. We take care of ourselves. While the fighting continues nearby we can not help but reflect on the unknown fate of our fellow USAID workers who live in some outlying areas that were extremely dangerous even before this Tet offensive. I also think about the invitation I had to spend Tet in Hue with the forester and his family. We have no reports on what is happening in Hue or the I Corps area but we suspect it is not good.

Three of our residents volunteered to ride shotgun on a fuel truck to the fuel depot to obtain diesel for our generator. They followed the truck and while waiting their turn to load they picked up some items at Meycord MACV BOQ PX. I used that

PX and snack bar when I lived at the Park Hotel. Then they waited until sniper fire subsided before they could get the truck to the depot. They told us US soldiers were lying in the street and firing at snipers who were shooting at them from a row of houses along one of the waterways. It was reported that gasoline tankers and diesel trucks were being fired on and set afire. Our group was fortunate to load fuel and return safely. We have started the generator and now have power in our apartments.

Sniper fire continued all day. There appeared to be three trouble spots within two blocks of us. Green Berets on top of the MACV Communications Center next door searched continuously with glasses and sniper scopes. Some 101st Airborne snipers came to our building in the afternoon and holed up in one of the apartments where they scoped the neighborhood for snipers. A sharp fire fight went on one block from our building. From cover behind our wall we could hear the sound of automatic weapons and the crash of concussion hand grenades.

Meanwhile, with martial law declared, Vietnamese idiots (or VC?) on their scooters and in cars continued to travel both ways on our one-way street. Again I can not understand why action was not taken to permit only military traffic. One of the Public Safety Officers who is a tenant in our building leaves each day and works with his precinct policemen. He provided us with graphic firsthand information on the battles taking place in Cholon. His eyewitness reports of fighting in the streets and roof tops were very vivid. Many of the SVN policemen he had trained and was working with have been killed in the last few days. Before Tet he attempted to obtain some M-60 machine guns and sub-machine guns for his men. He finally succeeded in making arrangements through the US Army at Long Binh (12km away) to obtain the necessary guns and ammunition, but the SVN police captain refused to send any of his men with the American advisors to obtain these items until after the Tet celebration. Many of those policemen will not be alive to participate in another Tet.

Today an attempt was made to pick up the helicopter that crashed nearby. A large double rotor Chinook helicopter hoisted the badly damaged craft above the trees, then dropped it to the ground and flew off on another mission. Several hours later we saw a semitrailer drive by hauling the remains of the twice-crashed helicopter. At frequent intervals during the day we saw Chinooks and huge Sikorsky flying cranes carrying damaged helicopters back to Tan Son Nhut.

Considerable aerial bombing took place all day in and around the city. It appears we are destroying the city to save it. Try to rationalize that policy if you can, we can't. Most of the selective bombing within the city is done by the Vietnamese Air Force in their propeller driven Sky Raiders. Bombing along the Saigon

River and Tan Son Nhut Airport is done by US jets. Huge columns of smoke could be seen raising to 25,000 feet from throughout the city. Helicopters continue to whirl overhead searching for snipers. FACs (Forward Air Controllers) fly low and slow over the city observing, directing and acting as a decoy to attract sniper fire. We saw tracers come up at one in the vicinity of the cemetery which is four blocks away. Later we learned several VC had been holed up there and came out at intervals to shoot at anything that moved.

Our friends across the street at the CID (Army Criminal Investigation Division) maintain tight security at all hours. I guess it is comforting to have them nearby. I witnessed two Jeep loads of heavily armed soldiers escort a civilian who was wearing a flak jacket (probably a high level CID or CIA person) into this building. The soldiers manned their machine guns and automatic rifles and appeared to mean business. However the women in the nearby RMK/MM (civilian engineering firms) apartment building momentarily diverted their attention. They were dressed in skimpy bikinis and waved at the soldiers, who had quite a time keeping their minds on their mission, whatever that may have been.

I took photos of the girls and the soldiers who were in the Jeeps. While I was talking to the soldiers about 10 Vietnamese men in civilian clothes came walking down the street carrying small bags, which could well have been satchel charges. The Vietnamese smiled and gave the soldiers friendly glances as they walked by. They appeared to be very pale, nervous and shook in their boots. But after all who wouldn't be nervous with three heavy machine guns, 14 automatic rifles and two grenade launchers trained on their backs and following them as they disappeared down the street.

As I mentioned, the traffic in our area seemed to have gone crazy and was speeding both ways on the oneway street and crossing the next intersection at high speeds. Within our narrow range of visibility we witnessed a number of accidents at the junction of Cong Ly and Phan Dinh Phung Streets. Two US Navy vehicles tried to occupy the same space at the same time and the occupants were carted off in an Army Jeep ambulance. The passage of ambulances never ceases, day or night.

We had interesting traffic. After all there was little else for us to see in our little part of the Tet battlefield. US Ambassador Bunker went by with his entourage of guards. SVN President Thieu and his protective cover went roaring by. Tanks and APCs traveled in both directions. Some of these vehicles appeared to be shot up quite badly, but the GIs aboard waved at us and gave the thumbs up signal. They looked tired and dirty from their extended duty.

At many times on this Friday, the sun was blotted out by

huge columns of smoke. The air was filled with all types of aircraft and we shuddered as we witnessed a number of near misses in the crowded skies overhead. We saw rockets being fired into the Airbase area and near the river bridges. Air Vietnam had just purchased a new 727 jet and attempted to land at Tan Son Nhut. Apparently it was denied landing privileges and waved off because we saw it bank sharply and head in the direction of Bangkok. The airport remained closed to all civilian and troop carrier military flights. On this day it appeared we were winning very few battles. There were reports the VC had captured and were holding out in the Cholon Childrens' Hospital.

About 1600 hours I came down to check with our guard. At that time an American woman, a Vietnamese servant woman and three young children came to our gate. They pleaded to be let in. I had our man on guard duty check ID cards and let them through the gate. A friend of mine came in at that time and we learned heavy fighting had run these people out of their home, which was about one half mile from us. They said VC dressed in black ran through their streets every night and had made their latest foray at 1300 hours that afternoon. The American woman said there were young girls carrying rifles and firing right along with the men. She said there were thousands and after one attack they changed clothes and mixed with the local people. The fighting in their neighborhood had been heavy that afternoon and many people were killed and wounded. She said the VC fired from rooftops, alleys, homes and stores at the ARVN, police and US Forces. Many helpless citizens were caught in the heavy crossfire. During a lull in the fighting this American woman, Antoinette D. Jordan, had taken off with her servant and three children. The girls were children of the Vietnamese woman who owned the house Jordan rented. She said they escaped with thousands of others from that Saigon battlefield. They traveled by foot as far as the Free World Assistance Headquarters (I passed this many times on the way to the pistol range and the PX.) Some Filipino soldiers had driven them to a nearby apartment where Jordan had friends and fellow workers. Mrs. Jordan, a grandmother, works for OICCN (Officer in Charge of Construction, Navy). She contacted several of her friends and asked them to take her and the Vietnamese in, but they all refused, on the basis they had no space to spare and were out of water. We knew they were out of water because people from that apartment building have been packing water in buckets from our supply.

These refugees had been told there were empty apartments in our building. There were none. My neighbor and I agreed to take them in for the emergency. But when we looked at his place it was a veritable pig pen and so cluttered that there was no room for any of them. It is unbelievable how a single man living in a two bedroom apartment, with two baths, a large living and dining room and kitchen could have so completely occupied and cluttered it. But he did. I took the whole outfit, the American lady, servant and three children and a tiny Chihuahua dog the girls could not bear to leave behind. The girls are 13, 15 and 16. I relinquished my bed to Mrs. Jordan. With my two large ottomans, and six from neighbors along with borrowed sheets we arranged a place for all of them. My couch is long and comfortable and suitable for me. They had nothing to eat for two days so I promoted dinner for them at a neighbor's apartment. I donated the Vietnamese food Mrs. Hai had previously given me and the Vietnamese seemed to enjoy this the best.

My guests were badly shaken by their experience. It required aspirin all around to get them settled. The little dog must have been shook up also because it shivered and jumped every time it heard rifle shots and explosions. When it settled down it was quite friendly. I learned the Vietnamese do not bother to house break their dogs . . . such are the fortunes of war!

The lack of having a trained generator operator resulted in a surge of power that blew TV tubes and water pump fuses. The water pump is required to pump water from a ground level reservoir to a roof top reservoir. Replacing the fuses put the pump back in order but TVs were out of commission. TV programs were all Vietnamese or very old US rejects, which I could do without.

I was on guard duty from 0200 to 0600 hours. My neighbor living in the large mansion behind the wall and row of coconut trees was General Edward G. Lansdale. He had been with the OSS in WW II and in and out of Vietnam since 1954 in various roles, such as over turning the government, running a covert dirty tricks program, vague advisory jobs, currently said to head a special counterinsurgency team and who knows what else. I have never seen him since he comes and goes at night in a limousine with black windows. I have seen some fierce looking Philippine guards around his house. They never smile. I learned earlier in the day from one of the Public safety men that Lansdale would have visitors and to be careful about any shooting in that direction.

As I walked along the wall that night I heard a noise just over Lansdale's fence. In the light of one of the many flares being dropped from planes overhead I first saw a black finger, then a black hand grasped a wire on top of the fence and this was followed by white eyes and the black face of a grinning Negro. I asked "who are you?" He replied, "we are the Screaming Eagles and I have a company of men behind me who have their weapons trained on you". I assured him I was friendly. I learned he was from the Washington, DC area also. I looked up at one of Lansdale's upstairs windows and in the black opening I could see the darker hole of the business-end of a heavy machine gun. I squatted down at my post, near the wall and watched the street, content to know we had good support from the 101st Airborne. The night was peaceful in our block. There was much action from Puff the Magic Dragon off in the distance. Mortar fire continued for most of the night and heavy artillery fire could be heard off towards the river and Bien Hoa. Flares provided good light and as they drifted down a moving mosaic was created by the shadows of the trees. It was really quite pretty.

I went back to the couch at 0630 hours. My refugees were tossing and turning. One of the children was mumbling Vietnamese in her sleep. The little dog had forgotten who I was and created quite a fuss when I came in but no one was awake.

(continued)   3 February 1968
Today is Saturday and we remain confined to quarters. The news you have been receiving must have filled you in about the battles all over Vietnam. The deaths of MPs, our military forces and the enemy have been very high. Our resident police advisor tells us bodies have been stacking up at the 17th Field Hospital. That is where we go for shots and treatment. Civilian and military hospitals are overflowing with wounded. Civilian casualties have been heavy in Saigon and all over the country. Today military airlifts are starting to evacuate wounded to the Philippines and Japan. American dead are reported to be piling up at the Tan Son Nhut mortuary. But in the heat of battle many reports may be exaggerated. But I can assure you a serious war has been taking place, all around us.

Last night three VC were killed only two blocks away from us. One of the drivers that picked up our Public Safety Advisors told us about it. He said the three bodies are still lying there and there are bodies in the streets and on the sidewalks all over the city. One unconfirmed report stated that the initial VC attacks in the Tan Son Nhut area hit a BOQ and wiped out 67 officers. Shortages are beginning to be reported. The Public Safety Advisors tell us ammunition supplies are running low for his men. As the Vietnamese depend on day-to-day purchases of food at the markets, with the exception of rice, they will be in serious condition if the siege continues. Most produce, meat and fish come in daily from outside the city and of course all commerce has been disrupted.

At 0900 hours the big sister of the girls and the landlady of my American guest arrived. She brought food and expressed her gratitude for taking in her family. Her name is Kim and normally she is employed as a hostess at Maxim's French Restaurant. She said the fighting in her area had broken all windows in her home and she must return to protect the place

371

from looting. Later she sent a Filipino employee of OICCN to give us a large bag of rice, apples and a case of chocolate milk.

If these battles continue I will run out of typing paper. We have just run out of water. The VC is reported to have destroyed a section of the main city water line. It may turn out to be only a blown fuse. People in our building are frantically running up and down the halls in search of water. I heard one poor creature moan "how will I wash my hair?" What most people do not recognize is that we each have a good supply of water in our hot water tanks. Besides that I have a case of beer and four imperial quarts of whiskey. At least we can go down fighting! It is 1430 and all is quiet. My refugees are spread out, all over the apartment taking naps. I am going to take a beer and a book up on the roof and catch some rays.

Later:

We received confirmation the VC did indeed blow up the main water line. We will be without water for several hours.

Another report was just brought in by one of our resident Public Safety Advisors . . . the VC have captured several American Jeeps which are armed with M-60 machine guns. In one instance it was reported that two Americans were sitting in the front seats and two VC dressed in ARVN uniforms were in the back seat. This has resulted in the military patrols being very cautious. One such Jeep just stopped across the street and I saw the guards jump behind trees and hold their rifles ready when the American Jeep approached. This incident has everyone very jumpy.

I came down from the roof to find Chi Ba (servant lady) preparing rice. I gave her a large can of chicken chow mein to add to our supper.

Our resident Public Safety Advisors are our best sources of information. They come and go at all hours. We learned more about the fighting at the Palace, just two blocks away. The VC holed up in an unfinished apartment building next to the Park Hotel, where I had lived for four months. From this location the VC could look direct into the police and military barracks on the Palace grounds. From this strategic location the VC fired rockets and directed heavy machine gun fire into the Palace grounds for 36 hours. If Mr. Passey had continued walking to work that first day of the fighting he would have walked right into the middle of the action. The VC had captured a heavy US .50 caliber machine gun and used it to riddle several Jeep loads of US MPs and ARVN reinforcements. Our Public Safety man said in our precinct alone eight policemen were killed and 26 wounded. The VC kept firing continuously and the police and military stayed low figuring they would burn out the barrels of their machine guns. But the VC kept this up for 36 hours before eight VC were killed and 12 captured. In this partly finished building they found the VC had packed in thousands of rounds of ammunition, including rockets and grenades. In addition to having one of our machine guns they had five AK-50 heavy Chicom machine guns. It was obvious that preparation for the Tet attack had been under way for some time. I recall there was a family living in a makeshift lean-to at one end of this partly constructed building. We are told that after the fighting everything in this area is in shambles.

After dinner I went up on the roof and could see heavy smoke and fires all around the city. We heard on the radio that everything was under control . . . the question was "whose control?" Sniping could be heard from the direction of the Embassy and the cemetery. After dark I saw our CIA man depart in the direction of the US Embassy with his infrared sniper scope equipped rifle.

One night the Personnel Manager of one of the leading civilian construction organizations got very drunk while visiting the women next door to our building. At 2330 hours he left the building and apparently could not find his car which was parked across the street on the sidewalk, behind two trees. He had parked it outside of a locked area contrary to orders. He ran up to our gate, where I was on guard duty, and looked frantically right and left and then apparently scared to death he took off running down the street as fast as he could. The street was completely deserted except for road blocks and other guards. He turned the corner two blocks away and disappeared. The women he had been visiting came to us and pleaded for one of us to go after him. We refused. But I did go out and move his car out of the street. One of the women told me this same man had been robbed and beaten at least twice while he was drunk. None of us had much sympathy or cared what happened to him. I learned later that several of the American casualties were similar drunks or the know-it-all types who stepped in where angels feared to tread. One US Lieutenant almost lost his life trying to rescue such a casualty near our USAID office.

(continued)   4 February 1968

Later Sunday morning conditions appeared to return almost to normal. There was no shooting or bombing and traffic increased. Kim, sister of my refugee teenagers, came at 1000 hours and said it was safe and they could all return home. Arnold Snowden, SCS Engineer, and I drove them home. We were required to go fully armed like all the other US vehicles. I rode shot gun with an M-2 carbine sticking out the window. We passed through a number of police and ARVN road blocks. We had a few anxious moments when we were bottled up in a traffic jam. When we finally arrived at their place we could see the windows in all of the homes in the area were broken and the walls pockmarked with bullets and shrapnel. A burned car remained in the street. Their home was a short distance off the main street, down a narrow lane. It was down these lanes they had watched hundreds of VC run back and forth on their way to various targets. We were invited to come in the house and walk down the lane further to one of the Buddhist Pagoda which was used as a headquarters by the VC. We said thanks, no thanks, and got out of there in a hurry. On the return trip we took a different route and passed through even more road blocks formed by sandbags, concertina barbed wire and bulldozers. At the road blocks we noticed all of the machine guns and rocket launchers were pointed in the direction we had just come from. We both were relieved to reach the safety of 192 Cong Ly.

That afternoon large double rotor helicopters could be seen shuttling 105mm artillery pieces and ammunition to various points in and around the city. Air America helicopters could be seen flying from Long Binh Army Base to nearby police headquarters. They were carrying ammunition to the police.

During the afternoon we could see propaganda leaflets raining down over the city. One leaflet has a photograph of VC bodies near the US Embassy. The other one showed captured VC prisoners. Stories of atrocities committed by both sides reached us on many occasions. It was reminiscent of the allied propaganda against the Huns during WW I. With the present credibility gap as wide as the Grand Canyon it is difficult to sort facts from fantasy and fiction.

Late in the afternoon we heard the sound of gunfire at our nearest street intersection. One of the returning Public Safety Advisors told us there was a dead woman lying in the middle of the street. Just before dark there was the sound of automatic weapons and rifle fire just outside my balcony window. It is wise during such an exchange of fire not to rush to the window or balcony. I do not know what that exchange was all about. During my 0300 to 0500 hours guard duty the street lights went out for awhile and made things interesting. There was intensive bombing and gunship action around us during the night. Many flares were dropped in our area and helped improve visibility. Other than that the night was rather uneventful.

Strict enforcement of a 1400 to 0700 hours curfew for Vietnamese and one of 1900 to 0700 hours for US military and civilians was effective in reducing traffic. The military were authorized to shoot at any vehicle that was not a police or military vehicle. Of course there was an open season on anyone caught out on foot during curfew hours. Our instructions were not to fire unless someone tried to come through or over the fence.

Our garbage pile on the parking area near the generator house continues to grow. Food cans are decreasing in volume but beer cans and whiskey bottles are increasing. At night during guard detail we occasionally hear a cat working over the garbage. As yet the rats have not moved in but it will not be long before they arrive. Some of our lazy tenants dump their garbage in the hallway. I have carried several bags of it down. Except for two men who refuse to come to meetings everyone goes up and down the stairs several times a day and could easily dispose of their own garbage. The two men who never show only talk to us through their closed door. They are absolutely petrified with fright. Yesterday two of our tenants took their maids home to a rough part of town and have not returned.

We have noticed during the past few days the traffic has been significantly reduced and the air is pure and clean for a change. There are very few of the polluting cyclos on the road. we believe most of the drivers of those vehicles are probably charter members of the NLF (National Liberation Front of South Vietnam or VC).

(continued)   5 February 1968

Today, Monday, we were notified to return to our offices after 0800 hours and only with an armed escort. I am the armed escort for our group. We returned with no difficulty. The traffic was very light as none of the Vietnamese civilians have been ordered to return to work. The lack of Vietnamese drivers created a problem for those who ride the buses and rely on chauffeurs.

As we entered the USAID building the Marine guards asked everyone if they had weapons or cameras. I told them I had a revolver in a shoulder holster and a Canon. Several of the Marines were visibly startled when I said "a Canon" and did not relax until I showed them the Canon was a camera. They held my camera for safekeeping because no one is allowed to have a camera in the building. This is to prevent photographing classified documents. I had to remove the shells from my revolver but the Marines allowed us to keep weapons. They said they were happy to have the additional firepower. The guards had all entrances well covered and there were additional Marines in our building. All were armed with automatic weapons such as Beretta sub-machine guns, AR-14s, etc. The additional Marines were eating C-rations and sitting or sleeping on the floor or couches. They looked dead tired and were bitter about the loss of their buddies in the attack on the US Embassy. These Marine guards are an elite group and members of the largest contingent of any overseas US Embassy guard detail.

At work we naturally were preoccupied with the past days' events and exchanged stories about our experiences. Flamm, Metcalf and Walter Pierce (our latest addition to the forestry team from SW Georgia) reported heavy sniper activity. From the roof of their tall hotel they could observe many of the first day's mortar attacks and fighting at the Palace. They said the police cleared the downtown streets several times by spraying the area with automatic weapons fire. Barry's group were called on to go to a nearby Vietnamese civilian hospital to give blood. Barry said civilians were lying all over the place, on the floor and several to a bed. Some were seriously injured but he did not hear a single cry, even from wounded children. The number of civilian deaths and injuries are unreported, but according to a reliable source they are heavy. The VC gained a strategic advantage during the Tet offensive by effectively demonstrating the Allied Forces were unable to protect the people, even in Saigon.

This was the day consumer prices skyrocketed as merchants began to extract their toll. A small loaf of bread costs the equivalent of 50 cents, eggs are 20 cents each and even more critically for the Vietnamese the price of rice jumped as high as $3.10 per pound. This will seriously affect the poor and working class who have nothing stocked ahead. Most people spent their meager funds on clothes for Tet, luxury foods, flowers and fruit for family, guests and offerings for ancestral altars. Homeless refugees continue to pour into Saigon from adjoining areas and many residents have been displaced by the massive destruction. It is impossible to have bombing, rocket and mortar fire in urban and village areas without destroying many homes, businesses and people.

About 1100 hours we received notice that work hours would be extended until 1630 hours except for those who needed to go to the Commissary for food. Many people were reaching seriously low food levels and needed to visit the Commissary. Special approval was granted to employees of USAID, JUSPAO and CORDS to go to the Cholon commissary between 1400 and 1800 hours. We were advised to avoid local markets, both due to the prices and lack of security. Apparently someone in their wisdom decided it was safe for civilians to go into the Cholon battlefield to purchase groceries. Someone else must have thought this through, or checked on the security in Cholon and found it lacking, because about noon the order was rescinded and everyone ordered to clear the USAID building immediately and return to their billets. It was pathetic to see the poor US civilians clutching bottles of booze they had picked up at the small nearby Meyercord BOQ PX. Oh, yes, many were in serious condition because they had run out of cigarettes!

On returning to our building I went to the roof to observe a very large column of smoke we had seen on our return trip. It appeared to be a large fire raging in the middle of the city near USAID 1 and 2. Apparently we got out just in time. I could hear two different fire fights in the general direction of our office building. People returning to our apartment building a short time later said they had seen two bodies lying in the middle of the street near USAID 2. Marines and MPs cleared our office buildings quickly, locked up the facilities and had taken defensive positions around the buildings and motor pool.

The large fire continued to rage all afternoon and night. many homes must have been destroyed, adding to the large number of refugees and their hardships. The refugee problem is growing but I find no great concern being expressed about the problem being created in Saigon. These additional homeless people will add to an already impossible situation.

With the aborted commissary trip many people in our building are critically short of food. Mr. Passey was down to a few crackers and a piece of bologna. Others took advantage of an offer to purchase a case of C-rations at $7.50 per case of 12 units. That evening Snowden fed three of us USDA-types.

After considerable griping and hearing all kinds of excuses we managed to find eight people who were willing to pull guard detail. I was on from 0230 until 0500 hours. Except for isolated shots at the nearby cemetery we had a quiet night. We have a Public Safety Advisor (CIA) who has a rifle equipped with a night sniper scope. I presume he is working over the cemetery. Bodies are reported to be piling up over there. In order to hold out for so many days the VC must be well dug in and are relying on supplies and ammunition moved in before Tet.

I went to the roof several times that night and from a safe location I observed Puff the Magic Dragon make pass after pass at an elevation of about 800 feet over Cholon and the race track. Solid streams of tracers could be seen flowing from the mini-guns on the airplane to the ground, just like a waterfall. It was awesome! Occasionally the waterfall reversed itself and could be seen reaching up for Puff and the flare ship. I could see the fire was not contained and continued to spread. Fire fighting in Saigon under the best of conditions reminds me of old silent movies featuring Keystone cops. Under our present conditions fire suppression is impossible. Panic among the civilians in those fire areas must be terrible.

(continued)   6 February 1968

Today, Tuesday, we were advised to return to work on a 0900 to 1400 hours schedule. Again travel to work was only under armed escort (me). Those of us from the Forest Service are disturbed because management does not make a decision to utilize the available manpower to assist refugees. I told our Assistant Director that under similar conditions the Forest Service would organize its manpower to handle this situation. I cited examples of what we had done in critical fire situations, rescue operations, floods and hurricanes. But he said he had no contacts with the Embassy and surprisingly, he had less information than our group about the status of the siege of Saigon. Our information comes from personal observations, reliable sources and over American Forces radio. We suspect he has been sitting alone in his apartment with his head in the sand or elsewhere.

Barry and I drove from our office to USAID 2 to send a telegram to the sea coast town of Vung Tau in an effort to retrieve Director Muoi. He had gone there with his family for Tet and had been unable to return. Barry and I had been invited to spend this past weekend with him on the coast as his guests, but fortunately or unfortunately we were unable to go. His office requested our assistance in getting Muoi and his family back to Saigon. We told them he probably did not have high enough priority to have this request approved. But we agreed to try and we sent the telegram.

While we were at USAID 2 which is CORDS (Coordinating Office of Revolutionary Support. i.e. CIA) Headquarters we purchased a case of C-rations at inflated prices of $8.75 or $1.25 higher than the cost the day before. As USAID employees we had to pay for this ourselves, but employees of CORDS, which are actually USAID employees administratively assigned to CORDS, are given C-rations free of charge. I plan to file a complaint with the Embassy on this gross discrimination.

Traffic to USAID 2 was more chaotic than usual. Military traffic, ambulances and civilians scrambling for food and refugees carrying or pushing carts with their meager belongings made an incredible sight. Seeing this mass of humanity struggling for scarce supplies and existence was unbelievable. We could see people lined up to purchase food, gasoline and kerosene for fuel. Police, military MPs and civilians tried unsuccessfully to direct traffic. I saw a small horse pulling a cart doing an outstanding job of wiggling through the traffic. The poor little horse was undisturbed by the frantic humans around him.

Our next trip was to the American Embassy to deliver the classified document I was supposed to have delivered last Wednesday. We found a route by the Central Market that bypassed the traffic jams. The market area was practically deserted since reports were circulating that the VC planned to destroy it. With the exception of two streets, the downtown area was also deserted. All shops were closed and boarded up. Even the normally busy black market was gone. Come to think of it, perhaps the VC operated it and they are now busy elsewhere! We drove past the Catholic Cathedral and could see where the VC had planted a small explosive charge in an effort to destroy the Madonna statute. She was scorched but still standing. The VC had set up a machine gun emplacement in this area for a short time. Today it was well guarded by ARVN soldiers.

We reached the new American Embassy. This ugly concrete fortress is placed in an attractive area of stylish French-colonial homes. As expected it was well guarded. We could see two places where the wall surrounding the compound had been breached. These gaping holes resulted from the explosion I heard at 0300 hours on the beginning of the siege of Saigon. The face of the building is covered with an equally unattractive device called a sun screen, actually it is a reinforced concrete rocket screen that stands about three feet out from the walls of the building. It was designed to deflect rockets and other explosives hurled at it. The screen lived up to design specifications. We could see evidence of where five or six rockets had slammed into it. On the top story, just under the helipad landing area, were a number of holes in the building. This damage occurred when the VC were shooting at descending helicopters. Decorative planters, marble facing around the entrance, patio floor, doors and iron entrance gates showed damage resulting from rockets, grenades and bullets. All windows in the structure are made of aircraft plastic therefore we saw no shattered glass. Much debris was being carted away by workmen, while others were repairing damages. This reminded me that the Embassy people were more interested in taking care of their own image problems rather than providing for destitute refugees.

In the reception area a number of Marines were on duty. As we entered they carefully checked our ID cards. I delivered my report to the office where I was to have had a meeting last week. Outside that office window a rocket had blasted through the sun screen and bowed in the plastic. That would have been a disruptive intrusion if it had occurred during our meeting. I could see very few people who normally work in the Embassy. Many offices were vacant, contrary to statements that we were back to normal. Outside we examined the locations from which the enemy fire had originated. The ground was still littered with cartridge casings and unfired bullets, mostly from M-16 ammunition, and grenade pins. I picked up a few for souvenirs.

Our two Vietnamese employees returned to work this morning. Mrs. Hai's family was safe. She said the VC had raced up and down the narrow streets leading to her home. That is where I searched in the pre-Tet darkness to locate her house. The VC had broken into a neighbor's home but had just searched the house and departed. Mrs. Hai's father and family vacated their home due to intense VC activity which was focused on the nearby US installations. They were safe in the home of friends. Mr. Thuong's home and family were safe. Here too there had been large numbers of VC. His wife's family home in Cholon and all their possessions were destroyed in one of the fires we had observed. I gave Mr. Thuong my surplus rice and asked him to give it to someone who could use it.

At 1300 hours we were ordered to clear out of our offices and return to our billets. Departure was quite orderly. People just marched down the stairs and out of the building. A battalion size force of VC is reported to be about two miles from here. Numerous pockets of VC are located throughout the city. Stricter curfew regulations have again been ordered.

Other reports tonight:

I fed three men who were running low on food. I have a case of C-rations and a supply of kipper snacks and grits to tide me over. I should be able to hold out for some time. My friends do not think much of my choice of food. I admit to you I am getting tired of the same old thing. But it is filling and beer helps ease it down. We hear there are 50,000 tons of rice in Saigon warehouses and on ships at the docks. That will feed many people if it can be distributed safely and quickly.

I read in the *Stars & Stripes* there are no real problems in Saigon and everything is under control. At the moment you could ask, "under whose control?" The VC have not taken the city but they have rather effectively paralyzed and tied up great sections of it. I would say that is far from being normal. From my comments you can gather there are many urgent tasks to be done. I hope in the coming days our skills and services will be used more effectively. These country-wide attacks are certain to curtail most of USAID's normal programs. If we could get out in the forest we could get some work done. After all the enemy has deserted the forests and is in the city, villages and hamlets.

Remember I wrote about our water supply being cut off? The water came back into the system that evening but none arrived at our side of the building. All night the water was pumped to the other side. The next morning as I went off guard duty the answer to the water problem became clear. One of the men had deserted his apartment and had left the faucets wide open. I found water running in the hall, down the steps and through the floor to apartments at lower levels. Later four of us solved the problem when we broke in and turned off the water.

Our garbage pile now grows at a slower rate, however the amount of beer cans and liquor bottles remains constant. We have a high percentage of heavy drinkers in our building. We would have been happy to have kicked some of them out the gate and turned them over to the VC. One evening while I was on guard duty, one of our resident drunks came down to the parking area, started his Jeep, turned on the head lights and headed for the locked gate. I stopped him. He wanted to go to the Hotel Embassy bar. He would never had made it past road blocks and dug in VC. He kept edging his Jeep toward the gate. One of the Public Safety men talked to him but he insisted on being let out. I brought up my carbine and said I would shoot out his lights if he did not turn them off and park his Jeep. He muttered something uncomplimentary about foresters and complied.

From my limited perspective this lengthy letter covers 10

days of the Tet Offensive. General Giap or whoever planned this operation will go down in history. The siege and defeat of the French forces was engineered by General Giap in 1954 at a place called Dien Bien Phu in North Vietnam. This was the end of the French Colonial Period. The sequel has been Tet 1968. Perhaps this is the beginning of the end of the American Period. Time will tell. Now I will put this letter in the mail. I hope that you and our friends will find it of interest. There are no commercial flights out of Saigon and it may take several days or weeks to reach you by the military route. Now I return to guard duty.

Chapter XXVI

Military & Other Battles

13 February 1968

The war continues tonight all around Saigon. There was a very large battle at 2000 hours. This one was located southwest of here, just beyond Cholon. Our troops must have detected a large enemy force coming or going because since that time our planes have been dropping tons of bombs and napalm and shooting rockets. Helicopters and Puff have been pouring machine gun fire into the area. The tracers could be seen between the bomb bursts and looked like waterfalls of fire pouring down from the sky. The expenditure of ammunition is beyond belief. I understand ammunition is now being flown in from the US at a terrific cost!

Since the roads have been secure, huge convoys of trucks have been bringing ammunition into the city from the large nearby Army camps. Overhead is a protective cover of helicopter gunships. For a while the VC had road blocks set up on all roads leading into the city and we were effectively cutoff from our ammunition supplies. Now, at least during the daylight hours, roads are fairly secure and convoys keep rolling in. All drivers are wearing protective flak jackets. These vests are made out of nylon plates and are heavy and terribly uncomfortable in the hot weather. I noticed during the peak of the fighting most of the American troops were wearing vests. Unfortunately there is a shortage of these life savers and not all of the troops are equipped with them.

Curfew remains in effect. We are confined to quarters from 1900 to 0800 hours. In the area surrounding our apartment building the Vietnamese curfew is from 1700 to 0800 hours. In the remainder of the city the Vietnamese curfew is from 1400 to 0800 hours. This has reduced the opportunities for the VC to move around. During open hours traffic is heavy. I suppose that many of the travelers are enemy scouts checking our defenses. The traffic is crazy, worse if anything. With many of the streets blocked, especially those near hospitals, military installations and police stations, everyone has taken to driving either way on one-way streets.

My food supply is low, but I have not considered it safe to drive to the Cholon Commissary and I do not have time since I have been assigned to an Emergency Warden Program at the Embassy. I am able to purchase a few items at the small

382

downtown Brinks PX. However no matter how you fix Kraft Dinner, Spam, sardines or kipper snacks they come out about the same every time. I enjoy kippers and Spam once in a while, but after 10 days anything gets old. My diet is not well balanced, but I am not starving. One of my neighbors brought me a loaf of bread and some fresh vegetables. Now I have some variety to go with my kippers and Spam. Today I got in line at the Embassy with Vietnamese employees and purchased 10 kilos of Thailand rice for the equivalent of 23 cents per pound. Before the US set up an effective distribution system rice was costing three times that price. Prices of fresh meat and produce have gone beyond the reach of anyone. Last week even the PX boosted the price of Spam from 42 cents a can to 55 cents. How about that?

As I mentioned earlier the family of Mr. Thuong's wife, lost their home and possessions in the fighting and resulting fires. One day he brought me several hundred piasters and asked me to purchase meat and fish for his father-in-law's family. Thuong said they were hungry and could not buy any of these items. We refused his offer of money and all of us on the forestry team pitched in and bought him a large amount of meat and fish. There was so much in fact that Mr. Thuong could not begin to pack it home on his bicycle. Martin Syverson loaded Thuong, bicycle and all the food in the Jeep and drove him home. Sy said it was an interesting trip. They were detoured by a number of road blocks and had to pass through narrow, spooky alley ways. These passages had been filled with VC not too many days earlier. But they made the trip safely. The next day Thuong made several bicycle trips to his wife's family and delivered food. They were very grateful. They are camped in a school with hundreds of other refugees and living under terrible conditions. They expect to be there for several months. The school system has ceased to function. I plan to give Thuong's relatives some of my Spam and sardines. Mrs. Hai's shortage was milk for her children. We picked up a supply of powdered Milk for her. At this time none of us feel a bit guilty about diverting PX purchased food items to these people. None of us have accepted any money for the donations.

Another group which has felt the pinch of the hostilities and curfew is the bar girls. They are about to starve out. The GIs had been moved to camps out of the city and during the fighting the GVN closed all bars and nightclubs. Moving GIs out of the city was wise, but their remoteness resulted in long delays in organizing counterattacks in the initial hours of the battle for Saigon. Armored cavalry dispatched to save Tan Son Nhut Airport from being overrun came from over 30 miles outside the city. They stormed in at top speed to join troops that had been airlifted by helicopter. From all radio accounts there was some very fierce fighting at the airport. Rescue forces were unfamiliar with the area and when they reached the airport in the dark they had difficulty in deciding which direction to go. But fight they did, and through their efforts the airfield suffered only minor damage. Attacking VC forces could have demolished the place but they apparently wanted to take it intact.

Heavy fighting has gone on from day one in the vicinity of the Phu To Racetrack, which is in the center of Saigon. I passed this facility going to and from the pistol range. On our first trip to the range we had driven through this entire area to become familiar with it. There are still pockets of VC in that neighborhood. About three times each hour the radio issues a warning to keep off the access road and not to approach the racetrack. Yesterday, Walter Pierce, our newest sawmill technician, was contacted by CBS TV. The TV people had been in the area and talked with a small group of soldiers who had been fighting there from the beginning. One of them was Pierce's son. He told them his father was working on the USAID forestry team. CBS decided this would make a good story and asked Pierce to go to the area with them. I gave him permission since he was provided good protection. I gave him one of our cameras and a supply of film to take with him. When Walter was over two hours late in returning, several people, including Barry, about had a fit when they learned I had approved this arrangement.

Fortunately Walter returned safely with a story to tell. Walter had given his son the camera to take his photograph when the protection force gave Walter orders to leave immediately. As he rushed away, Walter threw his son some film. The son yelled back that he would take some more photographs and return the camera to him in a few days. Tonight one of our government Miranda cameras is on the front line. The son is expected to be relieved in a few days. CBS reported this group of US soldiers have killed over 800 VC in the last 13 days and have been recommended for some kind of award for holding out against such odds. Many of young Pierce's buddies were killed or wounded in the battle.

Presently there is no way for us to carry out the forestry program. The airport and the Directorate of Forest Affairs is down to a small work force. The forester who went to Hue to visit his family can not be located. Several other foresters are missing. The USAID office is completely disorganized and demoralized. Meanwhile we are functioning as well as can be expected. Metcalf and Pierce have taken on some emergency jobs around town. Barry has helped out in several ways, including spending several hours restocking the shelves in the Commissary. Syverson has been holding down the office and is working with the Directorate on plans to accelerate lumber production to meet reconstruction needs. I have been assigned as an Emergency

Warden and have developed a "USAID Employee Locator System". The location of USAID employees was nebulous during the Tet offensive. No one, that is no one, knew where employees were located. Now we have all their locations plotted and someone designated as a warden to be in charge of each apartment, hotel and groups of houses. We have a system and people who can be contacted in the event of an emergency. There is still no plan for evacuation of civilians. While all this seems rather academic and logical, no planning or no system had been prepared during all the years we have been involved in this powder keg. Obviously during the past several days it would have been impossible to get people to the airport let alone evacuate them. The solution is to set up assembly points and evacuate them by helicopter.

The office I have been working in at the Embassy contains punched out plastic windows and the walls have been peppered with bullets. Crews are at work refurbishing the place. The breeched outside wall still has large holes that were blasted through during the initial attack. Bullet holes in the planters, entrance way and doors have been patched. A crew is at work repairing rocket holes in the sun screen. Other local employees of the Embassy were filling five kilo bags with rice for sale to employees who wanted it. This was where I purchased my rice today. This Thailand rice is rated poorly by the Vietnamese but today everyone seemed eager to purchase what was offered.

The Embassy is now bristling with guards. Some are filling sandbags and others are manning dozens of sandbagged bunkers on top of the Embassy buildings and around the grounds. The GIs formed a chain of manpower and were having a great time throwing sandbags to one another and up steps to the top of the rear garage. There must have been fierce fighting in the garage area since the place is riddled with bullet holes. Poking from the bunkers are .50 caliber heavy machine guns. The VC or anyone else would have an extremely difficult time entering the place under present circumstances. On the other hand I find myself wondering if the people running this show are very smart. According to *TIME* magazine all the intelligence reports indicated there would be trouble during Tet and the Embassy was a likely target. *TIME* reported there were five guards on duty the morning of 30 January, *NEWSWEEK* says there were three. Regardless, our Marines did well to hold out for reinforcements in spite of the poor odds. Remember Pearl Harbor?!

Across the street from the Embassy sits a bullet riddled, blood stained, vintage Citroen French automobile. Tires are flattened, windows shattered and blood stained seat cushions are evidence of what happened to one or more of the attackers. The driver's job was reported to be one of killing US guards at the gate. Just before 0300 hours he drove past the Embassy, turned around, drove to the side of the street opposite the gate and parked. When the first shot of the Tet offensive was fired by that driver, a Marine guard riddled the car with bullets and killed the driver. The battle at the Embassy lasted until 0830 hours.

That is all I have to offer for tonight. I still have quite a lot more to cover and will continue this letter later. It is 2230 hours and bombing and machine gun fire can still be heard off in the distance. Fortunately, for us at least, the action is not in our back yard, as it was last week. Now I will set out my boots, fatigues, carbine, pistol and ammunition where they will be handy just in case, and hit the hay.

(continued)   15 February 1968

Today is Thursday and I find it difficult to be believe I was at home with you less than a month ago. Actually it seems like I have been away for months.

Some 13 USAID/CORDS people stationed from the Delta to Hue are either listed as dead, missing or captured. Eight were injured in the fighting. Some of these were men I went through training with in Washington and others have worked with me here in Vietnam. In view of the ferocity of the fighting it is a wonder there were not more names listed. I place the responsibility for the deaths, losses and captured USAID employees directly on USAID. It is unfortunate but true, USAID had no plans or intention to prepare for any such emergency or lookout for the welfare of their employees. We Forest Service types are better prepared to look after our own welfare and safety since safety has always been a part of Forest Service activities. I am convinced, if the truth were known, that many of the deaths and injuries impacting US civilians and military personnel in Vietnam have resulted from accidents and so called friendly fire, and not from hostile action. Horseplay and lack of concern for safety is very apparent among the young GIs.

The numbers of dead and wounded on both sides have been significant. I gave blood today at the dispensary in response to an urgent appeal for blood for critically wounded Vietnamese citizens. More are being wounded and hospitalized each day. Two forestry team members gave blood last week at one of the Vietnamese hospitals. I could not get out at that time and was unable to give until today. My blood was drawn by a large Negro Corporal by the name of Davis. I have given many buckets of blood over the years and Davis did as good as job as I have ever had at any blood center. Davis, from the 17th Field Hospital, said the hospital was swamped and worked around the clock during the early days of the fighting. He said they provided initial treatment and then seriously injured military personnel and US civilians were evacuated to Japan or Hawaii.

While I was giving blood a Marine Captain came in with the last two Embassy volunteers. One of the Marine donors passed out but rapidly recovered under the TLC of the nurse in attendance. Refreshments included chocolate chip cookies sent in by US wives living in Taiwan.

Special approval had been given to civilian wives living in safe haven locations to come to Vietnam during the so called peaceful period of Tet. Several hundred of these wives were in Saigon for Tet. Somehow most of them were evacuated during the height of the fighting. The wife of one of the Public Safety Advisors in our building had much to worry about. Her husband was in on the assault made on the unfinished apartment building containing the 20 VC who were firing on the Palace and the Korean Embassy. He lost eight of his policemen and some 20 were wounded. He was pinned down for a long time by heavy machine gun fire. He kept up this activity for days and his wife experienced this first hand. But he kept coming back safely. Fortunately the serious fighting in his precinct was over by the time she returned to Bangkok.

Hourly radio announcements warn us to keep out of wide sections of Saigon. Other recent messages have canceled University of Maryland classes, language lessons and a number of other events. We are not returning to normal very rapidly. You will recall all the talk about the tremendous progress and improved security. I was asked about the security situation many times while I was on home leave. I stated some progress was being made in various pacification programs, but the security was certainly not improving. There were many outright lies about the security situation in the villages. This recent action proves the VC can strike at will. I assure you the troops that attacked Saigon did not come from Cambodia. I will tell you about that some day in the future.

The curfew for us has been relaxed and is now from 1900 to 0700 hours. Vietnamese still have a 1400 to 0700 hours curfew. Their absence continues to disrupt all vital services. We now have an adequate supply of water since the main line was repaired. We have fuel for our generator. Garbage trucks started removing mountains of trash that were creating a real mess. Our accumulation of garbage really was not as bad as New York City during a garbage haulers strike, mainly because we did not have much to eat or throw away, except for beer cans and whiskey bottles. Of course there has been no clean up in the areas seriously damaged by the fighting. Homes of tens of thousands of people were reduced to rubble by uncontrolled fires and bombing. Many of these homes were shacks but they provided shelter.

The lack of sanitation for the homeless is a serious problem and with it comes a serious health threat. Thousands are being vaccinated for one thing or another. Thousands of refugees are camped out along roads and streets in Cholon. The lucky ones are housed in trucks and schools. There are no estimates when schools will reopen. Schools are also being used as registration centers. An intensive effort is being made to register all males over 15 years of age and issue them new registration cards. I notice an intensive effort is being made to check the ID cards at all police and military roadblocks. Jails are full and overflowing with curfew violators.

Ambulances are in use continuously. Hearses serve a dual role . . . they transport both dead and wounded. On several occasions last week Barry was in the vicinity of the central market and the Saigon Hospital. Each time he was in those areas he saw truck loads of wounded and dead being hauled to and from the hospital. He witnessed one truck piled high with bodies and sitting on the tail gate were a number of wounded, bloody children and adults.

The radio and newspapers report "things are returning to normal very rapidly". However in talking with some of the Vietnamese who live in outlying sections of Saigon I learn that fighting and shooting goes on all night around them. The VC are either trying to escape the city or coming in to reinforce those still fighting. They must be taking an unmerciful beating. Artillery fire, bombing and helicopter gunship fire goes on continuously. There are areas in and surrounding Saigon that look like a no-man's land.

I continue working at the Embassy. One of the people I met there is in USAID's Industrial Development Program. Yesterday he inspected the textile plant. This one plant employed over 2,000 people and was one of the country's most productive and profitable industries. That US$9 million plant has been completely destroyed by bombs and fire. Other vital facilities are seriously damaged. Seventy percent of one paper plant was destroyed and one we work with is reported to be heavily damaged.

At the Embassy I talked to a group of USAID employees who lived near the boundary between Saigon and Cholon. When the fighting started they locked up and went to one of the hotels. Early this week they returned and found ARVN had first broken in searching for VC. According to their neighbors some of their possessions went away with ARVN. Then neighbors moved in and looted. When the USAID people returned they called for a crew to board up and secure the building. Four guards were put on duty. The USAID people returned yesterday and found only two guards and the remainder of their possessions missing.

Early this week the CIA sniper who lives in our apartment building finished his bloody business at the nearby cemetery. He operated from the roof of our building and a nearby one which overlooks the cemetery. In five nights he and his .270 rifle equipped with a Starlight scope disposed of some 20 VC as they emerged from graves and underground chambers to snipe.

Newspapers resumed publishing this week. At first they consisted of one page and now two are permitted. In Vietnam newspapers are heavily censored by the government. No editorials are permitted which might incite the readers. News is three to four days old. The articles you send me from *TIME* and other sources are all we are permitted to see. They make Tet sound pretty rough. And it was! The change that evolved from the streets of flowers, well-dressed people and the peaceful sounds of Tet was dramatic. Lives were changed and lives were lost. Streets were cleared of people and vehicles. The air was cleared of the choking blue exhaust fumes. Neon lights that cast a glow all over the city each night were cut off. Bars were closed and commercial airline traffic disappeared from the skies. Instead the skies were filled at all hours of each day and night with airplanes, helicopters, bombs, bullets, rockets, smoke and propaganda leaflets, all of which I mentioned in my previous letters. Leaflets continue to be dropped urging the VC to surrender. Some leaflets carry photographs of dead, wounded and captured VC. Others contain statements from captured VC urging their comrades to give up. More samples of these are enclosed. At times the streets, parks and roof tops were littered with a blizzard of propaganda paper.

Some of the battle damaged buildings within a five block radius of our apartment building will bear scars for years. The guard house at the Palace entrance and the unfinished building across the street that held the VC attackers now look like something out of a war movie. All windows in the building are broken. Two floors are blackened from the results of flame throwers and holes are punched all over the building from grenades, heavy machine guns, rockets and artillery fired from tanks. The Palace is set well back from the street and I am unable to determine what damage was done to the structure. The fighting which involved the VC and forces at the guard house lasted 36 hours. At least 50 or 60 people were killed in that one action alone, including two Jeeps loads of American MPs who came racing to the rescue when trouble was first reported at the Palace and the Hotel Splendid Hotel BOQ. All this action occurred in an area that I have traveled through at least twice each day since I have been in Saigon.

Guards and US Army are heavily armed but it appears in the past two days everyone, except the Americans, is lowering their guard. Many of the civilians working for USAID are armed, including many whom I suspect should not be since they do not know how to safely use what they are packing. Yesterday one of our group was at the crowded Commissary and saw a civilian showing off his ability to handle a pistol. He dropped it on the concrete floor but fortunately it did not go off. I carry a .45 automatic pistol in my brief case and have it loaded when I am traveling outside the office. The talk around town is that the Vietnamese government is planning to arm the citizens so they can protect themselves. If that comes to pass this really will be a dangerous place and provide the VC with an additional source of weapons and ammunition. I heard one report during the peak of the fighting that the VC did not need to receive fresh supplies of arms and ammunition because the Americans kept them well supplied. It is true they are armed with our carbines, M-16s and other weapons. Rumors were thick during the fighting and we were told to ignore them. Come to think about it, all the unspent bullets and empty shell cartridges I noticed in the street and on the floor of the blood stained car in front of the Embassy, were for our .30 caliber carbine, .223 M-16 and .45 automatics.

Recently before our guard detail started, one of the neighbors was burning trash and it sounded like an automatic carbine opened up. It was only exploding firecrackers. This noise alerted all the nearby guards and half the people in our apartment went running down the stairs to witness the action. In doing that they violated my cardinal rule . . . keep away and go in the opposite direction. If this had been an attack many people would have been zapped. I stayed in my apartment.

With no Vietnamese drivers many of the Americans are using motor pool vehicles or driving their own personal cars. Americans have been given permission to use US gasoline pumps if they are using personal cars for official purposes. I see many taking advantage of the offer. I bet the GAO auditors and inspectors who are assigned here on a permanent basis are going crazy trying to account for use of US supplies.

Theft and looting continues in many areas, many of which are not very remote. One of the men working with me at the Embassy parked his personal car in front of the Embassy. When we went out to drive to lunch we saw that some thief had pried open a wing window and stolen the radio out of his new Datsun. Imagine if you can, his car was parked in an area within sight of the Embassy, where several hundred soldiers and Marines are on guard and others are coming and going at all times and you will have some idea what we have to cope with in Saigon. If I wore false teeth I would have to keep my mouth shut for fear someone would steal them. We have to be alert and on guard at all times.

It appears we will return to a regular schedule next week. We are working on a plan to accelerate lumber production to support reconstruction needed to house thousands of refugees. Progress in the pacification program has stopped in all parts of the country. Nationwide damage must be tremendous. We have not heard any reference to secure areas during the past two weeks. A real tragedy has affected a number of Embassy and USAID workers. They are actually heartsick, honest true, cross my heart . . .! Bombs (undoubtedly dropped by a non-golfer flyer) have created traps in the center of the 12th, 14th and 17th greens. What a crazy world this is!

I appreciate receiving your valentines and letters expressing concern for my safety. I am being careful, really. I do not deviate from safe routes. I move quickly between my office and the apartment. I do not go to the Cholon Commissary because I consider the area too dangerous. My photo expeditions have ceased. I am not interested in taking photographs of dead bodies. One of the USAID men was seriously injured the first day of the fighting. He went to the Palace to take photos of the action. I will leave that to *LIFE*, Associated Press and UPI. Please save *LIFE* magazines for my historical collection.

We cannot accurately predict the outcome of the Tet offensive. In my opinion the US should concentrate on winning the war and reduce the political constraints imposed on the military. But under the present situation and leadership that is impossible. The other alternative is to get out. The USAID program does much good, but I fear money and effort for so-called pacification programs is going down the drain faster than it can be poured in. The poor Vietnamese people are suffering as a result of this very complex combination of factors. The tragedy of it all is there appears to be no workable solution. This is going to make a very sad, sorry chapter in the history of our country and all mankind.

Please continue planning for the trip to England in the Spring. I will need a trip by then. I am required to spend at least seven days in the US out of my available home leave. That means I can have just over two weeks in England. I suggest you plan to arrive early and spend a week or two in London and nearby areas before I arrive. Then we can take off for Wales and places in the country on the day following my arrival. There has been no indication the US will limit our stopovers in foreign countries in spite of LBJ saying that government travel to foreign countries is being curtailed.

As I stated earlier it seems like I have been here for months since I was last with you. Quebec seems to be a hundred years away and almost unbelievable. Undoubtedly it will be weeks before we can return to our field work and resume our projects. At this time it all looks pretty futile to me but perhaps things may improve. You can be certain of one thing, I am not going to take any chances. Good night, dears.

16 February 1968

In my last letter I commented on security. The following

message from the Embassy and note from Martin Syverson illustrates how we maintain security in Saigon.

"Local Police and Military Police have orders to shoot on sight anybody caught out after 1900 Hours" . . . US Embassy

Added pencil note: . . . "Better be home by 1900 or else!" . . . /s/Sy

I was!

(continued)   19 February 1968

Today the radio, TV and USAID Special Bulletin report provide us with an optimistic assessment that the situation is returning to normal. You may be interested in my observations regarding the present state of normalcy.

It is now 2100 hours and I have just listened to the news over the radio and watched the US Ambassador's press interview on TV. I had a Vietnamese TV shop adjust my set and it now works fairly well. I decided to go to the roof for a breath of fresh air. The night is very pleasant, with a light cool breeze and a few scattered clouds. Overhead I spotted the Echo satellite reflected among the crowded skies. Planes of all varieties are crisscrossing the sky tonight. There are flare ships, Puff, helicopter gunships, cargo ships, jet transports and fighter bombers. In every direction, 360 degrees around the city, brilliant parachute flares light the sky and ground area. Mortar and artillery fire are heavy in the direction of Tan Son Nhut Airport, Bien Hoa Airbase and along the Saigon River. The radio reported dozens of sampans were detected converging on Saigon via the River, tributaries and canals. Many carrying VC rocket ammunition were sunk.

My stay on the roof was short. Six of us talked about the events and our next home visit or R&R trip. Everyone wants to be somewhere else. We talked about USAID programs and where do they go from here? We discussed the need for a serious review and speculated on which programs might be curtailed. About that time someone tossed a hand grenade in the vicinity of the American Embassy. Some thought it was a mortar but I have thrown enough grenades and fired enough mortars to recognize the difference. Simultaneously, while we were standing there talking, automatic weapon fire opened up. When tracers skipped over our roof at a height of less than 10 feet I decided it was time to retire to my apartment and start this letter. I was just about run over by apartment dwellers who were on their way to the roof, two steps at a time, to see the excitement. That is an effective way to get a one-way ticket home in a flag covered aluminum box!

Yesterday morning and again this morning the VC unleashed a rocket and mortar attack on Tan Son Nhut Airfield. One rocket landed several hundred yards away and only shook the place. An early Sunday morning attack destroyed planes and the Air Force Chapel. This morning at 0600 hours they hit the Pentagon West and the Commercial Air Terminal. A young soldier had just completed his one year tour of duty in Vietnam and was on his way home, when he was killed before he could board the plane. Some 35 others were wounded as a rocket blasted through the roof and made shambles of the interior. Seats and counters were torn out. Debris and glass partly covered the blood on the floor. This scene was described to me by a friend who arrived shortly after the attack.

It would be difficult to count the number of times the airport has been opened and closed in the last two days. R&R trips, visitation to the US or trips to visit wives in safe haven countries continue.

It is extremely difficult for us to learn exactly what is going on, except for what we can see ourselves and hear from eye witnesses. The Vietnamese press is so strangled with censorship that we learn little or nothing from that source. This leaves us with the *Stars & Stripes* newpaper, American Forces Radio and TV, *TIME* and *NEWSWEEK*. The latter two sources are doing the best job of reporting the facts. *NEWSWEEK* is doing this at the expense of losing their key man.

Everett Martin, *NEWSWEEK* Bureau Chief in Saigon was deported last month by the GVN for his critical reporting. The articles you have been sending me are the type we do not see. The cartoon of Herblock depicting Westmoreland's words . . . "we have turned the corner" . . . and showing him running into a Viet Cong was appropriate, as was Conrad's of the *LOS ANGELES TIMES* shown in this week's *NEWSWEEK* . . . LBJ is sitting up in bed, holding his hand over the phone and saying . . . "what the hell's is Ho Chi Minh doing answering our Saigon Embassy phone?" You will be pleased to know the firing over in the direction of the Embassy has now quieted and I presume we are still manning the phones.

It is a problem these days to know just who you will meet when you round a corner. I continue to carry a pistol in my brief case in the event a need arises. We completed the locator system and we are now checking out the system in USAID 1, just to polish off the rough corners. Among other assignments I have at the Embassy is the task of processing requests for firearms. We have more requests than we have guns to issue. There appears to be much fear and uncertainty concerning who is protecting whom. My greatest fear is the gun in the hands of the civilian who does not know how to use it. Out of approximately 100 requests I have screened and processed, including personal interviews, I have approved only three requests.

I had one request for a weapon from a person who lives in an eight unit apartment complex which had the following inventory of weapons: four .45 cal. semi-automatic pistols, one .38 cal. revolver, one M-14 automatic rifle and one .45 cal. sub-machine gun. I did not believe the 8th occupant needed a weapon.

One woman who lives in an apartment near our office wanted a pistol. She had never used one but wanted something for protection. She said there was only one entrance to her apartment and she was afraid they would sneak up on her. I would not issue her a weapon, but I allayed her fears by telling her that most of the VC went barefooted and she could scatter a box of tacks on her steps and scare them off. I am not sure she fully appreciated my humor. She did welcome my suggestion to contact the housing people and ask them to move her into an apartment hotel until things quieted down. She like that idea and went off to make arrangements with the housing people.

Two women came to me with their application for a riot shotgun. Neither of them had ever used a shotgun and knew nothing about the operation of a shotgun. I doubt it they knew which end the pellets came out of . . . "Permit Denied!"

I took a .45 semi-automatic pistol away from a young man. He came to me one day and pulled his pistol out of his waist band and said the thing goes off unexpectedly. The magazine was in place, he had a round in the chamber, the hammer was cocked back and all he needed to do was to release well known safety devices common to the Colt .45 and touch the trigger. It is a wonder he did not shoot off some vital part of his body before I took the gun away from him. Marine guards shook their heads at some of my clients and were happy I had the job of issuing weapons rather than themselves.

People living near USAID 1 and 2, the US Embassy and the residence of Ambassador Bunker are not quite sure they fully appreciate their status neighborhoods. I know we have mixed emotions at our apartment building where we are surrounded by the CIA, CID, Navy Intelligence and near the cemetery, Palace and the Embassy.

I have many other items to write about in this letter but I will stop now and make a cup of hot chocolate and watch "Mission Impossible" on TV. In a way that program depicts these days in Vietnam.

(continued)   20 February 1968

Again tonight the sky is filled with planes, flares and tracers. Over the noise of the air conditioner and the sounds of the radio and typewriter I hear bombing and mortar fire. All day we could hear the crash of artillery fire which seemed to be in response to the lower pitched, but equally deadly sounds of mortar and rocket explosions. In less than one hour this afternoon I heard at least nine rocket explosions in the direction of Tan Son Nhut

Airport. The streets of Saigon are completely deserted tonight and our street seems unusually dark. Around us we can occasionally spot the dim glow from a cigarette being smoked by one of the guards at the General's house, MACV Communications Center or the CID across the street. Everyone seems to be alert for whatever may come next.

(continued)   21 February 1968
Morning again! The war was quite noisy last night. Tan Son Nhut was hit several times. This morning we hear Saigon is ringed by the enemy on three sides. The radio just reported an imminent attack on Saigon is expected. Someone must be reading the wrong script because this has been going on now for 22 days. Now off to work.

I rode to work in a friend's Jeep. We had not gone far when an old man on a bicycle abruptly turned in front of us as we slowly started up at a signal light. The Jeep ran over the back wheel of his bicycle and bent it. We sat there until a policeman reluctantly came to investigate the accident. He directed my friend to back up in order to extricate the bicycle. Then he indicated we should drive around the corner and wait. My friend was for running, but I held him back. After 15 minutes, with me serving as interpreter, the damages were settled on the spot for VN$200 (less than US$2). Everyone was happy with the settlement. The man can have his bicycle repaired for about VN$100 and he will pocket the savings. That is probably more than he could have earned by working all day. Of course he may have to share the settlement with the policeman. Who knows?

Noon . . . same day:
The latest USAID Bulletin announces a golf tournament will be held on Washington's Birthday. How is that for priorities? You recall I mentioned there are bomb craters in several of the greens. Now it will be easier for the golfers to stick their heads in the sand. There is actually more concern around here about the bomb craters in the greens, and the fact that the golf course has not been watered or the grass cut for over three weeks, than there is for the plight of the poor homeless refugees!

At USAID a Vietnamese friend and employee came to work limping and with a large bandage on his elbow. He told me on Sunday an American had run into his motorbike and knocked him to the ground, damaging both him and his bike. The American turned around and drove off without stopping. My friend said in the 13 years he had worked for the Americans he had learned to get the man's license number. In his injured condition and pinned down by his bike he recorded the license number. I hope the police catch that driver and throw the book at him. There are many similar accidents like this each day, some are fatal and the Americans speed off. The newspaper carried stories about hit-and-run accidents just before Tet. Now the Vietnamese are far from being good drivers, but they do show some courtesy to others. But this is not true of the offensive American drivers. We are warned continuously to comply with Vietnam's driving regulations. Of course no one has thought of translating these regulations and requiring each applicant for a license to at least read them and sign a statement indicating they have read and understand the regulations. The curb lane on all streets and highways is reserved for bicycles, motorbikes and horse carts. But just yesterday I saw a GI driving a truck in the curb lane with the air horn blaring at a poor old man on a bicycle. The GI truck forced the old man against the curb and he fell off his bike. I could not see if he was injured.

Arriving at work this morning I checked in at my office and was prepared to go over work plans with the crew. Instead of Mrs. Hai's cheerful greeting and beaming smile, I found her eyes full of tears. The other men had not noticed she was upset. I asked her what her problem was and she told me three VC were captured in their neighborhood last night. The VC carried secret documents which indicated they were going to attack her section of Saigon today. Between sobs she told me her children were at home alone. Her husband was at work at the electrical company. I offered to have one of our men drive her home. I also offered to let her and the family move in with me. She said she would work until noon and then go home to the children. Later in the morning she called me at the Embassy and told me her husband had found a place where they all could go for safety. Now that is a very sad story. This man and his wife fled from North Vietnam in 1954 to escape the Communists. They came south to the safety provided in Saigon. Now they are on the run again, but where do they run? The faith and confidence these people have in the Vietnamese government and the Americans to protect them must be near the breaking point.

Another Vietnamese employee in our office told me this morning he was going home at noon and move his family. He lives in Cholon near the Race Track where fighting has been going on continuously. He says word is out that the VC are going to attack again. He says the VC were detected yesterday setting up antiaircraft guns in his area. Many planes pass over this area in their final approach to Tan Son Nhut. He said American planes fired many rockets and heavy machine gun fire into his neighborhood last night. We hear continuous rumors about what the VC is going to do or not going to do. It is difficult to sort out truth from speculation.

Today the Information Warden Emergency Center where I have been temporarily assigned was on the receiving end of rumor after rumor. Most reports concerning the VC can not be verified, but in this war of nerves people are shook up. I have been flooded with requests for guns today. We exhausted our supply of .45 semi-automatic pistols. We do have a good supply of rifles and shotguns. I screen applicants carefully and interview them in person. Fortunately most of the people are honest in revealing their experience with firearms. You would be surprised to see some of the folks who want a weapon. Many have never used a firearm and would not know how to load or fire it and some are deathly afraid of using one. These applicants are not issued a weapon but I am doing well in counseling them about the dangers of firearms to the inexperienced.

This is not a happy place. Yesterday at noon one of the USAID men went home for lunch, apparently in good spirits. About 1400 hours he and his girl friend were stabbed to death. Neighbors report much noise and arguing came from the apartment and then complete silence. One of the men working in my temporary office and a friend broke in and found them both dead. My friend could not return to work until today and he is still visibly shaken by what he saw in that room. The deaths are being considered as a murder and a suicide.

Another friend returned to Saigon after visiting his family in Bangkok during the initial fighting. Tan Son Nhut Airport was under heavy fire at the time and all arriving and departing passengers were restricted to the base for 24 hours and put to work. My friend injected formaldehyde into some of the hundreds of dead GIs whose bodies were flown into Tan Son Nhut from around the country.

Another friend was requested to go to the holding-morgue and try and identify the body of one of his USAID colleagues. After looking at hundreds of bodies and pieces of bodies he could not identify any as the USAID employee. That friend has not been the same since that horrible experience.

The number of wounded and dead arriving at hospitals increases and then decreases for a few hours. Admissions are said to be normal and then there is a large increase. At one time the hospitals were surrounded by hundreds of refugees and family members of those taken to hospitals. Small children were the most pitiful sights. They were crying and could not understand what had happened to their parents. The GVN has moved refugees and children from hospital areas and relocated them to other sites. There must have been some tearful sights when the children and others were physically moved. A large orphanage is located along one of the routes I take in walking to work. I notice a large influx of orphans. It is a heart rendering sight to see their little hands clutching the bars of the iron fence and their troubled little faces peering out at you.

Schools remain closed all over South Vietnam. Many school areas are being used to house and care for thousands of refugees. The Revolutionary Development workers, those trained to handle the pacification programs in the country, have been pulled into Saigon and other cities to help with refugees. This means the countryside has been abandoned by these workers, who in most cases were the principal defense forces for the villages and hamlets. It is difficult to imagine what has happened in rural areas while major attention is focused on the cities.

It is time for me to return to work. I have been called back to the USAID office for a program review this afternoon. The USAID hierarchy has been reviewing the forestry program and the rumors are that the program and personnel are in for a large reduction. Whatever that means, since we are small already. I have mixed emotions about what may happen. I would hate to leave now when we have made substantial breakthroughs and progress, but it could happen. Certainly there is a need for reductions to achieve some real economy and improved management of the entire USAID program. The forestry program has had little recognition or encouragement from top level USAID management both here and in Washington. I expect an uphill fight to hold on to the program. We have presented justification and supporting documents for program realignment and continuation. We can, for example, provide and improve the production of materials needed for the vitally needed reconstruction. Forestry related activities, including timber receipts, have the potential to provide the GVN with a major source of revenue. We achieved a major breakthrough just before Tet in one province where we legalized cutting operations and made sawmill operators and loggers happy by giving them legal permits. This eliminated major sources of corruption and heavy payoffs at three levels. This action alone was instrumental in reducing the costs of lumber by over 50%, a savings that can be passed on to the poor Vietnamese consumer.

(continued)

Later, the next day at noon . . . 22 February 1968

Yesterday, before my program review meeting, I received word that USAID/Washington had recommended abolishing the forestry program. We are told that President LBJ has directed Ambassador Bunker to reduce the personnel in the US mission by 50%. One of the Assistant Directors of Agriculture (not our boss Assistant Director Overby) informed the USAID hierarchy that the forestry program could easily be eliminated and make the grand savings of some $200,000 (that is grand? out of a total agriculture program of $20 million!) I did not learn of any other program reductions or eliminations. The elimination of one entire program may look impressive and could show that USAID/Agriculture was cooperating all the way, with LBJ. Our Boss Overby was not around when this decision was made, he was visiting his family in Manila.

I have mixed emotions about the decision to eliminate the forestry program. I have believed for some time, long before Tet, that the entire USAID program in Vietnam needed a careful review concerning program priorities, progress and management. We came to do a job for the Vietnamese people and we have done it and made significant progress. Some of our major breakthroughs include salvaging war damaged timber, eliminating unbelievable amounts and sources of corruption, legalizing timber cutting operations and thereby bringing substantial revenue to the GVN. We have improved the operations of hundreds of sawmills. When we came to Vietnam there were less than 10 in operation. Now there are several hundred. We have helped modernize and motivate the Directorate of Forest Affairs. I dislike thinking about pulling out now and leaving the new Directorate staff to flounder on their own. All of us on the forestry team have the same objectives and thoughts. Frankly we are discouraged.

USAID's management style is difficult to comprehend. The decision to eliminate our program was done without offering us an opportunity for rebuttal. The people involved in making this decision obviously had not reviewed our recommendations to modify the program and redirect our priorities to meet the country's pressing needs, both financial and structural. Our recommendations for redirecting the thrust of the program had been put together by Syverson in a well documented proposal while the rest of the team were working on other emergency, Tet-related programs.

Today I feel like the commander of one of the Special Forces Camps which has been overrun by the enemy. The events leading up to the loss of a Special Forces Camp can be traced to someone on the inside opening the gates or sabotaging the operation. The proposed elimination of our program was an inside job, the enemy is USAID/Washington. They have never recognized the value of forestry in any of their overseas programs. "Forestry takes too long", they say. In my judgment USAID is interested only in short term results, those that have a quick political payoff for the agency. USAID has yet to recognize the value of the forest resources of Vietnam or the forestry program. To USAID, as it is to the US military, "it's that damn jungle that gets in everybody's way". USAID fails to recognize the forest resources are one of the most valuable renewable natural resources to be found in this tiny country. USAID chooses to ignore what the war is doing to the forests, let along the people. You have read about the US military bombing, the monstrous Rome plow land clearing machines and defoliation. USAID fails to recognize the value of protective forest cover on soil and water resources. In a country receiving rainfall that ranges from 40 to 160 inches per year, forest cover has a special significance. The Montagnard's slash and burn, or shifting cultivation as it is called, results in a tiny portion of the forest being lost. To counter all these pressures we are pushing salvage of war damaged timber wherever possible, encouraging reforestation, promoting selective cutting and protection and improved management of the most valuable old growth stands, forest nurseries, teak plantations and mangrove.

The Directorate has been known as a source of extremely profitable positions. Jobs were coveted, bought and sold for high, and I mean "high" equivalent US$ values. The job holder was able to recoup his payoffs through kickbacks obtained from sawmill owners and loggers. In such cases the GVN treasury never saw a piaster of timber taxes. We helped surface this problem in one province and encouraged the Director and Minister of Agriculture to initiate a program to eliminate corruption. How did we do that? We worked with and gained the confidence of sawmill owners and encouraged them to take the initiative and report actual demands for kickbacks to a responsive new government. The program showed promise of being expanded to the economically distressed and war torn I Corps after Tet. There are other provinces in need of priority attention to eliminate corruption. We find the forest industry people, if encouraged, will speak out for themselves, and keep us on the sidelines, where we should be. One of our program responsibilities was to reduce costs of lumber for domestic consumption. Eliminating corruption simply reduces costs by 50% or greater!

As I reflect on USAID priorities in Agriculture, which are rice and protein production and land reform, I agree that forestry does not contribute directly to these programs. However we can provide thousands of jobs and make a major contribution to the rebuilding of tens of thousands of homes and businesses that have been destroyed by the war. This certainly could not be accomplished by USAID's credit and cooperative people, or extension specialists or irrigation engineers. What about land reform? That is one of the great unreported failures!

Thinking our challenges through with members of our forestry team was a good exercise and we came up with a number of excellent ideas for a reoriented forestry program that could contribute to the welfare of the Republic of Vietnam. Today (Happy Washington's Birthday!) when I was called in for a program review I had good handouts and visual aids and gave a convincing presentation to the USAID/Agriculture people. This resulted in our proposal being transmitted and hand carried to the USAID/Vietnam brass. At least we have convinced the local USAID/Agriculture people that perhaps after all there is merit to the forestry program that they may have overlooked. It will be interesting to see the outcome of today's meeting. I sincerely hope we do not become another casualty. USAID has too many failures here and around the world. From a practical standpoint the Tet offensive provides an ideal opportunity for USAID to fall back, recognize the inevitable and utilize the Asian concept of saving face. The fact is, other than carrying on some public health measures and food distribution, USAID is not doing much productive work at this time. In fact much of AID's total efforts never reach the desperately poor people it was intended to help. An old peasant once told me the USAID workers were like mosquitoes . . . "they sting you, they talk a lot, but change little".

Speaking of survival, it is fascinating to watch the maneuvering and scurrying taking place among the professional USAID people to protect their jobs and coveted way of life. In the end I am confident their efforts will perpetuate management problems inherent in all USAID programs. Some of these people certainly could not survive long in the real world!

Now it is time for me to return to the Embassy to wind up the emergency project. I expect a large exodus of Americans soon. If it comes, rumors indicate we can expect to be flown to Australia and New Zealand. I am flexible and can either continue the forestry program or meet you next month in Sidney and then to New Zealand and Tahiti. I'll keep you posted on developments.

I came close to completing my emergency project but there were so many calls and requests for weapons that I did not finish today, perhaps tomorrow I can wind it up. The pressure to arm US civilians is heavy and I have quite a chore to screen the requests. More people are moving from outlying villas into hotels and the wardens are having difficulty in keeping records on moves.

Fighting was heavy around Saigon today and again tonight. The number of wounded and dead being brought into the hospitals has increased in the past two days of fighting. While I was washing dishes a nearby explosion knocked over some of the dishes I had stacked for drying. A friend was here watching TV and he took off at a run to go to the roof to check on the source of the explosion. That is a very dangerous practice but people are like the tiger, they are attracted by the sounds of explosions.

I listened to Radio Hanoi last night. They report each evening on fighting in and around Saigon. They precisely described the action, damage and injuries caused in attacks on Tan Son Nhut Airport Sunday and Monday. Radio Peking does the same thing except they harangue the American presence to the point of complete boredom.

One of the men I know works at the medical depot near the Phu To Racetrack. He says artillery was shot over their heads into the section of Saigon between the airport and the racetrack. Helicopter gunships fired machine gun fire into the area throughout the day. Jets and Skyraiders bombed and fired rockets into the area for hours. Many VC are reported to be dug in and were detected setting up heavy gun emplacements.

Barry Flamm returned from the Philippines today. He traveled to Manila with the first plane load of Philippine evacuees from Vietnam. He said the turnout of relatives, friends and the curious at the Manila Airport was tremendous. He was the only American on the plane and a CIA agent whisked him through customs and immigration formalities in five minutes. He found his trip to be very productive.

Again security precautions have been tightened in Saigon. Recently troops and guards seem to have regained some of their alertness. This indicates more action may be expected. All American troops, even announcers on American Forces TV, are dressed in fatigues and appear to be ready for action. American MPs have suffered heavy losses in the Tet offensive and they appear to be exhausted.

Walter Pierce contacted his son and the boy's commanding officer gave him a 48 hour pass. The 20 year old young man looked aged, tired and haggard. He has been through more than anyone should be expected to experience. He feels deeply about the loss of one of his closest friends in the early hours of the battle at the racetrack. Walter brought him in from Long Binh Camp which is out across the River and filled him with steak, eggs and milk. He had been eating nothing but C-rations for days. Prior to Tet he had been on search and destroy missions in swamps and rice paddies around Saigon. Since then he was involved in 20 days of fighting at the racetrack. One of the writers for the USAID publication interviewed Walter and his son for a news article.

We have seen a tremendous increase in flies in Saigon. Before the Tet offensive flies were uncommon in the city, in fact none of the hotel rooms or apartments have screened windows. The garbage and human waste have been accumulating and providing breeding places for the flies. Refugee camps are literally pest holes. USAID and Vietnamese public health people are trying to reduce sources which breed flies. Mass inoculations are being administered all over the country. Air America cargo planes are in the air continuously carrying vaccines and emergency supplies to all parts of South Vietnam.

Scenes around the streets:

Americans packing guns is a common sight. I saw one young USAID employee with a .45 pistol tied down to his thigh, ready for a quick draw. The frontier mentality and characters such as this one are a greater threat to the peace and security than anyone. I had one ask me the other day if I could help him let the hammer down on a .45 automatic pistol. This is the second instance I have been called on to help a novice "gun slinger".

The last time this young man tried to release the hammer the gun went off. Picture this if you will . . . he had a round in the chamber and the hammer back while the pistol was in the holster. Such carelessness gives me the creeps. We carry ours in our brief cases, or I can carry a .38 revolver in a shoulder holster when I wear a jacket. Either way our weapons are within easy reach and we know when and how to use them effectively and safely. Fortunately we have not had to use them for any purpose other than shooting targets.

The PTT, Post and Telegraph Office, has been mobbed on numerous occasions. People are frantic to send messages and receive word concerning relatives. The most frantic ones are those with relatives in Hue, where heavy fighting continues after 22 days. That beautiful ancient imperial capitol is being torn apart block by block to save it. I am thankful I did not go to Hue with that forester during Tet.

Employees of the Directorate have suffered. There have been unofficial reports of deaths and injuries among the personnel. There is no word on the fate of people in the Cantonment (Regional) Office in Hue. Throughout the country, homes of employees have suffered damage or complete destruction. The home of the Regional Chief in Dalat has been destroyed. I recall visiting his very nice home in the center of town.

GI vehicles are taking a beating. I see them driven around town with bullet holes through the windows and bodies, fenders are damaged and some Jeeps are sagging as the result of broken springs. I have seen vehicles carrying tremendous overloads of troops and ammunition. Gas caps are missing and noisy engines and transmissions provide evidence they are not being maintained.

Yesterday some ARVN troops appeared at one of the USAID warehouses and dragged two Vietnamese employees of USAID into the street and practically stomped them to death. ARVN soldiers are brutal to prisoners. We suppose these two were suspected of being VC. But who isn't? The VC or VC sympathizers have infiltrated every niche of Vietnam society.

This letter contains some of my recent observations. One of the clippings you sent me from the *Washington Post* indicated Saigon was returning to "normal". Now you can judge for yourself what is "normal".

Your welcome air mail letters are coming through in just five days. We are told regular mail and parcels in transit to Vietnam are bogged down. The radio reports a backlog of 150 tons of mail has been put aboard surface ships and will arrive in Vietnam sometime in March. Normally all mail is sent by air. However all air transport is being used to ferry arms, ammunition, medical supplies and aluminum caskets.

That is all for this letter. Good night.

Sorry we have had so much trouble balancing checkbooks. Having two checking accounts, one in Arizona and the other in Maryland can be confusing. We have too much of a time lag between writing checks, reporting and reconciling the balances. I came up with a solution! I have sent you all the checks and current stubs I have here. You can balance everything from there. I really have no need for a checking account at this time.

Gosh, what have I done? A prediction has just come true! My Mother said you would get it all someday and clean me out! You have just done it! Now you have everything. Please don't leave me now!

I receive your letters regularly and they are very dear to me. They add a touch of reality to this topsy-turvy world.

I continue to be careful and exercise great caution while the war is going on all around us. I am well and eating regularly. Some of my food gets a little monotonous, but it is filling. I miss the good BOQ food. But even that has changed. They now serve one standard meal. There is no variety and no choice. But few customers are complaining.

The victims of the latest fighting are pathetic, but there is no complaining. More victims are being created as dead and wounded continue to be brought into the hospitals. There has been some reduction in hospital admissions. However the fighting goes on all around us. Tan Son Nhut is showered with rockets and mortars every night. I usually go to bed between 2100 to 2200 hours and hear a few rounds go off, some are close enough to rattle the windows. Then around 0300 hours the noise of explosions wakes me for the remainder of the night.

Damage in Saigon suburbs is heavy. Several hundred homes were destroyed yesterday. People are really suffering. With the "need to destroy the cities to save them" all I can ask is, "who needs an enemy when they have friends like us?" Conditions in the urban areas are terrible. The VC go in among the people and we or ARVN blast them out. In doing so many VC and innocent people are killed and wounded. VC who are not killed, move out and later come back and the process is repeated over, again and again. The situation looks hopeless and has been for some time, but as you say in your letters, no one will admit it. Our side does not want to admit defeat or admit we can not drive the enemy out. Neither side will give in and the GVN says they will never seek or form a coalition government. Meanwhile cities, villages and hamlets are being reduced to rubble and the total number of dead and wounded increases.

I firmly believe I am doing what is right. We have helped the people find jobs and we have eliminated much corruption which was plaguing people and industries. The job has been difficult and challenging enough without having to fight the

USAID system for survival of the forestry program. USAID has given us instructions to phase out in Fiscal Year 1968. To date the US Department of Agriculture and the Forest Service are leaving the fighting for program survival to me. After all, USAID invited us into the country in the first place. The Forest Service and USDA will not be criticized if our program is eliminated. It will be due to the inept managers (if you can call them managers) and administrators of USAID. We have made our position clear, meanwhile we are busy helping rebuild the nation.

Our friend, Pete Stilley of Flagstaff, is okay according to my informants. He is doing an outstanding job of helping refugees in Danang. Military operations are very active in Danang and all over I Corps.

My last letter dated 19 to 22 February was mailed without stamps. With no stamps available in Saigon we are advised to send our mail without any. I marked the letter "No Stamps Available In Saigon ". Hope you receive it, if not let me know and I will send you a copy. Please do not be concerned about our income tax returns. I will have six months after I return from Vietnam for filing. Gene Jensen can help the girls with theirs.

I was surprised to learn your brother Bill Sanders has the farm up for sale. But that is his decision. It must be very difficult to be a farmer these days. I hope he has something left after the mortgage and bills are paid. It is sad to have the farm go out of the family after three generations of Sanders' ownership. I shall always have pleasant memories of the farm and your family.

My lunch break is over so I will sign off and return to work.

4 March 1968

Today is my first, and probably last, anniversary in Vietnam. I arrived here just one year ago today. Outwardly, my normal haunts and the surface of Saigon look about the same. There are some holes in the buildings, some buildings are missing, and some have fire scars. Today there are other changes worth recording in this letter.

On my arrival one year ago I wrote about the war going on around us. I commented about flares being dropped at night and a fire fight 14 miles across the Saigon River. In my first letter I wrote about the sound of the multiple "whumps" of B-52s dropping strings of bombs 40 to 50 miles from Saigon.

Last night was a typical evening and tonight sounds like a repeat performance. My windows rattle as the slam of an explosive concussion hits them. It sounds somewhat like your home on the farm in Clinton when the south wind slammed against the house on a bitter cold winter evening. At times dishes in the cupboard rattle. The bombing is not 40 to 50 miles away tonight, it is now taking place in and around the edges of the city. North, south, east and west, 360 degrees around us, we can see flares and hear continuous explosions. The VC are either trying to cross one of the major Saigon River bridges or blow them up. In contrast to one year ago, rockets and mortars are now being fired at us. Tan Son Nhut Airport and surrounding areas are a battlefield every night. Two nights ago ambulances shrieked past our apartment again and again on their way to the 17th Field Hospital. The Army reports "the poor devils have us surrounded". As reported in *TIME* magazine, 1 March 1968, there is a question as to just who the poor devils are. Reports reaching us over the American Forces Radio, TV, local papers and *STARS & STRIPES* are optimistic and in all recent attacks "the damage is reported to be minor and casualties light". Some of us wonder if the reports are true. What we have seen appears to be major!

Another difference that can be readily observed on my first anniversary is the attitude of the people. Some US tempers are getting short and we hear frequent, loud and bitter arguments. Very few people are able to leave Saigon for outlying areas. No planes are available for civilian transport. Demands on aircraft and pilots for transport of food, military supplies and medicines are punishing. You have read about the food shortages in Hue. The flow of vaccines to Tay Ninh is continuous and efforts are being made to combat a black plague epidemic. That was one of the last areas I worked in just before Tet. In the outlying areas the water supplies and distribution systems are reported to be damaged. Electrical facilities are down or the supply of diesel fuel to keep generators running is exhausted. In Saigon most facilities are working. Fresh vegetables, when available, are sold at a premium. We have been cut off from the daily flow of fresh vegetables from Dalat. Rice that can be harvested is piling up in the Delta country since it can not be transported. Since there are few rice storage facilities in the IV Corps area there will be much loss and spoilage of this vital commodity. Pork producers in the Delta are liquidating breeding herds. Some pork from nearby sources is being marketed locally. Today Saigon is isolated, as far as food supplies are concerned, except for limited produce from local suburbs and that which is brought in by air.

Now that Director Muoi has returned from Vung Tau to Saigon he is very security conscious. Since his return he was been working on the security features of his home. He has purchased all the GI sandbags he could obtain from the black market. He has strengthened the shutters and doors on his house. He is in the process of constructing a bunker in his back yard for protection of his family and himself. His home is near the Directorate.

Many of the Vietnamese people have lost faith in the ability of their government or our troops to protect them. The northerners who are now working as civil servants in their government and those working for US agencies are the most concerned. Tales of VC activity and the killing and fighting in Hue have been widely circulated. In that once beautiful city the VC had complete dossiers on civil servants and northerners. They were ruthless in rounding them up and killing them. Mr. Muoi has had sketchy reports from the regional and district offices. At this time it appears that all are accounted for except one man who is reported missing. One forester survived by hiding in a hole in his back yard for 26 days. He saw VC and North Vietnamese soldiers moving around his home each day, but he stayed put and survived.

Of the three FHA men who had dinner with us just over a year ago, each has recently received bad news. Mr. Deetz from Minnesota received word his son was missing in action. His son flew large cargo helicopters out of Danang. He flew supplies into Hue during the heaviest fighting. The load was delivered and the helicopter was monitored out to sea, that being the only safe route. Radar network monitored him as far as the Danang sector. However Danang radar never locked onto them. They either went down at sea or crashed in the forest covered hills north of Danang. Mr. Deetz has been in Danang this week and talked with other members of the squadron. There is no new word. The helicopter has not been sighted. Mr. Deetz is deeply concerned and holds on to the hope that the crew may be hiding out or captured.

One of the other men received word that his son was in some kind of serious trouble. The son had lived with them in their safe haven quarters until he returned to the US. A third man went home on emergency leave when his wife was found to have cancer.

I have a great feeling of relief each time I hear from you and learn all is well, that is everything except for your horrible Washington weather.

Mrs. Hai is worried about her family. They are still living in the same insecure area. She tells me the VC have all of the northerners marked for execution. People are given a meaningless trial before being executed.

We have had several farewell parties during the past few days for departing USAID people. In our Agriculture group two have completed their tours and departed. One of these was an economist who has been here for six years. I asked him if he could identify areas of progress. He said conditions were in far worse condition now than they were when he arrived.

On the way to one of these parties, which are held on a Saturday afternoon due to the 1900 to 0700 hours curfew, I saw three US soldiers in a Jeep towing a trailer filled with many cases of beer. The soldiers looked hot and dirty. They were armed to the teeth. A large Negro Sergeant was standing in the

Jeep with his hands holding onto a machine gun. The grin on his face seemed to say "this is ours to enjoy, let them try to take it!"

All around Saigon we can see parachutes hung up in trees. These come from parachute flares that are dropped over and around the city each night. These are common sights around Army outposts and Special Forces camps. Now that Saigon is a battlefield we look like a back country outpost.

Emergency controls are in effect concerning vehicle use. US soldiers can be seen driving Esso gasoline trucks. Vietnamese drivers are not permitted to enter fuel storage areas near the river and they can not enter the airport. Military convoys roll through the city continuously. Most are carrying supplies but some are carrying ARVN and US troops. Many of the streets in Saigon are sealed off, while others have temporary barricades which can be moved aside to permit official cars to pass. We are now experiencing a great deal of trouble in getting into Tan Son Nhut Airbase. Entry requires not only an individual ID but a special vehicle pass.

The normally happy-go-lucky University of Saigon students look depressed these days. Recently I was waiting for an officer at one of the BOQs and from the Jeep I watched a young student sitting across the street from me. He was trying to study but he was obviously upset about something. They have a right to be disturbed, classes have been discontinued, the University is closed and many students have been drafted into the Army. Others are working on emergency rehabilitation projects. Come to think of it that student may have been spying on the BOQ for the VC. Who knows?

We are now actively engaged in a large refugee housing project in the Cholon area. Thousands of homes were destroyed in this area and refugees are living in schools, tents and shacks in refugee centers. The area we are preparing for them is an old French military reservation. With the Directorate of Forestry Affairs we have been working out the details for lumber procurement. Lumber is being purchased by the Ministry of Social Welfare. The Minister of Refugees and Minister of Social Welfare were elated to find someone to help them with these details.

Wouldn't you know it? We had a problem at the outset of the project. The woman who was awarded the contact for lumber turned out to have six illegal sawmills, all were operating without a license. Her prices were the most reasonable and we worked with Director Muoi to legalize the operations. It was one of those deals where the woman was denied licenses so she operated illegally by making heavy payoffs. This is typical of some of the corruption we have uncovered. Director Muoi is anxious to clear up these cases as rapidly as possible. He has the full support of the Minister of Agriculture. We are now making progress in the elimination of corruption in three areas, Saigon, Tay Ninh and Ban me Thuot.

Walter Pierce, our newest millwright from Edison, Georgia is handling the lumber shipments as they arrive at the refugee resettlement area. He has developed a program and teaches refugees, Buddhist priests, Boy Scouts, university students and Korean soldiers how to cut and prefabricate housing components. Our biggest challenge on this project was to move from the talking stage into the planning and execution stages. I fear many projects in Vietnam are bogged down and no progress is being made for the same reason. Coordination is an unknown word as various US and Vietnamese outfits spin their wheels. But now at least we are at work doing something beneficial. We have the project well organized. Pierce is working at the refugee site and Syverson is working with the Directorate on lumber production and procurement. We fear we will have a serious problem when the available supply of logs is exhausted. I am working on log procurement and that is a problem. No roads are secure and we find that the fighting has destroyed or heavily damaged many logging trucks. We considered using military trucks on a back haul basis, but they are running the wheels off of these trucks for high priority military missions.

Barry Flamm and Walter Metcalf went to Ban Me Thout to evaluate the damage to the city and to the sawmills. They returned with their report and recommendations on what was needed to get badly damaged mills back into production to help rebuild the devastated city. General Cole, Chief of Staff of CORDS, asked for our help and said he would provide the funds to purchase the necessary sawmill equipment. Operators are ready to rebuild and return to full lumber production. The Directorate is making all of the operations legal. The Director and Minister of Agriculture approved our recommendation to replace, at no cost, the volume of logs and lumber that were destroyed in the fierce fighting. And we are ready to provide the technical assistance. But now the "he clerks" in CORDS (we had them in the Forest Service too) are staying up nights figuring out how they can stall the operation. Tomorrow we have a meeting with the General, perhaps we will find out who is running the show. That should be an interesting confrontation. As soon as we receive clearance Flamm and Metcalf will return to Ban Me Thout.

I continue using every means at my disposal to keep USAID from phasing out the forestry program, since it does not contribute to rice and protein production or land reform. It is a challenge to be carrying on a fight on at least three fronts at the same time. By the way, in addition to clearing the air on the streets of Saigon, the Tet offensive stopped defoliation. The spray planes have been assigned to military supply runs. I am conducting an analysis of the impacts of the defoliation program on the forest resources of Vietnam. It is a real accomplishment to have the forests recognized as a valuable resource rather than "that damn jungle that gets in everyone's way". My work is classified as TOP SECRET at this time. I will have more to write on the subject of Agent Orange in the future.

Last Saturday night and Sunday morning explosions originating from Tan Son Nhut Airport were frequent and noisy. It was impossible to sleep through so many barrages. I got up about 0400 and did my housework, laundry and ironing. I learned later that while I was doing these chores four Americans were killed and 35 wounded at the Airbase.

Sunday morning I walked downtown. At the Catholic Cathedral I stopped to examine the Madonna statute. The VC tried to destroy her. They cut through a protective barricade of barbed wire and set an explosive charge at the base of the statute. The sapper either did not know how to do the job properly or the good Lord was protecting her. She suffered only powder burns.

Beggars are thick along the sidewalks. Some are professionals I have been accustomed to seeing for the past year, but many were new. At one of the art galleries the GVN has set up a display illustrating VC death, destruction and atrocities. On display were burned statutes of Buddha, broken candlesticks from a family altar, a sewing machine and other familiar, badly damaged items. Some gruesome looking photographs of death and destruction were on display. The place was crowded with spectators and ARVN guards. I decided that this was a good crowd to avoid and I departed!

My next stop was at the APO to purchase stamps. A hand written sign stated "No 1, 2 or 10 cent stamps". I waited a few minutes for the window to open and sure enough there were no 5 cent stamps either. I helped the postal non-commissioned officer add "5" to the sign. That is the reason I keep sending you letters marked "No stamps available in Saigon". I hope you receive these bargain letters!

While on my walk I observed several commercial jets take off from Tan Son Nhut in a gradual spiral up over the city, until they reached a safe altitude and then headed in the direction of Bangkok, Manila or Hong Kong. On that day all airplanes were spiralling over the city in a gradual descent and then a final steep dive into the "safety" of the airport.

My last stop was the US Embassy. I left a message for one of the supporters of our forestry program. The place bristles with alert guards, both Marines and Army. I could see at least seven heavy machine gun emplacements on the roof and grounds. When I saw this show of force I recalled that only three, or was it five, guards were on duty at 0300 hours that first morning of Tet. I thought again about Pearl Harbor. At this time it appears we are covered.

(continued)
Later, Saturday evening   9 March 1968
The war in and around Saigon has continued all week. Last night I detected something different . . . it was strangely quiet. I went up on the roof and there was not one flare in the air or the sound of any explosions. It was a rare interval after almost 40 days of continuous battle sounds.

It is difficult for us to learn from the radio, TV or newspapers what is going on at this time. Apparently both sides are resupplying, licking their wounds and preparing for the next round. Attacks are being made here and there. Enemy losses are always the heaviest and ours the lightest. Again we wonder if truth has also been a casualty of this conflict. Two nights ago the VC hit the nearby Bien Hoa Airbase with a rocket and mortar attack. Some rockets hit the pilots' quarters. A USAID man from Bien Hoa was at the airbase and he told us casualties among the pilots was high. The radio made no mention of an attack in that area.

Ten days ago I wired USAID I Corps at Danang and offered to come up with Director Muoi to help get forest industries at Hue, Quang Tri and Quang Nghi back into production. Yesterday I received the following response:

INFO FROM QUANG NGAI, QUANG TRI AND HUE INDICATE LOGGING AND MILLING OPERATIONS NOT POSSIBLE UNDER PRESENT SITUATION. WILL ADVISE WHEN POSSIBLE AND YOUR ASSISTANCE INDICATED AS NECESSARY. END

The agriculture advisor from Danang came to my office and gave me the story of I Corps. All five provinces are a battle zone. Vietnamese and Americans' main job is to keep alive. Help is provided refugees, but conditions are terrible. These proud people, most of them farmers, have been taken or blasted out of their homes and put in make-shift refugee camps. Sanitation is bad and cases of the dread diseases cholera and plague continue to show up but are generally kept under control. Refugees are located either in the mud or under the boiling sun. At some camps rice and other inadequate food supplies are thrown off trucks to frantic refugees, just like feeding livestock. Although the Vietnamese are operating the refugee camps, few hearts and minds are being won by America under these conditions.

The reports I receive from Hue, from men who have been there, from what I read in *TIME*, *NEWSWEEK* and your clippings, makes it sound like that once beautiful city has been through Hell . . . and is Hell! I wanted to help in some small way as did Director Muoi. He agreed to give the operators free timber

411

and we were prepared to help rebuild the sawmills but now that is impossible. At least Syverson and Flamm were able to visit and see historic Hue before it was destroyed (to save it?).

Quang Tri is completely surrounded by the enemy and nothing is being done there except to watch and wait. The sawmill in that city has undoubtedly been destroyed.

Nowhere in I Corps is progress being made. Rice is rotting in the fields or consumed by insects or rats. No farmer dares leave his home to work the fields.

Before coming to Vietnam and since being here, I have read much about the history of Vietnam. While it is unbelievable, we have followed the French down some of the same roads, the same paths, used the same tactics and fought over the same fetid rice paddies in I Corps. We have suffered the same defeat and find ourselves in the same quagmire. The French made a serious mistake by holding out, actually they trapped themselves, in North Vietnam at a place called Dien Bien Phu. They were defeated in a final battle for that bomb scarred valley. That was the end of almost a century of French domination of Vietnam. Now we find ourselves in that same position, trying to protect a political-military symbol at a place called Khe Sanh near the DMZ. Our Marines are pinned down there and many are being killed and wounded each day. And for what? I am no military strategist but you do not do well in stopping infiltration of an enemy when you are surrounded. The prevailing attitude of our leaders is "the poor devils, they've got us surrounded". I admire our brave Marines who have the courage to remain there under 24 hour a day bombardment, but what else can they do when they are ordered to hold. Holding out in one of those camps which is under continuous direct fire must be nerve racking, if not terrifying. Khe Sanh is similar to the Special Forces Camp at Tra Bong where we set up the sawmill. I have seen it from 5,000 feet. There is an airstrip, trenches, underground bunkers and watch towers. Only the communications bunker is really secure from direct hits and space in the common bunker is limited to a few Marines.

That is the bleak picture of I Corps. It is doubtful we will return there. When we do, if we do, there will not be much remaining to build on. It will mean starting all over again.

Our efforts to find materials to help rebuild Ban Me Thuot advance slowly in spite of great odds. The US and Vietnamese in Ban Me Thuot want to rebuild. The General, Chief of Staff for CORDS, wants to rebuild. But the General's staff and USAID are dragging their feet. I was under the impression when a General gave orders to his staff it was up to them to carry out the orders. Not so! We made our recommendations. The General approved them and gave orders to his staff. Now his civilians, Colonels, Majors, Captains and Lieutenants are doing their best to stop the effort. USAID has all the materials in its warehouses we need to put the mills back into production. However to date we have been unable to pry them loose. I stated several months ago that USAID and the US military have enough idle equipment in their warehouses to win this war, win this peace and wage another war and win another peace. However, we foresters are now using a technique that was sometimes effective in "appropriating" material from neighboring national forests and ranger districts. We have the General's blessings and as soon as we pin point the required equipment we will move and move fast to requisition it. This will not be an easy task. Imagine if you will the chore associated with finding and sorting through hundreds and hundreds of acres of storage areas . . . bins, boxes and crates tucked away in hundreds of warehouses. It is absolutely unbelievable to see the materials that we have moved to this tiny corner of the world! And believe it, if you will, there are few people who know where specific items are located. You never saw such a paradise for treasure hunters! And that is where we come in! Pierce and Metcalf are real experts when it comes to finding and "appropriating supplies". They had their training in the hills of Kentucky and the redlands of southwest Georgia.

We have made good progress on our refugee housing project in Cholon. That has given us much satisfaction and high visibility for the forestry program. The lumber is flowing in from the lady's (now legal) mills. Pierce has done well in handling the lumber shipments and training the prefabricators. I am going to be with him tomorrow in the good clean dirt, work up a sweat and blow off steam that has accumulated this week as I dealt with bureaucrats.

The dry season is with us in Saigon and temperatures under the boiling sun are in the mid-90s. Tomorrow the temperature is expected to increase. Up on the roof tonight I could see huge thunderheads over the South China Sea. Cloud to cloud lightning was mixed with artillery fire and flares. It was a pleasant sight.

The people of Dalat have asked for our assistance in resolving a GVN bottleneck in issuing sawmill and logging permits. They need lumber for reconstruction of the many homes damaged and destroyed in the Dalat battles. Dalat is much like other provinces where we have not taken action to investigate and encourage the loggers and mill owners to report corruption. As soon as we begin work with the local operators and gain their confidence, they open up and "blow the whistle". I took action this afternoon and had Director Muoi secure approval for the Minister of Agriculture to clean up recent forestry corruption in Dalat. A telegraphic order went out to the Regional Forester of that area directing him to prepare permits for each sawmill and logging operation and have them ready for issuance on Tuesday. Arrangements are made for Martin Syverson and Assistant

Director Hiep to be there to meet with the local operators and issue permits. Local foresters and the province chief may become slightly ill, because they have been "bleeding the industry dry" and requiring payoffs and kickbacks as a condition for operating illegally. You can believe me when I say this is not "penny ante" stuff! This is one of the rare cases where specific steps are being taken to clean up corruption at the provincial level. It takes much courage on Director Muoi's part when he takes on powerful military province chiefs. Muoi makes the equivalent of a few dollars over a US$100 per month. Over the years the equivalent of many thousands of dollars have been drained from the pockets of sawmill owners and loggers.

USAID is still scheming on how to reduce the number of USAID/Vietnam employees. One day we hear all of us will be gone by June. Then after a little salesmanship we hear only two or three of the forestry group are leaving, then it is four. The numbers change from day to day. Recently we have found a ground swell of support for the forestry program. No one knows who will win out in this maneuvering. I know for certain, in the long run, the losers will be the Vietnamese people. They have little to look forward to.

Occasionally people ask "when are you foresters leaving"? I smile and quote Mark Twain after he read his own obituary . . . "I fear the report of my death has been greatly exaggerated!" We hope when the smoke clears part of the forestry program will survive.

ODDS & ENDS:

The radio continues to caution us to keep off Plantation Road and the area south to Cholon. This morning there were mortar attacks in those areas.

We have seen and heard much publicity about the M-16 rifle. The radio just now is advising GIs to keep magazines clean and not to use oil because it collects dirt. They are told tracers are pretty but they foul up the gas ports and lack the punch of ball ammunition. The M-16 is a beautiful rifle but I am appalled at the condition of some soldiers' rifles.

I am well equipped with a number of protective devices. I now have a .38 cal. revolver, a .45 cal semi-automatic pistol, a .30 cal. semi-automatic M-1 carbine, a .30 cal automatic M-2 carbine, a few grenades and a fully equipped field surgical kit. These would not be considered to be standard forestry instruments by the Society of American Foresters, but they are useful in Vietnam. None of us go looking for trouble, in fact we go out of our way to avoid it, but we believe in being prepared. The surgical kits were delivered to me for issuance to the wardens. When two Vietnamese Embassy employees came in carrying these kits the Marine guards just about had "mule colts". You see the kits are just about the size and shape of the deadly dynamite satchel charges used by the VC. We find the Marines are a little jumpy both here in Saigon and in I Corps.

The most frequent warning we now receive over the radio is for drivers to check vehicles for attached bombs. We have no trouble checking our (trusty/rusty) Jeep because it is full of inspection holes. In fact it would be difficult to find a place strong enough to hold a bomb. USAID's Jeeps came from an USAID reconditioning center in Belgium. We understand a number of USAID employees were separated at that facility for fraud. Seems like the Vietnamese are not the only ones to get in the fraud act. But after all the Jeep has served us quite well.

That's all for tonight.

11 March 1968

It now appears the forestry program will be reduced to one person. If it comes to that I have recommended Barry Flamm remain for another 18 months tour of duty to manage the forestry program. This is the proposal USAID/Vietnam has presented to USAID/Washington. USAID/Washington says close out the forestry program in June 1968. To date the Forest Service has been silent and is leaving the negotiations to me.

There is much to be done here but the USAID people are stumbling, bungling and stalling most projects. USAID people are most concerned about their own future and major reductions in all programs. As a result progress in most program areas is slow or non-existent. I have never seen so much foot dragging and procrastination. I have mixed emotions about leaving. I want to help the Vietnamese people but I am tiring of fighting the USAID hierarchy.

Other than that all is okay here. I have been terribly busy in the office and have no plans for going out of town.

It now appears we can get out of here in any direction we select. Rather than being shipped to Australia against my wishes, I want to plan my own departure and make reservations. I will plan to arrive in London two or three days after you all arrive. That will take care of any delays I may experience en route. I need your estimated dates of departure in order to make my reservations and have the travel people obtain my tickets. Your British Airlines deal sounds unbelievable, I suggest you grab it. You will be a rich reservoir of knowledge on the British Isles. Besides having you along will be so much better than a brief case full of notes. Please have the girls check with the American Automobile Association concerning International Drivers Licenses. They should study up on international traffic signs.

If our program is reduced my plans will be flexible and I will not have to comply with those home leave rules. We can spend more time in Britain than we originally planned. I will keep you posted on USAID's final decisions.

I have no birthday card for you, Cindy, but Happy 17th Birthday on 17 March. Be good to your mother and ask her to give you a check from me.

22 March 1968

This has been a rewarding and a discouraging week. Rewarding in that the forestry program is going well. Three of our men are making good progress on the refugee housing project in Cholon. We are making continued progress in cleaning up corruption and receiving good support from Director Muoi and sawmill owners. We are gaining high level US Embassy recognition of the serious impacts the war is making on valuable forest resources.

The discouraging part of the week concerns the proposed program reduction. I worked hard to retain the entire forestry program but it may be reduced to one forestry advisor. Subject to the approval of Chief Cliff and Forest Service Director of International Programs Clark Holscher, Barry Flamm will remain for another tour of duty. I am trying to place Syverson and the two millwrights in the Refugee Program. We may not succeed in transferring them.

I informed USAID and USDA I want to return to the Forest Service. I will be prepared to leave on or about 10 May. I made my reservations with Pan American Airlines this morning. I have a feeling time is running out for me and I had better get out while I can. Time is also running out for the people of this little country. I have little reason for optimism as long as we continue to pursue the present course of military and political action. I view Robert F. Kennedy's entry into the scene as a ray of hope for the world. Although I am not certain the world can stand our friendship and protective support much longer.

I have a depressed, let down feeling when I think about leaving before our work is done. But I have made the decision to go and it is time. I hope my decision is supported. I want out!

I plan to leave here on 10 May for the following places:

10 May To Bangkok, Thailand
14 " To Calcutta, India
15 " To Katmandu, Nepal
20 " To Calcutta
21 " To Istanbul, Turkey
23 " To Frankfort, Germany
1 June To London & a happy tour of the British Isles
24 " To Washington, DC, home, debriefing and reassignment

If there is any delay in receiving official travel authorizations, I will fly direct to London. I hope time permits the trip I have scheduled. I need time to return to normalcy. I am not certain I am prepared for the culture shock of returning

416

o the real World. USAID and the US military spend much time orientating new arrivals to the perils and culture shock of Vietnam. Then practically overnight when the tour of duty is done we are dumped back into a culture that is strange to us and does not understand us or what we have experienced. I expect to cover this subject in my debriefing.

Best wishes for a good spring vacation at Isle of Palms. I will be thinking of the fine time you will have in South Carolina. That was a wonderful idea of having a trailer set up and waiting for your arrival. You two drivers, listen up! Be careful driving and take care of your Mother!

I am putting this letter together at the Cercle Sportif pool. I must close now and have lunch at the club restaurant. Fortunately this place did not suffer during the siege of Saigon. Being operated by local Vietnamese, with French support, it probably is owned by VC interests.

I am in better spirits now that decisions have been made. I will push ahead and finish jobs I want to complete before I leave.

30 March 1968

This will be a short letter. I want to get Cindy's tax return in the mail.

I just returned from a trip to Quang Ngai. The situation looks even more hopeless than ever in I Corps. There is no security, no one can get out to do any productive work. Attacks are expected everywhere, at any time. The Army sweeps through and clears an area. There are heavy losses on both sides, bodies are counted, the VC is pushed out, the VC returns and the cycle is repeated again and again. There are hundreds of plague cases in Quang Ngai. If this sounds like the Middle Ages, it is! I see more hatred in the eyes of the suffering local people. The French were hated in Quang Ngai Province and now they appear to hate Americans even more for the suffering they are experiencing.

Our volunteer from the IVS (International Volunteer Service), forester John Chitty is quitting and ready to pull out because there is no security in Dalat. Up to now John, with his ability to speak French and Vietnamese fluently, has been able to travel where the US Army and ARVN fear to go. He wants to work for the Forest Service. I will recommend him for placement.

I can hardly wait to get out and meet you in England. I wish we were there now that it is spring. But time will pass rapidly and we will soon be together again. Working helps pass the time and all of us have been out on projects this week. None of the other USAID programs seem to be doing much of anything at this time.

Cindy, I will attach separate instructions on how to handle

your Federal Income Tax Return. The 0.75 cents Maryland refund is not worth the effort of filing. You should check with Gene Jensen, but my recommendation is to forget it. Melissa, your withholding statement shows no tax was deducted and you earned less than $600, therefore you are not required to file a return.

In regard to our tax returns I will file for an extension. For the record my Foreign Service grade is FC-2 which is equivalent to a GS-16. The current salary $21,636 + 25% for hazard pay + SMA + my cost of living allowance. All of which comes to approximately $50,000 per year.

Gwendel, Happy Anniversary today! You never remember the date but I always do.

31 March 1968

I have just returned from the TSN Heliport where I put Barry Flamm and two ecologists from the states on a helicopter which will take them to Tay Ninh and a Special Forces camp near the Cambodian border. They will be examining heavily defoliated areas.

Back in my apartment I had ironing, dusting and floor cleaning to do. An old Vietnamese lady who cleans the halls of the building has been after me for some time to hire her to do my cleaning and washing. Today I decided to hire her for VN$3,000 or US$25 per month. Melissa, this is very inexpensive, but after all she provides only the services listed above.

Gwen, thanks for the birthday card. I look forward to our celebrating in England. In preparation for the trip you all must have a smallpox vaccination and a tetanus booster. Call the State Department Medical Department and tell them you will be meeting me and these services will be provided free of charge.

You should plan on bringing enough money to England to support me in the style that I am accustomed to. Please proceed with plans for visiting your sister Leela and family in early August, that is before the 15th, because of Cindy's hayfever. My ragweed hayfever has not bothered me for many years, perhaps she will outgrow her allergy also. While not knowing what I will be doing on my return to Washington, I presume I can take leave during that period. I will have a new Hong Kong suit when I return. I ordered it last week at the PX where Hong Kong tailors have a shop.

As you know I have accumulated many additional hours of annual leave time while working in Vietnam. Working on Saturdays and holidays has added many hours of compensatory time. I have used no annual or sick leave, only compensatory time. With the special authority federal employees are given for working in Vietnam, I should be able to carry over a maximum of 360 hours of annual leave until I retire. That should be worth a lump sum payment of over $5,000. Not bad! But I earned every hour of it!

Sounds like you and Betty Carroll had a good visit. She is really a character but we both love her dearly. It has been a pleasure knowing Betty and Franklin since our time together, seems like a 100 years ago, at Happy Jack, Arizona. I recall Betty's feelings about ants. She would not like the ones we have here, but I agree with her thoughts in every respect concerning Vietnam. The entire world would be better off if every human in Vietnam, including us and all the other ethnic groups, disappeared and the ants took over. The situation looks more hopeless with the passing of each day.

You are incorrect about the 15,000 poor, sad and unemployed prostitutes on at least two counts. There are 20-30,000 and they are not unemployed or even underemployed. Many work on a full time basis in my apartment building and every US leased facility in Saigon. I have not yielded "chua" (yet). I plan to hold out past 10 May.

We visited my Vietnamese refugee family a couple of weeks ago. They are getting along fine. Fortunately the big sister salted away lots of money from her job at Maxim's and she and her sisters have not had to hustle.

Last week, Bill Abbott, the Assistant Administrator of the International Agriculture Development Service was in Saigon to check on the phase out. He says Chief Cliff is waiting to write me until dates of departure are definite. He says the Chief has a good job to offer me, as his assistant, whatever that means. Then with a twinkle in his eye he would say no more, whatever that infers? He says Chief Cliff wants me back and agrees on leaving Barry Flamm in charge. Mr. Abbott sees no problem in setting my departure for 10 May and Assistant Director Jerry Overby agrees. It appears I am on my way out. I am what locally is known as a "short timer". As the GIs calculate their time, it is "so many wake up calls".

My efforts to retain the forestry team is gaining momentum. Several, high level requests are in for up to three of the men to stay. Lew Metcalf would also like to remain but I prefer to see Walter Pierce get in his year, especially since his son is here. Pierce worked very hard to get here and was turned down many times but his persistence paid off and he was finally approved for the appointment. He has been a strong addition to our team and his efforts on the refugee housing project have attracted strong support for the forestry program.

The temperature is increasing each day. Yesterday it was 97 degrees. The sun is bearing down mercilessly. But I enjoy this kind of weather. When I am in Saigon I normally have a lunch of papaya and two small delicious bananas while sitting beside the Cercle Sportif pool. That is a feature and attraction I will miss when I leave.

During the past two weeks we have been busy in Saigon and the field. Some of our trips would be considered out of the ordinary when compared to the usual forester's work program.

Martin Syverson was in Dalat to help facilitate reconstruction of that beautiful city. The local forester was withholding delivery of permits for cutting timber. However he did find the time to issue cutting permits to a few chosen individuals, i.e. those who provided kickbacks. USAID advisors in Dalat attempted to resolve this problem themselves but were unsuccessful. Sy went up with his counterpart, Mr. Nguyen Van Hiep, and quickly resolved the problem. Permits are now issued to sawmills for cutting and the refugees have been given special permits to obtain government timber at cost. We are making an effort to eliminate the minimum cost for the refugees but the Minister of Agriculture is dragging his feet. He does not want to provide free use permits and set a precedent.

Between the stories Sy brought back and those I obtained from IVS forester John Chitty, it apparent that Dalat experienced a very difficult time during the Tet offensive. The filtering process (censor) kept peaceful Dalat out of the news. Being so peaceful Dalat had very few troops for protection before Tet. You will recall my descriptions of the beautiful city and pine covered Central Highlands. Dalat was a R&R center for rich Vietnamese and Chinese and the VC. Dalat was off limits for most Americans. It was an area where both sides had the philosophy of live and let live. But with Tet, accommodations changed. VC troops from other provinces came in and hit the city hard. They occupied the center of the city. Losses to local troops and their American advisors were heavy. VC intelligence was outdated because they hit the old MACV headquarters which had been relocated several months earlier. John Chitty told me he was pinned down at his house by VC who occupied nearby houses. Their objective was the precinct police station. Knowing this the police fired at everything that moved. ARVN troops with armored vehicles were called in to drive them out. Chitty watched a brave ARVN Lieutenant lead his troops in an attack, but as they neared the VC lines the troops deserted him. He in turn had to fall back. Then the ARVN troops and vehicles emptied their weapons firing at nearby houses, completely missing the houses containing the VC.

Some VC hid out in a refugee settlements after the initial attacks. Between bombing and artillery fire, refugee areas were demolished. They found no VC casualties. The same situation occurred in the area containing fine French-style villas. On a beautiful ridge above Dalat a few VC holed up in one of the houses. The Vietnamese Air Force saturated the area with bombs and rockets and did tremendous damage. Facilities not damaged or destroyed were occupied by ARVN soldiers who looted everything of value.

In downtown Dalat the VC occupied the theater which is near the central market. They were in a part of the theater building capped by thick concrete and invulnerable to attacks by helicopter gunships. From this protected site the VC killed every person who showed themselves. A US teacher went by the area before he knew it was occupied. As his Jeep went by his Vietnamese companion was shot and killed. When the teacher drove out of the danger zone he found four slugs imbedded in the back of his flak jacket. Bodies laid in streets near the theater and the market for days under the hot sun. Finally on the fifth day of the siege, Montagnard rangers from Ban Me Thuot were brought in to ruthlessly drive out the VC.

With the exception of a few isolated patches, thousands of acres of pine plantations and newly established pine seedlings were destroyed by fire. That was tragic loss of valuable timber and watershed cover. Many Montagnard villages in Dalat Province are now occupied and controlled by the VC. Chitty has no opportunity to carry on the forestry program. Therefore he has asked for release from IVS and applied for work with the Forest Service on one of the national forests. He has completed many fine projects and will be leaving shortly. So goes it with beautiful Dalat.

Since writing to you last I returned to Tay Ninh and Quang Ngai. Common factors to be found in both provinces were fear, destruction and plague.

On our flight from Saigon to Tay Ninh City we flew above the clouds at 5,000 feet. The trip in a Porter turbo-prop normally takes about 20 minutes. After flying for 40 minutes I wondered what was up. I looked around and down through holes in the clouds. I could see peaceful farmland beneath us. There were no bomb craters, no defoliation and no fire scars. Off in the distance to the southeast I could see Black Virgin Mountain, a prominent landmark just outside Tay Ninh City. Our good old Air America pilot was lost. I pointed out the mountain to him and he agreed we were slightly off course. We had violated Cambodian airspace and were well across the border. Fortunately Cambodia has no fighter aircraft and we got out of there, fast!

Approaching the city I could see where bombs and artillery fire had cratered most of the fields. In practically every direction I could see where tanks, APCs and trucks had churned up most of the poor, overworked farmers' fields. Landing in a Porter is scary, since these things land in the middle of town on a runway that is less than 100 yards long. On takeoff the pilot cranks the flaps down, throws power to the engine and then points the nose up and takes off in less than 100 feet. Amazing flying machine, the Porter, I am impressed with its performance!

My mission to Tay Ninh was a follow up trip to verify that the sawmill owners had received permits to legalize their operations. I was in Tay Ninh prior to Tet with Director Muoi. Accompanying me on this trip was Ms. Elizabeth Landeau an industrial specialist from Bien Hoa. She has a very strong Russian accent and is a whirlwind to behold, especially when she talks over an aircraft radio with one of the Bien Hoa Filipino radio operators. Both VC spies and I are confused when they try to follow a conversation between those two. Ms. Landeau is an ardent supporter of the forestry program and has been pressuring USAID and the Embassy to support and continue the forestry program.

We had received reports from Tay Ninh concerning innovative new methods of kickbacks and payoffs. However we were pleased to find the mill people were very happy with the new arrangements and had no problems with local officials. We also detected a very positive change in the attitude of the local forestry and province officials.

The meeting with the sawmill people, local forestry and province officials is worth noting. The meeting was carried out through interpreters. I spoke a few sentences and then listened very closely to be sure the interpreter was translating my words and the responses from others correctly. Working through interpreters can be frustrating because many times they think they know what you are saying and even worse they may have their own ideas about what they think you want to hear and they put it all in their own words. I had an advantage because I could understand the local people fairly well and the interpreters knew that.

So what was the problem? I have described some of the noisy meetings I have attended, but this one was even more exciting. We helped form the Tay Ninh Sawmill Association. The meeting was held in the home of the President of the Association. His home was beside a busy main unsurfaced road. Every few minutes a convoy, or a tank company or a fleet of APCs (armored personnel carriers) would go racing by. The noise and dust were terrific. Added to this was the frightening sound of pistols, rifles and machine guns being fired in the front yard and you have the perfect recipe for bedlam! After we all hit the deck and dived under tables and chairs we learned the firing came from a group of civilians being trained in civil defense. This was about as dangerous as confronting the VC but we all survived and completed the meeting.

We visited one of the sawmills that had been destroyed in the Tet fighting. The owner was a former Army officer who invested his life savings in the mill. When the VC attempted to hit a nearby district headquarters they were driven off. They took refuge in the mill where between attacks from helicopter gunships and dive bombers, the man lost his mill, home and a substantial amount of lumber and logs. He and his surviving workers had pulled together scrap lumber, roofing and constructed shacks for families who had lost their homes. They were in the process of trying to piece the mill back together, but it looked hopeless. The mill owner and his wife had been to province headquarters and Saigon trying to file a claim and obtain a loan. All they found was disappointment and a run-a-round in the bureaucracy. Ms. Landeau told these folks she would take action and have them back in operation in a short time. I recall the statement, "never under estimate the power of a woman!" I am confident when she gets rolling on this crusade this man will receive his loan or perhaps an outright gift of money and equipment just to get Ms. Landeau off the backs of the GVN officials. I have often thought if this woman ever gets captured by the enemy, and she goes were angels and tanks fear to tread, the war could be over in a hurry, just to get rid of her!

The people of Tay Ninh City have been expecting new attacks every night for a week and they were nervous and on edge. Most of the local residents had bunkers in spaces adjacent to their homes. City streets and areas next to police, US and province headquarters were encased in concertina barbed wire. We have made friends in Tay Ninh, but hatred and mistrust of Americans runs high for many reasons. I could see the hatred in their eyes. They blame us for their suffering.

Convoys roar through the picturesque, narrow streets at all hours, day and night. On one of my earlier visits to Tay Ninh the Senior US Province Advisor was pulling his hair because a US tank had just run over a Lambretta carrier loaded with nuns, nurses and old women. Most of the passengers were killed or seriously injured.

During one of my recent trips to Tay Ninh American advisors were trying to calm a group of irate farmers. GIs in a passing convoy had tossed lighted matches at a farmer's ox cart which was carrying dry rice straw and a large log slung underneath. The farmer and oxen were badly burned and the straw, cart and oily log were destroyed.

CIA agents had antagonized both sides. Two CIA characters made frequent nocturnal calls beyond the inner defense perimeter to visit Vietnamese girl friends, allegedly to gather intelligence information. The CIA types committed a series of errors. They established a pattern when they made these trips on a regular basis. The women they visited were also the girl friends of the local VC. One dark night as the CIA men drove past the cemetery where the VC laid waiting in ambush, a solid wall of machine gun fire stopped those visits forever.

Then there are the poor loggers who are harassed by VC and GVN officials. The VC required loggers to carry money, food, medicines, ammunition and other supplies for their use. Loggers were forced to make payoffs and kickbacks to ARVN, police and until recently, local forestry and province officials. Loggers are hated by ARVN, Special Forces and other US military forces. FACs fly over logging areas on the lookout for loggers who may have strayed beyond designated "safe-areas" into the so-called "free-fire" zones. H&I (artillery term for harassment and interdiction) fire is directed at the "free-fire" zones at any hour, day or night. The amount of artillery shells and helicopter machine gun and rockets fired into these "free-fire" areas is astronomical and costly (artillery shells cost US$50 to $100 each). Lately the bombers have been dropping napalm on the loggers in addition to anti-personnel bombs. One of my trips to Tay Ninh came just after 28 loggers had been killed and 30 wounded from H&I fire ordered by a FAC observer. Wounds resulting from this massive fire power are just horrible. The results of H&I fire can also be seen around every one of the 50 Tay Ninh sawmills. There are piles of ruined band saw blades. I could collect bushels of large, very wicked looking, razor sharp, case-hardened shrapnel fragments that have been chopped out of logs.

Some day I will provide more information about being out in the "boonies" which is called "War Zone C". This hateful area covers what was formerly an excellent forest reserve near Tay Ninh City. (Then there are "War Zones A and B".) "C" is a wicked, much fought over area, where no quarter is given by the VC, NVA, Special Forces and our mercenaries, the Cambodians, Montagnards, Nungs and Vietnamese irregulars. We have been involved in "C" on many occasions. Barry Flamm has visited all Special Forces camps along the Cambodian border in this area and obtained significant information regarding the defoliation program. In doing this he has slept in rat infested bunkers. Once he had to dive headlong into a bunker during a mortar attack, only to have others dive in on top of him. In measuring forest inventory plots to check impacts of repeated defoliation he and a Vietnamese forester were surrounded and provided security by a Special Forces machine gun company of 100 men. While doing timber inventory work he has been chased by VC in dense tropical foliage and impenetrable bamboo thickets where he could not see more than a few feet in any directions. Fortunately the thickets impeded everyone and he was able to keep a jump ahead of enemy forces while obtaining important information. Colorado State University and the Society of American Foresters told us forestry was all fun and games. They told us we could hunt, trap, fish and live in a cabin in the woods. They didn't tell us about all the hazards associated with forestry!

We were scheduled to have a helicopter pick us up for a reconnaissance flight over one of the defoliated areas near the Black Virgin Mountain and Cambodian border, but someone had a higher priority for the chopper and it never arrived. While waiting for our flight back to Saigon a medivac, "Dustoff" helicopter, arrived with a load of wounded GIs. One GI's leg ended in a stump which was covered with a bloody field dressing. His boot with the rest of his bloody ankle and foot were beside him. Other wounded soldiers appeared to be in equally serious condition. The temperature in Tay Ninh that day was 100 degrees.

I could fill more pages with stories about Tay Ninh, but they will have to wait for another time. That's all for now.

Chapter XXVII

G-O-O-D-B-Y-E, V-I-E-T-N-A-M

3 April 1968

I hope all goes well on your end regarding London trip planning. I will send you information on my hotel reservations en route as soon as Pan American Airline confirms them. I plan to check with them later today.

We are all as busy as cats on hot tin roofs. Believe me when I say tin roofs are hot in Vietnam! Barry Flamm has returned following several days in the camps along the Cambodian border. Walter Pierce is on the Cholon refugee housing project. I am taking Martin Syverson and Lew Metcalf to the airport in a few minutes. They will be leaving for the Central Highlands. I am involved in revising the revised budget and work plans. Both Barry and I are preparing special classified, top secret reports.

Got to run to the airport now.

4 April 1968

This is just a quick note before I leave for work. I am enclosing a list of the hotels where I will be staying during my travel from here to London.

We still do not have official confirmation on who remains in Saigon or the departure dates for the others.

Our Assistant Director Jerry Overby goes to Tokyo today to meet with Secretary of Agriculture Orville Freeman. Martin Syverson will be on the same flight to attend an agriculture fair. His wife will join him for a tour of Japan before she departs for Portland, Oregon and he returns to Saigon.

I plan to shop for a suitable Vietnamese painting for my new office, wherever that may be. Do you want one?

9 April 1968

I hope this brief letter reaches you at Isle of Palms (w/cc to Bethesda). Sounds like you may be having wonderful weather in South Carolina. I wanted to tell you about the good news I received yesterday from Deputy Chief M. M. Nelson. He says the Chief and staff have approved me for the position of staff assistant to Nelson. This will return me to National Forest Administration. I am pleased. I hope you will be also, since no move is involved.

I have returned to Saigon following several trips into various parts of the country. I have only two more scheduled trips to make before I complete my work plans. I will leave Vietnam on 10 May.

I plan to keep free of VC and any "hearts of gold" for the duration.

The following memorandum was received from the Forest Service:

Forest Service
Washington, DC 20250
April 8, 1968

6130

To: Mr. Jay Cravens
     Forestry Team Leader
     Vietnam
From: S. Fine
     Washington Personnel officer
     Subject: Status Change
          FOR OFFICIAL USE ONLY

We have been asked to formally transmit an offer to you following your return from Vietnam this summer.

The position will be that of a GS-15 Forester vice Leon Thomas who is retiring in April. A copy of the job description is attached for your review. The job will be slightly different from that described as additional coordinating responsibility in Budgeting, Travel and Manpower Ceiling and PPB (program, planning and budget) work is added. Also the responsibility involved with the Multiple Use Advisory Committee is included in the job.

We will protect your salary to a step in GS-15 that is at least equal to the rate you are now receiving. Since you are now receiving $21,632 per annum you will be placed in Step 7 at $22,082. This results in an increase of $450 per year.

We will be obtaining Department approval on this action as soon as possible. Please notify us as soon as possible if this assignment is agreeable to you.

Prior to your return we will be contacting you on further paperwork involved in your assignment.

Enclosure

Note: I accepted the offer by return mail!

14 April 1968

This letter will pick up my story at the point I concluded my 31 March letter. I have been out on trips during the past two weeks and that will explain the time lapses in my letters.

I wrote about being out in the Tay Ninh "boonies" or if you prefer, "boondocks". One of the fillers on American Forces

Radio is given by Cris Noel. Cris is a sultry voiced, luscious looking discjockey. To encourage GIs to brush their teeth they are told a freshly cut piece of bamboo makes a good tooth brush. In closing the announcer adds "Cris Noel brushes her teeth, even when she is out in the boonies".

American Forces Radio and TV have numerous fillers, including how to keep the M-16 rifle operating, the importance of taking malaria pills, a warning not to congregate at bus stops, check your vehicle for devices designed to remodel you and your vehicle, give your buddy a ride, etc. Advice is given about saving money, not to spend piasters and reviews of news headlines of 20 to 30 years ago. Speaking of advice every soldier of World War II is familiar with the posters (you probably remember them from Camp Carson) of the beautiful girl standing on the street corner by the lamp post waiting for the boys to come by . . . the caption in that war read "you can't tell a girl by her looks". In this war the poster shows a sloe-eyed beauty, a GI holding an M-16 rifle, with a bulldozer and a Jeep nearby . . . caption reads "know your equipment". All these little items are fillers on TV. So much for that.

I made a return trip to I Corps and Quang Ngai the last week of March for the purpose of getting the Tra Bong Sawmill back into production. Since Tet, air travel has been spasmodic and very undependable. Planes are being used on emergency projects and by VIPs. Ranch Hand C-123 pilots tell me they are now hauling trash around the country. There are not as many planes now as there were before Tet. The explanation of the latter, while not admitted, is obvious. I had a Class 2 Priority on Air America for a 1430 hours departure. Arriving at the terminal at 1400 hours I was told a mistake had been made and the special flight arranged for me to Danang had departed at 0900 hours that day. I slowly counted to 10 in Vietnamese and back again in English, and with much patience and speaking Vietnamese to one of my friends in Air America I arranged a 1500 hours departure to Danang, with the bonus of a drop off at Quang Ngai. That meant I did not have to go to Danang, work my way into town and then talk my way onto the morning courier flight to Quang Ngai.

We had a smooth flight to Quang Ngai in a Beech Vopar (twin turbo-prop jet) and arrived at 1730 hours. I obtained a ride from the airport into the city from an AMA doctor. You recall my description of the chuck holes with intervening strips of asphalt. Now the chuck holes are tank traps. Quang Ngai had changed from pre-Tet days. Some of the changes were visible and obvious to me, others were described by the doctor who gave me the ride.

Additional airplane carcasses at the airport provide vivid evidence why there is a shortage of airplanes. This provided further evidence of the improved accuracy of Tet snipers. Narrow roads and city streets were further restricted by miles and miles of concertina barbed wire and steel posts which lined roadways and restricted access to homes and businesses. Children could be seen playing in and around the tangled masses of barbed wire. Several gaps in the wire were evidence of the furious fighting and bombardment. The doctor told me the number of dead and wounded was more sickening than anything he had witnessed during WW II and Korea. The body count of VC and NVA was reported to be very high. Remind me someday to tell you about conducting a war by body count.

The doctor told me the aftermath of the Tet attacks was equally bad. Caring for the wounded and dying was a major task. Patients were discharged far too early for their own welfare to make room for more serious cases. Disease followed in the wake of battles. He said they had treated over 200 cases of plague at the Quang Ngai hospital during the month of March. Just that day he had detected a case of very contagious, usually fatal bubonic plague (pneuo-variety). The plague victim had come in undiagnosed and put in bed with a man with a broken leg. Due to shortages, the sharing of beds in Vietnamese hospitals is a common practice. The plague victim was coughing up blood. The doctor was still shaking his head and saying he could not believe a situation could bring two patients together in the same bed and to ultimately share the same fate. The doctor said not a day went by without him saving many people from dying. You can imagine what it is like over vast parts of this country where there are completely inadequate (non-existing) doctors, medical facilities and supplies.

Since this was to be my last trip to I Corps before leaving Vietnam, I made the best of it. I thanked a few people for their support. And I read the riot act, both volumes, to both Vietnamese and Americans for dragging their feet and failing to support the Tra Bong Sawmill operation. I am now able to interject a few choice Vietnamese words for emphasis. To make a long story short they agreed to get on the ball, which included moving in the elephants which had never been delivered, help develop markets for lumber, arrange for backhauls to air lift lumber over existing VC road blocks and to arrange for a GVN loan to the timber cooperative. Lumber markets should be no problem because the amount needed for reconstruction is immense.

I scheduled a helicopter for transportation to Tra Bong where I expected to stay for several days to see operations under way but an ambassador-type character had a higher priority and took off with some VIPs and my helicopter on a show-me trip. I was stranded in Quang Ngai City with no promise of another helicopter for a week.

Before dark I walked to the MACV compound where I had always eaten when staying in Quang Ngai. It was covered in sandbags and many of the barracks buildings were completely destroyed. I hope the FACs and other military advisors who lived there were in bunkers at the time of the attack. I stayed at the guest house of the Province Senior Advisor (CIA-type). His girl put me in the so called bullpen with two Korean medical supply technicians. All of the rooms in the more secure bunkhouse were taken. I had learned enough Korean to greet them and establish cordial relations. The bullpen was okay, had comfortable beds, toilet and shower, air conditioning (which did not cool), and millions of malaria bearing mosquitoes. Overhead protection against mortar and rocket fire consisted of a thin corrugated tin roof.

I slept well that first night in spite of nightly forecasts of enemy attacks. The mosquitoes were bad enough. There was no protective netting but thanks to liberal doses of repellent I made it through the night. Sleeping, sweating in 100 degree night-time temperatures and rubbing on mosquito repellent every few hours is not my idea of fun and games.

When the PSA returned the next day he had a fit when he learned that an American had been permitted to sleep under the tin roofed affair. He insisted I move into one of the more secure downstairs rooms in the secure building. According to emergency instructions I was told in case of an attack to run to his house or go to the bunker on top of the more secure guest quarters. One night it was so hot (over 100 and no air conditioning) and the mosquitoes were so thick, I went for a walk inside the compound's high walls. I checked out my two designated escape routes and found both locked and barricaded. It would have required a dynamite satchel charge for me to have retreated to my assigned post. I talked to the Vietnamese guards and learned they would be pleased to make room for me in case of an emergency. Again my luck held and there were no attacks. Three days later as I departed Quang Ngai the local people were piling on additional sandbags and digging their bunkers deeper.

While waiting to be placed on the Air America manifest, I went to a nearby Vietnamese guard post which was beside the PSA's and Province Chief's offices. There I saw a Vietnamese soldier sitting in a tightly cramped position in a shallow hole, under the boiling sun and covered only by interlaced barbed wire. I learned the soldier had deserted his post the night before and was working out his punishment. Vietnamese prisons have a similar punishment in what are called "tiger pits". The Vietnamese can be very brutal.

My Air America return flight to Saigon was canceled, but I went to the airport and hitched a ride on a C-130, 4-engine turbo jet cargo plane. My transportation priority also includes use of military aircraft. The crew was a friendly group of career Air Force officers and a Sergeant crew chief flying out of Clark Airbase in the Philippines. Fortunately they provided me a seat in the front office. I was glad because the cargo bay had the sickly sweet smell of death. They had hauled 23 Marines out of Khe Sanh the day before in rubber body bags to their last stop in Vietnam, which was the Tan Son Nhut mortuary. Before we took off for Nha Trang we loaded 50 Vietnamese and GIs in the very foul smelling cargo bay. The temperature at the airport at that time was 110. I am certain that the cargo bay was 130 degrees. Not a pleasant place, Vietnam.

The crew of the "Gone Goose" told me they had made two trips into Khe Sanh that month. They said it was like flying into the barrel of a canon. The first trip was a supply run and it was touch down, drop and go. The next one carried in Marine replacements and evacuated the 23 Marine bodies. In spite of a great number of patches the "Gone Goose" flew like a dream.

During the flight the crew chief brought forward a 14 year old Vietnamese boy from the cargo bay. The copilot let the boy sit on his lap, taught him how to read instruments and let him fly and slowly turn the plane as we flew south, about a mile off the coast, to Nha Trang. I took a photograph of the boy flying the plane. The plane commander asked me not to show that photo to their group commander. The boy of course was thrilled. I learned from talking to him that his parents were sending him from Quang Ngai to Nha Trang to attend school. All Quang Ngai schools have been closed since before Tet. Don't you know that boy's classmates will think he is the biggest liar in South Vietnam when he tells them he flew one of the Air Force's large cargo airplanes!

Nha Trang was even hotter! We loaded cold Cokes, JP4 fuel and about 100 more GI and Korean troops. Passengers in the cargo bay were jammed in like sardines. Of course there were no seat belts! It was a great relief to take off and climb to 8,000 feet where it was pleasantly cool. We flew through valleys in the sky surrounded by tremendous, towering white cumulus clouds. The huge plane was not disturbed by unstable air masses. It was an impressive flight, especially since I was permitted to sit up front in the office. We returned safely to Tan Son Nhut and I hitched a ride into Saigon before the curfew.

My next two trips are clouded with the official cloak of Top Secret. Both trips were made on separate Sundays and took me to widely separated areas in South Vietnam by military aircraft. I will be free someday to discuss many happenings, but not at this time. Again my luck held, the missions were completed and we returned safely.

I was scheduled to be in Tay Ninh last week but I decided since I was a short timer I would ground myself for the duration.

Later I learned on one the days I was scheduled to be in Tay Ninh, the VC set up a mortar position in one of the lovely French rubber plantations and shelled the city. Three shells landed just outside the place where I normally stay in Tay Ninh. It has been that way for months, trouble is just ahead of me or just behind me. Gwen, that is what I meant when I wrote in one of my recent letters . . . "time is running out for me". It is time to go. Frankly I must admit I will be happy when my Pan Am plane lifts off from Tan Son Nhut one last time.

After weeks of uncertainty the forestry program is not folding. Three advisors will remain. The amount of work to be accomplished will certainly decrease. But this work force of experienced forestry advisors can provide Director Le Van Muoi and his staff with important assistance while supporting various USAID, CORDS and military programs. I am pleased and relieved with the results of our efforts.

During the past two weeks, before I arbitrarily grounded myself, I made two trips with Mr. Muoi, Mr. Hiep and Ms. Landeau. She has really taken to the forestry program, especially in III Corps. My colleagues tease me and say she is doing it all for me. But I can tell you it is above and beyond the call of duty to spend 12 to 15 hours a day in the field with her. She pumps for information, prods and pressures high level province chiefs and senior American advisors, like the legendary John Paul Vann, into supporting the forestry program and individual projects. I must admit we have taught her well and she can prepare a technical forestry program request or report as well as Flamm, Syverson or myself. I give her a great deal of credit for she has been an important factor in our effort to retain the forestry program.

One of our trips took us to Binh Long province to size up the forestry situation. We made a low level reconnaissance flight in a Huey helicopter for a close view of some fine old growth timber stands. The pilot refused to fly over one area which he said had a reputation of being too deadly for Hueys. This was one of the areas where Barry Flamm had experienced problems just the week before on the ground. The pilot told us he had brought six American bodies out of there just that morning. I told him as one short timer to another I respected his judgment and besides he was driving. Our flight carried us over some beautiful, friendly country containing a number of Montagnard villages. We spotted several, out of control, war-related forest fires and some lush green rubber plantations. The French-owned and operated plantations are politically sacred and protected by a surrounding 5km buffer strip which keeps all defoliation, artillery and bombing activity well away from the rubber trees. Of course the buffer strip does not keep the enemy out. These places are deadly as far as American GIs are concerned. The enemy has huge base camps hidden in the plantations and can come and go at will. I now fully agree on Gene Jensen's hostile attitude towards DeGaulle and his countrymen. Such are the perils of a war run by the politicians.

During a break in our flight we dropped down in one of the secure villages for lunch. If the surroundings can be overlooked, you can enjoy a good, inexpensive Chinese meal in these tiny village restaurants. The flies are thick, mangy looking dogs scavenge on the filthy floor under your feet for food before it falls through the cracks in the floor to grunting, sway-backed, pot-bellied, stinking Vietnamese pigs rooting, fighting, squealing and snorting under the building. Cooks and waiters glare at American customers for any number of alleged or real problems we have created. But ignoring all this, the food was delicious, especially when drenched with the flavorful, odoriferous nuc mam. In one of those places Barry Flamm can win all the honors for eating the most. He can put away eight bowls of rice, in addition to all the other dishes. He even surprises the Vietnamese who can put away prodigious amounts of rice.

Following lunch we visited a local sawmill. The owner had met with us earlier in the day and reported he had absolutely no problems. He did not expect us to come to his mill. When we looked at logs scattered around the mill yard we could not see, contrary to Vietnamese law, one legally stamped and scaled log. The owner and the local forester looked like they had been caught with their hands in the cookie jar, up to the arm pits! Mr. Muoi disposed of the problem on the spot. I understand the forester is now in some army training center, far away up north. Everywhere we take Mr. Muoi and his staff we find similar problems. But, ever so slowly, we are making progress in identifying and cleaning out new sources of corruption. In looking at the situation from the perspective of the Vietnamese employee who is involved in such illegal practices, we find their official pay and rice allowance are grossly inadequate to provide for their family. If a Vietnamese forester is fortunate to have an official vehicle, his allowance will not provide enough for petrol to get him out of town. These are some of the reasons for accepting payoffs, i.e. it is "you scratch my back and I will scratch yours" sort of thing.

By far the greatest problem we found in Binh Long province was the lack of security for loggers. In fact they had more problems from local ARVN friendly forces than the VC. The ARVN forces found great sport in shooting into logging areas and the payoffs they extracted from loggers and log haulers were far greater than VC demands. The VC is very clever in not antagonizing the local population. This fosters hatred on the part of local people towards the GVN and Americans. Action to reduce loggers' problems has been taken up with high level

Vietnamese and US military channels.

On our return from Binh Long I noticed Barry Flamm was becoming visibly thinner. I insisted he go to the 17th Field Hospital for a checkup to determine the cause of his weight loss, stomach cramps and dysentery. Tests revealed he has amoebic dysentery plus a few miscellaneous, odds and ends of choice Southeast Asia parasites. The amoebic could be traced back to one hot day in the Philippines when he took a drink from a cool high mountain stream. The treatment for this ailment is said to be long and unpleasant.

Between trips I was called to a meeting in Bien Hoa with the Commanding General of III Corps. General Mearns, a former National Forest grazing permittee from California, needed our assistance. The subject of our meeting must wait for a later discussion. Before the General could see me I waited until he concluded a meeting that was in progress. I noticed a Jeep parked outside that carried a 4-star flag. This could have been the vehicle of Generals Westmoreland or Abrahams. It turned out to be General Westmoreland. He came out first, of course, walked up to me, and said "my name is Westmoreland, what is yours, sir?" I told him my name was Jay H. Cravens, Forester with USAID. He told me he and General Mearns had discussed my coming and they appreciated my offer of assistance. We had a brief conversation ("censored") and he opened the door of General Mearns office, introduced me and then departed. That room was filled with a "galaxy of stars". I counted six of them on various generals' collars. It was an interesting day and started us on another project that was to be completed before I departed Vietnam. A discussion of this subject must wait for peace-time or some other suitable occasion.

The next trip, unrelated to the above affair, took Director Muoi, Assistant Director Hiep, Ms. Landeau and me to the province capitol of Ham Tan. This province extends from the South China sea coast, across the coastal plains and hills to the Central Highlands. This province has everything, including fisheries, agriculture land in the coastal sands and upland valleys and forests that extend from the coast to the highlands. The war has damaged both people and natural resources. But work and life goes on for the Vietnamese and Montagnards. The province chief, a Colonel, offered us the use of his US provided helicopter and he and his US Colonel advisor accompanied us to see what they could do to help improve the forestry situation. We flew over much of the forested area and obtained a good perspective of the extent of the forests and war damage. The US pilot was another short timer and we did not fly at low levels. We landed at two district headquarters and were briefed by the district chiefs, Vietnamese Captains, and their US advisors. We received much information concerning forestry problems. Having Director

Muoi with us provided the opportunity to take action where it was needed and within his authority.

As we lifted up and out of one village we made a low pass over a beautiful, clear mountain stream. Twenty feet below us beautiful, young, bare-breasted Montagnard maidens waved at us from the middle of the stream. I was sitting back with the port side gunner where the safety belt permitted me to lean out and take a good look. I had the Miranda camera set at 1/1000 of a second and equipped with a 125mm lens. The lens opening was properly set but I must have had, what shall I call it, "buck fever". I failed to take a single picture as we roared past at 110 knots. The person who said "war is hell" clearly had Vietnam in mind.

I have more to write about this trip, but first I must prepare for my final field trip. After that one I shall complete one last long letter. The war goes on in Saigon. Yesterday they blasted the TV station, minutes after I passed it in the Jeep. Today I have heard other nearby explosions which jarred the building. I must be ready for whatever comes next.

16 April 1968

I hope your trip to South Carolina went well. Now you can begin serious preparations for the trip to London. My reservations en route are confirmed except for Munich and Nepal.

My last general letter will describe my recent trips. A number of them have been on classified missions and the telling of those stories will be delayed. Mr. Muoi and his staff have been with me on the non-classified field trips. Mr. Muoi asked me to remain as his advisor, but I told him Barry Flamm had been selected and I must return to the Forest Service. He accepted the decision. He and Barry work well together and I anticipate no problem. Barry departed yesterday for 24 days home visit. In view of the uncertainty about who remains and who goes, I have been unable to obtain approval for the other men to depart on a home visit. Official word on personnel has not arrived, but we expect confirmation any day.

After taking a malaria pill on Saturday I became ill. My stomach was upset. I have lost 10 pounds in the past week. I am convinced my stomach problem is not the same as Barry's amoebic. I have experienced stomach upsets from malaria pills on at least two other occasions. The medics tell me this happens to about 30% of the people taking malaria prophylactic medication.

Lunch break is over and now to return to work.

21 April 19

Just 19 more "wake ups" and I hope to be on my way out of here. Recently I have been working seven days a week and I am

435

tiring. It will be a relief to be on my way, have time to rest and reflect on the past 15 months of my life. In the days ahead I will wind up my work, sort, pack, purchase gifts, ship my household goods, attend a farewell dinner as guest of Mr. Muoi, update inoculations, take a preliminary departure physical examination, have a property check-out, etc. On the advice of experienced Foreign Service people I will have my final physical examination and clearance from the Washington State Department Medical section, rather than the 17th Field Hospital.

Tomorrow Pierce and I accompany Director Muoi and three of his men on a trip to Phu Quoc Island. That is one of the spice islands of the Orient and is located in the Gulf of Siam off the coast of Cambodia. We plan stops at Ha Tien near the Cambodian border, fly over Ca Mau Peninsula which is the extreme south end of Vietnam and make a final stop at Can Tho in the Delta. This last trip will provide me with complete coverage of South Vietnam, from the DMZ on the 17th Parallel to Phu Quoc Island on the south. We plan to return on Wednesday evening, the day Syverson is due back from Japan.

I have been confirmed by Pan Am for a room in London w/o private bath at the Regent Palace Hotel, Piccadilly Circus. I am scheduled to arrive there on 1 June from Frankfort at 1735 hours. When you arrive I will move in with you and share bed and bath, if I may. It seems like a 100 years since we have been together.

I have worked at the office all day. It is now 1730 hours and I am going to the Splendid BOQ for dinner. I am reducing my food supply at the apartment now that the end is fast approaching. If the VC hits us again, I may lose more weight. I have had severe dysentery for one week and have not been able to sleep well. Pills are ineffective. I plan to see a doctor when I return from the field trip. I must say it certainly is a reducing experience!

I am glad I hired a maid. It provides more time for work and some relaxation. She does excellent work on the apartment, my clothes, shoes and dishes. I seldom see her since she comes after I go to work and is gone before I return in the evening.

24 April 1968

I was pleased to receive your letter of the 17th and learn you had a good trip to and from South Carolina. I am glad you escaped the DC riots.

I returned from the field trip late yesterday. We were scheduled to remain for another day but the VC chased us out of one area. Here again I was just ahead, or just behind trouble. Now my field trips are concluded. If I can survive Saigon for the next 16 days I will have it made. There is continual talk of more trouble in store for Saigon, but that may just be a part of the terror tactics and war of nerves to keep everyone on edge. I will have more information on these items in my last general letter.

I saw the medics today and obtained a different medication to help clear up whatever is causing my dysentery. Tests discovered no bugs or amoebas. I was told I should recover in a day or two.

We certainly are having unique problems with our checking accounts. At the moment I have just over $300 on hand. I have donated money for refugees, orphans and church relief, but none for unemployed bar girls.

I am sending you silk cloth and pepper from Phu Quoc Island. The cloth is smuggled in from Thailand. If the pepper is damp please spread it out to dry. For several centuries pepper has been produced on that island for export to the West. Please advise me what bank I can write a check on to purchase $500 worth of traveler's checks for my travel to London. I do not plan to spend that much but it will be well to have it with me.

I guess you are relatively pleased with my new assignment with the Forest Service, especially since we do not have to transfer. I hope the next move will be to a regional forester position in the West. The Forest Service and USDA should be pleased when they receive confirmation that Flamm, Syverson and Pierce will remain on the forestry program in Vietnam. That is a major accomplishment since we were a complete write off at one time.

Your plans for England, Wales and Ireland sound good. I am looking forward to driving the back roads and visiting pubs and inns. Yes, I heard about London Bridge "falling down", to Arizona. Strange isn't it?

28 April 1968

I appear to have recovered from my gastric ailment and I am gaining strength. I was literally wrung out.

I worked at the office until 1000 hours this morning and then drove to the Cholon PX for supplies. My new Hong Kong suit has not arrived. They expect to receive it this coming week. There is much talk about another VC attack on Saigon. I do not intend to return to the Cholon PX, the place is just too dangerous. I will ask someone to pick up the suit for me. I now understand how everything comes into perspective when the GIs talk about "getting short" or becoming a "short timer". You become more cautious. I can not help but remember the GI who was killed in an attack on Tan Son Nhut Airport just before he stepped on the plane to go home for the last time. So until it is "wheels up" on 10 May "cerre la guerre"!

I returned to the apartment and spent the remainder of the day sorting slides and wrapping packages for mailing to you. These are items I do not want to take around the world with me

437

or ship with my household goods.

I now eat most of my meals at the Splendid BOQ. They serve a field ration mess. The food is satisfactory and less expensive. Breakfast is 27 cents and supper 55 cents. In this war the military still serves "SOS" every day for breakfast. In my war "SOS" was creamed chipped beef on toast. In this war it is creamed hamburger on toast. It tastes about the same. I have a few emergency rations just in case we are tied down again. I expect to survive nicely.

Tan called me Saturday and said he heard I was leaving. He invited me to visit him at the Presidential Palace. He will send a car for me tomorrow. It will be good to see my old friend before I leave. I know he recommended me for the Distinguished Service Award from the Ministry of Agriculture. He was a good friend.

5 May 1968

Sunday morning and now it is only five more wake up calls before I pull out. The remaining days will be filled with all work and no play. Recent demands on my time have been extremely heavy. In addition to work there is the USAID sign-out procedure which requires running around all over the city to obtain clearances on this, that and other things.

I have spent several evenings packing my gear for shipment. I selected two paintings and purchased more gifts. These are carefully packed. The movers came yesterday for my foot lockers and wooden boxes. I am now down to USAID's linens, utensils and items I can leave in the apartment when I depart. I have given my old clothing to one of USAID's Chinese drivers who is large enough to wear them.

The new suit has not arrived. So much for the good service provided by Hong Kong tailors. "So sorry, it may arrive in the next five days".

Last night four members of the forestry team had a delicious dinner at Director Muoi's home. Next week will be occupied with meetings, reports, lunch on Thursday for all the USAID and Directorate forestry group and dinner in the evening with the Assistant Director of USAID/Agriculture and his staff.

Now I am off to the BOQ for a 27 cent breakfast and then return and devote the day and night, if necessary, to conclude my last general observations for "Dear Gwen & Girls". It promises to be a long and I hope interesting letter.

Following the helicopter trip over the mountain stream, Montagnard maidens and the forests we returned to Ham Tam for a meeting I had scheduled with local sawmill operators and charcoal producers. With our encouragement they "let it all hang out" and told Director Muoi about their problems. Essentially the problems were the same as we had discovered elsewhere. There were large payoffs, no permits, marketing problems and no security or protection from airplanes, helicopters or artillery. It was encouraging to see a glimmer of hope radiating from the bright eyes of those operators. At last they had found someone from Saigon who could help them. During the meeting we had Coke and ice, made from questionable water sources. I did not want to offend our host by refusing the ice so I gritted my teeth and drank. By that time it made no difference, the Chinese food was reacting so I had little to lose! Following the meeting we were invited into the back room of the office for a feast of the largest lobster, and flies, that I have ever seen. Fortunately the lobster shells kept the flies off until we were ready. I ate faster than the flies and enjoyed the feast. By that time I was accustomed to having dogs underfoot and pigs that could be seen and smelled through large cracks in the floor.

At the conclusion of a seven course meal, we were herded into Jeeps and taken to the officers' club for beer and (ugh!) angel food cake. While there we learned our helicopter had been dispatched on an ammunition run. On the return trip an engine malfunction had caused the helicopter to crash into the roof of a house. Heavy damage was done to the house and the chopper but fortunately no one in the house or in the crew was seriously injured. All I could think of at the time was "there but for the grace of God go I".

You might have expected the helicopter crash to have stranded us in Ham Tan, but that did not occur. Ms. Landeau's arrangements for Air America aircraft are a wonder to behold. They always come through for her. The pilots and operators (CIA) do not want to incur her wrath. Come to think of it, she well could be an employee of the CIA. That thought had never occurred to me, until just now!

We dropped Ms. Landeau at Bien Hoa and departed for Saigon. I am always impressed by the smooth handling of the extremely heavy traffic at Tan Son Nhut Airport. The Vietnamese air controllers and their FAA advisors handle everything from the slowest, to the world's fastest and largest airplanes on two parallel runways. There are always several planes landing and taking off at the same time. The fastest aircraft must be the courier aircraft that makes daily flights between Saigon and Washington's Andrews Air Force base. It breaks the sound barrier and creates a sonic boom both coming and going. There must be a line of flying tankers strung out across the Pacific and US to accommodate this plane. I have never seen it and do not know what it is. Frequently I can spot an SR-71 "Blackbird" in one of the TSN concrete revetments. When this Top Secret reconnaissance plane is on the ground it is always surrounded by a large contingent of heavily armed Air police.

Back in Saigon I had to clean up and drive the other three members of the forestry team (Flamm was in the US) to Mr. Muoi's for an excellent Vietnamese meal. The tremendous meal was served by a servant woman. Mrs. Muoi remained in the kitchen supervising the preparation and only appeared to receive our greetings and expressions of appreciation at the end of the meal.

Picking up and returning Syverson to his apartment was a major challenge. He lives near Tan Son Nhut airport. Driving through all the barricades and concertina barbed wire was a demanding task. Roadblocks were set up every few blocks. Many military convoys were on the move that night. I saw one convoy of 14 Jeeps and trucks full of US MPs racing toward me (going the wrong way) on my street. I did not choose to argue who had the right-of-way and just before we met, I pulled into my apartment driveway, a few minutes before curfew. Artillery and helicopter gunships were very active that night. It appeared flare ships were dropping enough flares to outline the entire perimeter of Saigon, including the suburbs. I believe all this show of force was for the benefit of VC and NVA spectators. Regardless of our tremendous fire power enemy forces apparently can hit us any time and any place they are willing to invest their manpower and ammunition resources. Most of their ammunition was brought down the Ho Chi Minh trail on bicycles. That may be difficult to comprehend, but that is how it is with the dedicated other side. Sometimes I think we chose the wrong side when it comes to highly motivated people with honesty, dedication and leadership! But then we might have corrupted even them too with our billions of US$s.

I am down to one M-2 automatic carbine, 150 rounds of ammunition, two cans of Spam and 5 kilos of rice. I expect that to last until 10 May. I have disposed of all the other weapons and ammunition. And I have consumed the last of my kipper snacks! Instead of selling my .38 cal. revolver shoulder holster, perhaps I should have kept it for use with my .38 in the Washington, DC combat zone!

That night was a noisy one. It was difficult to sleep. Artillery was fired throughout the night and incoming mortar rounds started falling about 0430 and kept up until about 0600 hours. Mortar explosions rattled my windows but none landed nearby. American Forces radio gave out with the regular "G-O-O-D M-O-R-N-I-N-G, V-I-E-T-N-A-M" (I will not really miss when I leave) and had no news about any of the evening's military activity. Following breakfast at the Splendid BOQ I drove to the office to wind up a few affairs. I seemed to be the only one working, other than the boss. At noon I walked downstairs and found the Marine guards had the place locked up tight and were letting no one into the building. The guards told me there was heavy sniper activity in Cholon and at each of the bridges entering Saigon. That report changed my plans. I was to have driven a friend to the pistol range. Since the range is deep in Cholon, I drew him a map and left him a note on his desk advising him to go there only after peace is declared. I returned to my office, packed my work and drove back to the apartment in haste through all but deserted streets.

During the remainder of the day flights of jets and gunships crisscrossed the sky. Nearby I could hear sounds of rocket and machine gun fire. Troops raced both ways on our one-way street, just like Tet. One of our building residents came in from Tan Son Nhut and said the airbase was an armed camp, with heavy sniper fire being directed in from all around the perimeter. It appears my last week will go off with a bang! When I arrived over a year ago the area around Saigon was under fire. And they are still at it with increased fury, both in and around the city. That's progress?

My last field trip carried me deep into IV Corps, the Delta country and to Phu Quoc Island with Director Muoi, Mr. Hiep, Mr. Qui, Mr. Phuong and Walter Pierce. We had a special flight arranged on one of those beautiful Porter Pilatus, turbo prop airplanes. This was my first low level flight to that part of the Delta. I observed irrigated fields of green near the many forks of the mighty, silt laden Mekong River. But most of the area was burnt brown from the dry season, fires and defoliation. I saw fires burning in rice stubble and the mangrove forests. Near the coast the large, VC infested U Minh Forest had many fires which have been burning for weeks. Damage appeared to be heavy. Military activity prevents any fire control. All local foresters can do is watch it burn, day after day, until the rainy season suppresses fires.

Our first landing was at Rach Gia on the Gulf of Siam. Like so many places, there are two airports, one in the city and one in a near-by village. We landed at the village airport and the pilot immediately took off for another assignment. We were met by a driver and all six of us climbed on board a vintage Jeep and headed for the forestry office.

The countryside was fertile farmland and well irrigated. The local people have an ingenious way of using the tide and a series of check dams to irrigate these flat coastal areas. We crossed one of the many canals we had seen as we flew into the area. The extensive canal system was started many decades ago by the French. Each Vietnamese family was required to devote many days each year to construction and maintenance of the canal system. These are the highways of the Delta. Most of the water craft we saw were a motley, run-down mixture of motorized sampans and "African Queen" type of derelicts. Sampans are driven in these shallow waters by outboard motors which have an extremely long, nearly horizontal shaft which drives a propeller several feet behind the vessel.

All of the water craft flew a small SVN flag which indicated, at least in the daylight hours, their sympathy was with the government. I saw several heavily armed Vietnamese and US Navy crafts on patrol. American Navy advisors, assigned to "Operation Game Warden", waved back at us. I have noticed the American fraternity grows closer together out in the "boonies". The job of the Navy "game wardens" in this area is to provide security and patrol hundreds of miles of canals, tributaries and the main Mekong River. They have hazardous duty traveling on these narrow waterways. Much of the mangrove forest cover and many fruit trees have been killed by defoliation and fire in an effort to eliminate ambush cover. The US Navy also has "Operation Market-Time", another frustrating and dangerous job, which consists of inspecting thousands of small fishing and transport sampans that ply South China Sea coastal waters all the way from the DMZ to the Gulf of Siam and Phu Quoc Island.

We called on the province chief, a Vietnamese Army Major and learned their greatest need was a dependable supply of logs. He said they needed authority to import Cambodian logs since logs from Saigon and Phu Quoc were too expensive. Our next stop was at the office of the local forestry chief. He had a staff of 10 men. For an area with no forests and few sawmills that appeared to be a well manned organization. It is a mystery what they all do. Next we visited local sawmills which appeared to be well stocked with large, beautiful, unmeasured logs. You guessed it! The logs came from Cambodia! We found mills to be well maintained and properly aligned. These are manufactured and maintained from local materials. It is a pleasure to see those old French-designed band mills cut a 1/12-inch kerf and peal off clear board after clear board. Vietnam has much labor and these labor intensive mills require much manpower. This is one of the reasons we have resisted USAID's pressure to develop a sawmill program which imports highly efficient, difficult to maintain, American sawmills. Hand labor consists of rolling logs to the saw, pushing the saw through the stationary log and carrying the large boards to lumber stacks for drying. Most of the mills had adjoining boat works. Here they hand crafted beautiful small and medium sized sampans for use on the canals. I could have purchased one which would have made a beautiful duck boat for the equivalent of US$30.

We had lunch in a weird Chinese restaurant. It would have rated a Class Z (minus), if rated by Coconino County's friendly public health officer. But with continuing dysentery, I had little to lose. I made the mistake of being escorted through the kitchen to the washroom. That just about did me in. That terribly hot and dirty kitchen could have come from "Dante's Kitchen in

Hell". A skinny, nearly naked Chinese cook was cooking all parts of the hog, dog, chicken and duck over a charcoal fire. His array of goodies looked just about like the voodoo stands that fascinated me down in the Amazon country. The flies were making good progress in packing off the food. When I reached the washroom and where the dishes were rinsed off, I decided my hands were not so dirty after all. It was not an appetizing set up!

I refused ice and drank hot beer. I had seen the urinal where the ice was stored. When the food came everyone dug in with gusto. Vietnamese can really make food fly when they dig in with chop sticks. I occasionally dropped some on the floor but that really didn't matter. The floor sweepers consisted of a pregnant bitch mongrel dog and another with a litter of puppies hanging on to her pendulous tits. They cleaned, so to speak, the floor as fast as we spilt food. As I looked around at this scene I burst out laughing, I can not remember ever laughing so hard in my life. My hosts thought I was enjoying myself. They could never have understood my thoughts at that moment. I was thinking about our good old county sanitarian, Ed Stanfil who made the Kaibab Lodge dispose of sugar bowls because they were unsanitary. So much for that unforgettable culinary experience!

Our pilot met us at the airport on schedule. After receiving a briefing concerning the use of life vests and ditching procedures we took off for Phu Quoc, some 75 miles out in the Gulf of Siam. We flew over small islands, some of which were inhabited and others which looked like the desert islands in story books . . . places where you imagine you would like to be shipwrecked with someone. These jewels had small brilliant blue lagoons, snow white beaches and coconut trees swaying in the breeze. We passed over another fascinating group of small islands called Iles des Pirates, where old time, and present day, pirates must have had a delightful time raiding spice laden sailing ships. Our flight did not cover the shortest distance because we were required to follow a prescribed flight path to avoid Cambodian airspace. This reminded me of the Tay Ninh flight!

Phu Quoc Island is about 30 miles long and 15 miles wide at the widest point. The south end of the island has a VN navy base and a prisoner of war camp. There are said to be 20,200 VC on the Island. Some 20,000 are in prison and 200 are on the outside harassing everyone else. Flying north from the prison camp we passed over an abandoned rubber plantation and thousands of hectares of pepper, spice and herb fields. As we neared our destination at Duong Dong a sudden rain squall moved in from Cambodia and we circled around the storm for 20 minutes until it lifted. This additional flying gave us a good look at many ship wrecks and stands of old growth timber on the north end of the island. There is a good supply of standing timber but the lack of roads and VC activity limits development and utilization.

We were met at the airport by the local forester and some of his staff. As the Jeep had no top we waited under cover until the rain stopped. The heat was very oppressive, even for me. The drive through the fishing village was most interesting. The predominant odor, which was much stronger than the fetid open sewers, was fish. Tons of fish are caught in the waters off Phu Quoc. Some is iced for shipment to Saigon but most of it is salted and made into the pungent fish sauce called nuoc mam.

We could see hundreds of people mending nets, salting fish and working on a wide assortment of boats, which defy description. Naked children splashed in the fresh water ponds remaining after the rain. We could see beautiful young girls in every direction. Police had apprehended two Thai fishermen in Vietnam territorial waters and were leading them off to prison. The handsome Thais had broad smiles on their faces but the look in their dark eyes indicated they could cut your throat in the blink of an eye. They were dressed exactly like 17th century pirates, complete with head bands, baggy pantaloons and colorful sashes, all made of beautiful Thai silk.

Small children with monkeys on the shoulders passed by the front of the unimposing Hat Lam Vu district forest office. The District Chief was busy and we held a brief meeting with the local foresters. They were all far from their homes in Saigon and very unhappy to be stationed on this remote island. They have only one sawmill which they serve very well. But with little else to do they resort to the inexpensive, ample supply of opium, gamble with cards and play with beautiful young girls. (The girls are beginning to look beautiful to me, so it is a good thing I have only four days to go.)

We took a walk to the mouth of the river and watched numerous fishing boats come and go. The little engines with their "putt, putt, putt" did not appear to be too seaworthy. But the waters were abounding with fish, in fact we saw them catch huge nets full of fish only 200 yards offshore. At the mouth of the river was a shrine and a statue of the Virgin Mary called "Duc Ma". We sat there for a while watching women across the river mending nets and children playing in the sea. The high temperature and humidity were brutal. Since Walter Pierce and I had no mosquito nets for use on this malaria infested island everyone decided to stay at the coastal bungalow. This well weathered, picturesque structure had a multiple peaked roof of some sort of French architecture. The inside was fairly clean. Antique French, pull chain plumbing fixtures were served by huge water jars, of the size Ali Baba could hid in.

I was the first to go for a swim and then the entire group joined me. The water was warm, clean and really delightful and relaxing. At that time all I could think of was the saying of Kaibab Ranger Pat Murray, "I wonder what the common people are doing?" At that time Arizona, New Mexico and Bethesda seemed to be a million miles away on another planet. The beach was a coarse sand, made for wading and swimming in a gentle surf. I thought the place could be paradise if there was just peace and security. Further up the coast I could see hundreds of thatch covered huts, hootches as the GIs call them. The homes were not fancy or particularly clean, but as I watched the children run and play in the surf I thought they were much better off and lived better lives than the street children of Saigon and Cholon.

On our stretch of beach an old man and two boys loaded an ox cart with sand. The old man beat the oxen unmercifully to get them to pull the heavy load of wet sand up a slight incline. This man worked until dark and I saw him at it again at daylight.

After the swim I washed off by dipping fresh water from one of the huge wooden water jars. Then I put on my black pajamas and looked just like the local folks, Revolutionary Development workers or the VC. My skin is about the right color but the round eyes and height betray me.

The District Chief came by for a visit and a meal. He was an Army Major, smooth talking, and luxury oriented. He is reported to be doing very well administering the people of the island. The fact is he is "getting fat" in his systematic bleeding of the entire population. One logging contractor, for example, refused to begin operations when he could do no better than share 50% of the profits with the District Chief.

Dinner was one of those all out affairs. It was prepared in both the bungalow and a thatch covered hut out back. Every dish contained cuttlefish, which is the well known specialty of Phu Quoc. We were served large shrimp, fat oysters, fish of several varieties and fresh vegetables. The food tasted good but I discretely refused the oysters because earlier I had seen them being harvested at the outfall of a fetid sewer. There were few flies since it was almost dark. The kitchen and cooks would not have rated even a Class Z (minus) license.

After dinner we sat on the porch, smoked Cuban cigars and watched lights come on from many fishing boats. Those just offshore were netting cuttlefish. Some fishermen used fire but most of them used gasoline lanterns to attract cuttlefish. About 15 miles out we could see the lights of other fishermen who were after the larger ones.

About 2100 hours one of the local Special Forces men came by to give us the good word. They had heard two Americans had arrived and they wanted us to know they had received word the VC was going to attack in force sometime during the night. He said in case of an attack we should either go to the District

Chief's compound 300 meters away or to the Special Forces compound which was 1,000 meters down the coast. The District Chief had 100 men protecting him and the 12 Special Forces had 400 local troops. I thanked him for the information and his concern. After he left I told Walter Pierce it would be suicide to try and reach either place in the event of an attack. In the dark there would be too many trigger happy gunmen for us to go to either compound. I could just see myself running towards one of those places in the dark of night in those black pajamas. We both agreed the old man's sand pit, 100 yards away, would be the place to go. I spent the night under mosquito netting that had mysteriously appeared. A faint breeze failed to penetrate the netting and it was hotter than Hades. With my .45 for a bedfellow, I slept well, except for frequent trips to use the facilities.

At daylight I was awakened by swayback Vietnamese pigs, snorting, rooting and stinking under my window. I walked to the shore and looking out over the water I could see the cuttle fishermen coming into port. While I was sitting beside the sea a small boy came by carrying his baby sister hung over his hip in the typical Vietnamese style of carrying babies. That practice makes them slightly bowlegged. The baby girl had her ears pierced and bits of dirty string had been inserted to keep the holes open. The little boy turned out to be quite a talker when he learned I could speak Vietnamese.

Before leaving the shore I had attracted a dozen bright eyed children. We had a good time talking. While there we saw a school of small fish, the kind caught and used in making nuoc mam, about 200 yards offshore. From nowhere, it seemed, a fishing boat swooped into the area and before I knew it they had dropped a net in a circle around the fish. Then a dozen or so men came running from the village with buckets, jumped in the water and swam out to the boat. They all pitched in and hauled in the heavy net loaded with millions of the tiny fish. I suppose the extra helpers took their pay in a bucket of fish.

We visited the sawmill, a typical French-designed horizontal band rig. None of the logs were stamped or scaled. I guess there was a shortage of foresters! The largest investment on the island appeared to be the nuoc mam factory. The factory consisted of 50 large wooden vats. The vats are 10 feet high and 8 feet in diameter at the top. Each of these vats costs VN$200,000 (VN$118 = US$1), which is quite an investment. Believe it or not the tiny 2-3-inch long fish are gutted and the eyes removed by dozens of women workers. Fish are packed in vats a layer at a time, then salted, then a layer of fish, more salt, etc. until it is filled. This mixture sits ("rots" is a more suitable term) in the boiling hot sun for a year. At the end of that time the remaining liquid is drawn off. The first drawing is No. 1 luxury stuff. The smell around this factory and the final product is indescribable.

Before leaving we were taken across the river by boat to the local market. Fish, fruit, vegetables and nuoc mam were for sale. Business was brisk but the flies were about to carry off the food faster than the buyers. We stopped at a refreshment stand and each of us had a refreshing drink of iced lime. I drank this against my better judgment, but gad it was hot! I bought five liters of No. 1 nuoc mam to take back to my friends in the Directorate and USAID office. In fact everybody bought some. I hated to think what the CIA pilot would do to us if any of those bottles broke in flight. He would probably dump us and the airplane in the Gulf of Siam. There is a local Saigon saying that there are two sure items to be used for catching a beautiful Saigonese girl . . . "one is a bottle of No. 1 Phu Quoc nuoc mam and the other is a Sears Roebuck catalog" . . . enough of those kinds of sayings!

Our Air America Porter buzzed the village at the appointed hour and we all hopped into the old worn-out Jeep, crossed the rickety bridge and headed for the airport. As we departed beautiful Phu Quoc Island I thought of the people in this small insignificant speck in the Gulf of Siam. No one, other than the District Chief, would ever become rich, no one would ever suffer from the cold and there would always be plenty to eat. I concluded they were pretty well off after all and besides they would not die of carbon monoxide poisoning, and related lung problems, like the people in Saigon.

Our next stop was to be Ha Tien, which is located on the Gulf and the Cambodian border. It was only a short distance but again to avoid Cambodian airspace we had to go the long way around. We flew over vast stands of mangrove which were interlaced with miles of canals. The Directorate of Forest Affairs puts a portion of stumpage receipts back into the transportation system. They hire people to construct and maintain canals for improved access into the forests. Ha Tien was a pretty little fishing village at the mouth of a very muddy river. In fact the silt could be seen for miles out in the Gulf from this one source.

We buzzed and circled Ha Tien for 20 minutes. In the town we could see troops on the move and witnessed air strikes on outlying areas. When no one came to the airport to meet us we got the message that something was amiss and we headed for safer parts. We were running low on fuel and were required to stop at Rach Gia for jet fuel. This required landing on the short designated airstrip in town. Actually the landing strip was a city street. Police detected our problem and closed the cross street. We landed with only a few minutes of fuel remaining. Refueled we went on to Can Tho, one of the principal province capitols in the Delta. We quickly adjusted our schedule after Director Muoi checked in with the province officials and learned that attacks were expected throughout the Delta. We concluded our business quickly and returned to Saigon just before curfew. That is the story of my last field trip in vietnam.

Now for today's (5 May) events. The battles in and around Saigon have grown in intensity throughout the day. There have been many casualties. Three newspaper reporters, including one representing *TIME*, were killed near the Cholon PX. Many police have been killed and injured. Three helicopters were shot down within 1/2 mile of our apartment building. It is now 2100 hours and for the past two hours mortar shells have been falling not too far away. The chatter of gunships has gone on continuously, especially the "burrrpppppppp" of the Gatling guns in the C-47 Dragon Ships. I am going up on the roof for a while and watch the action before concluding this letter.

Several battles are taking place tonight in and around Saigon. In one part of Cholon it sounds like they have called in tanks. Skies are full of planes and flares. I witnessed two gunships pour streams of tracers into the Phu To racetrack area.

This will be my last letter of this type from South Vietnam, that is unless the airport is closed next Friday. I am glad I have taken the time to record my observations of this forester's unique experiences. I wish I could say life is improving for the Vietnamese people. That is the reason we made a family decision for me to come here. At this time that goal has not been achieved. Our American efforts have resulted in vast areas of Vietnam being laid waste. Tacitus, that ancient Roman historian and orator, could have predicted the outcome in Vietnam when he wrote . . . "They made a wasteland and called it peace". (Karnow, 1983) We have created a desert and destroyed valuable resources, cities, villages and hamlets to "save them". Many people on both sides have suffered and died. Now we are trying to call it "peace". No way! At this time there is no "light at the end of the tunnel" and no end is in sight for this tragic chapter of the world's history. Often I think my experiences belong in "*Alice in Wonderland*". Who knows what/who will pop out of the rabbit hole?

7 May 1968

The battle of Saigon goes on in and around Saigon with unabated fury. Fortunately the areas between my apartment, USAID and the BOQ are secure. I come and go in relative safety.

My one last problem is that Hong Kong suit. I will go to the PX in Cholon if the area is secure. I will carefully check out the security situation before I go. Please leave the worrying to me. If the suit does not arrive you will need to bring my gray suit. I will have a total weight allowance of 64 pounds on the return flight from London. That should give us an adequate allowance for any items we purchase.

I have my airline tickets and everything completed except for picking up green dollars and traveler's checks. The last obstacle could be the route to Tan Son Nhut airport. It was under fire all day yesterday and closed to all except military traffic. So keep your fingers crossed and I may be on my way come 10 May.

I do not feel too sharp but I believe I have overcome my dysentery and gained back a few pounds. You were correct about the last minute activities, they have been hectic. I decided to consign my household goods to Smith Mayflower in Washington, DC with instructions to deliver on or about 27 June. That is better than having it dropped in your lap just before you leave for London. The government will cover any storage costs.

I hope your travel arrangements are complete and you can settle down to packing. I will call you from Frankfort for any last minute messages. In case of any mixup you can meet me at my London hotel or at the American Embassy in London. Congratulations! I am pleased to hear all check books are in balance! Please keep them that way!

I look forward to healing any wounds on the good old 1953 Chevy station wagon. I have no desire to purchase a new car. I even look forward to doing the lawn, windows and plumbing. Most of all I look forward to being with you, taking you places and doing things together. Perhaps we can make up for some of the lost time.

9 May 1968

Dear Gwen & Girls:

Well I have come down to the wire! As usual I am working late tonight in the apartment on reports. The past few days have been hectic in more respects than one.

I had a hunch I had better stay out of Cholon. I gave Martin Syverson the money to pay for the suit and mail it to you. He and Walter Pierce braved the battles and went to the PX. The suit is in Vietnamese customs along with the last regular shipment from Hong Kong. It may be delivered to the PX today.

Fighting in the past two days has been more severe than during Tet. More bombs and napalm have been dropped on Saigon than during the first few days of Tet. Fighting goes on day and night. Refugees are streaming out of the Cholon area. I do not have the time or compassion to help them at this time.

Saturday a friend of mine moved into an apartment I almost selected when I was first looking at apartments. The residents were under fire all day Sunday until they were evacuated by an armored car at 1630 hours that day. Their building was riddled with .50 ca. machine gun and rifle grenade fire. This was all friendly fire directed at the VC who had the place surrounded. My friend told me the residents of the apartment building crawled under beds for protection. They had a harrowing experience. The warden system we set up and the radio we provided were all that saved them. The US military refused to go in and rescue them because of the heavy casualties they expected to suffer. After the residents were evacuated, ARVN drove out the VC, and friendly troops moved in and completely looted and emptied the apartments.

Today one of the USAID offices was evacuated due to heavy fighting. Fires raged all around the south and western parts of the city. All fires were started by our napalm, bombs and artillery fire in an effort to drive out the VC and save the city.

Mrs. Hai and our other North Vietnamese USAID people are frantic about current developments. VC are found throughout her neighborhood which is near Syverson's. They are kidnapping people and killing them and as many GIs as they can find. The radio continues to announce many VC are being killed and friendly casualties are light. Censorship is utterly ridiculous. It reminds me of fiddling while Rome burns. The city is being torn apart at the seams but you would never know it by reading the newspapers or listening to the radio. Last week when the TV station was blown up, the radio reported the action was 10 to 15 miles outside Saigon. The locals say the TV was destroyed because the programs were so poor.

One more wake up and tomorrow, if the airport is open, I am gone! If not, I will remain until other arrangements can be made.

It was a sad day today when I said goodby to my USAID Vietnamese friends. My maid and Mrs. Hai, our secretary cried, as I bid them farewell. Mrs. Hai is such a good person. I shall always remember her.

When I catch up with myself and I am away from the noise and tension of battle and the pressure of this phase out I will write to you. I miss you all so much and look forward to our meeting in London.

Love, Jay

* * *

In spite of continuing mortar and rocket fire, the route to Tan Son Nhut Airport was open and commercial flights landed on 10 May 1968 for rapid unloading, refueling, loading and departure. As directed, I ran to the plane and as we were taxiing at a high rate of speed to our takeoff runway I saw a luggage tractor pulling trailers loaded high with 13 aluminum caskets to a waiting Air Force cargo plane. Those remains of 13 GI's were far from being the last to leave Saigon. The war would go on for another seven years and many more aluminum caskets would be shipped out from Tan Son Nhut for the long journey to

Arlington and other cemeteries to be received by grieving next of kin. Before it was over the names of those 13, along with those of 50,170 more dead and missing, would be carved into the black granite V, known as The Wall, or officially as the Vietnam Veterans Memorial imposed on the peaceful green Mall in Washington. Flights of fully loaded, huge C-141 hospital planes would head out continuously for military and veteran's hospitals around the world, where shattered bodies and minds would be treated. The sad chapter of American history recording our shameful involvement in Vietnam was not yet ended. I was grateful to be leaving on Air Vietnam's crowded coach class (substituted for Pan Am at the last moment) even though it had a faint detectable odor of nuoc mam.

As we spiraled upward to gain safe altitude, I had a good look at sprawling Saigon. We crossed the Saigon and Mekong rivers to land at Phnom Penh, Cambodia and on to my first rest stop in Bangkok. I had very mixed emotions to be leaving Vietnam. Images of many of the people I had met raced through my mind. I recalled the progress we had made, the challenges we had faced and the multitude of problems remaining for the Vietnamese and the Americans. I could not resist recalling the sad history and the tragic present circumstances being experienced by the Vietnamese people, all of them from the Chinese border to Phu Quoc Island. At that stage the future for Vietnam and the United States did not look promising.

There were moments when I wished I was somewhere else, but my Vietnam tour of duty was memorable, a once in a lifetime experience. Of our group only Lewis Metcalf had no visible health problems. One of the really good guys, tireless, productive Walter Pierce, died of leukemia which was judged to be connected with exposure to Agent Orange. Agent Orange also impacted Jack Shumate, Barry Flamm's and my successor, who lost motor control of his muscles and was given a disability retirement. Martin Syverson suffered a hernia and Barry Flamm fought amoebic dysentery for well over a year. I had the normal health problems related to the tropics and lost part of my hearing. All of us experienced the effects resulting from prolonged use of malaria suppressants. Fortunately none of us contracted malaria, hepatitis or one of the dreadful diseases, but all of us were deeply affected and will never forget the swirling red dust which covered everything and the blood, green forests and fields, heat, smoke, and the never ending noise we lived with day and night. I will never completely forget the smell of herbicides, rotting fish, fermenting fish sauce (nuoc mam) and human waste. As I think back to the day of my departure, and until the day I die, I can clearly see in my mind the 13 flag draped aluminum boxes carrying GIs to their final resting places. Also in my mind I can see the tragic endless procession of refugees wandering aimlessly, seeking refuge with no place to run. No one, military or civilian, wives, husbands, children, mothers, fathers, brothers, sisters and friends, who was a part of our country's poorly conceived and ill-fated involvement in Vietnam can ever forget the Vietnam experience.

Vietnam Epilogue

I returned briefly to Vietnam in 1973 to evaluate the forestry program. At that time some of our war prisoners were coming home. But the war just was not going anywhere and, in spite of the peace conference, there was no resolution in sight. Politicians were looking for a way out . . . looking for someone to blame for America's failures. Author William J. Lederer stated it succinctly . . . "The people who have botched things up are your neighbors, your friends, your relatives, your countrymen. They are ourselves." (Lederer, 1960)

Significant progress had been made through the forestry program. Where there were only five active sawmills in the entire country when we arrived in 1967 there were over 500 operating as a result of our efforts. I thought back to the criticism USAID/Washington heaped on the forestry program when they were trying to scuttle it. I was also reminded of author William R. Corson's critical comments concerning USAID when he wrote in 1968: ". . . we pay each of the five forestry experts assigned to CORDS/USAID an average annual salary, including fringe payments, of $38,000—this in a country where there has been no timbering for the last six years." (Corson, 1968) It was one battle after another.

My 1973 journey in Vietnam took me high over the heavily forested central highlands. I examined some of defoliated areas and could observe recovery of some species of trees. We landed at such places as Ban Me Thuot, Kontum and Pleiku. My CIA, Air America helicopter pilots refused to fly lower than 5,000 feet due to the likelihood of unfriendly fire. By that time the US forces had pulled out and military operations turned over to the war-weary, poorly motivated ARVN troops. Former US military camps, airfields and fire bases had been abandoned and were in shambles. Everything not tied down to the earth had been stolen and removed. Weeds, brush and trees were sprouting and reclaiming some of those forlorn, forgotten sites.

Local Vietnamese officials and sawmill operators I contacted were apprehensive and anticipating the final drop of the curtain of doom, now that the US military had deserted them. Men, women and children stared at me with hatred in there eyes; obviously I represented the disastrous impact our American presence had brought to them and their precious land. In Ban Me Thuot I felt an enemy presence and there were no American GIs to protect me. My hair bristled and stood out on the back of my neck. It was very spooky and I was glad to be on my way. I did not know that the presence I felt was the North Vietnamese Army behind the nearby hills and just around the corner. At that time the NVA was massing for its final assault on the highlands.

In 1973 the Vietnamese people were benefiting from the improvements we had made in forestry and the forest industries. Some of the same problems, such as the pervasive, wide spread corruption, that existed before we arrived, were still there and probably prevails today. In 1973 USAID was promoting the sale of large South Vietnam timber concessions to Japan or whoever was interested. I was able to stop this movement when I recommended against it. I described countries throughout Southeast Asia that had exported logs and jobs and left nothing but devastation, economic disaster and misery behind.

In 1975 near the end of America's bitter involvement and before "peace" was declared, I helped pull some official strings, and extract some of the Vietnamese foresters and their families to the United States. The USAID forestry program held out until 23 April 1975, a week from the close of the final chapter of the United States traumatic involvement in Vietnam, a real tribute to the people who succeeded me. I can visualize myself standing on the roof top of my apartment on Cong Ly Street on that final day, 30 April 1975, of our country's ignoble retreat from Vietnam. I would have had an unobstructed view of the US Embassy. I could have seen the harrowing roof top helicopter evacuations. I might have heard the anguished cries of the loyal Vietnamese employees of the US Government who we left behind to cope with the uneasy peace, imprisonment, reeducation and the unknown perils that awaited them. My efforts over several years to help our Vietnamese secretary, Mrs. Hoang Minh Hai, who served us so well from 1967 to 1969, leave Vietnam were unsuccessful and frustrated by the US Orderly Departure Program. Those State Department bureaucrats, operating in the comfort of Bangkok, never acknowledged one of my many requests for her departure from Vietnam.

My tour of duty was unique. During my 18 months of travel throughout the 44 provinces of Vietnam I confirmed what a military officer told me over dinner in a Saigon BOQ shortly after I first arrived in Vietnam . . . "It is a shame what we are doing to this lovely country" . . . I could now add . . . and to the people . . . and to America. If we had just studied the 3,000 year history of Vietnam and profited by the study of that history, and thought things through, how different life might have been for the Vietnamese and America. But we didn't and we were doomed to repeat history and what Chinese, French and Japanese invaders, had experienced. As my administrative assistant and interpreter Tran Cao Thuong wrote in a final letter to me . . .

"You are really living in a crazy soil with a crazy government, a government *not* for the people, *not* by the people, and *never* for the people! Please feel with my situation!" Someone said that Lewis Carroll could not have written Vietnam . . . it would have been too unbelievable for *Alice in Wonderland*. After 18 months I could relate to those thoughts.

What were my last thoughts as I departed from Vietnam for the last time? Why did we invest so much blood and treasure in an area so remote as Vietnam and of so little apparent significance? Some day I must give thought to writing about my unique Vietnam experience, perhaps I could shed a bit of light on this complex question. Yes, the Hong Kong suit caught up with me at the airport on that last May day in 1968! I received a "0+" score on my language examination . . . not enough for that pay raise, but it did open many doors, minds and a few hearts. "Xin loi" (I apologize), and G-O-O-D-B-Y-E, V-I-E-T-N-A-M. I am gone!

BOOK FOUR

The Team

The 1960s and 1970s saw many bright, well educated young professionals and technicians enter the Forest Service. Ed Cliff retired in 1972 and John McGuire became Chief. Leadership during these decades was confident and well organized. Field personnel were highly motivated and the many users of the national forests were well served. Where completed, multiple use plans and environmental assessments provided sound guidance for more sensitive management of resources, including endangered species of plants and animals. Management improved further through the environmental decade of the 1970s. "Forests For the Future" was the beginning of a number of well thought out, imaginative programs. There was increased concern for maintenance of adequate raw material for industry from national forests and private woodlands. Put into action, research results resulted in improved forest management and utilization techniques.

The Forest Service became involved in the innovative Forest and Rangeland Renewable Resources Planning Act of 1974 (RPA) which set the stage for improved national forest planning and programs promising more cooperation with the states and private land owners. As the public became more involved in the planning process each interest group wanted more of what they perceived to be most important.

After 25 years of serving as the living symbol of forest fire prevention at the National Zoo in Washington, DC, the original Smokey Bear retired in May 1975. A new living symbol of Smokey Bear, also a black bear from New Mexico, had been on the job for four years as an understudy, and took over his new role.

Poorly designed and implemented timber harvests on national forests in West Virginia and Montana and the carefully designed and orchestrated efforts of certain preservation groups, resulted in foresters losing their image of wise land managers and being cast as timber people or loggers. Environmentally leaning federal judges were liberal in granting injunctions and rendering court decisions which stopped a number of viable timber harvest programs. We lived in a litigious society where it became standard procedure to "sue the (federal) bastards". The 1976 National Forest Management Act directed how national forest plans and timber harvests should be made and how public interests should be brought into the process. The Forest Service could not make a plan or a timber sale without having it immediately appealed. The Forest Service became involved in what promised to be decades of conflict.

The 1980s brought Forest Service leadership changes and the new leaders did not heed Gifford Pinchot's seventh maxim or advice to Forest Service people: "Don't try any sly or foxy politics because a forester is not a politician". The Chiefs made headlines. Natural resources, and the people managing them, became embroiled in one controversy after another. As the Forest Service was planning for the year 2020 it may have lost touch with rapidly evolving contemporary changes. Certainly it faced its greatest challenge as I returned from South Vietnam.

Chapter XXVIII

Welcome Back To the Real World

The trip around the world and time in Britain with the family helped me make the transition from Vietnam to the real world. It was good to be home and move into a pleasant routine with family, friends and work. Gwen had adapted to Bethesda and enjoyed activities involving the girls and their girl and boy friends. I was accepted back into the family by all. We encouraged the girls and their friends to spend time in our home and they all seemed to be there continuously.

Melissa had done well in college and waited until I returned home to marry Bob Humphrey of Phoenix. Fortunately Bob escaped Vietnam and served his Army time and was released. They moved to northern Arizona where Melissa worked for the National Park Service at the South Rim of Grand Canyon National Park and Bob was a lookout-fireman on the Kaibab National Forest. They loved the West and stayed on in Flagstaff to attend Northern Arizona University.

Cindy had a number of fine girl friends. Sue Maynard was one of them and was like a third daughter to us. Then there were the boy friends. The ones who stayed home and out of the Vietnam draft were more or less normal. One fine young man, with everything going for him, went to Vietnam. He returned completely changed and unable to cope with the real world. His brains were scrambled when he was discharged from military service. The causes were the same as I had observed in Vietnam. He went off the deep end when he lost close buddies in the fighting. He became a "space cadet" and drifted off into a world of his own.

Myrl and Gene Jensen and his elderly mother provided Gwen and me friendship that had grown stronger over the years. We were family. Gene was a commissioned officer in the US Public Health Service and we enjoyed the benefits of their membership in the USPHS Officers' Club near us in Bethesda on Old Georgetown Road. We had many wonderful lobster boils and bushels of steamed clams. The shellfish were fresh, juicy and good, but they couldn't be otherwise, since Gene was responsible for the USPHS Shellfish Program. Within a few months of my return, Gene accepted an Intergovernmental Personnel Act (IPA) transfer and moved to the State of Virginia where he was responsible for Virginia's water program. We had many delightful visits to their lovely home in Charlottesville. Later on they moved to a similar position for the State of Kansas and lived in nearby Topeka.

I had my debriefing sessions with USAID and International Agriculture. We made progress in gaining recognition of the forestry program. The respective groups supported my philosophy of helping developing countries help themselves. Their attitude had changed and they embraced my idea of helping improve the host country's techniques and "home-made" machinery and putting masses of unemployed to work rather than imposing our way of doing things and loading them with our complex, while more efficient, equipment they could not afford or properly maintain. Basically I understood the reality of producing short term benefits. Short term benefits are what most politicians were looking for and the bureaucracy survives by innovation and being responsive to the political process. This fact of life is not well understood by many resource professionals. But it is a fact of life in public service as well as the private sector. In my closing comments I commended USAID and the State Department for the outstanding orientation and training we were given before going to Vietnam. We were well prepared going in and there was no culture shock, however I told them they were terribly derelict in providing no safety training in Vietnam and not preparing returnees for what awaited them back in the real world. They took my recommendation "under advisement", where it remains buried to this day.

My decision to have my USAID final physical examination performed by the State Department Health Service, rather than the 17th Field Hospital, was a good one. They were able to clear up most of my ailments. Their intensive tests verified I had lost a significant amount of my hearing due to repeated exposure to explosions, jet engines and other high level disabling sounds. Annually I was called back for Federal Employees Compensation evaluations for the next five years. I was provided a hearing aid and a lump sum settlement for my hearing loss. But ever since 10 May 1967, when I left the noises of Vietnam behind, I have the sensation of continuous ringing in my ears and that promises never to go away. I have learned to live with the ringing and hear fairly well unless there are conflicting sounds such as electric motors, engines, jet airplanes going overhead and in crowded situations with many people talking at the same time. Gwen says I can hear what I want to hear and credits much of my hearing difficulty to inattention. It is so good to live with such an understanding woman!

I was welcomed back to the Forest Service and had an interesting job as Assistant to the Deputy Chief for the National Forest System, M.M. "Red" Nelson. I was responsible for budget coordination for all of the NFS program areas and served as secretary to the Secretary of Agriculture's Multiple-Use Advisory

Committee. I attended all Chief and Staff meetings and was very much back in the Forest Service mainstream. I learned much about operations at the Washington level. I was impressed by the importance of taking prompt action, preparing timely and responsive replies to incoming letters and meeting due dates. I quickly learned referral letters carrying a red margin came from the White House. In those days when President Nixon's power hungry musclemen, John Ehrlichmman and H. R. Haldeman, yelled "frog", we all jumped. I learned the sensitivity of the Forest Service was much greater in the Washington, DC area and appeared to diminish as it passed down the organization ladder to the ranger district level. In my judgment that was a paradoxical situation since 99.9% of the Forest Service's problems at that time occurred at the district level rather than in Washington. I would have to give that more thought. The assignment was good training for me and I learned a great deal. I commuted to work by bus and did not have to leave when many meetings broke up in the late afternoon by the call of "car pool". Frequently I found myself remaining to assist Chief Cliff on some urgent matter after the staff had departed. Gwen quickly adapted to my devotion to the Forest Service, after all that was nothing new for her and the girls.

In those days the Forest Service was a close knit team and human resources were highly regarded and appreciated. We all respected our tremendous leader Chief Cliff. The Chief's staff, Regional Foresters, Directors and the entire field organization worked well together as a smooth functioning team. We all knew where the Chief stood on issues and where we stood with him. We understood where he was leading the Forest Service and we all went down the road together. Art Greeley, the gentlemanly Associate Chief and Red Nelson gave me a number of interesting and challenging assignments. I had considerable traveling to do, but by then I was a seasoned traveler.

I gave many talks and slide shows about our forestry program in Vietnam. I recall one luncheon meeting lasted until 1500 hours. Most of the participants wanted to see all my slides (and that was a lot!) and then kept me after the slides were shown answering questions. I was pleased by their interest.

During this time I recommended Barry Flamm for the "Outstanding Young Man (under 40) in Government". My nomination was approved and Barry returned from Vietnam to Washington to receive the Flemming Award. Barry and I co-authored a widely read and quoted article on the "Effects of War Damage on the Forest Resources of South Vietnam". (Flamm & Cravens, 1974)

Gwen and the girls continued to have good health. However occasionally Gwen had a kidney flare up. Tests were negative and medication controlled the problem. Our family life was good and we enjoyed the girls and their friends. We encouraged the girls to bring their friends to our home for entertainment instead of running around all over the metropolitan area . . . and the DC area was a combat zone in those days. The girls had longhaired boy friends who played in a band. To avoid being too square, I let my Vietnam-style haircut grow out.

Until you have been under the same roof, poor roof, with electric guitars, horns and drums you can not comprehend or appreciate what noise really is. Some of the music must have approached the level of Vietnam's battlefield sounds. One summer evening the young people gathered at our home for an all out, well behaved music session. All went well until neighbors, two blocks away, complained to the Montgomery County police. However nearby neighbors sat on their lawns enjoying the music and sent the police on their way, to catch real criminals.

My colleague Vern Hamre was still in the Chief's office. Earlier we worked together as Assistant Directors in Watershed Management. Vern and his wife, Pete, spent many days and nights in Bethesda at the National Institutes of Health Hospital caring for their son Jon. We were all very sad when Jon died of leukemia. Shortly after the loss of Jon, Vern was offered a Deputy Regional Forester position in one of the western regions. At that time deputy jobs were pretty much dead end positions and I had told Deputy Chief Nelson I was not particularly interested in being considered for one. I was always a great admirer of Vern and his outstanding ability and when he was selected I detected a change in the wind. I told Red Nelson I was interested in one of those Deputy positions. He pulled on his ubiquitous golden brown meerschaum pipe and with a twinkle in his eyes said, "changed your mind, did you?" Within a short time I was offered a lateral move, with no promotion as usual, to the Eastern Region Deputy Regional Forester position in Milwaukee. Most of my transfers involved moves with no promotion. The only exception was the promotion to Vietnam and when I returned I was demoted back to a GS-15, but at a higher salary, less taxes of course. It really did not make any difference by this time in my career since I had reached the maximum allowable Federal Civil Service salary level.

For a change, Gwen and Cindy were more or less ready to move. Wisconsin was closer to Gwen's family in Minnesota. We made a house hunting trip to Milwaukee and put everything we had saved, down on a 1896 Victorian on the east side of Milwaukee at 2732 North Shepard Avenue. Efforts to pry part of the money we needed for the down payment and closing costs out of my Mother in Cedar Rapids, Iowa were as usual unsuccessful. She still planned to take it all with her. This move placed us in our 23rd house since we had been married and was no different than the others. It cost us dearly. We were over the weight allowance and transfer payments at that time were grossly inadequate to cover total moving costs. Gwen had always wanted a big old home and fell in love with Milwaukee's eastside, which was near Lake Michigan, in an area of large old homes, near the Downer Avenue shopping area, University of Wisconsin-Milwaukee campus for Cindy's school. A bus line gave Gwen ready access to shopping and libraries.

Before moving we visited the Jensens in Charlottesville. On the return trip to Bethesda Gwen and Cindy spotted an auction in process. I swear, on a dark and cloudy day, they could see an auction 10 miles away! I tried to divert their attention and get them to look in the opposite direction, but it did not work. We stopped and before long we paid $100 for a very large pump organ (the kind used in old fashioned churches). Recognizing the organ could not be loaded in the small Volkswagon Beetle sedan, Gwen made arrangements for its transport to Bethesda. The Milwaukee house accommodated all the "junk" furniture (really fine antiques) we had picked up over the years in abandoned mining camps, city dumps and second hand stores in Fort Collins, Flagstaff and Salt Lake City. While in Milwaukee on the house buying trip we visited the Salvation Army secondhand furniture store on North Avenue and purchased a large black walnut dining room table and seven velvet covered matching chairs for $99. The "new" 15 room house was to be filled in a short time.

Chapter XXIX

The Lands Nobody Wanted

On the move to Milwaukee we transported the normal complement of treasures, including cats, plants and the girls' friend, Sue Maynard. Sue decided to come and live with us. The girls found jobs in the nearby shopping area. Gwen was busy fixing up the house and spent a great deal of time with the girls. When not traveling I worked on the house. It needed much work inside and out. I was glad the Forest Service had trained me to be a plumber, painter, carpenter and electrician. Work on the house was my therapy. The remainder of the time I traveled. The 20 state Eastern Region required much travel and then there were the frequent trips to the Washington Office and training sessions. I made it my practice during my Forest Service career to select at least one major training session each year. This helped me grow professionally and enabled me to keep up with new management trends.

Having spent much of my career in the West, Washington, DC and Vietnam, I found Region 9 to be a new and different challenge. I was impressed by the work of the people who had built the Region out of abused, cutover and burned over brush patches. Other than a few thousand acres of Public Domain land in northern Minnesota and Michigan's Upper Peninsula the 10 million acres of National Forest System land had been purchased. What a challenge Clare W. Hendee and those hundreds of early builders of the region had! They selected, appraised and purchased "the lands that nobody wanted", (Shands, 1976) and built an infrastructure and organization to protect the land and resources. Management came later. Those early builders were giants among men and I developed a great respect for the legacy and the lands they passed on to us. With this background we gave priority to land purchase and added to the national forest land base. We had a long way to go since the Hoosier National Forest, for example, had less than 25% federal land ownership within the proclamation boundary. There was much for everyone to do.

Not all Forest Service people were impressed with the Eastern Region. Some of the western bred foresters, especially the seven western Regional Foresters, looked down their noses at the 10 million acres of poor little R-9. But we had a lot going for us and we were in a building stage . . . adding land, constructing recreation facilities. The abused, cutover timber land was becoming more productive each year. That greatest of all conservation programs, the Civilian Conservation Corps, had given us plantations, roads, recreation and administrative facilities. The saddest chore we had in 1969 was to begin closing some 13 Job Corps Centers, only two centers were retained in the entire region. I could never forgive President Nixon for his decision to close most of the centers. In addition to curtailing a creative program of training under privileged youth of America, we had excellent programs to terminate, people to place or separate and facilities to dispose of or sign over to some other public agency. In Wisconsin I signed over the Clam Lake Job Corps Center, including facilities and equipment to the University of Wisconsin System.

I went to the Region as Deputy to Regional Forester George S. James. George was an excellent teacher and very astute and sensitive about public relations and the political and legislative processes. He turned the natural resource programs over to me. But my number one job was to get to know the Forest Service people, where they lived, what they were doing, their needs and problems. I quickly learned that George and I could do very little ourselves and that it was up to the other Forest Service people to get the jobs done and serve the public. The western regions really had it simple. They had responsibility for Forest Service programs in only one, or two, or three, or four, or five or even 11 states at the most to cover. We had 20! The geography and travel for all of us in the Region was a killer. We traveled by commercial airlines and in our two twin engine Aero Commander aircraft with our outstanding pilots. For the next seven years I spent much time in airplanes and airports. I sometimes believed I was doomed to spend my life in airports. But while there I always had one or two briefcases full of work to complete. I spent many, many hours outside of my regular work hours reading and learning about the region, the Forest Service people and USDA programs.

We had many public pressures and hordes of the public found the eastern national forests had what they were looking for in the way of yearlong recreation, scenery, wildlife, timber, clean water, minerals, oil and gas and wilderness. Residing within the boundaries of the Eastern Region was 53% of the nation's population. At times we were certain all of them were searching for a piece of, or niche, in one of the national forests. Large numbers of the public looked to the national forests as a welcome refuge of peace, relaxation and escape from the crowded cities. They took for granted the fact that their national forests were there and being well cared for.

What do the American people see when they visit a national forest? They see green expanses of open public land they can enjoy. Often they see evidence of use they do not understand or appreciate. A road under construction, or any development project and timber being harvested are not always a pretty picture. Some see this as deterioration of the environment. The forest manager explains that the impact is transitory and the end results will be a fine facility for public use, or a new crop of timber for future generations. An increasing number were saying "we do not want development or that kind of timber harvest!" Rebellion was in the air and some of us "omnipotent foresters" (Behan, 1966) and other resource professionals failed to listen and get the message.

Regional Forester James told me he had a gut feeling there was trouble brewing in West Virginia and he assigned me the task of going over to meet with the public and local Forest Service people and see what was going on in the hills and hollows of the beautiful Monongahela National Forest. Our family knew and loved that area as the result of many vacations we had spent in the Seneca Rocks and other lovely areas.

On that first official trip to visit Elkins, WV I met the new young Forest Supervisor Tony Dorrell and attended a barbecue with the local people. I went, all full of enthusiasm, energy and a desire to tell those people what a wonderful job the Forest Service was doing. Guess what the Izaak Walton League and the state legislators barbecued? Why me of course! They did not like our timber practices and they told us everything that was possibly wrong with our timber management practices, especially clearcutting. I thought back to that day in 1948, on the Kaibab National Forest, when I was introduced to the Forest Service Manual. I had read the 1897 Organic Act and at that time questioned our cutting of trees that did not meet the criteria of being "large, old growth, dead and dying trees". But I had been put in my place and told the 1897 Act was an ancient, outmoded law, and besides that we had years of research to back up our practices. Guess who got sued for violating that law? It was the Secretary of Agriculture, Chief of the Forest Service, et al. I was to soon to become the most sued "et al" in the federal service, because times were changing.

George James retired early in 1970 and I was promoted to Regional Forester. There was no increase in pay, just increased authority and responsibility. We had outstanding people and I delegated much authority to highly competent people. The outstanding John A. Sandor, Assistant to Chief Cliff, transferred to Milwaukee to become Deputy Regional Forester. John and I had a strong top quality staff and we all worked well together as a team. In addition to the tasks of approving work plans and budgets, I continued concentrating on human resources. I had to meet Forest Service people throughout the widespread region, become familiar with their work and the problem issues. This involved continuous travel to 14 national forests. I literally lived out of a suit case.

The Monongahela was only one of those national forests, but it demanded a great deal of my time. I had to learn what was going on and what had gone wrong in West Virginia. I had to meet the public, the users and get to know our Forest Service people, including the outstanding Forest Service research scientists and the state and private forestry experts. I spent a great deal of time looking at our cutting practices and engineering work. I had to agree with some of our critics. The clearcutting did not look good.

The Monongahela was a brush patch when the Forest Service acquired it for a few dollars an acre, almost half a century earlier. It was given fire protection and then management. Timber management started slowly and consisted of individual tree selection, essentially taking out the trees passed over during early cuts. It was really economic highgrading. The more valuable trees from a timber and wildlife standpoint were the intolerant species, trees requiring full sunlight to reproduce and grow. These species were suppressed and disappearing. Small research cuttings made earlier, demonstrated that small clearcuts achieved this objective and stimulated intolerant species.

But we created problems on our way to federal court. The small clearcuts responded so well in obtaining regeneration of desirable new trees that our timber people made the cutting areas larger. Five acres looked good, so 50 acres, or 500 or even 1,000 would be that much better. Release of intolerant species required cutting large trees and small poles and pulp size trees of less desirable species. But there was no market for the small stems and 15 to 20 cords per acre of felled material was dropped to slowly decompose in some very scenic, sensitive areas for all to see. Foresters were so proud of what they were doing they had attractive signs erected around the Forest explaining that a timber harvest was really no different than cutting a field of corn. It could be cut, replanted and would grow back. Try telling that to a 75 year old state legislator who as a small child sat under large oak trees with his daddy, hunting turkeys. One fall he returned and his favorite tree was gone and in its place was a field of stumps and logging debris. He could see one of those oaks growing in his lifetime and 60 to 80 crops of corn. There was no comparison in his mind. The people of West Virginia did not like clearcutting. They particularly did not like to see lovely hardwood stands being clearcut and replaced by plantations of disease free, fast growing white pine.

We supported the besieged Monongahela Forest people and the Chief backed us. Generally what we were doing was biologically and silviculturally sound. We were the professionals. We knew what was best in the long run. The only problem was the local people did not like it and we were poor listeners. We changed our cutting practices. Clearcuts were reduced in size, some were eliminated, others placed in less sensitive areas. Utilization was improved and we did not waste any merchantable wood. We used more shelterwood cuts, which created only small openings in the Forest, but these were enough to obtain increased growth on the residual trees and regeneration of new desirable species. But the local people would not acknowledge we had changed. They had us on the run and on the defensive. So we entered a period of injunctions and court battles. Federal District Court Judge Maxwell ruled against us. It did not matter that cutting on private lands within and adjacent to the Forest was disastrous in all aspects. The Society of American Foresters and State wildlife biologists supported us, but the timber industry sat on its hands while we were being crucified. Judge Maxwell's decision to stop clearcutting was overruled by the Court of Appeals. While we won on an appeal and the Supreme Court refused to hear the case, that was not the end. The opponents were successful in changing the federal law. This resulted in the National Forest Management Act. The important outcome of the Monongahela case was that we changed our timber management and utilization practices and improved our environmental sensitivity. We listened. We got the message, but it was too late, the environmental movement gained headway and the National Environmental Policy Act (NEPA) also became law.

Another critical environmental issue involved Sylvania on Michigan's Ottawa National Forest. The Forest Service acquired this lovely 25,000 or so acres of lakes and forest from an outdoor club owned by officials of General Motors. Forest Service acquisition plans called for intensive development of the tract to satisfy County officials who objected to taking this land off the tax rolls. A self appointed advisory group called Wilderness Watch, headed by a Green Bay dentist took us on. That involved a whole series of battles that approached the intensity of Vietnam's Tet offensive. But we listened, agreed and modified the plans to preserve more of this lovely area.

The Boundary Waters Canoe Area was another of our environmental battlefields. Shortly after I became Regional Forester and the National Environmental Policy Act became law, I called for a meeting in Duluth with the Superior National Forest team, my staff, and Attorney In Charge Jack Curtis and Attorney Jim Pfeil, to begin preparation of a BWCA management plan. We recognized we were heading for litigation and it paid dividends for us to have the expert Office of General Counsel lawyers with us from the beginning.

The BWCA had been around since the 1920s, and part of the one million acres was designated as Wilderness under the 1964 Wilderness Act. But the Secretary of Agriculture was permitted to make exceptions in this "one of a kind canoe area" and certain existing uses such as timber cutting, road building, motor boats and snowmobile use were continued in portions of the BWCA. It was a design for disaster for the administrators to be conducting these activities in a unit of the National Wilderness Preservation System. It was a "no win" situation. The area never had an integrated management plan. A new BWCA management plan had to go through the NEPA process. At that meeting I made a prediction. Regardless of what plan we came up with, the BWCA would eventually become pure wilderness and the non-compatible uses would be eliminated. It was just a matter of time. Meanwhile we would do what we could to permit appropriate timber cutting and provide access for less active older and handicapped users.

We improved the quality and appearance of the BWCA and produced suitable habitat for grouse and deer and food for the endangered timber wolf. Our effort to improve the quality of the canoe country experience were supported by an earlier administrative order which prohibited the taking of cans and bottles into the area. About that same time trappers announced they were moving into the BWCA to trap wolves. I directed Forest Supervisor Craig Rupp to issue an administrative order and close the area to the taking of wolves.

The management plan was completed and we chose to continue some of the consumptive uses. That led to a series of injunctions and prolonged litigation in the Federal District Court of Judge Miles Lord. The final court case lasted some 30 days and as the case progressed the Judge set up his ambush for the "responsible federal official" . . . that was me! I was referred to as that "entity from Milwaukee", an outsider. What they did not know at first, was that I had used that area since I was 12 years old and knew most every trail, lake and river. When it came my turn to sit in the witness chair, again as "et al", Judge Lord grilled me and literally roasted me. He made every effort to get me to admit we had violated the Wilderness Act. Our attorneys, Forest Supervisor Harold Andersen and staffman Bill Spinner gave me very strong support. But after five days I was glad to be delivered from the "right hand of the Lord". Even the attorneys for the Sierra Club and others, acknowledged they had never seen a federal judge try so hard to intimidate a witness. Judge Lord ruled against us, but again we won on appeal. In fact the Court of Appeals chastised Judge Lord for exceeding his judicial authority. End of BWCA issues? Never, the groups that did not approve of our plan worked to change the law and the BWCA was made "more or less pure".

The Eastern Region was an exciting place. What I enjoyed most was working with our Regional Office staff and the people on the individual national forests. They were the ones who performed the work and met the goals and objectives set out in our work plans and budgets. I really appreciated our human resources. I became familiar with most of the people, knew most of them on a first name basis. I recall one meeting we held in Milwaukee with 60 people from all 14 of the Region's national forests in attendance. The group which had organized the program asked me to say a few words. They were satisfied to have me say the words and then leave the program to them. Following my remarks the program chair asked everyone to stand and introduce themselves and tell what they did and where they were from. I interrupted the Chair and told him I would like to make the introductions. I started around the room and introduced 59 people from all over the region . . . Minnesota to Maine and West Virginia to Missouri . . . by their name. I told what they did and where they were from. To my embarrassment I stumbled over the last and 60th person, one George Webster, who worked in the Regional Office and one I knew well. I was forgiven for that mental blockage. I did not tell the group I learned that secret from former Chief Richard McArdle who was a whiz with names. But I kept trying because I recognized the importance of people.

Early in the 1970 Gwen became critically ill and was rushed to the Hospital. The problem was a huge stag horn kidney stone that had gone undetected in one of her kidneys until it flared up. The urologist in Bethesda had missed it. Gwen knew something was seriously wrong and was waiting in the doctor's office when she said it felt like someone shoved a knife into her back and her temperature shot up. Doctors rushed her to St. Mary's Hospital, near our home, and fought to reduce the terrific infection and her terribly high temperature before operating. The infection began destroying Gwen's blood platelets and they had to go for the source of the problem. The kidney was too damaged by that time to be saved and was removed. All that mattered was to save her life. She was deathly ill for days, but her tough Sanders' genes came forward and saved her. That was the second time in my career when the Forest Service came second to family. But Gwen recovered and life went on as usual.

Cindy and Steve Weston were married. Melissa and Bob were divorced. Melissa moved to California and Bob stayed in Arizona. Sue Maynard married Bob Fox and moved out of our home to Wauwatosa. Cindy graduated with a Bachelor of Science in Speech Therapy and went on to earn a Masters Degree, also at the University of Wisconsin-Milwaukee. We had lots of things going on, plus trips to England. One of them was "free", so to speak. Gwen won a contest and we had free (less an income tax deduction) round trip tickets on TWA to London. On these trips Gwen searched for her English, Irish and Scottish roots and began the preparation of manuscripts for the two branches of her family.

I devoted much time to the people in our 14 national forests.

Most of these were very positive relationships. Unfortunately there were personnel issues requiring adverse actions. In these cases we tried to be firm and fair and exercised as much compassion as possible. I had very strong support from Harry Halvorson, Betty Peschek, Jim Frey, Carl Webb and Jack Heintzelman. These were outstanding personnel management people and I appreciated all their good work.

The Eastern Region was an innovative place in those days. We had a very active program and we were rated tops in the Forest Service for meeting the Chief's objectives and goals. A spin off from the Monongahela case resulted in our working with the Southern Region to develop a land use planning system that was ultimately selected as the model for the entire Forest Service. We also pushed for the establishment of some kind of a wilderness system for the eastern United States, after all that was where the majority of the nation's population lived. Atlanta Regional Forester Ted Schlaefer and I were chastised by Congressman Saylor of Pennsylvania for being so bold as to suggest and select wilderness areas. After all that was a job for the Congress. (Congressional Record, 1972) But it was our people who did the "spade work" that created the Eastern Wilderness Act and designation of some lovely areas as Wilderness.

We had a prescribed burn get away one Spring on the Superior National Forest as a result of freak weather conditions. This slash disposal project also disposed of logging equipment, thousands of cords of decked pulpwood and logs and burned several thousand acres, including islands over 1/4 mile from the mainland, before it was controlled. Chief Cliff wrote me a strong letter of reprimand over that "Little Sioux" fire. To show he was serious, he sent every other Regional Forester a copy. The other Regional Foresters were sympathetic and recognized "there but for God go I". But I, and we, survived that set back and developed an even stronger prescribed burn program and wild fire control team under such very strong Fire Chiefs as Hoot Gibson and Ed Heilman and their exceptional staff people.

The Chief appeared to enjoy our innovation and moving our programs ahead. He seemed to agree with my emphasis on sophisticated under development of the land, rather than over development, and stressing the magnificent beauty of "the lands nobody wanted". However there were times when he shook his head and bit a new hole in the pipe stem of his ubiquitous curved pipe. One such occasion resulted from production of an innovative annual report. Most Forest Service reports at that time were dull compilations of how many acres were planted to trees, the number of trees pruned, acres thinned, board feet cut, visitors days of recreation use, miles of road and trail constructed and reconstructed, acres of range reseeding, etc., etc., At that time we had a problem. Our publics did not particularly like some of our work and we wanted to let them know we heard them. Working with our creative writer John Forssen, John Sandor and I came up with a publication called ". . . a little rebellion, now and then". (Cravens, 1974) This 33 page report had creative and eye catching art work by Beverly Jaquish, a tremendous message, few statistics, no charts and no tables. I put my signature to the following:

The conflict and contradiction of the human spirit constitutes the most inscrutable and important web of inter-relationships with which this nation's fledgling environmental awareness has yet to come to grips. For it is in the depths of mind and soul, in the rubbing of elbows on crowded subways, in the laughter and sorrow which 210 million people experience individually and collectively that the quality of life in this society ultimately depends.

We are not aware, certainly of all the relationships—nor even all the species—which exist in the natural world; but we lack only the asking of the right questions under the right circumstances before those secrets are unlocked. But the mysteries of the mind, the anguish and the anger that move people in and out of dark corners, so often escapes us in dismal speculation.

During brief moments, inspiration may visit a diligent curiosity: and out of that momentary vision, we construct endless systems to accommodate the homeless and the hungry, the impoverished and the uneducated; we create cities to precise specifications and measure the values inherent in our environment by computer code. But no sooner does the system emerge than the man for whom it was designed disappears.

Yet we continue, somehow persuaded that time alone will produce a workable system. We continue, failing for all our energy and expertise to realize that the system is the individual—changeable, unpredictable, but the system, nonetheless. And to build for anything more or less is, finally, not to build at all.

Only slowly are we learning the real value of alternatives.

/s/ Jay H. Cravens
Regional Forester

We were proud of our product and sent copies far and wide, including to the Chief and Secretary of Agriculture. All went well until the Chief and the Assistant Secretary got their heads together. You would have thought we had committed the cardinal sin and burned Smokey Bear. I was called in on the "carpet" and told such things as "this is no annual report, there are no statistics, no tables and few numbers. What are you trying to do? You are forbidden to issue any more annual reports until they are cleared by Washington." (i.e. the great white father's blessing, amen!) That did little to take the wind out of our sails since the accolades, editorial awards and letters of commendation from some of our most ardent critics flowed in to the Secretary, the

Chief and to our offices praising our ". . . a little rebellion . . .". The Chief got a new pipe stem and nothing more was said. I must admit this was not your normal federal report!

Before retiring in 1972, Chief Edward P. Cliff in referring to leadership in resource conservation, challenged a "leadership that is responsive to the needs and wishes of the American people". Chief Cliff had the respect of every man and woman in the Forest Service. He was the leader and spokesman for State & Private Forestry, Forestry Research and the National Forest System. He was respected as a world leader in forestry. We all were sad to have him retire. He was one of the great ones and an inspiration to us all. But "long live the Chief", and John McGuire moved up to become Chief and landed running. We had a orderly transition and again we all worked well together as a team and knew where the Forest Service was going. John was an effective leader and we moved smoothly ahead.

During my tour of duty in Milwaukee I had three excellent official foreign trips. I was a member of the USDA team attending the United Nations Food and Agriculture Forestry Committee meeting in Rome. Chief McGuire sent me since he knew I was interested in international work. After sizing up FAO, both FAO and I decided there were better ways to continue my career. I was designated as the coordinator of the US Delegation to the VII World Forestry Congress in Buenos Aires, Argentina in 1972. In 1973 I was sent to Vietnam by the Chief and the USDA to review the USAID Forestry program.

Gwen and I enjoyed our trip to Argentina. In advance of the Congress we saw the scenic west side of South America. We could have gone on an American Forestry Association guided tour. But we studied the countries, decided where we wanted to go, and went on our own. We first enjoyed Ecuador and got off the beaten path to visit some of the famous market towns north and south of Quito. We stood straddle of the equator north of Quito.

We visited Peru and spent time in Lima, Iquitos and a jungle camp on the Amazon, Cusco and Machu Picchu. We caught up with the AFA tour group in Cusco and many of them were as sick as dogs. They had not read up and followed the advice for surviving in areas above 11,000 feet. When Gwen and I arrived in Cusco, we went slowly to our hotel, laid down and rested for two hours, ate lightly of soup and bread and drank no alcohol. The tour group did just the opposite and suffered from debilitating high altitude sickness. We stayed over night in a tiny hotel at magnificent Machu Picchu and walked ancient Inca trails after the tour groups caught the train back to Cusco.

Next we traveled by train the full length of Peru to Puno, caught a taxi and drove by Lake Titicaca to the Bolivian border at Desaguadero and rode the interesting "smugglers' bus" to La

Paz. We traveled around southern Bolivia by bus to Cochabamba and returned to La Paz before rejoining Pan Am and taking off from the 12,000 foot high airport for Buenos Aires. Our takeoff seemed to take forever. We traveled over 8,000 feet on the runway before we slowly climbed up through that rarefied air. We did not make it all the way and were "bumped" for two enjoyable days in Asuncion, Paraguay at Pan Am's expense. We arrived in Buenos Aires just in time to join the US delegation at the Congress.

The Congress was a high point of my career travels. Gwen and I loved Argentina and enjoyed meeting people from all over the world. I kept busy making arrangements for the US delegation. One of my chores was to change money at the best rate for the delegation. I found I could double the official exchange rate with the proper contacts. My contact was the local, General Manager of Allis Chalmers, South America, who Gwen and I met on the plane out of La Paz. I must admit I was a bit apprehensive one day when I took our new friend $25,000 in green US dollars. He asked me to give him a little time. Three hours later I had the money (at double the official rate) and returned it to those who had entrusted me to exchange their dollars.

The biggest challenge I had at the Congress was to keep reminding Chief McGuire to address the representatives from mainland China as being from the Peoples' Republic of China, not Red China. There were some tense moments between Taiwan and mainland China representatives when it came to the placement of their flags and order in tree planting ceremonies, but we came out on top and the Chief made us all look good.

Among my other duties I helped Mr. Charles Nelson, President of the Nelson (tree) Paint Company organize a reception. The EL Presidente Hotel, where we were staying, went all out with the arrangements. Gwen and I never before or since, participated in such a magnificent affair.

Gwen and I had time to ourselves to eat in some of the tremendous Buenos Aires restaurants, enjoy the music and see some of their famous dancers. We visited the old town and purchased some inexpensive and colorful paintings of Gauchos and local scenes. I had responsibilities to complete for the delegation and Gwen returned to the States ahead of me. The evening she flew across the Andes to Chile was the same day a Uruguayan rugby team crashed in the snowy mountains. The 45 people who survived for 10 weeks were forced to cannibalize their dead companions. On departure I crossed those same magnificent 20,000+ feet Andes during daylight, unaware of the plane crash and the tragedy unfolding below me. A Chilean friend who I had met at the Congress, Hugo Bianco, appeared at the Santiago airport and presented me with five bottles of the finest wine, Chilean of course. Great people, those South Americans!

Back again in the real world my work progressed and I enjoyed it all, even the troublesome issues. One trip took me back home again to Indiana, to the Hoosier National Forest. While there I returned to Bloomfield where I was born. What I remembered in my eyes as a child was different. The house where I was born and lived in until we moved to Kansas was much smaller. The beech trees were larger, but of course the giant American chestnuts were dead and gone. My grade school building was replaced by an apartment building. The Court House and my Father's office building looked the same. I winced and lightly touched my bottom (which my Father had blistered) as I failed to find the "JHC" I had carved in the Post Office corner stone in 1930. The hill I slid down in the winter was not as high as I remembered. I also visited the cemetery in Worthington where my Grandfather and Grandmother Hampton and Aunt Madeline Kester are buried. I must return to Worthington one day and make arrangements for a new headstone for Worden Hampton, since his marker is spalling and his name will soon be obliterated.

I was detailed to the Chief's office in 1975 for nine months as a member of the Forest and Rangeland Renewable Resources Planning Act (RPA) Team. Deputy Chief Red and Mildred Nelson were on a world cruise and I "house sat" their town house in southeast Washington. It was a pleasant walk to work. For a portion of that lengthy period I commuted from Washington, to the Twin Cities to be in Judge Lord's Court and to Milwaukee to look in on the Region's business, our new Deputy Regional Forester Stan Tixier and Gwen. In Washington I was responsible for the policy portion of the program and had over 200 people throughout the Forest Service working under my direction. Our RPA team had offices in the sub-basement of the South Agriculture Building. We were confident they did this to keep us from jumping out the window. That was really "mission impossible", but our team got the job done and on time. Picture this, if you will, I was responsible for planning Forest Service policy and programs ahead to the year 2020. And I was the one, back in 1948, who was so presumptuous as to tell the Forest Service planners that long range planning (annual work plans) was a waste of time.

I have very pleasant memories of all the wonderful people I worked with in the Eastern Region, about the remarkable growth of our productive timber resources, and the improvements we achieved in engineering, land acquisition, minerals, range, recreation, water, wildlife and fish habitat, wilderness and personnel management.

Speaking of wonderful people, Ed Groesbeck and Gordon

Bade and I had many delightful summer trips during the 1970s and early 1980s. Ed, Gordon and I had worked together—they were my mentors in the Southwestern Region and had a great influence my early Forest Service career. Our favorite trips were to the BWCA. Being with those two gave me an exciting new perspective of the history and ecological succession of the "one-of-a-kind" BWCA. Both Gordon and Ed had begun working for the Forest Service on the Superior National Forest in the late 1920s. These two outstanding foresters retained in their brilliant minds a wealth of information about the area. We could see striking and visible changes and improvements that had taken place over the years on old logging and fire areas. It was obvious to me that working with nature the Forest Service people had done well and cared for the resources while serving the people.

Over the years our trips covered the BWCA from east to west . . . from Lakes 1,2,3,4, to Insula, Thomas, Fraser, Kekekabic, Ima, Ensign and Snowbank . . . that trip and its innumerable portages was a challenge. Other trips were made to Basswood and Crooked lakes. We paddled and portaged down Nina Moose River, through Agnes to La Croix and from Crane Lake and Loon River into La Croix. The companionship, the experiences and the gallons of delicious mojaka, fish stew, we shared on those trips, enriched the quality of my life. From time to time others joined our group. We added my son-in-law Steve Weston, brother-in-law Rod Gingrich and his son, Ed's daughter Barbie and husband Stan Wyche and Flick Hodgin, my Forest Supervisor on the Kaibab National Forest. Then there were the wilderness fly-in trips with Ed and Gordon into Ontario, a tremendous trip to Alaska by Marine Ferry, float planes and train and a wonderful horseback trip into Colorado's Mt. Zerkel Wilderness (Gwen, Cindy and Melissa were on that one). There must have been over a dozen of those great trips.

Shortly after Christmas 1975 and the New Year's holiday Chief McGuire called me and dropped the last shoe. He offered me the Associate Deputy Chief position, for National Forest System resources. The job would promote me to a GS-17 (at the same $36,000 of course). I told him thanks, but no thanks, I was not interested. I liked being where I was, enjoyed what I was doing, had a new Deputy to introduce to the region and since I would reach age 55 in March, I expected to become quite independent. I planned to remain in place in Milwaukee, retire and seek a university teaching position. The Chief thanked me for my views and said "fine, I expect you to be in Washington on the job next Monday morning at 0730 hours!" He added that I always had been quite independent. That was not a popular message to carry home to Gwen. However, she had no problem with my last Forest Service transfer and remained in Milwaukee. She reminded me we had moved 23 times and Milwaukee was

474

her last move. Since I was not quite 55 and eligible for optional retirement I went to my last Forest Service job on my own. Two items saddened me (1) leaving the wonderful people of the Eastern Region who had enriched the quality of my life and (2) in spite of spending up to $20 million each year on land acquisitions there remained the unfinished task of building the region.

My challenge to the people I left behind in the Eastern Region was to listen to those voices of "rebellion now and then" and carefully reflect on Gifford Pinchot's, Ed Cliff's and other forestry leaders' directions. The resources are there to serve mankind and "the good of the land must always defer to human welfare as the basis of judging goodness and badness. It is a comforting thought that there will be few, if any, real cases of human welfare requiring what is bad for the land". I left the region confident it was in good hands.

BWCA 1970 . . . Two of the Best "Timber Beasts"
Edward C. Groesbeck & Gordon Bade

Chapter XXX

Washington Revisited

The return to Washington, DC was an easy move. I attended the Regional Foresters' and Directors' meeting at the end of January 1976 and just stayed. I carried two suitcases and a brief case, flew to Washington National Airport and caught a cab to M. B. Dickerman's house in Alexandria, Virginia. I lucked out again and had another house sitting situation. Dick and his wife Mei were out of the country on an extended international trip. Now my Washington Office assignments had given me an opportunity to live in Maryland, the District of Columbia and Virginia. Commuting to work from Virginia was not exactly fun and games, but without family obligations I could go to work early and work late. I put in many hours for the brief time I was to work in the Chief's Office.

Working under Deputy Chief Tom Nelson my portfolio included all natural resources programs, including timber, range, fish and wildlife, minerals, soil, water, recreation and wilderness. While I had experience in all these fields, it was the outstanding program directors and their competent staff people who managed the programs. The Renewable Resources Planning Act provided funding to bring programs more in balance than they had ever been before. Up to that time we had 95% financing of our budget requests for timber activities, but Congress only met about 35% of our requests for programs such as fish, wildlife and recreation. RPA helped us go from programs that were badly out of balance to more realistic and effective multiple-use financing. We made good progress in all areas and were able to bring aboard on the national forests more of the specialists necessary to improve our management in all resource areas.

Our training programs were intensive throughout the country. This required much travel on the part of the program people and me. The most demanding tasks in the Chief's Office were the daily, even hourly, crisis situations. More frequently the crises resulted from political pressure being placed on our people by some interest group that wanted more land or a greater share of the resources. It was one critical issue after another. Livestock people wanted increases in permitted livestock use. Wilderness proponents wanted more set-asides of land for single purpose use. Timber interests wanted increased harvest levels or complained about volume estimates or some timber administrator's decision. Recreationists wanted more recreation facilities or services. Wildlife interests were concerned about wildlife/livestock conflicts and believed wildlife should have preferential treatment. Oil, gas and minerals interests wanted more opportunities to extract resources with fewer environmental constraints. The dam builders wanted more dam sites. We had a strong environmental ethic and effective procedures to evaluate these proposals. All activities were subject to NEPA requirements and we were the federal experts in preparing environmental analyses. After all we had 15 years experience in preparing multiple-use impact surveys.

We had an effective decentralized organization and when complaints reached the Chief's Office we made every effort to have problems solved at the ranger district and forest supervisor levels, rather than in the Washington political arena. However some project proponents would not accept a ranger's, forest supervisor's or regional forester's explanation or take "no" for an answer. They would go the political route. I had worked with many pressure groups over the years, but the one group which consistently went the political route to obtain what they wanted were the ski area proponents. Then it fell on us to resolve the problem when they came to Washington. Frequently it was not in the public interest to issue a permit for new or expanded ski developments. They would not listen to us. One such proponent was a high level political appointee in one of the nearby federal agencies. He went all out to exercise pressure on the Secretary and the Chief. But our excellent environmental analysis strengthened our position and we refused to approve his application.

Sometimes we had a short turn around time to explain or resolve a problem. I recall arriving at the office at 0700 hours one morning and the Chief called me into his office to resolve a range/wildlife conflict. The conflict involved the black-footed ferret and livestock use. Livestock interests advocated intensive prairie dog control to enhance livestock use. The Chief asked me to go to a prominent western senator's office at 0930 hours and explain the issues involved and our position, which was opposed to prairie dog control. At 0700 hours I knew nothing whatsoever about the elusive, rare, black-footed ferret. As far as I was concerned it was one of those critters you captured at midnight using a gunny sack and rubbing two sticks together. As on most all issues we had the expertise in the Washington Office. Within the hour one of our experts had briefed me on the relationship between the black-footed ferret and its principal food source, the prairie dog. We explained and resolved the problem to the satisfaction of the Senator and retained a suitable prairie dog ferret ratio. By 1100 hours I returned to my office to tackle the next waiting crisis. There was no end to them.

In spite of effective field people, a large number of

complaints reached the White House, the Secretary's office and the Hill. We seemed to go from crisis to crisis. Most of these situations resulted from individuals or groups who were never satisfied . . . they always wanted more. Wilderness advocates had an insatiable appetite for more single purpose set asides. The end seemed to justify the means and frequently the means were not completely honest, such as assembling a series of phony horror pictures depicting the results of poor logging practices or false numbers to cast an unfavorable light on national forest timber harvest activities. When we investigated and located some of those sites we found most of them did not involve national forest land. When we produced evidence to prove their claims were false, they just shrugged and said "we made our point, didn't we?" That was a most difficult situation since our people were doing quality work in the field.

My last week in the Forest Service gave me mixed emotions. In the absence of the Chief I served as Acting Chief. It was a busy week and the Washington Office team worked smoothly. We held staff meetings, I signed letters, issued orders and responded to the White House, Secretary's office and other political types. At one 0730 hours staff meeting I was told to expect a call from the Canadian Ambassador through the State Department. The previous night a North American satellite had detected the start of hundreds of lightning fires in the Province of Ontario. By 0745 hours we received the call and a request for 1,000 hand tools to supply Canadian fire fighters. Before 1000 hours our team had gone to work and we authorized the Boise Interagency Fire Center to send the tools. Before noon a US Air Force C-141 transport was dispatched to Canada with the tools. Within a few hours the White House, the Secretary's Office and our team received expressions of appreciation for our good work from the Canadian government.

My other fire related task that week was not a happy one. I appointed and dispatched a team to Colorado to investigate the circumstances related to the deaths of members of a Coconino National Forest "hot shot" fire suppression crew. These men were some of the nation's best fire fighters, but a west slope crown fire trapped and killed them, proving once again how very dangerous forest fires can be.

We had a strong team and all of the Deputy areas and staffs worked well together. While I loved the Forest Service and my years in the National Forest System, State & Private Forestry and Research (my international assignment), a job offer from the University of Wisconsin-Stevens Point caused me to reexamine my future. I must admit I was pleased to receive an offer to teach in Wisconsin. I could help prepare young professionals for work in the "real world". By that time I had 28 years of federal service, four years of military time and over two years of unused sick leave to add to my retirement base.

I retired effective 16 July 1976 with 34 years of creditable federal service. At that time I was a GS-17 with a good salary that was limited by law to $36,000 per annum. In October of that year the salary for my grade level was increased to almost $60,000. But I had no regrets in leaving. I had a good career, helped train and develop many Forest Service people, provided the public with goods and services and helped improve the quality of resource management. I would miss the fine people of the Forest Service, but I would only slightly miss the crises and conflicts that promised to increase. Since the Forest Service was people, I strongly believed I could continue serving my organization by training bright young people for promising careers in the Forest Service. I was ready to retire and go to my next career. I looked forward to working with bright eyed optimistic young students.

The Washington office staff organized a coffee reception for my retirement affair. It was well attended and I was presented a magnificent electric typewriter for producing my university lectures. Gwen was busy with family affairs in Milwaukee, but our good friends the Jensens attended the affair and were amused when someone asked if they were my parents. Since Gene and I are the same age that provided me with a chuckle. Moving out of Washington was easy. I packed my two suitcases and went back to Milwaukee, Gwen and the cats.

University of Wisconsin-Stevens Point Forestry Professor
and Students 1981 (UWSP Photo)

Chapter XXXI

The Joy of Teaching

Stevens what? What is it and where is it? I recall raising the same questions when I attended the Saigon memorial service for Dr. James Albertson and the other presidents of Mid-West colleges and universities who perished in a Vietnam air crash in March 1967.

Actually I made the decision to leave the Forest Service and teach in July 1948. I resolved at that time to have my life, family and career in shape to retire at age 55 and do something else. Over the years I counseled others to prepare themselves and their family to do the same. My resolution was firmed up over almost three decades of personal observations. I saw too many people who stayed on in the Forest Service too long. By too long, I mean 40 to 45 years of service. Some of those old timers remained productive and creative until the day they left the Forest Service. Unfortunately some of those retirees died shortly after retiring, having little else to do after giving their lives to the Forest Service. But in my judgment some of these senior people hung on too long and outlived their usefulness. The organization had to scratch its head to come up with something for good old Tom, Dick or Larry to do. I determined this would not happen to me.

I saw many bright young men and women come from fine colleges and universities into the Forest Service. They were highly motivated and well trained in the biological, engineering and other sciences but they could not cope with the real world. Academically they knew how to manage range, recreation, timber, soil, water, fish and wildlife habitat and the other resource activities. Many of the foresters, like me, were attracted to the profession for the opportunity to hunt, trap, fish and live in a cabin in the woods. They loved the woods. They were real tree huggers. Working with others and with the many publics was as unfamiliar to them as dealing with aliens from another planet. Their interpersonal relations rated a "D minus". Few had effective communication skills. Some failed during their probationary year, while others lasted into an unhappy mid-career. I had experience in the real world of forestry and the scars to prove I had been there. I could work with students and help prepare them for the real situations and tell them how to survive in any organization they selected.

During my last few months in the Forest Service I pulled together a resume. I began my job search by writing to the

University of Wisconsin-Madison and the University of Wisconsin-Stevens Point. I knew people from both of these schools quite well from my work with them while I was with the Forest Service in Milwaukee. I sent letters to the Deans of both universities. On receipt of my letter Drs. James Newman and Robert Engelhard of UWSP got their heads together, approached the faculty, Dean Daniel Trainer of the College of Natural Resources and the Chancellor about my interest in coming to Stevens Point. That very same day I received a telephone call from Dr. Newman inviting me to come to Stevens Point to put on a seminar and meet students and faculty. Actually the seminar was a method of putting a prospective faculty member on display for all to see and size-up.

In conjunction with an April 1976 trip to the Eastern Region I visited Stevens Point, met people and presented my seminar. By that time in my career, after receiving early training in Toastmasters, giving talks to various groups and conducting uncounted training sessions, I felt quite comfortable speaking to large groups and answering questions. The next day I had an appointment with dynamic, red vested, Chancellor Lee Sherman Dreyfus (later to become Governor of Wisconsin). It was a full two days and I returned to Washington feeling quite good about my prospects. Before the week was out I had an offer to become Associate Professor of Resource Management, with the opportunity to be converted to full Professor of Forestry with tenure within two years if the University and I were satisfied with my performance. There would be no requirement for me to conduct research and to publish or perish. My job was to teach and advise students. My 10 months teaching contract offered a second salary increment based on teaching experience. And that amounted to 28 years of Forest Service experience and two years of graduate teaching. Within two years I had 32 years of teaching experience and was the only non-Ph.D., full professor, with tenure, in the University of Wisconsin System.

I had only a few weeks of retirement before I reported to the University. Stevens Point is located in central Wisconsin and 160 miles from my front door in Milwaukee. Being predictable, Gwen remained in Milwaukee. It took me a while to convince the faculty that Gwen would come to Stevens Point for social affairs but would reside in Milwaukee. In view of this situation I rented an apartment which was within easy walking distance to the College of Natural Resources. My teaching schedule for next several years permitted me to leave at 0900 hours after class on Thursday and return in time for classes at 0800 hours on Monday morning. Normally I returned to Milwaukee every other week and returned to Stevens Point on Sunday afternoon. Frequently I visited my aging Mother in Cedar Rapids, Iowa. I drove about 15,000 miles per year in all kinds of weather. The balance of the time I remained in Stevens Point and was available to advise students, prepare lectures and assist students who needed additional help with their classwork. Some of the Southeast Asia and African students requested help in order to understand the subjects. These students had good basic intelligence but their comprehension of English left much to be desired.

Shortly after I began teaching at the university I received a telephone call from Chief McGuire. He informed me Smokey Bear had died and his ashes returned to Capitan, New Mexico. That brought forth a flood of memories from the 1950 fire when I had held the tiny Smokey and the many times I had recited the story of Smokey Bear to Melissa and Cindy when they were small children. I contacted the local newspaper and helped them put together a story about Smokey Bear.

Preparing lectures was demanding during that first year. I used the Forest Service lesson plan approach. This consisted of developing an outline, including cues as when to use visual aids, when to smile, tell a joke and ask questions. At a glance these lesson plans provided me the information I needed. Students enjoyed my classes and I earned good student evaluations which led to promotions, merit increases and tenure. My classes ranged in size from over 200 in the Introductory Forestry class to 30 in Resource Administration, 40 in Integrated Resource Management, 100 in Natural Resource Policy and Law, 40 in Natural Resources and Public Relations, 20 in Forestry Labs and 12 in senior seminars. At the beginning I worked long and hard hours to prepare my lecture lesson plans. Once I completed a semester I had something to work from and improve for the next round.

My first semester was very challenging. In addition to the lesson plan preparation I spent time with students helping them understand the class work and counseling them on career opportunities. Then there were the weekly CNR faculty meetings with the Dean. Each faculty member was required to serve on a University committee. I selected Community Relations. All of this took time, but I soon became familiar with the College of Natural Resources, University operations, faculty, students and the community. I used my skill in recalling names and students were more alert when I called them by name. Students often reminded me it was important to them that I knew them on a first name basis.

Two months into my teaching career at UWSP I finally received a response to my University of Wisconsin-Madison employment application and letter. The Madison Dean stated he and the faculty had considered my application. While they were interested and recognized I could bring a new dimension to their program they had nothing to offer. That is a so-called prestigious University's response for, "we do not want a non-Ph.D.". I will always be grateful to Jim Newman, Bob Engelhard, Bill

Sylvester, Dan Trainer, the College of Natural Resources faculty and Chancellor Dreyfus for giving me the opportunity to come to "Stevens what?"

On my arrival I found most UWSP students and graduates preferred to remain and work in the Lake States. In my classes and advising I provided them with insight on the exciting opportunities to work elsewhere, particularly in the West and Alaska. Before long we placed students and graduates throughout the US. We had good cooperation from the National Forests in accepting volunteer, temporary and permanent placements. We had one of the best placement records for graduates of all the forestry schools in the nation. Other schools could not comprehend how we achieved 100% placement of our forestry graduates. The faculty and the able Mike Pagel in University Placement prepared them and I was pleased to do my part.

My Society of American Foresters involvement took on a different approach. During my first semester I accompanied Dr. Newman to a national SAF meeting. At that meeting there were no UWSP students or graduates and few other students in attendance. I decided to change that. On return I became advisor to the Student Chapter of UWSP SAF. We gradually built up our membership and at the peak we had the involvement of 200 forestry students in SAF activities. Each year we had a pulp cut where the students were given operational and safety training in making pulpwood sales, harvesting pulp, arranging transportation and selling products to one of the nearby paper mills. This was our major fund raising effort to finance part of the costs for students attending national meetings. We prepared students for the annual SAF meetings. We told them how to meet prospective employers and the importance of having a well prepared resume in their possession. My final instructions were a warning to the effect that at the meeting they could say "good morning" and "good night" to one another, but I did not want to see them talking to one another during the meeting. The point being that many foresters are more comfortable visiting with friends and people they know, rather than mixing and talking with strangers. But I informed them that their fellow students had no jobs to offer and there were many major employers of foresters at national SAF meetings. They got the message, they mingled, mixed and talked to others. This impressed the SAF members and we received many compliments concerning the fine UWSP students and their ability to communicate with complete strangers. That probably did more than any one thing to make the name UWSP well known throughout the forestry profession. The students put us on the map.

We usually took 30 to 40 students to each annual Society of American Foresters meeting. I traveled with them in University vans or chartered buses. Some of my involvement on such trips could be considered above and beyond the call of duty. Normally I traveled with the students in a small bus equipped with a tape deck and *loud* speakers. When you listen to heavy metal music, like that made by performers such as Pink Floyd, you have a major challenge. I recall attending one such non-stop musical concert for 17 hours from Stevens Point to Birmingham, Alabama and return. I thought my war shattered hearing would be gone forever. To add to my misery at the next University affair the students presented me with a Pink Floyd collector's album.

I did not intend to teach summers. Following my first academic year I was invited to accompany a "Partners in Americas" group to Nicaragua. We loaded two used school buses with hospital and educational equipment, 10 University students and five adults. It was an exciting 4,000+ miles trip. One bus broke down on a Texas freeway and we cooled our heels waiting for repairs to be completed. We had to talk ourselves and the equipment through the rigid Mexican customs. We drove to Mexico City, Vera Crux, across the Isthmus of Mexico to the west coast of Mexico. Then, like in Vietnam, we bribed and paid our way through El Salvador, Honduras and finally to Managua. I was invited by a Somoza government to prepare a natural resources evaluation of the country. In three weeks I became well acquainted with the problems and opportunities, the latter being limited. Nicaragua had some intestinal bugs that out classed those I had experienced in Mexico and Vietnam. But I survived and enjoyed my opportunity to learn more about this tropical country and the gracious people.

During our month long break between Christmas and the resumption of classes, Gwen and I made numerous trips to Guatemala. We loved the country and the people and brought back many pleasant memories, ponchos and magnificent hand woven cloth and blankets. Melissa accompanied us on one trip. We included Copan, Honduras, Tikal and Los Lisas. Copan and Tikal belong with the "World's Seven Wonders" but Los Lisas in the extreme southwest corner of Guatemala ranks with the foulest "arm pits" of the world.

The second summer I volunteered to teach at our summer camp in the Black Forest of West Germany. That was a delightful experience. We started students out with two weeks of training at our regular Wisconsin summer camp. Then we flew to Frankfurt and traveled by bus to the State of Baden-Wurttenberg's training facilities at the ancient castle known as Burg Hornberg, southwest of Stuttgart. There we inventoried and mapped the magnificent German forests, studied wildlife management and had frequent field trips to historic and scenic parts of Germany.

Gwen and Melissa joined me for a scheduled five day break in the German program. I rode in their rented car to Switzerland. Gwen loved the farms and accompanied an old Swiss farmer from his mountain pasture to the milking shed. She did not even bother the cows at milking time. I took that girl off of a Minnesota farm but fortunately her farm background was with her all her life. It rained in lovely, green Switzerland and I predicted as soon as we passed through a long tunnel we would be in sunny dry Italy. So be it! Gwen and Melissa thought I was a real weather expert. Really it was the well known rain shadow created by the Swiss Alps that enabled me to make such a reliable prediction. I left Gwen and Melissa in Milano and returned to Germany by train to the summer camp, while they continued their delightful travels through Italy.

I enjoyed every semester of teaching and was recognized by the students with an "Outstanding Faculty Member" award. On three occasions Gwen and I took students on UWSP Semester Abroad programs. We spent a full semester in Spain where we lived in a small pension (Hostel Lamar) with 25 students near the world famous Prado Art Museum. We loved Spain, the people, the food, the tappa bars and our frequent group trips around the country. On one such trip Gwen and I met Bob Hatfield, an English engineer with Whimpy Engineering working in Spain. Bob became a close friend over the years.

Our other two semester abroad trips took us to London where we stayed for three months each time. On all the programs I taught two international natural resource courses and administered the programs taught by local professors. In Britain we had another UWSP faculty member with us. We could take turns and travel around the country on alternate weekends. We enjoyed living in London and everything English. Gwen lectured on genealogy and had a unique opportunity to pursue her genealogy research and continue the search for her family roots. In Milwaukee, or any other city, Gwen had the ability to get lost going around one block. In London she could walk six winding blocks from our lovely Peace Haven Victorian Mansion where we all lived, take the underground into central London transfer to another line and finally arrive at the Barbican Station, walk five blocks to Royal Genealogical Library or one of the many record offices. And she could generally find her way back with little difficulty.

On all semester abroad trips we thoroughly enjoyed the students. However at times they could be a challenge. Some had never been out of Portage County, Wisconsin and when they were turned loose on their own in London, Paris, Munich, Rome or Madrid . . . it could be exciting! There never was a dull moment! We have as many tales and memories as there were students, 45 each time. In connection with all of the European Semester Abroad Programs we spent one month touring Europe by train or chartered bus and visited Belgium, France, Luxembourg, Germany, Austria, Switzerland and Italy. My favorite spots were Brugge and Salzburg. These were unique experiences and we were grateful to the students and UWSP for making them possible.

I had no intention to teach summers when I went to UWSP. I looked forward to having a nine months teaching contract and taking the summer off to travel or go on fishing trips with old friends. However things changed. The first summer I went to Nicaragua and the second summer to Germany. From that time on I was asked to teach at our summer camp at Clam Lake. By teaching six weeks I rounded out my salary to 12 monthly payments rather than 10. But the big reward was working in the field with students I had taught on the Stevens Point Campus.

Teaching at Clam Lake gave me mixed emotions. Clam Lake was one of the first Job Corps Centers I had visited on my return from Vietnam. As Regional Forester I had signed over the property, facilities and equipment to the University of Wisconsin System. I recall suggesting to the Chequamegon National Forest staff that they should condemn the small Job Corps buses. Now here I was using the same facilities and little blue buses. I enjoyed teaching forestry, surveying and plant identification to all of the CNR majors at summer camp. I tried to make those activities interesting for the forestry, wildlife, soils, recreation, resource management and environmental education majors. I spent a great deal of time after supper and on weekends helping the students learn their herbaceous and grass plants. I knew the plants well from my training at Fort Collins and as the result of the experience I gained in wildlife and range management work in the Forest Service. During my last few years of teaching the summer camp was moved to Tree Haven near Tomahawk, and much closer to Stevens Point. The land and facilities at the new field station were donated to the University and were a definite improvement.

Summer camp started immediately after final exams in May. As soon as our grades were turned in for the spring Semester we went to summer camp for six weeks. At the conclusion of each summer session I usually had a trip planned for Ed Groesbeck and Gordon Bade. We either went to the BWCA, Ontario, Alaska or Colorado. It was a good life and I returned home to Milwaukee often enough for Gwen to be glad to see me and then I left before she could get mad at me. Actually the summers, 30 days of school breaks and holidays, provided time for Gwen and me to do many things together such as our trips to Britain, Mexico and Guatemala.

I had good rapport with students and faculty during my 11 years at UWSP. However I grew weary of the tedious University committees very early. As an experienced administrator, I was often appalled when I looked at the administration of the University. If the organization and procedures were looked at by any rational, independent thinker, I am confident the verdict would be that the university administration could not possibly work. But it did work in spite of such archaic methods as faculty governance. I soon found a way to relieve myself of committee assignments. I volunteered to work at the University Advising Center. This was a program set up to advise the many undeclared majors, students who came to the University with no idea about what courses to take or what major to select. I was an experienced advisor and was sought out by forestry, wildlife and other majors about course work and career opportunities. I found the advising effort to be productive and I was relieved of the monotony of sitting through hours of unproductive University committee work.

While at the University we were encouraged to do outside consulting to keep up with the state of the art in our respective professions. I worked on a Minnesota Timber Resources Study project with the George Banzhaf and Company team. For almost five years, working through the Banzhaf office, I served as a technical advisor to the legal counselor of Dow Chemical on the Agent Orange Product Liability Litigation (MDL N. 381. United States District Court, E.D. New York, Sept. 25, 1984 as modified). I agreed to help Dow since they were a very responsible firm and had used every means possible to avoid dioxin contamination of the 2,4,5-T during production. By that time I was well informed on all aspects of the Vietnam Ranch Hand operation. I had access to more declassified documents and files, which confirmed my earlier conclusions that the US Government ran a "sloppy" operation. The day before I was to leave for the trial in New York, to serve as an expert witness, an out of court settlement was reached.

During my University career I served on the Board of Directors of the Michigan-Wisconsin Timber Producers Association. I was appointed by Wisconsin Governor Lee Sherman Dreyfus to Chair the Governor's Council on Forest Productivity. I was frequently asked to prepare suggested remarks for the Governor for use at some of our state-wide conferences and banquets. Governor Dreyfus was a quality speaker and no one could provide remarks suitable for his use. Seldom did he use more than the first sentence. The words just slid off his "silver tongue" with the greatest of ease. He embarrassed me frequently when during one of his major speeches he would add that one of the best things he did while being Chancellor at UWSP was to hire Cravens away from the Forest Service.

While at the University I enjoyed charter membership and served as a member of the Board of Directors of the Wisconsin Forest History Association. In this organization I could rub shoulders with some of the great timber men of Wisconsin like the first consulting forester George Banzhaf, timberman Gordon

Connor, Consolidated Paper's Stanton Mead and many, many others.

During these years Matthew and Benjamin were born to Cindy and Steve Weston. Melissa and Aubrey Serfling produced Ethan and Melissa came to Milwaukee a few years later and following a high risk pregnancy produced twins Delaney and Mackenzie. Elizabeth Weston was born in 1990. That made six delightful grandchildren and we enjoyed them all.

As I neared age 65, the University administration started making noises that sounded like, "it's time to retire". I was asked to stay on for a while, but accepted their "carrots" and was ready to go at age 66. The annual Society of American Foresters Student Chapter banquet was a surprise affair. Chief Max Peterson of the Forest Service (Chief 1979-87) came to honor me. He talked about my careers and poked a little humor my way. Max and I were regional foresters at the same time and we worked together in the Washington Office. Bob Engelhard's remarks to the several hundred people in attendance stated I had instructed and advised well over 6,000 students. Dean Trainer commented on me being a fantastic instructor and committed to the students. He said my major contribution was that I gave the students outstanding hands-on training. When I was called on for remarks I corrected the Dean and reminded him that he had warned me to keep hands off (of the women students). That had the audience rolling in the aisles. It was a very pleasant affair and I was presented with a handmade oak, schoolhouse clock.

My last semester was a Semester Abroad in Britain. On my return from Europe, at the end of Semester 2 1986-87, I retired after 11 years at UWSP. I retired from the teaching profession with very fine memories and a sense of real accomplishment.

Chapter XXXII

Now is The Future

After returning to the United States from the Semester Abroad in Britain I closed out the program's travel account and completed my debriefing with UWSP International Programs people. One of the major responsibilities of the international programs administrator is to account for all expenditures in foreign currency. Keeping track of daily exchange rates and expenditures for group expenses was a real challenge, but I was experienced and quickly closed out with the proper balances. Before I departed Stevens Point I went the rounds of the University and thanked faculty, administrators and staff for 11 good years.

After being away from home for so many years, some of it for prolonged periods time, Gwen had become accustomed to my being away and could get along nicely without me. Coming home full time was a challenge for both of us. She needed time to herself, to read, work on her genealogy research and divide her time between the girls and grandchildren. However my homecoming went smoothly. We did many things together and spent time with Cindy's family. I went to Melissa's in San Anselmo frequently and enjoyed brief periods of the good life in California with Melissa, Aubrey and the grandchildren. So Gwen and I both had time to ourselves.

Again I was not ready to fully retire and accepted an offer to become an Associate of George Banzhaf and Company. George and I had become well acquainted over the years and he frequently asked me to join the firm after I retired from UWSP. I was not a full time salaried person. I worked on projects which could utilize my experience. One of my first projects was a birdseye maple marketing study. This project required travel to Michigan and High Point, North Carolina, one of the world's major furniture manufacturing centers. I learned a great deal about birdseye maple and what drives the furniture manufacturing business.

I had the privilege of preparing and submitting the nomination of George Banzhaf to the Wisconsin Forestry Hall of Fame. It was my pleasure to attend the ceremony in which he was inducted and entitled to have his award displayed in the Wisconsin Forestry Hall of Fame. Unfortunately George became ill in 1987 and passed away within a few months. His forestry business was organized in 1929. He was the first consulting forester in Wisconsin and a tremendous person. We missed his easy smile and humor around the office.

Bill Banzhaf, as President of the firm, continued to manage the business, supported by the highly productive Sam Radcliffe, Mike Sieger, Tim Cayen and me. I helped Office Manager Cindy Johnson with some of the administrative duties.

It was during my first few months at Banzhaf's that John Sandor, Society of American Foresters Council member from Juneau, Alaska and other SAF members asked me to consider running for SAF National President. After talking it over with Gwen and Bill Banzhaf, who was also on the SAF Council at that time, I agreed to be a candidate for the position of Vice President of the 20,000 member professional forestry organization. Gwen welcomed the idea, since she knew it would keep me busy. One of my first trips as a potential candidate was to accompany SAF President J. Walter Meyers, Jr. to Alaska and become acquainted with some of the Tongass National Forest issues. We had a good reception from the Alaska Society and I was encouraged to run for the office. I was well known among federal SAF members and my work with students had given me exposure to a broader segment of the national membership. Again the students helped me in that respect. I had received the prestigious national SAF John Beale award for my work with students. Wisconsin had elected me to Fellow status and presented me with the State Society's John Macon award.

I was elected Vice President. At that time forestry and foresters were being battered by preservation interests and my position as President was to speak out for forestry with the "right stuff". We had a vacancy in the Executive Vice President staff position. I encouraged Bill Banzhaf to apply for the position. After careful consideration Bill decided he was ready for a career change, applied and was selected by the Council for the position. He was good for SAF and provided excellent support and strong national leadership. Sam Radcliffe became President of George Banzhaf and Company.

The next three years were busy ones. My Mother had mellowed over the years and grown closer, and more generous, to the family. She lived alone in Cedar Rapids until full time care was required. At that time she accepted an invitation from Gwen and Cindy to move to a nursing home in Milwaukee where Cindy worked as a speech therapist. She enjoyed being near us and we saw her frequently. She passed away peacefully in her sleep of a ruptured aorta at the age of 93 on 22 April 1989 and was buried next to my Father in Iowa.

I was on the SAF Council for three years, one year as Vice President, one year as President and one year as Immediate Past President. As Henry Clepper, Executive Director of SAF had told me years earlier . . . "be involved". I never expected to be involved at that level, but I did my best. If Henry had been alive, I think he would have shared my pride in serving the profession at the national level. It gives me a humble feeling to see my photograph hanging near Gifford Pinchot's in the Presidents' Room at the national SAF office in Bethesda. During my three years on the Council I traveled throughout the US and represented SAF at a number of Canadian Institute of Forestry meetings across Canada. I enjoyed my association with the Canadians and became a member of their professional organization.

Gwen and I traveled to Spokane by Amtrak to attend the annual SAF meeting. We lived it up and, for the first time ever, we had a bedroom compartment on a train. It gave me a great thrill to conduct the national assembly and address the membership in Spokane. I gave seven speeches at that meeting. I was very involved in Society business during my three years on the Council. I attended many State society meetings and gave short talks to the membership. Banzhaf's gave me strong support, such as office space, telephone and secretarial services. I devoted most of my time and energy to SAF. Thinking back, I gained more respect for former SAF Presidents, like the great Charles Connaughton, who had a full time job as Regional Forester. I do not know how he and the others were able to handle both responsibilities, but Charlie did both, and did them well.

Banzhaf's was the successful bidder on a Forest Service contract to conduct a series of nationwide land management planning workshops. We designed a program to facilitate and receive input from federal, state, and local officials and representatives of Indian tribes concerning their views of recently completed national forest land management planning process. My team consisted of two recently retired Region 9 forest supervisors, Jack Wolter and Wayne Mann. We had approximately 90 days to complete the workshops and prepare a final report. Within seven days of the completion of each of the nine workshops we were required to submit a draft report of the findings to the Forest Service contracting officer. We held the first meeting in Milwaukee as a test. We received approval to proceed. We then took to the air and held a series of one and two day meetings in Albuquerque, Atlanta, Denver, Juneau, Missoula, Portland, Sacramento and Salt Lake City. The participants had some excellent thoughts and these were documented and reported to the Forest Service. I am not quite sure what the Forest Service did with this valuable input. The Forest Service was slipping back into a 1970's mode of not being good listeners and continued to lose strong local grassroots support that had been painstakingly built up over the years.

George Banzhaf and Company was encouraged by the Forest Service to enter into a recruiting contract with the Pacific

Southwest Region. The objective of the contract was to help the Forest Service meet a federal court order to increase the numbers of women and minorities in the work force. We provided California's Region 5 with applications from well qualified women and minorities for a variety of professional and technical positions. We received weekly work orders to supply candidates to fill positions on the 17 Region 5 national forests and the regional office. We developed a worldwide recruiting network through the Peace Corps and the Department of Defense. Over a two year period we located hundreds of well qualified candidates, screened and processed their applications. Most of the applications we processed came from Returned Peace Corps Volunteers (RPCVs). After almost two years of work on the project we placed only five candidates. I donated hundreds of hours to help the Forest Service and these worthy candidates. According to my calculations I earned 28 cents per hour for my efforts.

There were reasons for our poor accomplishments. The Forest Service seemed to suffer from organizational paralysis. Nothing could be done within a reasonable time frame. Sometimes it took weeks, even months to make a decision on placements. Neither the candidates nor our office received any reasonable notification on the status of applications. Many of our well qualified candidates grew tired of waiting for the Forest Service to make a decision and accepted job offers from other federal and state agencies. For some unknown reason well qualified Native American males were not selected. Another reason for our lack of accomplishments was the ineffective management of the program by the court ordered "Consent Decree" group in the Forest Service regional office. The leadership in the San Francisco office of the Forest Service was ineffective. Regretfully, people in key leadership positions were not experienced or well qualified. After serving in a dynamic, smooth functioning Forest Service for almost 30 years I was appalled by the ineffectiveness of those groups. Their arrogance was unparalleled in the history of the Forest Service! We terminated the contract because we could not tolerate having qualified candidates dangling on a string for months. Following termination of our contract we continued to provide advice to qualified RPCVs and Department of Defense applicants on how to apply to other Forest Service regions and other federal agencies. In addition to the recruiting problems we had with Region 5, the morale of people in that once proud region was at rock bottom and most of the rapport and support that had been generated over a century at the grassroots level was lost.

Occasionally I did some real forestry work for Banzhaf. I surveyed and marked property lines, inventoried and mapped proposed timber sale areas and inspected cutting areas. One of my inventoried sales cut out within -2 1/2% of the inventory estimate. Pretty good for an old forester! I prepared reports and served as a technical advisor and expert witness for a number of law firms on a variety of court cases, including a logging equipment product liability case, wooden fence construction, and fire and timber trespass situations. Most of the time I helped out with administrative chores in the office. I was not particularly concerned about what I did, or if I was paid for my services, I just wanted to keep busy and contribute to the George Banzhaf and Company operations. In my spare time I began putting my thoughts together for this story.

From time to time I was invited back to UWSP as Professor Emeritus to give guest lectures, serve on the Forestry Advisory Committee and attend College of Natural Resources and SAF banquets. I continued to serve as a reference for former students and provided career advice. Stevens Point continued to produce well trained field foresters. From time to time Banzhaf used Stevens Point students and graduates on forest inventory work.

My schedule at Banzhaf's provided ample time for travel to Hawaii, Colorado and Arizona to hike with friends and family. Gwen and I continued to make frequent combination business and pleasure trips to London and other favorite parts of the UK, our favorite part of the world. I did well as a Premier Frequent Flyer on United Airlines. After some of my harrowing trips through Chicago's O'Hare Airport, I am not certain the perquisites of the Frequent Flyer program are worth the effort. Occasionally I have a bad dream in which I am traveling to the hereafter and a delay is announced. I am routed through Chicago's O'Hare Airport! Now that is a frightening thought!

The United Nations Food and Agriculture Organization in Rome asked me to participate in a short term African assignment. I almost got to Africa on a FAO assignment to Uganda, Kenya and Tanzania. I was selected for an Upper Nile watershed study team. I had my physical, started on malaria suppressants, upgraded my typhoid, yellow fever and cholera inoculations, had visas for entering those countries and spent hours reading about that part of Africa. Then just before my departure for Rome and Africa, someone in Rome discovered I was 70+. Seems like anyone working for FAO that is 70 or older must be approved at the highest level of FAO Rome. Therein I was lost in the awesome United Nations Food and Agriculture Organization bureaucracy. I never heard from them again. Perhaps it was for the best. But I always wanted to see and do something to improve conditions for the people and resources of Africa. But that was not to be. It becomes a tremendous problem for older people when others decide they are too old to do anything worthwhile. But it gave me time to climb more western mountains with Gene Jensen.

I retained my close ties with the workers and retirees in the Eastern Region and the Southwest Region of the Forest Service. I attended a number of picnics at Grand Rapids, Minnesota and mingled with the famous builders of the great Eastern Region. As a member of the Southwest Forest Service Amigos I attended many "Roundups" at Quemado Lake in New Mexico and assisted wagon Boss Ed Groesbeck and his son-in-law successor, Stan Wyche and ex-regional foresters Bill Hurst and Steve Yurich on the firewood detail and the cooking. We lost Gordon Bade and Ed Groesbeck, two of the greatest gentlemen and top quality foresters I ever met. The quality of my life was much better because of them and their families. I was privileged to deliver the eulogy at Ed Groesbeck's funeral. During this sad visit to Albuquerque I was able to have one good, final visit with two very ill old friends, Dahl Kirkpatrick and Norman Weeden, truly great associates of mine.

Time permitted me to serve on one of the United Nation's (UNESCO) Man and the Biosphere Committees. I was a member of the Temperate Ecosystem Directorate. This three year appointment provided an opportunity for me to work with some very high quality scientists from federal agencies, the private sector and academe. It was a great honor and privilege to help plan projects for the improvement and protection of some of the world's ecosystems. This resulted in frequent coast to coast travel for the two years I served that group. I also served as a member of the National Forest Service Museum Board of Directors and attended meetings in Missoula, Montana.

I have maintained contact with many of the young people who entered the Forest Service in the '60s, '70s and '80s. Some have done well and others have been frustrated in their careers. I have great confidence in the people of the Forest Service team. They know what to do, if they are just given an opportunity.

In retrospect as I look at the succeeding processions of federal administrations from President Truman, to Eisenhower, to Kennedy, to Johnson, to Nixon, to Carter, to Reagan, to Bush, to Clinton and whoever comes along, I would like to think that one of these four years we will have an administration which will pay some genuine attention to conservation programs. Our very capable resource managers can help our natural resources blossom and provide the quality amenities, goods and services our nation needs. While I am optimistic, I am reminded that Voltaire once said . . . "in a democracy the people get the kind of leadership they deserve". Surely we deserve better.

I have an original, unpatented, uncopyrighted statement I use to remind my listeners when they advocate setting aside more productive woodlands . . . *the greatest enemy of the forest is neglect*".

I have many good memories of my travels on *A Well Worn*

Path. Some of my fondest recent memories are about trips to England where Gwen and I enjoyed hiking through the Yorkshire moors and along Hadrian's Wall in the north of England. Then there are the trips to Arizona, Utah and Colorado where I am privileged to hike some 50 miles each year in the Arizona desert mountains in the early Spring and the high mountains in Colorado and Utah in the summers with my best old (same age!) friend Gene Jensen. Our enduring friendship with the Jensens and my life with dear Gwen and the girls have been the high points in my life as I traveled through time. It has been a good life. With that I end this story.

BIBLIOGRAPHY

Behan, R.W. *The Myth of the Omnipotent Forester*. Journal of Forestry 64,6:389-407

Carson, Rachel. *Silent Spring*. Boston: Houghton-Mifflin, 1962

Congressional Record, *E 3519*. Washington, DC: US Government Printing Office, 11-11-72

Corson, William R. *The Betrayal*. New York: W.W. Norton & Co., 1968

Cravens, Jay H. *A Little Rebellion, Now and Then*. Milwaukee, WI: USDA Forest Service, 1974

Cravens, Jay H. *The Sawmill in Vietnam*. American Forests, 9/67

Flamm, Barry R. and Jay H. Cravens. *Effects of War Damage on the Forest Resources of South Vietnam*. Journal of Forestry, 69,11:784-789

Highlights in the History of Forest Conservation. USDA, Washington, DC: US Government Printing Office, 1976

Karnow, Stanley. *Vietnam A History*. New York: The Viking Press, 1983

Lederer, William J. *Our Own Worst Enemy*. New York: W.W. Norton & Co., 1968

Martin, Everett. *Editorial*. New York: NEWSWEEK, 9/25/67

McArdle, Richard E. and Elwood R. Maunder. *An Interview with the Former Chief, USDA Forest Service, 1952-1962*. Santa Cruz, CA: Forest History Society, 1975

Morris, William, Ed. *The American Heritage Dictionary*. New York: American Heritage Publishing Co., Inc., 1975

Page, Rufus. *Forest Industries of the Republic of Vietnam*. Washington, DC: USDA Forest Service, US Agency for International Development and Directorate of Forests, Ministry of Agriculture, Republic of Vietnam, 1967

Pomfret, John E. *Founding of the American Colonies*, 1583-1660. New York: Harper & Row, 1970

Saigon Sunday News. *Editorial*. Saigon: October 8, 1967

Service, Robert. *Collected Poems of Robert Service*. New York: Dodd, Mead & Company, 1940

Shands, William E. and Robert G. Healy. *The Lands Nobody Wanted*. Washington, DC: The Conservation Foundation, 1977

Sheehan, Neil. *A Bright Shining Lie*. New York: Random House, 1988

Westmoreland, William C. *A Soldier Reports*. Garden City, NY: Doubleday & Company, Inc., 1976

Index
(+ indicates numerous references)

Regional Forester Jay H. Cravens 1972 (Pohlman Studios)

About the Author

Jay H. Cravens, a native of the Midwest, resides in Milwaukee, WI. He is active in the forestry profession. He frequently travels to the western United States and to England where he hikes well worn trails in deserts, the high Rocky Mountains and the Yorkshire moors with friends and family.

His reasons for writing his story were to describe a remarkable forestry career and provide clues for resource professionals to help smooth their paths through any bureaucracy. He also describes his unique experiences as a civilian forester working with the Vietnamese, U.S. military, CIA and a host of bureaucracies in war ravaged South Vietnam.

Jay is a graduate of Coe College and Colorado State University. His first book is written in a pleasing and often engaging style.